CONTEMPORARY ERGONOMICS 1996

CONTEMPORARY ERGONOMICS 1996

Proceedings of the Annual Conference
of the Ergonomics Society
University of Leicester
10–12 April 1996

Edited by

S.A. Robertson
University College London

Taylor & Francis
Publishers since 1798

UK   Taylor & Francis Ltd, 1 Gunpowder Square, London EC4A 3DE

USA   Taylor & Francis Inc., 1900 Frost Road, Suite 101, Bristol, PA 19007-1598

A catalogue record for this book is available from the British Library.

ISBN 0-7484-0549-6

Cover design by Hybert Design and Type, Waltham St Lawrence, Berkshire.

Printed in Great Britain by T.J. Press (Padstow) Ltd

Preface

Contemporary Ergonomics 1996 are the proceedings of the Annual Conference of the Ergonomics Society, held in April 1996 at the University of Leicester. The conference is a major international event for Ergonomists and Human Factors Specialists and attracts contributions from around the world.

Papers are chosen by a selection panel from abstracts submitted in the autumn of the previous year and the selected papers have the opportunity to be published in *Contemporary Ergonomics*. Papers are submitted as camera ready copy prior to the conference. Details of the submission procedure may be obtained from the Ergonomics Society.

The Ergonomics Society is the professional body for Ergonomists and Human Factors Specialists. Based in the United Kingdom it attracts members throughout the world and is affiliated to the International Ergonomics Association. It provides recognition of competence of its members through the Professional Register. For further details contact:

The Ergonomics Society,
Devonshire House,
Devonshire Square,
Loughborough, Leics.
LE11 3DW
United Kingdom

Tel /Fax. +44 1509 234904

Contents

Donald Broadbent
Memorial Address

ERGONOMICS IN ORBIT

H.S.Wolff

Professor Emeritus
Brunel Institute for Bioengineering
Brunel University
Uxbridge, Middlesex UB8 3PH

Much of Ergonomics is concerned with matching the forces due
to gravity to human comfort and ability. In orbit where
effective gravity is virtually absent, the rules are changed. The
human operator is also changed in terms of his range of sensory
perception, may in addition be under psychological stress and
on short missions, probably badly overworked. The paper first
outlines how different life on orbit actually is and why, and then
lists broad categories of engineering and ergonomic adaptations
which are required.

Terrestrial Gravity

A great deal of human activity is concerned with countering the effects of
gravity. By implication therefore, so is a good deal of the work of the
ergonomist, be it concerned with the design of seating, lifting, or body
posture. All evolution since the appearance of the first self replicating living
cell has taken place under the influence of the earth's gravitational field and
organs for sensing the direction of this field are incorporated in virtually all
organisms and thereby constitute an important element of the overall sensory
input. It is even possible, though perhaps not yet important to the work of the
ergonomist, that some organisms can sense the subtle variations of the
gravity vector imposed by the motion of the moon.

The effects of gravity as sensory input

In man, the organ which is primarily concerned with the detection of the
direction and magnitude of the orthogonal vector of the gravity field is the
otolith organ of the vestibular apparatus. This, like most accelerometers
consists of a flexible component, which carries a mass (crystals of calcium
carbonate), and which is connected to a position transducer which measures
the deflection of the elastic portion. If we were to think of this device as
having a positive output when the head is the normal way up, and a negative

output when it is upside down, it would have zero output under conditions of free fall (zero effective gravity). Angular movement is detected by the semi-circular canals, in all three planes, but this system is unaffected by the gravitational field.

However, the otoliths are not the only detectors of gravity, both pressure receptors in skin and stretch and proprioception receptors in muscles and joints, are sensitive to gravity because of the mass of the body and limbs. For instance if one were to reach up in order to adjust a control knob, one of the directions of movement is defined by the gravity vector. Lastly, the behaviour of objects around us, when they are falling, constantly remind us of gravity.

Gravity also plays a part in the distribution of blood and body fluid, because for instance whilst standing there is a hydrostatic pressure of over a metre of water gauge on the vessels in the legs, and a slightly negative pressure above the heart, **superimposed** on the blood pressure. The bones of the skeleton are sensitive to mechanical stress and those most concerned with weight bearing, such as the bones of the leg have a mechanism of adjusting their strength in relation to the imposed stress. If this stress is removed, the bones become weaker due to demineralisation and a condition very similar to osteoporosis results.

It is clear therefore that gravity is not only responsible for a mass of sensory input, but evolution has taken it into account when "designing" the body both structurally and in response to weight imposed stresses.

Conditions on orbit

When an astronaut is launched into space and relatively few minutes later finds himself in earth orbit, effective gravity **suddenly** disappears. It is not that gravity itself is changed much, because in low earth orbit the distance to the centre of the earth is barely increased, instead terrestrial gravity is counteracted very nearly exactly by the "centrifugal" force due to the orbiting motion of the space craft. This after all is why it stays in orbit rather than escaping from the pull of the earth or falling back onto it!

To my mind the miracle is how **little** actually happens. Before space flight became possible all sorts of prognostications set out to prove that man could not survive, all of which have been proven to be fallacious. Individuals have now lived under **"micro gravity"** conditions for over a year, in relative comfort, boredom apart, whilst admittedly taking certain countermeasures to reduce the longer term effects, such as bone demineralisation.

My conjecture is that in the 2 billion years since life first started a large proportion of the time was spent in water. It seems plausible to me that a good deal of the rules according to which life is organised were laid down at the time when organisms were practically weightless and that therefore the only problems likely to arise were concerned with weight bearing structures and hydrostatic forces. In parenthesis it should be said that being neutrally buoyant is NOT the same as being in a micro gravity environment; a ball bearing in the stomach of a neutrally buoyant diver would still roll about if the diver turned upside down, whilst this would not happen to a ball bearing in the stomach of an orbiting astronaut. Nevertheless immersion in water or even prolonged bed rest can reproduce some of the effects of prolonged exposure to micro gravity.

What does happen very quickly is a sense of disorientation which can culminate in "space sickness"; symptomatically similar to sea sickness, though immunity from the effects of the latter due not correlate with susceptibility to the former. This is thought to be due to the combination of fluid shift into the upper part of the body because of the contraction of vessels and tissues in the lower part which are no longer distended by hydrostatic pressure, and the confusing input from the otolith organ. This signals continuous free fall, a state of affairs for which the brain has no programme, because on earth a duration of more than a few seconds is likely to be fatal, and in any case the eyes contradict the evidence. However at the most three days later physiological adjustments have been made, the otolith input is disregarded and the individual functions more or less normally

Apart from the sudden disappearance of effective gravity, there are other environmental changes. Space craft by their nature are cramped, and the "day" in terms of alternation of light and dark becomes about 90 minutes. In US missions which tend to be of a length between 7 and 10 days, every minute of the astronauts time is carefully scheduled; this together with the very high motivation of the crew mean that every individual is under considerable pressure, intensified by communication with the ground which includes communication with principal investigators who want THEIR experiment done as well as possible.

Lastly there is the psychological state of the astronauts. Even allowing for the very careful selection, there must be some realisation of the danger, particularly during launch, that there might be a disaster. The odds for Shuttle launches have been put to be of the order of 1 : 100 (they are probably better now), and for many of the crew the mission represents the fulfilment of years of training.

As far as possible I have now set the scene; described the work place, the cockpit apart (which is in principle not very different from that of any large airliner in the US Shuttle), and the physiological changes which will have occurred in the crew.

Design problems.

What follows is more of the nature of a list, the lecture will illustrate how some of the problems have been solved.

In performing any action like turning a wheel, pulling or pushing, on earth we are reasonably well anchored by our weight, against which the purposeful forces can react. This is not so in micro gravity, and therefore tools and controls either have to incorporate adequate fixation of the astronaut, or preferably contain the forces within themselves. An example would be a simple spanner as against a two handled instrument in which the effect is achieved by squeezing two handles together.

The crew have a tendency to propel themselves through the cabin by giving an aimed push an floating to the destination. Not only must sharp corners be avoided, but all equipment must be designed to resist the impact of an over enthusiastic launch.

Because there is no internal UP or DOWN it is attractive for designers to use all surfaces for equipment or controls. Apparently astronauts do not like this as further contributing to disorientation and so far (perhaps not true for the NASA Space Station), there has been some pretence of having a floor and a ceiling.

Free liquid/air interfaces must be avoided, because apart from the weak wetting forces nothing will keep the liquid in the container if this is jerked. This means that all liquids must be contained, and every precaution taken that particles and droplets do not escape into the atmosphere. Droplets and particles will not sediment under gravity, so that even a sneeze will produce a cloud of droplets which remain suspended and dry producing bacterial containing particles which may be inhaled. Continuous filtration reduces this problem.

Eating, drinking, cooking , washing, showering and particularly elimination have to be designed to work without gravity, the micro gravity toilet is a major engineering project!

The layout of scientific and control equipment must take into account that convenience or necessity may require access from unusual angles, for instance two crew members floating above one another. Other equipment like glove boxes (with which the Institute has been particularly concerned), may require simultaneous immobilisation of the trunk, fine manipulative works with the hands, under good direct vision for appreciable periods, whilst preventing the work space from filling up with floating objects, spilled liquids, and toxic fumes.

Extra-vehicular activities require much more complex modifications from either terrestrial or diving practice, though because the suits are run on pure oxygen at well below atmospheric pressure, the problem of "bends" when making the transition from cabin to suit has to be prevented.

In longer missions equipment is required on which exercise can be taken to counteract at least some of the muscle atrophy and bone demineralisation. So far designers have failed to produce exercise machines or games which are fun, or failing that produce a useful output. This represents an interesting area for the lateral thinker.

The Future

The author firmly believes that there is a cultural dimension to manned space flight, which otherwise cannot be defended realistically on scientific grounds. If he is right, and the manned exploration of the solar system is one of the objectives mankind has, than the ergonomists of the future have interesting tasks in front of them.

Keynote Address

PEDESTRIANS IN THE ROAD ENVIRONMENT

Heather Ward

Centre for Transport Studies
University College London
Gower Street
London WC1E 6BT

Increasingly the need is being recognised for improvements in the road environment to enable people to have safe access on foot to homes, schools, shops, open spaces and other facilities. In order to achieve this we need to know much more about walking patterns of different groups of people and the risk of injury to them as pedestrians in different parts of the road network in an urban area.

This paper explores some of these relationships and draws on the results of a large study of pedestrian activity and accident risk undertaken in Northampton for the AA Foundation for Road Safety Research. The paper also draws out some of the implications for policy makers, both nationally and locally, and for those involved in road safety education and engineering.

Introduction

On an average day about three quarters of the population of a typical British town walk somewhere in the road environment, even if only to their car and from it to a shop or workplace. Over the period 1985/86 to 1992/94 there was an increase in the distance travelled in cars of 32 per cent for drivers and 28 per cent for passengers. Some of this increase can be accounted for by 35 per cent of women owning cars now compared with only 13 per cent in 1975/76. Over this same period there has been a decline of 18 per cent in the amount of walking, with the largest reduction being in the age group 5-15 years (Department of Transport 1995b). This is of particular concern in the case of children because there is evidence that their independent mobility is being reduced by parents as a response to fears for their children's safety on the roads (Hillman, Adams and Whitelegg 1990). Many older people also feel threatened by traffic because they find it more difficult with increasing age to see and hear the approaching vehicles. As their level of agility decreases with age they find it difficult to walk smartly across the road in order to avoid oncoming vehicles.

The quality of the road environment in which we walk is important and after decades of the motor vehicle being at the centre of the planning process more account is now being taken of the need for safe and pleasant pedestrian movement. This concern is not new and was expressed in 1963 by Buchanan in the Ministry of Transport Report *Traffic in Towns*

...walking is also an integral part of many other matters, such as looking in shop windows, admiring the scene, or talking to people. In all, it does not seem to be far from the truth that the freedom with which a person can walk about and look around is a very useful guide to the civilised quality of an urban area. (p40)

Over the last decade more attention has been focused on providing a safer and more pleasant environment for pedestrians. We have seen this in the attention given to town centres where many have been pedestrianised to provide a traffic free environment in which people may shop and meet.

Being out on foot is seen by some to be dangerous and to be avoided, especially after dark on quiet streets. People are concerned not only about traffic but also about being threatened by other people on foot, by those riding bicycles on the footway and by people harming or abducting their children. These issues are complex and need sympathetic treatment by planners and others responsible for the road environment if we are to succeed in encouraging more walking and fewer short trips by car.

Last year over 1100 people were killed as they walked or crossed the roads in Britain and a further 11 800 were seriously injured (Department of Transport 1995c). Nearly 3000 of those killed or seriously injured were over the age of 65 and a further 5000 were 15 years old or younger. These figures indicate that risk of injury to pedestrians is a big problem.

We cannot ban traffic from all the roads and streets in our towns but we can target areas in which we can reduce its impact on pedestrians. However, in order to do this we need to have a much greater understanding of the amount and pattern of activity which takes place every day in the road environment by pedestrians according to their age, gender and social background. We also need to know more about people who rarely if ever go out on foot at all.

This paper explores the patterns of activity on foot by the residents of Northampton. The data for which were gathered as part of a study of pedestrian activity and accident risk undertaken in 1992/3 for the AA Foundation for Road Safety Research. Issues will be discussed that relate to injury of pedestrians by vehicles and what contributions may be made towards a road environment which is both safer and more pleasant for travel on foot.

The study

Northampton was chosen in which to collect information about walking because it is a medium sized town which has a typical pattern of streets and housing types ranging from victorian terraces near the town centre, through between the wars council and privately built housing, to the modern development corporation and 1990s housing on the eastern and western outskirts. The age distribution of people who live in Northampton is typical of towns of its size so is its ethnic mix and number of cars per household. About 164 000 persons aged five years and over live in Northampton with females making up 51.5 per cent of the population.

A stratified sampling sample of about 400 households was chosen and every resident aged five or over was asked to fill in a diary of their activity on foot over a 24 hour period determined by the study team. In this way walk information was collected from 1037 residents. The importance of this approach was to sample across the different types of housing which in turn include the various road layouts which people walk alongside and cross. The main weakness of using this sampling method was that it concentrated on households and not on individuals who ultimately provide the basis for the exposure data so some gender and age groups were not sampled in quite the same proportions as found in the population of Northampton.

Respondents to the household surveys provided a wealth of information about their pedestrian activity but they could not reasonably be expected to recall all the characteristics of the road environment in which the walking took place; enough of their time had been taken up in asking them to describe the route they had taken for each walk and where they had crossed each road. The information about the road environment in which each walk took place was collected by a member of the study team who rewalked the routes described by the respondents and noted down the presence of different features of the road and roadside using a classificatory system devised to deal with all the information about each walk.

The methodology developed for use in this study owes much to earlier work of Todd and Walker (1980) but has been adapted for use in one town and for all ages of people. It can be used in many different situations to enable comparisons of relative risk to be made between people of different ages in different road environments either in the same town or region or between regions. It will also provide a basis for aiding the exchange of information about effectiveness for pedestrian safety of different policy decisions about the pedestrian environment and the education and training of road users.

Who walks when, where and with whom?

On an average day in Northampton about three quarters of the residents walk somewhere and between them they walk about 133 000 kilometres and cross about 636 000 roads. This is the equivalent of 825 metres for every man, woman and child aged 5 and over. They each cross about four roads a day which adds about another 30 metres to the distance walked. In general the number of roads crossed by a person is proportional to the distance walked which is what might be expected. For every person who did walk somewhere on their survey day, the average distance walked was of the order of 1.15 kilometres with about 5.5 roads being crossed.

One of the determining factors in the amount of walking is health and general mobility. Data from the 1985/86 National Travel Survey show that people with a physical or long-standing health problem that makes travelling difficult, make on average about half the number of journeys made by people without such problems (Oxley 1993). People were asked how often they had gone out in the previous week and also whether they had any health problems that affected their mobility. Eighty three of the people we interviewed reported a health difficulty, however nearly two thirds of these (60 per cent) went out on foot on the survey day. We interviewed 24 people who were registered disabled and of these, half went out on foot somewhere on the survey day.

Whether or not a car is available to the household may affect the amount of walking by household members. Nationally, about 87 per cent of 5 to 10 year olds who live in households where there is no car walk to school. For households with one car, about two thirds of the children walk to school but in families where there are two or more cars only 36 per cent these children walk (DoT 1995b). About three quarters of the people sampled lived in households having access to a car which is about the average for Northampton. We found that people living in households without a car walk further than those with a car; one kilometre compared with three quarters of a kilometre. But whilst they walk further they tend to walk less often. People who live in households where the head was not in regular employment and there was no car were appreciably less likely to walk than any other group.

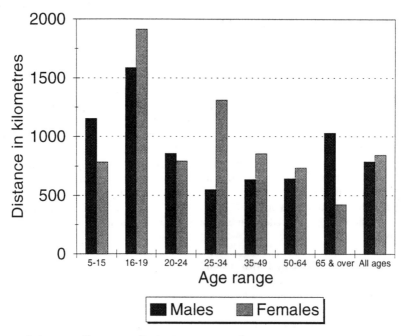

Figure 1 Average distance walked on an average day

Males and females do not walk the same distance in a day as, on average, females walk about 8 per cent further than do males. In distance this translates to about 60 metres per person for all females and about 100 metres for those who walk somewhere on an average day. The distance walked is not uniform across all age ranges and this can clearly be seen in Figure 1. Of particular interest is the amount of walking done by young women aged 16-19 years and 25-34 years in comparison with men of the same age. Male children walk further than female children with the 10-15 year old boys walking nearly twice as far as girls of the same age. Within the different walking patterns it is interesting to note that about 60 per cent of all walking is done within 500 metres of home and whilst boys and men venture further afield than do girls and women, females tend to do their extra 60 metres of walking nearer to home on the quieter residential roads.

Most walking is on streets close to home which is consistent with the finding that most walking is on residential roads. This may not be a profound finding at first sight but it does have relevance to wider issues of how people relate to the environment near to where they live and how educational material relating to the use of the roads should be targeted.

Accompaniment, or escorting, of children by an older child or adult is common, especially when the younger child needs to cross a busy road. In our study we found that the 5-9 age group were accompanied by an adult on 77 per cent of the times they crossed a road and a further 10 per cent were in the presence of older children. On main roads about 40 per cent of crossings made by the 10-15 age group are accompanied by an adult. However, on the local residential roads the level of accompaniment of these older children by an adult is only about 27 per cent.

The time of day the walks start is also interesting. In general, about equal amounts of walking are accounted for by walks starting between 07.00 and 09.59 and between 15.00 and 18.59. About half the distance walked is done between these hours.

On average, about 8 per cent of all walking is done in the hours of darkness. The 5-15 year age group do most of their walking starting in the two periods described above which corresponds to the times they are walking to and from school.

Hillman et al (1990) found that 98 per cent of junior school children and 76 per cent of senior school children are not allowed out alone after dark. This together with the findings of Hillman et al that increasingly children are being taken by car on the school journey lead them to identify five adverse consequences:

increased resource costs
constraint of adult opportunities
contribution to increased traffic congestion
removal of a routine means for children to maintain physical fitness, and
limitation of opportunities for the development of independence.

The 16-19 year age group do nearly a quarter of their walking in the evening after 19.00 hours but the largest amount, nearly 600 metres a person, is started between 15.00 and 18.59. This together with the finding that they venture further away from home on foot and cross more main roads than any other age group indicates they are perhaps travelling to the town centre where there is not only a greater concentration of main roads but is where the activities are located that attract young people in the evening.

Perhaps not surprisingly, people aged over 65 year do two thirds of their walking between 10.00 and 14.59. These older pedestrians have a walking pattern more similar to that of children than to that of other adults with about 85 per cent of the distance they walk being within one kilometre of home. The main difference being the number of main roads the older people cross. Children cross on average one main road every other day whilst the older pedestrian crosses about one a day. This is probably a reflection on where in the town the older people live and their lifestyle which includes visiting the shops which tend to be located on or near to main roads.

Car ownership and use is not distributed evenly across the age range of people in Northampton. From our data we could determine where the walk started from. About one third started from home, a third from another walk, a further quarter started from a car and in Northampton, only three per cent started from the bus. The 16-19 year olds and the older people were the greatest users of the bus with about 11 per cent of their walks starting from a bus journey.

The destination of walks also shows an interesting pattern across the age groups. The 16-19 year age group link fewer of their walks to a car than does any other age group which again highlights their independence and lack of personal transport other than their feet. By contrast, the 50-59 year age group link more of their walks to a car than any other age group. In general, about 30 per cent of the walks done by the people of Northampton are 50 metres or shorter but these represent only about one per cent of the total distance walked. Of these very short walks, about 44 per cent were to a car. Over two thirds of all walks are no longer than 350 metres and of these nearly a third were to a car whilst only three per cent were to a bus stop. The differences in patterns and amounts of walking by people of different ages, gender and personal circumstances briefly covered in this paper highlights the complexity of the task facing those responsible for planning new and improving existing pedestrian environments.

Where in the road environment are people injured?

The location in the road environment together with the distribution over time of accidents involving injury to pedestrians are of importance to those involved with making the roads safer and more pleasant for all people as pedestrians.

In Britain about 95 per cent of all pedestrian casualties are injured on roads where there is a speed limit of 40 miles/h or less. There is comparatively little walking in rural areas and alongside the faster roads which explains the predominance of the problem in urban areas. Because pedestrians are vulnerable to injury when struck even at low speeds by vehicles, the proportion of all injuries sustained by pedestrians that are fatal or serious is higher in urban areas than for say, occupants of cars who are protected by the vehicle shell itself. The proportion of pedestrian injuries which were fatal or serious in Northampton for the period we studied was 37 per cent which is somewhat higher than the national average of 27 per cent.

Crossing main roads is difficult for pedestrians of all ages with half of all injuries occurring there. These roads are especially difficult for those over the age of 60 years as nearly two-thirds of injuries to this group of people are sustained whilst crossing these busy roads. This age group tended to be more severely injured than would be expected from national figures with 57 per cent being killed or seriously injured on Northampton's roads. Up to the age of 60 years more males are injured than females.

Children and young persons between the ages of 5 and 15 years comprise 39 per cent of all pedestrian casualties. None was killed in the five year period studied but 89 were seriously injured. Whilst older pedestrians tend to be injured when crossing or walking alongside main roads, children of school age are injured closer to home with 68 per cent of their injuries being sustained on the residential roads. In this age group boys are injured more frequently than girls in a ratio of approximately three to two.

The fact that most walking is done close to home has already been discussed and in relation to this it is of interest to look at where the casualty was injured to see whether this is on the very familiar streets close to home or on those some distance away. The young children were injured close to home with 80 per cent being injured within 1 km and over half injured within 400 metres. Older children venture further afield and this is reflected in their injury pattern with 40 per cent of the 10-15 year olds being injured within 400 metres of home and two-thirds within one kilometre. The wide ranging 16-19 year olds tend to be injured at places more distant from their homes with only 17 per cent of casualties being within 400 metres of home. The older pedestrians have an injury pattern more like that of children with 73 per cent of casualties being within 1 km of home. These patterns are not surprising given where the walking is done and serve to reinforce the need for safe access to shops, schools and amenities for all age groups whether they live on a quiet or a busy road.

Pedestrian accident risk

Much is known about casualty rates and levels of risk for drivers and passengers of vehicles as estimates of distance driven or passenger kilometres have been collected for some time, for example as part of the National Travel Survey (DoT 1993). These casualty rates are expressed per 100 million kilometres. The casualty rate in common usage for pedestrians is expressed in terms of per 100 000 population as no current measures of exposure exist for walks of all lengths on different types of road by people of different ages.

We have estimated that the residents of Northampton walk 133 000 km on an average day and cross some 636 000 roads. All this activity results in about 0.43 pedestrian casualties a day. Using the casualty rate based on population we find the rate each day is 0.26 per 100 000 head of population which gives an annual rate of 95.4. Expressing this as a casualty rate where the level of exposure is taken into account gives 411 casualties per 100 million kilometres walked or 66 casualties per 100 million roads crossed. Whilst in reality it takes the people of Northampton nearly three years to walk this far it does provide a casualty rate which is expressed in similar terms to those for drivers and passengers of vehicles. For 1992 data the casualty rate per 100 million vehicle kilometres for bicycles was estimated to be 526 and for car drivers to be 34 per 100 million vehicle kilometres (DoT 1993b).

Estimates of the amount of walking indicate that females walk a greater distance and cross more roads whilst the casualty information indicates that males are injured more frequently than females. When the amount of walking is considered along with the numbers of casualties we find that females have a casualty rate of 263 per million kilometres walked, or 54 per 100 million roads crossed whilst the corresponding rates for males are higher at 386 and 82 respectively.

The addition of exposure to the pedestrian casualty information puts into a different order the risks of injury to people of different age and gender and these are shown in Figures 2 to 4. The addition of exposure data confirms our knowledge that the risk to children on all parts of the road network is higher than for adults but it brings new insights into the vulnerability of the older pedestrian. Young adults aged 20 to 24 years are also at higher than average risk of injury when walking alongside traffic or crossing the road. When teenagers leave school they appear to become much better able to keep themselves safe on the road as pedestrians. This 16-19 age group walks the furthest per person and crosses more roads yet the rate per kilometre walked or per road crossed is only one third that of children of school age.

The casualty rates per 100 000 head of population indicate that boys aged 5-15 years are about one and a half times more at risk of being injured as pedestrians than girls of the same age. When exposure is taken into account we find that boys and girls aged 5-9 have very similar casualty rates per kilometre walked but boys are

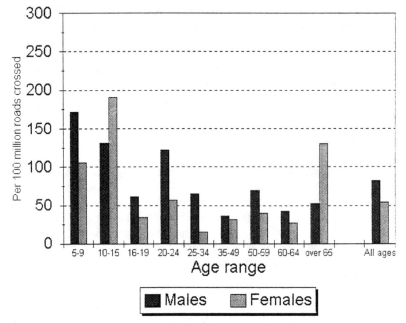

Figure 4   Casualty rates per 100 million roads crossed by gender and age

one and a half times more likely to be injured when crossing the road. Girls aged 10-15 appear to be less able than boys of the same age either to walk alongside or to cross the road safely. Indeed girls of this age have the highest casualty rates of any age group for both kilometres walked alongside traffic and for roads crossed. The increase in ability of females in late teenage years to keep themselves safe from injury as pedestrians is even more marked than for males of the same age.

Throughout adulthood up to the age of 64, females have lower casualty rates than males, the rate for young adult females being half that for males of the same age. Older female pedestrians are as much at risk as children aged 5-9 years when walking alongside traffic and about the same as 10-15 year old males when crossing the road. Older females are two and a half times more at risk of injury as pedestrians than are males of the same age for the same distance walked or number of roads crossed.

Just over half the pedestrian casualties occur on the main roads in Northampton but they only account for about 20 per cent of the total length of roads in the Borough and only 16 per cent of the walking of Northampton residents is alongside these types of road. For all residents of households aged 5 years and over the casualty rate per head of population indicates that the average person is just over twice at risk on the main roads as on the residential roads. However, when exposure is taken into account the relative risk is more than six times for walking alongside traffic and more than nine times for crossing the road.

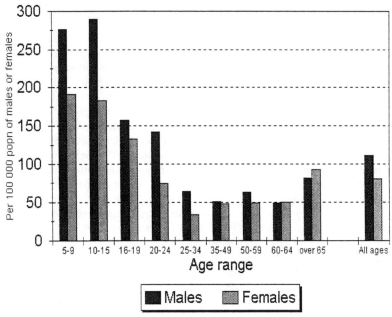

Figure 2   Casualty rates per 100 000 population of males or females by age

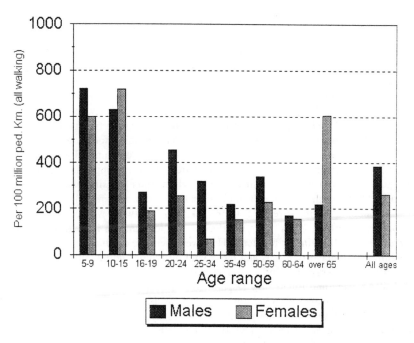

Figure 3   Casualty rates per 100 million pedestrian kilometres (all walking) by gender and age

Towards a safer road environment

The findings of studies of pedestrian behaviour have implications for policy decisions at both national and local level in regard to the provision of education and training for all road users, for enforcement, for the design and layout of new roads in urban areas, and the improvement of existing roads both in terms of management of both traffic and pedestrians and in changes to the road environment itself.

The study reported here highlights the need for education and training of people of all ages, especially the young. In order to increase the effectiveness of such programmes the awareness and knowledge of the problems faced by young people should be raised among those who are responsible for their safety such as parents and guardians, teachers and school governors. In many Local Authorities, Road Safety Officers are introducing in-service education and training programmes for these groups of people. Far too many children and young people are injured whilst they are crossing the road so the challenge is for educators, planners and road safety professionals together to identify locations and behaviours which can be targeted for remedial treatment. That children are walking less is of great concern to those responsible for promoting healthy transport policies and lifestyles. Children should be encouraged to walk more but this can only be achieved if the environment is improved to make it both safer and less threatening for children and their parents.

The older pedestrian is seen to be almost as much at risk as children and this is an issue which also needs addressing by education and training. The older age group has for too long been left out of the planning process and it is a group dominated by women as they tend to live to a greater age than men. These women tend to be frail and liable to severe injury when struck by a vehicle. They are much less likely than average to have a car thereby making it important both to understand their mobility problems and to work with them to make their trips to the shops, post office and elsewhere safe and with reduced threat from fast moving traffic.

It is encouraging that so many people with disabilities that make walking difficult do go out on foot. The need for sympathetic planning of the road network and transport systems is paramount to enable all people to be independently mobile. This is highlighted by the Department of Transport's Mobility Unit who have a commitment to 'achieving, where possible, a fully accessible transport system within a barrier-free pedestrian environment' (DoT 1995a). Planning for people with disabilities includes planning for the elderly as so many of them are hampered by reduced mobility. The main changes are the introduction of more dropped kerbs, tactile paving to guide and warn those with impaired eyesight to safer crossing places and improvements in the design of audible signals at pedestrian crossings.

People do one seventh of their walking on footways with cracked paving or badly worn surfaces. This is not a reflection on Northampton in particular as most urban areas suffer similar problems. However, there are implications for the ease of mobility for older pedestrians and those with difficulties going out on foot. Department of Trade and Industry figures suggest that up to ten times as many people attend Accident and Emergency Departments of hospitals with injuries sustained in falls on the footways and other transport areas as are injured in road traffic accidents (DTI 1990). Whilst it was not possible in this study to calculate casualty rates other than for road traffic accidents, investigation of other aspects of pedestrian safety are important when considering the pedestrian environment in its wider sense.

It is not just the older and younger pedestrians who are at particular risk as the study has highlighted the young adults, especially males. How to reach this group poses a challenge to educators at both local and national level. Young girls aged 10-15 years are at greatest risk of injury of any age group. This is interesting because they walk about 120 m less than the average person and about 150 metres less than the average woman. One possible explanation put forward by some is that these girls do not have the same experience of the road environment as boys of the same age because their movements are more restricted by their parents and the girls themselves do not venture out as much as boys.

Besides the age and gender differences the most important finding to emerge, and perhaps the one with the most far reaching implications, is that of the high relative risk of injury on main roads compared with local distributor and residential roads. In turn, the local distributors carry a higher risk to pedestrians than the residential roads. The problem is complex due to the interplay of function and use of these roads in providing routes for traffic to gain access to the commercial centre and to move goods and people from one part of the town to another. This conflicts with the needs of pedestrians to be able to cross the road safely without needing to walk undue extra distance or incur undue delay. Sixteen per cent of the walking and fifteen per cent of the crossings produce nearly 50 per cent of the casualties. Those with responsibility for the safety of the road environment need to work closely with the educators and Police to understand more clearly how to reduce the conflict between pedestrian and vehicular movement on this type of road in order to reduce risk to pedestrians.

The large differences in risk of an accident in daylight and darkness point to the need for further work to be carried out to increase our understanding of the difficulties encountered in darkness both by pedestrians and by other road users.

The amount of walking at different distances from home can provide input to the discussion on how planners and those responsible for the safety of the road environment enter into dialogue with people about different road safety problems and possible solutions to them. People's perceptions of their environment may well differ depending upon how familiar they are with it, which in turn may depend on the amount of walking and social interchange with neighbours. More work is needed to understand how people relate to different road environments.

This study has made an important contribution to increasing the understanding of the patterns of walking of pedestrians of different ages and gender in different parts of the road environment. Its findings can contribute to the discussion about the direction of future research into pedestrian safety and provides more information than was hitherto available to input to policy decisions about the safety of the pedestrian environment.

The study also provides a methodology to enable comparative studies to be undertaken to enable an exchange to made of more systematically collected information between regions or between countries. This in turn will enable good practice in relation to decisions about the pedestrian environment to be shared more widely and help us to put into perspective by the addition of exposure data the apparently poor pedestrian safety record in Great Britain compared with some other European countries.

Acknowledgements

The study was undertaken on behalf of the AA Foundation for Road Safety Research between December 1991 and April 1994. Mike Boyle and Brian Langer administered the study and Barbara Sabey was the Technical Adviser to the Foundation and her support, guidance and encouragement are recorded with thanks.

The text of this paper draws on material in the published report of the study and the contribution is acknowledged of the other authors; Julie Cave, Alan Morrison, Richard Allsop and Andrew Evans of UCL, and Carin Kuiper and Luis Willumsen of Steer Davies Gleave.

Special thanks go to the 1037 people of Northampton who gave their time to answer our interviewers' questions and who so conscientiously logged their pedestrian activity and to those who contributed in smaller ways by responding to our surveys.

Thanks are also due to the staff of Northamptonshire County Council, Northampton Borough Council, Northamptonshire Police, Nene College, Northampton General Hospital and to the interviewers who collected the household and personal walk information.

References

Department of Trade and Industry (1990) *Home and Leisure Accident Research. Twelfth Annual Report: 1988 Data,* (Department of Trade and Industry, Consumer Safety Unit, London)

Department of Transport (1993a) *National Travel Survey 1989/91. Transport Statistics Report,* (HMSO).

Department of Transport (1993b) *Road Accidents Great Britain 1992,* (HMSO)

Department of Transport (1995a) *Mobility Unit 1995-96 Overview.* (DoT)

Department of Transport (1995b) *National Travel Survey 1992/94. Transport Statistics Report,* (HMSO).

Department of Transport (1995c) *Road Accidents Great Britain 1994,* (HMSO)

Hillman, M. Adams, J. and Whitelegg, J. (1990) *One False Move... a study of children's independent mobility,* (Policy Studies Institute, London)

Ministry of Transport (1963) *Traffic in towns - a study of the long term problems of traffic in urban areas.* (Buchanan Report), Reports of the Steering Group and Working Group appointed by the Minister of Transport, (HMSO)

Office of Population Censuses and Surveys (1992) *1991 Census. County Report: Northamptonshire Parts 1 and 2,* (HMSO)

Oxley, P.R. (1993) Special needs: a fairer opportunity for mobility, in P Stonham (Ed.) *Local Transport Today and Tomorrow,* (Local Transport Today, London) 61-65.

Todd, J.E and Walker, A. (1980) *People as pedestrians.* Office of Population Censuses and Surveys (Social Survey Division), (HMSO)

Ward H, Cave, J.A. Morrison, A. Allsop, R.E. Evans, A.E. Kuiper, C. and Willumsen, L.G. (1994) *Pedestrian Activity and Accident Risk,* (AA Foundation for Road Safety Research, Basingstoke)

Cognitive Quality in Advanced Crew Systems Concepts

A TEST PILOT'S PERSPECTIVE ON ADVANCED AVIONICS

Wing Commander Ian Burrett RAF

Officer Commanding
Heavy Aircraft Test Squadron
DTEO Boscombe Down
Wiltshire UK

With the ever increasing sophistication of avionic systems, we are fast
approaching the situation in which industry can offer systems that,
individually, are manageable and highly capable but which are so complex
to operate together that aircrew workload can become intolerable; in some
areas we are probably already there. This problem is partly caused by a
tendency to procure sub-systems before fully considering how they will be
integrated into the cockpit; perhaps more importantly, the technology
required for full integration is only now bcoming available. Future
combat cockpits are likely to have some sort of Electronic Crewmember
which, in theory, promises to solve many of the pilot's problems.
However, there are a number of fundamental issues to be resolved. This
paper deals with some of these issues from a test pilot's perspective.

Background

Until the 1970s, aircrew had fairly rudimentary avionic systems at their disposal:
simple analogue instruments, stick and throttle, a weapon control system, a weapon aiming
system and, if they were lucky, a radar and some sort of navigation system. The crew had
very little information to assimilate and very few systems that required active control.
Unfortunately, the cockpits of most of these early generation aircraft were ergonomic
disaster areas so they could be extremely difficult to operate effectively, even though the
roles they had to perform were relatively simple. On the other hand, most aircraft had at
least 2 crew to share the workload. There were single-seat aircraft around, such as the
Lightning, in which workload was high, but manning and recruitment in the RAF was high
enough to assure a steady stream of more capable aircrew onto more demanding aircraft.
Against this background, the role of the test pilot was fairly straightforward.
Aircraft performance and handling qualities were considered paramount so flight testing
concentrated on aspects such as manoeuvrability, speed, maximum range and so on.

Cockpit design was, unfortunately, a secondary issue. Flying the Buccaneer, for example, was akin to driving a car with the fuel gauge embedded in the back of the passenger seat and the gear stick hidden under the driver's seat. Unfortunately, like many aircraft of this generation, the human was not considered early enough in the design process. By the time an aircraft reaches flight testing it is far too late to make substantial design changes. Therefore, a cockpit had to be considered acceptable if most instruments could be seen and all of the controls could be reached, given some manual dexterity. The test community could do little more than apply restrictions on the way an aircraft was operated to ensure that the impact of cockpit deficiencies was minimised, particularly if there was some risk of an accident. The Jaguar Navigation and Attack System is a good example: The original system was so poorly integrated on the aircraft that the pilot sometimes had to dedicate most of his attention to operating the system, at the expense of other tasks such as lookout and ground avoidance; this is thought to have contributed to a number of fatal accidents. However, it was almost 10 years before a more satisfactory system could be installed; in the meantime, a number of operating restrictions were applied to the system which reduced the aircraft's operational effectiveness.

Way Forward

The Ministry of Defence has made some progress in this area by putting human factors issues at the front end of the procurement process and other nations have made similar changes in their procurement procedures. Unfortunately, there is still a long way to go because it is still too easy to pay lip service to Human Factors by simply ensuring that **individual** avionic systems and sub-systems are easy to operate. We need to develop tools and procedures that allow us to assess the effectiveness of the **total** system, including, of course, the aircrew. In the UK, the research community is now very active in this area.

The operators also have alot to contribute because we need to be able to state exactly what our systems must be capable of; we can hardly expect industry to demonstrate the effectiveness of an avionics system if we cannot describe in detail what the mission is. This is a difficult problem because we can no longer predict the threat we face next year, let alone in 15 years time when current research programmes mature.

The aircrew community will also need to change its philosophy if it is to cope with the demands of future cockpits. To date, as systems of increasing complexity have been added, aircrew have adopted a 'can do' attitude, content if a sub-system contributes only a small increment to his or her total workload; that is what they are trained for and paid to do. However, this can only go so far. For example, EF2000, one of the most sophisticated aircraft in the world, has been designed from the outset with Human Factors as a priority. However, the cockpit workload will still be very high in some scenarios because there are simply so many systems with which the pilot has to interact; each of these systems increases total workload to some degree but very few could be considered to **share** the intellectual workload. I would suggest that the workload on current generation aircraft represents a limit. On future combat aircraft, aircrew will have to dispense with the 'can do' attitude and demand avionics systems that will take on a much greater share of tasks that have previously been in the human's domain.

Current aircrew would blanch at the idea of simply relinquishing their authority to the aircraft avionics (and with good reason, as discussed later). However, they might accept that they could **share** workload with, and accept advise from, an Electronic Crewmember; in my view, they will have to. However, there are many fundamental issues raised by this concept which will have to be resolved. Some of the main issues are discussed below, from a pilot's perspective.

Command

Perhaps the most important issue is whether the pilot should still be in command of future combat aircraft, or whether primary authority should be given to an Electronic Crewmember. There is probably not a fighter pilot in the world who has any doubt about the answer to this - aircrew must be in command. The reason is essentially the same as that for having a human in the cockpit at all. Modern computers, particularly Knowledge-Based systems, offer a quite remarkable leap in decision making capability, but we cannot directly compare these machines with the human brain. We cannot even agree on a model of human intelligence yet, so we have no way of producing an electrronic equivalent (an alternative perhaps, but not an equivalent). Therefore, I believe it will be some considerable time, if indeed ever, before an Electronic Crewmember could take the intellectual lead in aircraft cockpits. There are some experts in Artificial Intelligence who would disagree with this view but they are a long way from convincing the end users.

There are also strong political arguments for retaining human command of combat aircraft. Commanders will continue to feel unable to trust electronic systems to act autonomously for the foreseeable future so they will continue to apply restrictive rules of engagement; for example, pilots may have to visually identify a target visually before attacking, to check that the sensors have correctly identified that a target is hostile. Whilst such restrictions can prevent systems from being used to their full potential, this is a small price to pay for minimising the risk of civilian casualties or unnecessary collateral damage which is one of the prime considerations on the modern battlefield. These rules of engagement are also the politician's final means of controlling events on the battlefield and politicians are even less likely than aircrew to entrust this control to computer systems. Thus, even if we develop intelligent, predictable and trustworthy avionic systems, this is more likely to translate into an increased confidence of success for the human electronic team, rather than a relinquishing of control.

There is another important area in which electronic systems are found wanting. Computers cannot replicate the ingenuity of the human mind that enables us to respond to unexpected situations in novel and unpredictable ways. For example, the ability to grasp an opportunity, apply our ingenuity and creativity, and determine a course of action unique to the individual and situation - a course of action that a computer might consider illogical. This is a great strength because it allows us to surprise an adversary, which is one of the key principles of war at all levels from grand strategy down to a brief air combat skirmish.

Only when this capability can be achieved by a computer do I feel that we should consider replacing the human in the cockpit. Until then, our efforts should be directed towards improving the way information flows to and from the human, providing him or her with timely advise that can be easily assimilated and taking on many of the mundane tasks which currently take up alot of the pilot's attention. There is a perfectly adequate brain in the combat cockpit already so, rather than trying to replace it, we should be trying to improving its effectiveness.

Crew Composition

The development of future advanced avionics will have to cater for a number of factors in the composition of the crew, particularly if an Electronic Crewmember is to be considered part of that crew.

First, the composition of a crew can have a marked affect on the way an aircraft is operated, from routine tasks such as fuel management, right up to major tactical decision making. Essentially, each member of the crew adapts his technique to suit the experience and capabilities of the other crewmembers. Future avionics may need to do the same.

Second, even a constituted crew (that is to say a crew who routinely and frequently operate together) can function quite differently from one mission to the next, depending on fatigue, distraction, motivation and so on. To what extent should the avionics cater for this vast array of human foibles?

Third, aircrew we can expect to be flying combat aircraft in 15 years time will be very different to those operating today. We need to be developing systems that will be used by people who are currently still at nursery school, not for us die hards who can barely use a computer and secretly wish we still flew Spitfires.

Fourth, we will probably have to accept that the average experience of future combat aircrew will be much lower. The days are long gone when people joined the RAF with the single ambition of flying fighters until they were dragged off to a rest home; today's student pilots are already talking about which airline they want to join once they have had a few years fun. As a result, we could find that most front-line pilots have only been operational for a couple of years.

Fifth, by the same token, we could be forced to reduce the time spent in training in order to maintain our front line strength. Therefore, the cockpit must be designed so that aircrew can become fully operational very quickly. We may also need to standardise cockpits as much as possible so that aircrew can transfer between aircraft with minimal conversion training.

Finally, with fewer aircraft types in the inventory, we cannot accept a demanding aircraft and expect to be able to cream off the most capable aircrew to fly it. Future aircraft will have to be designed to be flown effectively by pilots who, today, would only just get through training.

Crew Numbers

The question of how many crew we will have in future combat aircraft is currently being hotly debated in many Air Forces. On the one hand, a single-seat aircraft is normally cheaper to procure and cheaper to run. Proponents of single-seat aircraft argue that, with good enough avionics, it should be possible to perform many roles effectively with a single-seat aircraft, even though the same roles used to require two crewmembers. However, others will argue that, given equally sophisticated avionics, a 2-man crew will always be more effective. This is probably true, but with limited financial resources, the real question is whether, say, 200 2-seat aircraft would be more effective than 250 single-seat aircraft. At the moment we do not have the answer to this, largely because of the points made earlier that we need to develop means of measuring performance and we need to define more accurately the missions that future aircraft will be expected to perform.

Crew size has a direct relevance to two aspects of advanced avionics. First, the human factors research community has an important contribution to make in determining whether the avionics we can expect to be available will, indeed, enable one man to operate effectively over the future battlefield. Second, the flow of information and team dynamics of 2-seat aircraft is quite different to that of a single-seat aircraft so the development of future avionics may have to follow 2 separate paths at some stage.

Trust

Finally, I would like to make a few points about the issue of trust. The first point is that, notwithstanding the concerns about future experience levels, we should not over-estimate the degree of trustworthiness we require of future avionics or aircrew. Aircrew are invariably highly motivated and, more importantly, have a highly developed survival instinct. For example, the Tornado routinely flies at low level in cloud using the Terrain Following Radar (TFR). It would be a bold pilot who trusted this critical task to the electronics alone but the operation is safe because the pilot monitors the system, ready to override it if necessary. It would be an even bolder pilot who attempted to fly at low level in cloud without the TFR. In other words, trust is earned by the **combination** of human and machine, not by the machine, or the human, alone.

On the other hand, if future avionic systems are to **share** the workload, we will have to reduce the pilot's involvement in the control loop for some tasks. This will require a detailed knowledge of the risks involved, coupled with some difficult decisions about the risks we are prepared to accept. As a general rule operators will expect a very high level of confidence in potentially life threatening situations, so we would expect to exploit the capabilities of both human and machine. Conversely, for routine tasks, in which the consequences of a mistake are 'tolerable', we will allow the machine to make decisions autonomously. Between these extremes, we may need to develop a range of predictable automation levels. Operators will have to resist the temptation to be conservative in their definition of 'tolerable'; we will have to relinquish as much control as we dare if we are ever to meet the objective of sharing workload.

Conclusion

This paper has outlined some of the major issues that need to be considered by both systems engineers and human factors experts involved in the development of advanced avionics for future combat aircraft. There are many others. The effort, and cost, of solving all of these over the next few years will be enormous and, indeed, will probably only be achievable through international collaboration between the major research centres, the aeronautical industry and the test flying communities. Most important, the front-line needs to be prepared for a radical change in the way combat aircraft are operated.

COGNITIVE COMPATIBILITY: ADVANCED DISPLAY CONCEPTS FOR FUTURE AIRCRAFT CREWSTATIONS

Dr. John M. Reising
Kristen K. Liggett

Advanced Cockpits Branch
Wright Laboratory
Wright-Patterson Air Force Base, OH

David C. Hartsock

Veda, Incorporated
5200 Springfield Pike
Dayton, OH 45431

With the rapid advancement of digital avionics technology and the development of electro-optical devices for cockpit use, designers can easily overload the pilot with information which may or may not be necessary for the task at hand. Initial indications are that pilots perform well with the first generation of these new devices, but how about the next generation? This may depend on how well the pilot and the cockpit communicate. Displays that provide a mental model of the workings of the aircraft system in question, advanced interfaces between the pilot and aircraft, and the ability to summarize meaningful information are keys to this compatibility.

Introduction

Digital avionics technology is rapidly becoming a key factor in the success of modern aircraft. The effects of digital technology are especially being felt in the cockpit where electro-mechanical (E-M) control and display technology is rapidly being replaced by electro-optical (E-O) devices in both military and civilian aircraft cockpits. For example, both the new Boeing 777 on the commercial side and the F-22 on the military side are intended to have E-O displays (active matrix liquid crystal displays (AMLCD)) that cover the most important portions of cockpit real estate. Also, with the advent of such technologies as touch sensitive overlays and eye control, the same physical devices serve both as control and display, blurring the previously held careful distinction between the two. A third factor facing the pilots of modern aircraft is the unprecedented amount of automation in the cockpit (Hughes and Dornheim, 1995). As the conversion continues, there is concern as to the pilot's efficiency with this equipment. "Somewhat paradoxically, machines that can do more, and do it faster, provide the basis for systems that are increasingly demanding of the human operator, particularly in terms of cognitive requirements" (Howell, 1993, p.235). Initial indications are that pilots perform quite well with the new equipment; however, the display formats are first generation -- reproductions of E-M instruments. How well will the pilots interact with the second generation display formats, more innovative controls, and increasing levels of automation?

The answer to the above question may depend on how well the information transfers back and forth between the pilot and the cockpit. Another term for this transfer is cognitive compatibility which Selcon defines as "The facilitation of goal achievement through the display of information in a manner which is consistent with internal mental processes and knowledge, in the widest sense, including sensation, perception, thinking, conceiving, and reasoning." (S. Selcon, personal communication, Dec 1, 1995). The purpose of this paper is to translate this definition into cockpit applications and provide practical guidelines which can be used by the crewstation designer.

Guideline 1: Conform to the Pilot's Mental Model

Mental models play an important part in the efficient operation of systems (Wickens, 1992). Since direct views of the inner workings of a system are often not possible, e.g., the flow of electrons inside the avionics system, displays are a major means of conveying information on the operation of a system. The closer the display formats conform to the pilots' mental model, the more beneficial they will be. Pilots form a mental picture of how a system should work (at a top level) and base their trust in the system according to how the system conforms to this picture or mental model. "A mental model is a representation formed by a user of a system and/or task, based on previous experience as well as current observation, which provides, (most if not all) of their subsequent system understanding and consequently dictates the level of task performance" (Wilson and Rutherford, 1989, p. 619).

Three ideas have been underlined in the above definition to stress its key aspects: representation, understanding and task performance. The pilots' representation leads to their understanding of the system which in turn leads to their performance with the system. For example, if the pilots' mental model of a fuel system pictures the flow valve lever in line with the flow when the fuel is moving and at right angles when the flow is shut off, then that is the way it should be portrayed. It is not important that the valves are electronic and do not have a flow valve handle to turn. An example of not conforming to an operator's mental model is illustrated by the use of reverse notation on early calculators. To add 3 + 2 = 5 instead of punching the keys in this order, the task had to be performed in the following order: 3 then 2 then +. Needless to say many operators had difficulty in using these calculators.

A cockpit display format called the Pathway head up display (HUD) (Reising, Liggett, Solz, and Hartsock, 1995) is an example of matching the display to the mental model because pilots are, in reality, tasked with flying a commanded path in space. Current HUD symbology (Figure 1) requires the pilot to keep the pitch and bank steering bars centered to achieve flight on the commanded path. Conversely, for the pilots to stay on the commanded path when using the Pathway HUD format, they simply fly down the roadway -- a clear mapping of the task and the display used to achieve the task. The roadway is made up of a continuous string of path blocks drawn in perspective, representing 45 seconds of flight into the future (Figure 2). The format incorporates a velocity index displayed in the shape of a small aircraft, called the follow-me aircraft. The follow-me aircraft is drawn to fly along the left side of the pathway at an altitude equal to 150 feet above the desired altitude. It always flies the

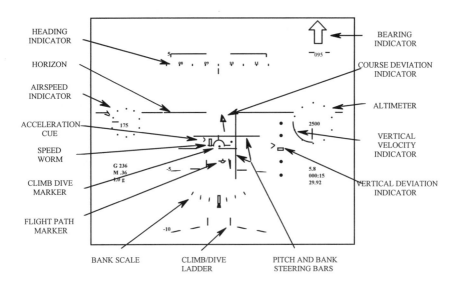

Figure 1. Standard HUD Symbology

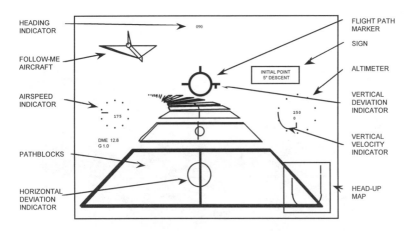

Figure 2. Pathway HUD Format

perfect commanded path at the correct airspeed. To fly the commanded path displayed by the symbology, the pilots only need to fly in an echelon formation on the right wing of the follow-me aircraft. This places the pilots approximately on the centerline of the course. "Road signs" alert the pilots of profile information such as navigation points, glide slope steepness, exact route changes, and a brief description of that change.

A comparison between the traditional HUD format and the Pathway HUD format showed that there was a significant difference in pilot performance -- subjects performed better using the Pathway HUD format than the standard HUD symbology in all cases. Pilots comments attributed the advantage of flying the Pathway HUD format to the fact that they could see their route in the form of a roadway from their present position to a point 45 seconds into the future. This allowed them to anticipate necessary control movements. The Pathway HUD format was described as instantaneous situational awareness.

Guideline 2: Make the Interface with the Crewstation Transparent

When pilots communicate with their team members in aircraft such as the Royal Air Force's GR-1 Tornado or the US Air Force's F-15E Strike Eagle, they frequently use voice, a very natural means of communications. Unfortunately, when pilots communicate with the aircraft's onboard computers, they are often forced to wade through numerous levels of indenture to reach the appropriate command. However, new interface devices have lessened this problem. Touch sensitive overlays and voice controls are two means to achieve easy communications. Boeing's new 777 uses touch sensitive cursor control devices so that the Captain and First Officer can achieve easier interaction with the multiple AMLCDs on the flight deck. Voice control is also becoming a viable means of pilot interaction with the cockpit (Figure 3). Some recent experiments with a connected word recognizer have shown that it is possible to use conversational commands and still achieve 99% recognition accuracy (Barry, Solz, Reising, and Williamson, 1994). The ultimate goal of the conversational commands is

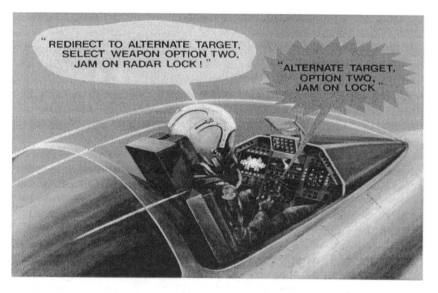

Figure 3. A Pictorial Example of Conversational Commands in the Cockpit

to emulate the interaction of the GR-1 and F-15E crews. An example of a conversational command is shown in Figure 4. The pilot has the capability with multiple paths to say four different phrases that mean the same thing. The pilot is not required to remember one specific phrase ("brittle" speech) to accomplish a task as in

Figure 5. As long as there are alternate paths to obtain the same goal, conversational commands are possible.

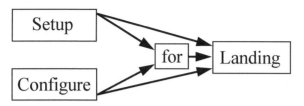

Figure 4. Multiple Path Voice Command

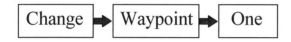

Figure 5. One path voice command

Guideline 3: Present Summarized Information

Even though an efficient means of communication exists between the pilot and the avionics system, it does not mean that information always flows in a clear and concise way between them. In modern military aircraft cockpits, pilots very often suffer from data overload, and with the inclusion of information from off-board sources, this overload problem may get worse. The designers of the displays can solve this problem by allowing the avionics system to present only summarized information. Icon based display formats, supplemented by text when necessary, are a very efficient way to achieve this goal. Steiner (1989) presented pilots with system status information in both text form and through icons. For very simple displays the pilots performed equally well with either type of display; however, as the displays became more complex, as measured in bits, the icon based format was clearly superior. The pilots comments also supported the fact that icon based formats were easier to interpret and gave better situational awareness. As an example of additional work in this area, Way, Hobbs, Qualy-White, and Gilmour (1990) developed a series of crew alerting system status displays. Figure 6 shows a summarization of a fuel pump failure. In this display, the

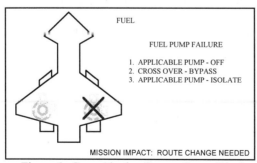

Figure 6. Crew Alerting System Status Display

problem is depicted with icons, instructions for solving the problem are outlined, and the mission impact is stated.

Conclusion

E-O displays, coupled with new cockpit control technology, have the potential of significantly helping pilots efficiently interact with the crewstation while maintaining adequate situational awareness. However, this technology by itself is no panacea; in fact, if not implemented in an intelligent manner, it could become a detriment to the pilot. The designer still needs to spend the majority of the time figuring out how the subcontrol modes coupled with the myriad of possible formats, "play" together to present the pilot with a clear picture of what the aircraft is doing and how to change its subsystems if required. The new technology is a two edged sword -- it offers the designer virtually unlimited freedom to present information to the pilot; on the other hand, it also gives the designer the opportunity to swamp the pilot in data. The cognitively compatible glass cockpit will be the key to making the technology the pilots' friend, rather than their foe.

References

Barry, T., Solz, T., Reising, J., and Williamson, D. 1994, The simultaneous use of three machine speech recognition systems to increase recognition accuracy, In *Proceedings of the National Aerospace and Electronics Conference*, (IEEE).

Howell, W. 1993, Engineering psychology in a changing world. In Porter and Rosenzweig (eds.) *Annual Review of Psychology* (Annual Reviews, Inc.) 231-263.

Hughes, D. and Dornheim, M. 1995, Accidents direct focus on cockpit automation. Automated cockpits special report, part 1. *Aviation Week and Space Technology*, 52-54.

Reising, J., Solz, T., Liggett, K., and Hartsock, D. 1995, A comparison of two head up display formats used to fly curved instrument approaches, In *Proceedings of the 39th Annual Meeting of the Human Factors Society*, (HFS).

Steiner, B. and Camacho, M. 1989, Situation awareness: icons vs. alphanumerics. In *Proceedings of the 33rd Annual Meeting of the Human Factors Society*, (HFS).

Way, T., Hobbs, R., Qualy-White, J., and Gilmour, J. 1990, *3-D imagery cockpit display development*, Technical Report WRDC-TR-90-7003.

Wickens, C. 1992, *Engineering Psychology and Human Performance*, (HarperCollins Publishers Inc.).

Wilson and Rutherford. 1989, Mental models: theory and application in human factors, *Human Factors*, **13**, 617-634.

DECISION SUPPORT AND TASK NETS

Erik Hollnagel, M.Sc., Ph.D.

Principal Advisor, Man-Machine Interaction Research Division
OECD Halden Reactor Project, P. O. Box 173, N-1751 Halden, Norway
e-mail: erik.hollnagel@hrp.no

"Intelligent" automation is feasible only if the target function can be clearly described and if the interaction between automated and human tasks can be ensured. The traditional approach is based on information structuring, which assumes that decision making is an orderly process. An alternative is proposed which uses task nets to describe the decision elements. Task nets represent the functions of the joint cognitive system in decision making, and make minimal assumptions about the nature of human cognition.

Introduction

One of the most important consequences of automation is how it has changed work with technological systems. In general, the introduction of automation has progressed in "bottom-up" fashion, going from repetitive manual actions, via simple control tasks, to more complex activities that involve "thinking" rather than "doing". Due to the capabilities of modern information technology, we are rapidly facing the challenge of "intelligent" automation involving cognitive tasks such as monitoring, planning, and decision making. This challenge creates problems that cannot be solved simply by applying the principles of already known solutions.

Whenever a function is to be automated, two questions must be answered. First, how the target function can be described in sufficient detail to enable automation. This must include a description of all the possible situations or conditions that can occur, since the adequacy of the automation otherwise cannot be guaranteed. Second, how an appropriate interaction between the automation and the remaining (non-automated) parts of the system can be ensured. As tasks move from simple manual tasks over basic control tasks to decision making, both questions become more difficult to answer. There are, for instance, no descriptions of decision making that have the details needed for extensive automation. Providing the proper coupling or interaction between the automation and the user also becomes more difficult. Since the operator clearly cannot be excluded from the loop, the question is how tasks and responsibilities shall be shared, how information shall be provided that enables the operator to perform appropriately, and how the operator can control the automation rather than be controlled by it. The success of "intelligent" automation requires that effective answers are found to both questions.

Decision Making: The Ideal And The Real

Decision making is often described as a formal process of reasoning where the decision maker first identifies the alternatives, then compares them, and finally selects the optimum one. While few experts seriously believe that this is how people actually make decisions, the normative description has nevertheless influenced most of the popular attempts to model human decision making which commonly depict a process divided into a number of stages. Although models may vary in level of detail the basic principle is always the same: that decision making can be adequately described in terms of a limited number of steps that are carried out in a prototypical sequence.

Although this type of model has been adapted by the mainstream of ergonomics and cognitive psychology, it is not universally accepted (Bainbridge, 1991; Hollnagel, 1993). More interestingly, some classical descriptions of decision making propose quite different paradigms. An interesting alternative is the "muddling through" description of decision making (Lindblom, 1959). Here decision making is seen as going through the following stages: (1) define the principal objective; (2) outline a few, obvious alternatives; (3) select an alternative that is a reasonable compromise between means and values; and finally(4) repeat the procedure if the result is unsatisfactory or if the situation changes too much.

It is not hard to argue that the latter description corresponds better to what people actually do. It is therefore worrying to note the discrepancy that exists between psychological or cognitive models of decision making and accounts of actual decision making. In the former, decision making must comply with the inherent constraints of the information processing model; in the latter, the control of decision making depends on the context. The two types of description provide different answers to the questions of what the target function is and how the interaction should be designed. According to the "ideal" description the target function is given by the prototypical version of decision making as information processing and the interface between humans and automation can accordingly be based on the principles of information structuring. According to the "real" description the target function must be based on an understanding of how people cope with the practical demands, and the interface must be based on the principles of task nets. Since the design of interaction and interfaces has more often been driven by theory than by practice, it is little wonder that ergonomics and automation design seem to face such serious problems.

Natural And Constrained Decision Making

In real life people regularly make decisions in an informal manner by looking for the first acceptable solution. This is not as haphazard as it may sound, since the search can be facilitated by preparations of various kinds. An example is mission planning, where the possible conditions are examined in advance and the potential situations are evaluated (Amalberti & Deblon, 1992). From this basis a set of responses can be defined that are readily available and likely to succeed. A variation of that is formalised training, where appropriate responses are taught and, hopefully, learned. In both cases, by preparing for the potential situations that may occur, actual decision making is made faster and less resource demanding.

Work in a natural environment requires a mixture of skilled actions, established routines or procedures, and more or less complex decisions. Here decisions occur as a natural part of the way in which the situation develops or as part of a context. Decision making is only one of several activities that the operator must carry out, and decisions are made whenever the situation requires it. This means that the natural decision making is prepared, that the context and the conditions are known, and that the decision is integrated in the overall pattern of activities. The need to make a decision is not suddenly forced upon the operator, but can be anticipated and fitted into the natural flow of work.

In contrast to that, constrained decision making is separated from the flow of activities, and is often the only thing the operator is required to do. Work in a constrained environment is confined to a specific range of activities, usually thinking rather than doing. Consider, for instance, highly automated systems, such as glass cockpits or NPP control rooms, where decision making is a specific and partly isolated activity. According to the premises of system design, there should be a well-defined transition between situations where the system is in charge and situations where the operator is in charge. This goes both for manual actions and for complex decisions.

In reality, however, the transition between automated and human tasks is often ill-defined. At the upper range of automation we find situations where either the systems has failed or which have not been anticipated in the design. Such situations are left to the operator, who in particular is given the responsibility of making the required decisions. System design may assume a clear division between automated and manual functions where the human operator takes charge when the stipulated conditions are met. But in practice operators often interfere when they perceive that there will be a need for intervention or when they no longer trust the system - even though it may not be the situations dictated by system design.

The design of "intelligent" automation usually presents the operator with several inconsistent principles. On the one hand he is not supposed to intervene in the process while the automation works, and may in many cases be directly prevented from doing so. Yet the interface and displays often maintain the illusion that such intervention is possible, by providing information that enables the operator to follow the developments of the process. On the other hand, when operator intervention and decisions are required they are usually not well supported by the interface. Natural decision making requires a continuous display of information. Constrained decision making requires a comprehensive presentation of all the relevant information at the beginning of the situation. The reality is neither one nor the other, but rather a blurred situation that is caused by the seemingly forgotten discrepancy between the real and the ideal.

Decision Support Through Information Structuring

In the design of intelligent decision support the adage has been: (1) to provide the right information, (2) presented in the right way and, (3) at the right time. Unfortunately, this is easier said than done. To see why this is so I will briefly examine each of the three conditions of information structuring.

In order to provide the right information it is necessary to know what the situation is and which information the operator requires. This exercise is fairly easy to do in retrospect, but is much more difficult to accomplish prospectively, as discussed by Woods et al., (1994). There is a significant step from *post hoc* analysis to being able to specify in advance which information is required. To accomplish that it is necessary to identify all the potentially important situations and describe their information needs. Quite apart from the fact that this is susceptible to the n+1 fallacy, it creates the problem of reconciling the design of n specific displays with the goal of maintaining global situation awareness. The balance between the two is difficult to strike, particularly as current display technology forces the use of piecemeal or fragmented information presentation.

In order to present the information in the right way one must know both what the situation is and what the user needs. The latter presents a very difficult problem, since the user's state of mind can only be predicted in a rather general way. The preferred solution is to apply some type of user model, either based on stereotypes or specific theo-

ries of human information processing (e.g. May et al., 1993). The basic problem with
this is that the variation between and within users and domains is too large and that it is
impossible to anticipate the specific needs of a specific user at a specific time. The reported successes of various systems are usually due to human ingenuity and adaptability
rather than the qualities of the information presentation. In practice the alternative solution - designing a display as well as possible and training the users to understand it - has
more leverage and is therefore often chosen.

Finally, in order to present the information at the right time, it is again necessary
to know what the situation is. In this case it is fortunately possible, in principle, to find a
solution. The right time means that the information should be presented neither too early
nor too late. Most dynamic processes can be described in terms of limited number of
states, and the knowledge of the transition between the states can be used to determine
the right time for information presentation. Similarly, the time constants of the process
can be used to synchronise information presentation, at least in an ordinal sense. In most
cases an approximate solution is therefore possible.

In relation to "intelligent" automation, information structuring is an attractive
idea, although it has been very difficult to turn into reality. The main problem, and possibly also the main reason for the lack of significant results, is that information structuring requires that the situations - and in particular the transition between automated and
human tasks - can be analysed in advance and that there will be only insignificant differences between what has been anticipated and what will actually happen. Information
structuring also assumes that a person will respond in a predictable way to a given input,
an assumption that has been reinforced by the apparent simplicity of information processing models.

The ease by which specific situations can be analysed has made the problems
seem simple, but experience has shown that generic design principles cannot be derived
from an accumulation of particular solutions. Information structuring can be a powerful
technique if both the target function and the interaction are well defined. But this is precisely where the problems lie for "intelligent" automation and information structuring is
therefore not useful as a general principle. Information structuring furthermore tends to
favour prosthetic solutions, because of the need to prescribe what the information should
be and what the user will need. But this does not work well when decision making is
"muddling through" rather than rational choice.

Decision Support By Task Nets

An alternative to information structuring is task nets, which are based on the
principles of cognitive systems engineering, and in particular the concept of a joint
cognitive system. The purpose of a task net is to ensure that the joint system, i.e., the
operator together with the "intelligent automation", has performed all the necessary parts
of a decision. Task nets are based on the assumption that it is possible to define the elements that are sufficient to make a decision, but that is neither necessary nor useful to
prescribe a specific order or sequence in which they must be accomplished.

The elements of decision making can be derived either inductively or deductively
(e.g. Hollnagel, 1984). If we consider the description of real decision making, such as
muddling through, at least two alternatives must exist before a selection can be made;
the same is the case in rational decision making. Starting from the final choice the decision process can be traced backwards through various paths that correspond to generic
descriptions of decision making. The task net describes the elements of decision making

based on a goals-means relation. As shown in Figure 1 such a description can encompass formal as well as informal decision making. The pragmatic requirement is that some path has been followed through the task net, but not that it is one particular path.

The task net can be used to specify the minimal requirements to a decision element. Each element serves to accomplish a specific purpose or goal, and similarly has a specific set of pre-conditions that must be met before it can be accomplished. Thus "select alternative" can be carried out only if some alternatives have been found. Therefore, if it can be determined in a situation that alternatives exist, then it is unnecessary to invoke earlier parts of the task net. The alternatives may be the result of immediately preceding steps or the outcome of preparations and expectations, but all that matters is that they are there.

A further advantage of the task nets is that a specific element may be taken care of either by the "intelligent" automation or by the operator - or even by both. If some of the elements are performed by automation, the task net will show those which must be done by the human. For instance, the identification of the base state, i.e., the background information about the process and the initial conditions, may be automated via rule-based queries. Similarly, once the base state and the current state have been described the overall goals may automatically be defined, e.g. from plant specifications and operating criteria. Automation can also be made to depend on the conditions such that a decision element is carried out by machine "intelligence" only if enough details are available. Conversely, if the available information is insufficient for the automation to work, the operator may take over using the human abilities for pattern recognition or reasoning by association.

Finally, in task nets the organisation of the decision elements, i.e., their temporal sequence, is less important as long as the final choice is made in time. The representation in the task net shows the logical rather than the temporal structure between the decision elements. Task nets therefore do not impose a specific way of doing things or a fixed allocation of functions in the way that an information processing model does. Altogether, task nets have the following advantages for the design of "intelligent" automation:

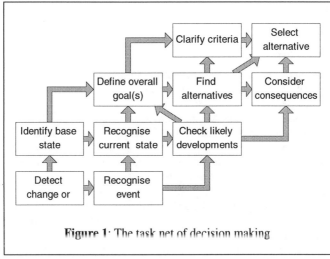

Figure 1: The task net of decision making

+ They can be used to identify the decision elements that can be automated, given a specific domain and application. The automation need not be rigid, but represents a possibility rather than a necessity.

♦   Task nets allow multiple solutions or multiple paths to a decision, and do not pre-scribe a specific path. They therefore allow a flexible distribution of the decision elements, according to the current conditions. Depending on time and capability, a decision element may be accomplished by "intelligent" automation or by the human.

♦   The "intelligent" automation may conceivably be parallel, i.e., several decision elements may be addressed at the same time, thereby improving the throughput. The specification of pre-conditions for decision elements can also be used to structure the situation.

The purpose of task nets is to ensure that the joint system has not missed any significant elements when the decision is made. Although there are multiple paths through the task net, not every path is correct. Thus if an alternative is selected based only on the recognition of the event, it is reasonable to characterise the decision as deficient. The task net can thus serve as a check list for the joint system. The specific implementation of this will depend on the domain and the nature of the tasks. It is not inconceivable that the complete monitoring of the task net can be done by the automation itself, e.g. using techniques from plan recognition or dialogue structuring. It may also be done by the operators, or via a supervisory function. In some industries, operators are trained to follow a set of rules when making a decision; there is no reason why the same practice cannot be applied to "intelligent" automation in a joint system. The task net can be used to answer the question of what the target function is, and how the interaction between automated and human tasks can take place. It recognises that human decision making more often is a "muddling through" than an orderly, rational process. The task net, finally, does not include any strong assumptions about the operator's behaviour or psychological "mechanisms", nor does it impose a specific order of working.

References

Amalberti, R. & Deblon, F. (1992). Cognitive modeling of fighter aircraft's process control: A step towards an intelligent onboard assistance system. *International Journal of Man-Machine Studies, 36,* 639-671.

Bainbridge, L. (1991). Mental models in cognitive skill: The example of industrial process operation. In A. Rutherford & Y. Rogers (Eds.), *Models in the mind.* London: Academic Press.

Hollnagel, E. (1984). Inductive and deductive approaches to modelling of human decision making. *Psyke & Logos, 5(2),* 288-301.

Hollnagel, E. (1993). Modelling of cognition: Procedural prototypes and contextual control. *Le Travail Humain, 56(1),* 27-51.

Lindblom, C. E. (1959). The science of "muddling through." *Public Administration Quaterly, 19,* 79-88.

May, J., Barnard, P. J. & Blandford, A. (1993). Using structural descriptions of interfaces to automate the modelling of user cognition. *User Modelling and User Adapted Interactions, 3,* 27-64

Woods, D. D., Johannesen, L. J., Cook, R. I. & Sarter, N. B. (1994). *Behind human error: Cognitive systems, computers and hindsight.* Columbus, Ohio: CSERIAC.

COGNITIVE COMPATIBILITY: A CONCEPTUAL FRAMEWORK FOR COGNITIVE QUALITY IN ADVANCED CREW SYSTEMS DESIGN

Robert M Taylor

DRA Centre for Human Sciences
Farnborough, Hants GU14 6SZ.

Recent problems with advanced aircrew systems have led to renewed interest in the cognitive quality of human-systems judged in terms of the performance of cognitive functions. Currently, there is a lack of formal methods specifying system cognitive requirements and for quantifying compliance in a systematic manner. Cognitive compatibility (CC), is believed to be a key determinant of system cognitive quality. This paper describes the background and approach to the development of CC-SART, an experiential measure of CC, and discusses some of the implications for systems design.

Requirements capture failure

We believe that the root of many of the problems of advanced system design lies in the general lack of understanding of how to capture and to specify users requirements for new systems with advanced technology. Current methods are not consistently reducing systems-induced "human error" or systems failures, or routinely reducing user workload or enhancing user situational awareness (SA). More specifically, we feel that the main weakness with current methods is their failure to capture user cognitive requirements.

Systems can only be as effective as the quality of the system knowledge applied in the process of system design. In designing systems with high levels of complexity in system functioning, it is increasingly difficult for the designer to maintain an appropriate focus on the role and needs of the human user. The experiential and knowledge gap between the designer and the user of complex systems is widening, and this chasm is becoming difficult to bridge. Differences between old and new systems are also widening, limiting the ability to transfer design solutions between generations of technology.

In general, new systems produce more data faster than previous systems. Also, the time available for users to perform information integration, assessment, and to make decisions is reduced with advanced systems compared with earlier simpler systems. Coupled with increasing automation, this changes the nature of users' tasks. The nature of users' tasks is becoming more thinking than doing, and more attentionally demanding. Such tasks include situation assessment, planning and decision-making. These user tasks involve relatively high level, cognitive functions, perceptual skills, and human memory and reasoning processes. The systems design requirements for such thinking tasks, and their associated cognitive functions and processes, are more difficult to define than physical tasks, and they change more rapidly with the changing operational environment.

Knowledge acquisition methods

Experience with knowledge acquisition methods in the creation of "intelligent" decision-support systems has revealed that the required procedural and perceptual knowledge is difficult to recall and to verbalise, and for the engineer to access, comprehend, and represent. Understanding and specification of the cognitive requirements of tasks is needed to fully anticipate system effectiveness during the design process when performance of cognitive functions is a major system task variable. Furthermore, the introduction of new technology to aid the user in the performance of cognitive functions, such as decision-support systems, and concepts for the human-electronic crew, will require definition and resolution of the cognitive task requirements, if this technology is to be integrated into systems in a systematic manner.

System cognitive quality

Assessment of Human System Integration (HSI) effectiveness, and of the success of human-centred design strategies, can be approached through consideration of the resultant system cognitive quality. Cognition is normally taken to refer to the part processes of the mind involved in human knowing such as conscious and unconscious reasoning processes, understanding and judgement. Hosting human cognition are mental or psychological processes. The performance of high level cognitive functions, such as situation assessment, planning and decision-making, and in particular the setting and maintenance of mission goals, is a key human contribution to advanced human-machine systems. Cognitive quality can be considered as the quality of performance of these cognitive functions, judged both in terms of the attentional or cognitive effort expended, which since cognition (and attention) is a scarce resource, should be minimised, and in terms of the benefits of the performance achieved.

Training is the traditional method of reducing cognitive effort in systems operation. Training is costly and it places a load on human memory; its effectiveness depends on the ability to recall the learning. Thus, any reduction in systems operation training need due to systems design can be considered as an index of increasing systems cognitive quality. By the same token, cognitive quality can be considered as an indicator of HSI success. Cognitive quality can be simply described as the conformance of the system with the specified cognitive requirements. This conformance can be achieved through consistency and compatibility with cognitive requirements. It is intended that through consideration of cognitive quality issues, progress can be made towards a better understanding of human problems in working with advanced, computer-based systems.

Cognitive compatibility

We have sought to take further our understanding of the relationship between cognitive quality, SA, user knowledge and systems design requirements by investigating the nature of cognitive compatibility (CC). CC is considered as the degree of consistency, congruency, and mapping between tasks on the one hand, and internal mental processes, knowledge and expectations on the other. Thus, CC is probably a key determinant of system cognitive quality.

Compatibility Types

Concern with compatibility in systems design originates from the ideas of Arnold Small on stimulus-response compatibility, presented to the first meeting of the Ergonomics Society in 1951, and the subsequent work of Fitts and Seeger (1953). McCormick and Sanders (1982) define compatibility in relation to human engineering design, as follows: " the spatial, movement, or conceptual relationships of stimuli and responses, individually or in combination, which are consistent with human expectations." In their taxonomy of different types of compatibility, conceptual compatibility is described as referring to conceptual associations, intrinsic in the use of codes and symbols, or culturally acquired associations.

More recently, Wickens (1984) has discussed the importance of compatibility (or congruence) between levels of representation which form the basis for understanding systems, namely the physical system (which is analogue), the internal representation or "mental model" (which should be analogue), and the interface between the two. Compatibility between the real system and the mental representation is a matter of training, which can be facilitated by anticipation in design. Andre and Wickens (14) use the notations S for stimulus, C for comprehension or cognitive understanding, and R for response, suggesting that S-C mappings are concerns of cognitive compatibility and S-R mappings are concerns of physical compatibility. Intuitive pictorial formats, icons, direct manipulation interfaces, and schema-based display formats are practical examples of the application of the principle of cognitive compatibilty in systems design.

Measurement of cognitive compatibility

A measurement of conformance is needed for the systematic improvement of CC. The range of candidate measurement approaches include performance-based metrics, physiological indices, memory probe measures of knowledge, and subjective self-report ratings. Our initial focus has been on the development of subjective measures, primarily for reasons of ease of implementation both in the laboratory and in the field, following the example of workload measurement (i.e. SWAT, NASA TLX) and our work on SA measurement using SART. Our experience indicates that validated subjective techniques can have both practical utility and predictive power (Taylor, Selcon, and Swinden, 1994))

Measurement of unobservable cognitive states, not directly available for analysis, presents practical and theoretical difficulties which affect the validity and reliability of the data. Establishing the validity and reliability of subjective, self report CC ratings presents particular problems. The validity of subjective assessments rests on the fundamental theoretical assertions that degrees of CC are experienced at levels of awareness, that awareness is mediated by working memory, and that the associated mental activity leaves a memory trace (Baddeley, 1994).

CC-SART Development

In order to investigate the subjective dimensions of CC, we have sought to develop an experiential measures approach, using a task created to manipulate CC variables, based on the following working model for types of CC (Taylor, et al, 1995):

$$CC = K((Md)(Sp)(Mv)(Cn)(Ns)$$

(CC is Cognitive Compatibility score; K represent constants; Md is modality compatibility; Sp is spatial compatibility; Mv is movement compatibility; Cn is conceptual compatibility; Ns is not yet specified compatibility e.g. social dimensions).

Task environment

The task was a highly abstract, computer-based simulation of flying an aircraft in tactical situations. Subjects were required to provide directional (left/right) responses to a multi-modal display of situational information. Information on other aircraft locations was presented visually and auditory, and subjects were required to make orienting responses (fly towards or away from) in relation to those locations. The presentation of the information was designed to provide correlated and uncorrelated task cues demonstrative of varying Md, Sp, Mv, and Cn compatibility. A total of 60 CC task situations were created demonstrating degrees of compatibility and incompatibility, with an equal number of correlated and uncorrelated task cue combinations.

Construct elicitation

In the first phase, 30 non-aircrew subjects were presented with sub-sets of the CC task situations, and personal constructs associated with CC were elicited using a Repertory Grid procedure, similar to that used to develop SART. Subjects were guided by a broad working definition for CC, namely "Cognitive compatibility of advanced aircraft displays is the facilitation of goal achievement through the display of information in a manner which is consistent with internal mental processes and knowledge, in the widest sense, including sensation, perception, thinking, conceiving and reasoning." They then provided subjective ratings on dimensions of the elicited constructs for 22 of the CC task situations. 56 construct dimensions (32 unique) were elicited and rated in this way. Guided by the strength of correlations of the ratings with task response times (RTs), and by the inter-correlations of the subjective ratings, the set of construct dimensions was reduced to 22.

In a second phase, 16 task situations, chosen to represent a range of difficulty, were presented to 20 non-aircrew subjects, and ratings were obtained on the reduced set of 22 construct dimensions. From analysis of the ratings obtained, and of their association with the task RTs, and from consideration of the structure of the semantic associations, 10 construct dimensions were identified as characterising the main CC variability in the task. The ratings of these 10 construct dimensions appeared to be organised in three main statistical clusters and three principle components dimensions. Following group discussions with subjects who provided the personal constructs, the three principal groupings of personal constructs were identified as Depth (or level) of Processing (DoP) - naturalness, automaticity, association, intuitiveness; Ease (or difficulty) of Reasoning

(EoR) dimensions - straightforward, confusability, understandability, contradiction; and Activation of Knowledge (AoK) dimensions - recognisability, familiarity.

In the third validation phase, the full set of 60 situations were presented for rating to 30 non-aircrew subjects, in a balanced experimental design. Semantic compression was used to define the 13 rating dimensions in terms of the meanings of associated constructs. Analysis of the results showed significant main and interaction effects on RT's and ratings of the compatibility variables, and provided evidence for the validity of the derivative primary constructs EoR and AoK. However, the DoP dimension generated a different pattern of ratings than the group sub-constructs. Subsequently, this dimension has been redefined as Level of Processing (LoP), to clarify the intended meaning. The resultant set of 3-d, 10-d, and composite 13-d rating scales are the basis of a new experiential measure of CC, designated Cognitive Compatibility - Situational Awareness Rating Technique, or CC-SART, available in paper and disc form (Apple Mac Hypercard).

Principle Psychological Dimensions
The principle psychological dimensions of CC can be defined as follows:
a . Level of Processing: The degree to which the situation involves, at the lower score level, natural, automatic, intuitive and associated processing or, at the higher score level, analytic, considered, conceptual and abstract processing.
b. Ease of Reasoning: The degree to which the situation, at the lower score level, is confusing and contradictory or, at the higher score level, is straightforward and understandable.
c. Activation of Knowledge: The degree to which the situation, at the lower score level, is strange and unusual or, at the higher score level, is recognisable and familiar.
A simple additive formula could be used for a single index from the ratings:
$$CC(c) = AoK + EoR - LoP$$
On the basis of this work, can simply define CC as follows: "...ease of processing with appropriate expectations". Further validation of CC-SART is being conducted through application of the measurement technique in experimental studies of crew system design concepts for improving cognitive quality and the performance of cognitive functions.

Cognitive requirements specification

The capturing of system requirements, and their incorporation into a system specification, is a key activity in the life-cycle of any product. The current approach, which is to integrate post specification through progressive prototyping and acceptance testing, will not suffice for future systems in which cognitive functionality will be shared between the human and machine system components. We are seeking to provide a generalisable framework for specifying cognitive requirements of systems which describes the scope of what needs to be known during the system design process. This framework will need to formulated in terms which can be translated into practical guidance for an integrated engineering solution to the cognitive requirements problem.

The psychological dimensions of CC incorporated in CC-SART have similarities with the Skill, Rules, and Knowledge (SRK) taxonomy or model of human performance (Rasmussen, 1983). Level of Processing intuitively equates to skill-based behaviour; Ease of Reasoning to rule-based behaviour; and Activation of Knowledge to knowledge-based behaviour, or intelligence. The SRK model considers humans as goal-orientated, and categorised into three phases. These phases encompass performance stretching from early knowledge-based behaviour (unfamiliar situations but goal awareness), to rule-based behaviour (derived from previous experience and with relatively slow use of feedback from the work environment), to skill-based behaviour (performance is smooth with little conscious control). The SRK performance distinctions have parallels in many other approaches to the categorisation of human behaviour. For example, the SRK model can be compared to Fitts' three skill learning phases of 'early - cognitive', 'intermediate - associative', and 'final - autonomous' (Fitts & Posner, 1967).

We propose to examine the efficacy of the cognitive compatibility taxonomy on which the CC-SART is based, coupled with the SRK model and other related taxonomies, for cognitive requirements specification, and for the development of protocols for a function-based, cognitive task analysis. Assessment of compatibility with cognition will aid in the capture of human cognitive schemas. Compatibility assessment will also aid in the evaluation of the share of total system cognition to be in residence within the machine. The use of validated CC assessment tools, such as the CC-SART, could offer substantial assistance in the design of the context sensitive, Cognition Adaptive Crew Station of the future. Such a crew station will be designed with the aim of promoting efficient HSI and, as a resultant, promoting individual, mission, and system related SA.

References

Andre AD and Wickens CD. 1992. Compatibility and consistency in display-control systems: Implications for aircraft decision aid design. Human Factors, **34**, 6, 639-653.
Baddeley A. 1993. "Working Memory and Conscious Awareness", In Collins AF, Gathercole S, Conway M, and Morris P (Eds),"*Theories of Memory*" (Hove, Erlbaum).
McCormick, E. and Sanders, M. 1982. *Human Factors in Engineering and Design*, (New York: McGraw-Hill).
Fitts, P.M., & Posner, M.I., 1967, *Human Performance*, (Belmont, CA: Brooks/Cole).
Fitts, P.M. and Seeger C.M. 1953. S-R compatibility: Spatial characteristics of stimulus and response codes. Journal of Experimental Psychology, **46**, 199-210.
Rasmussen, J., 1983, Skills, Rules, and Knowledge; Signals, Signs and Symbols, and Other Distinctions in Human Performance Models, IEEE Transactions on Systems, Man, and Cybernetics, **SMC-13**, No.3, May/June.
Taylor, R.M., Selcon, S.J., and Swinden, A.D. 1994. "Measurement of Situational Awareness and Performance: A Unitary SART Index Predicts Performance on a Simulated ATC Task", In "*Human Factors in Aviation Operations*". Fuller R., Johnston N., and McDonald N. (Eds),3. (Aldershot, Avebury Aviation).
Taylor R.M., Shadrake, R, Haugh J, and Bunting A. 1995 . "Situational Awareness, Trust and Cognitive Compatibility". In "*Situation Awareness: Limitations and Enhancement in the Aviation Environment*". AGARD Conference Proceedings. (AGARD, Neuilly-sur-Seine). April (In press).
Wickens C.D. 1984. *Engineering Psychology and Human Performance*, (Columbus: Merill).

COGNITIVE QUALITY IN ADVANCED CREW SYSTEM CONCEPTS: THE TRAINING OF THE AIRCREW-MACHINE TEAM

Iain MacLeod

Aerosystems International
West Hendford, Yeovil
Somerset, BA20 2AL, UK

Training the human in complex skills is expensive and difficult. Training a human team in the complex skills associated with the teamwork needed to operate advanced systems is even more expensive and difficult. To construct a team consisting of human and machine members is currently impossible but is desired. This paper looks at some future problems inherent in the promotion of quality in, and training of, the conceptual human-machine team.

What is an Aircrew-Machine Team?

A team can be described as a set of two or more people or system components who interact, dynamically, interdependently, and adaptively towards a common agreed goal. In a team specific roles are assigned and the methods of interaction are agreed, trained, practised and modified as necessary through work practice and experience. Therefore, effective teamwork relies on a developed and synergistic goal directed effort by all the team members. In the case of an human-machine team, synergy can only be achieved if::
- the machine elements of the partnership can be made situationally and tactically aware and assist and advise the human in a timely manner conducive to the effective completion of the aircraft mission - a quality Electronic Crew Member (Taylor & Reising. 1995);
- the human elements of the partnership trust and aspire to teamwork with the machine.

The training of the aircrew - machine team is currently taken as the training of the aircrew to operate the aircraft. The aircraft design is seen as fixed and can only be changed through design modifications applied to part or whole of the aircraft. Thus, it can be argued that all existing operational Human-Machine System (HMS) are flawed systems in that they are designed to consist of a human plus a machine with only the human being trainable.

System design should aim for a system performance that is greater than the sum of the performance of each of its component parts. A man-machine team should strive to be a 'Joint Cognitive System' (Hollnagel and Woods, 1983) in that the image of the human to the machine and the machine to the human should support their synergy.

A tenet of this paper is that future advanced HMS will be largely flawed unless they are specifically designed to allow the performance of effective full system training within their role, are contextually aware, and are flexibility in their approach to diverse situations. To reach any maturity, the concept of the aircrew-machine team must aspire to the development and maintenance of a high quality of teamwork. Such teamwork must be understood, specified and trained prior to its improvement through application and practice.

This paper will approach the issues of aircrew-machine teamwork and training from several different perspectives, namely those of:

- team situational awareness;
- tactics and strategies;
- perspectives and influences on HMS team qualities;
- analysis, direction, control and supervision in the HMS team;
- HMS ab initio and continuation training;
- requirements capture and HMS specification

What is Team Situational Awareness?

Situational awareness (SA) is associated more with complex 'open skills' than with 'closed skills' (Poulton, 1957), in that its sustenance requires a good deal of interaction and appreciation with its operating environment. SA is a necessary and intrinsic property of applied skills. It is also a product of the quality and experience of skill application, but is variable under the influences of certain personal, individual and organisational factors.

Team SA involves a joint management and sharing of information, and the collective projection of that information into a future context. The SA possessed by a team not only relies on the knowledge and skills of the individual team members, and their conception of the perceived working environment, it involves team agreement on the understanding of that environment. This shared agreement implies training, a within team continual awareness of each others' proficiencies and limitations plus a collective appreciation of the standard of current team performance (MacLeod et al, 1995).

Considering future HMS teams, it is suggested that good synergy between human team and machine involves the establishment of extended joint cognitive HMS where physical interfaces are not serious barriers to smooth and proficient system performance.

Moreover, SA supports the maintenance of trust. Trust is essential to sustainable teamwork. Trust is promoted by awareness of the predicaments of both the team as a whole and of the individual members, as well as awareness of the quality of team performance towards shared goals, through a consensus on the means of their achievement (see Taylor & Reising, op cit).

Team SA relies heavily on the maintenance of common goals. This maintenance can only be achieved by a team through agreed endeavours, a willingness to co-operate, and a consideration of the needs of fellow team members. These agreed endeavours must be based on team training promoting a shared knowledge of tactical and strategic rules, procedures, and guidelines devised to support the team role within their working environment. Tactics and strategies will be discussed next.

Why Tactics and Strategies?

A tactic is defined as an arrangement or plan formed to achieve some short term mission goal. The goal may be an end in itself or serve as a stage in the progress towards a later mission objective. A strategy governs the use of tactics for the fulfilment of an overall or long term mission plan. Tactics are normally recorded as formal written procedures. When learnt they reside in human memory / cognition. Usually, human work experience with a system leads to modification of any formal system related procedural tactics.

The human perceives the world through the interpretation of information gleaned by the senses. This perception can be achieved through direct observation of the world or by the use of a man machine system's interface with the world. The use of interpreted information is governed by learnt tactics and strategies. These tactics and strategies are tuned through training and experience and are governed by the human roles within a man machine system and the human's interpretation of these roles.

A good quality tactic supports operator task routes to assist skilled performance, is an assistant to the application of all levels of aircrew skill, and allows a flexible and adaptive approach to a changing and potentially hostile environment. In an HMS, the performance of tactics and strategies should be enhanced by system equipments designed to aid the human to select and interpret information contained in the working environment,

Formal Tactics and Strategies were seen as representing an organisations recommended methods of achieving a desired man-machine system performance. They also currently provide a bridge between an operator's performance, assisted by the application of cognitive tactics and strategies, and the control requirements of the engineered part of the man-machine system, as represented by equipment operating procedures.

The different forms of tactical and strategic plans, and equipment operating procedures, should all be as closely associated as possible for the efficient operation of a man-machine system. Knowledge of the equipment operating procedures, and formal tactics and strategies, allows the operator to apply their skills through a formulation and application of cognitive tactics and strategies. Thus cognitive tactics and strategies representing the operator's conception of the means required to achieve man-machine system goals.

Tactics and strategies are continually mediated by both the information that the operator already possesses and information gleaned from the working environment. Currently all HMS performance is under the directed influence of human cognitive tactics and strategies. In the future, cognitive tactics should be devised for work throughout the joint HMS team.

For the conceivable future, only an operator or team can direct an HMS towards its given goals. Too often formal tactics and strategies are devised late in the design of a man-machine system and represent a method of equating problems in the design. However, by explicitly approaching how effectively a system should and might be used by the operators, at as early a design stage as possible, it is suggested that the whole design process and final product must benefit (Macleod & Taylor, 1994).

With reference to an HMS team, there must be perspectives on human teamwork that also can be applied to HMS teamwork. A brief examination of some of these perspectives follows.

Perspectives and influences on HMS team qualities

It has been proposed that there are seven 'Cs' of good military team performance: Command, Control, and Communication; Co-operation, Co-ordination, and Cohesion; and Cybernation" (Swezey and Salas. 1992) These seven 'Cs' are grouped as follows:

- Command, Control, and Communication (C^3);
- Co-operation, Co-ordination, and Cohesion (CO);
- Cybernation.

Thus teamwork is considered through inter-team interactions (C^3) and within-team interactions (CO). Cybernation, a term used to refer to highly skilled team performance where that performance far exceeds the sum of the skills of the individual team members, will be argued as a necessary aspiration of the aircrew-machine team.

A manifestation of modern advanced technology is the increasing emphasis in design on efficient HMS communications with many HMS related agents outside the commonly accepted boundaries of the aircraft HMS. A major concern in the C^3 area is to provide the support needed to allow necessary team co-ordination and structure to exist between system agents in order that the mission goals can be gained and the results disseminated. Thus the emphasis on C^3 is not only on the capabilities of the communication bearer (i.e. the physical form of the means of communication) but on the influence of C^3 on team interactions and performance. For example, the poorer the support to control afforded by the design of the HMS, the greater delays and uncertainty in the information presented to the operator and the HMS's operating authority. Poor quality information will adversely affect an operators use of tactics and strategies and their maintenance of appropriate cognitive modes, or necessary conditions for cognitive operations, to effectively support HMS direction and control. [1]

Inter team performance comprises goal directed behaviours by a team during task performance. The team behaviours depend on co-operation, co-ordination and cohesion between the team members to convert the individual behaviours of the team members into the required team behaviour. The required team behaviour will be guided by teamwork as influenced by the formal tactics and strategies applicable to the HMS

Cybernation occurs both within and between teams and represents the high degree of competency that can be achieved when the collective performance of the team is greater than that possible through sum of their individual skilled performances. The means of achieving cybernation is not only through the selection and training of the individuals for the team, it is through effective team training. Cybernation is also strongly influenced by such as organisational mores, inter-team empathy, within team concern for individual members, team spirit, feelings of 'belonging' and friendship to name but a few.

One of the main influences on the degree of teamwork possible is the size of the team with relation to the form and amount of work it is required to perform. Particular team structures and sizes are frequently argued under cost constraints, the argument being primarily supported by statements on the forms, levels, and efficiency of current methods of

---

[1] Hollnagel describes modes of control as Strategic, Tactical, Opportunistic and Scrambled, For explanation see Hollnagel 1993.

automation of system functions. It appears that automation is being increasingly incorporated into HMS, not to promote a net gain to safe system operation and performanc but to promote a net saving in the monetary life-cycle costs of the system.

Unfortunately, though HMS functions are derived under the 'solid' logic of systems engineering, they fail to fully encapsulate the cognitive logic and functions of the essential human component(s) of the system, the one form of component essential to the performanc of system related analyses, direction, control and supervision. The frequent result is an ove committed team both with relation to their role within the HMS and their maintenance of good teamwork. It is suggested that a more human centred and multi-disciplinary approac is required to formulate the principles for future HMS design.

Heil (1983), in a particular suggestion to cognitive psychologists that should arguably be applied to most disciplines, stated that practictioners:

" .. might conceivably profit by coming to recognise which of their commitments are due not to a clear perception of the nature of things, but to gratuitous philosophical prejudices that serve no purpose save that of forcing one's thoughts into certain narrow channels and sparing one the intellectual labour involved in coming to see thing s differently. "

Analysis, direction, control and supervision in the HMS team

In military operations, the use of HMS is normally for a particular purpose or missior In aircraft terms a mission might consist of one or more flights. Each mission and flight ha to be planned in detail to meet a strategic aim. This proactive mission planning allows pre flight rehearsal of the HMS tactics with relation to the expected flight. Once airborne, the proactive planning acts as a reference template to short term reactive planning. Reactive planning is necessary to equate unforeseen perturbances to the proactive mission plan, and direct and control HMS performance towards the mission objectives.

The need for reactive planning is dependant on human perception and analysis of mission situations. The initial perception of information is influenced by such as HMS design, the environment, the aircraft situation, the vagaries of human attention and the knowledge and efficiency of the adopted tactics. The feedback from situation analyses is used as a basis for the redirection of the aircraft to meet mission goals, or to control the aircraft to improve the HMS performance towards its goals, or to maintain supervision of t aircraft HMS and its situation.

At present, the human is required to perform all the above forms of task without any true assistance from the other components of the HMS. This is because the other components have no situational awareness or applied knowledge of the prescribed tactics and strategies. Current HMS systems act as designed and as used by the operator. HCI principles of display are evoked to support the quality of information presentation to the operator. Further, the diverse type of information produced by such sub systems is a mixture of machine interpreted situation related data, collateral data such as obtained from flight and mission planning, and ephemeris data presenting short term information on fligh or sensor performance. The se diverse types of information should not be confused.

The HMS operator is forced to analyse and associate all perceived information. Incorrect interpretations and associations must be avoided (some can be fatal). Interpretations can be dependant on the nature of the environment, on the HMS situation a mode of operation, on the types of information, on correct association of diverse information, and on the idiosyncrasies of equipment design or the human operator.

If a quality HMS team is to be allowed by technology and design, methods must be developed to allow the human and the machine to assist one another as full team members. The difference that must be maintained between man and machine is that the human must always be in a position to lead the team. However, the machine must have the pertinent knowledge and capability to assist the human in a timely and appropriate fashion with rega to the tactical situation. The machine should be 'aware' of the quality of human performance and adapt its performance and advice accordingly; possible examples are the proffering of additional assistance to the normal if an operator is seen to lack skill, attentio or is being distracted from a task or situation by a current performance of an over onerous activity.

As an example, a paper map can be considered as an inflexible but important source o collateral information assisting aircrew planning and anticipation. Moreover, good map design sets appropriate expectations which prime perceptions and direct attention to the critical features and important relationships in the environment.

Understanding of the current situation, the ability to think ahead, and to make effective plans are key to aircrew operational performance. Therefore, what is basically needed is an adopted digital map technology that is an integral part of the HMS team, that helps the aircrew anticipate and manage cockpit and cabin workload, that enhances aircrew and team situational awareness, and supports aircrew planning, decision making, and weapon system control, all in an adaptive, flexible, and intelligent manner (Taylor & MacLeod, 1995).

The HMS team may consist of many components, both human and machine. In the multi operator case, machine team components will have difficulty in proffering advice and adapting to the performance of one human operator. However, there are additional problems if the machine components are required to team with several operators at once. Problems that might occur include questions as to which human operator is currently HMS leader, or which human is currently under performing with relation to task and their interaction with other crew members. Some early suggestions to the solving of such problems by machine have been indicated by Novikov et al (1993).

In association with the above will be a need to keep all team members aware of the performance of the system as a whole, of the quality of their own contributions, and the contribution of others. Such a machine supported feedback is essential to aid the maintenance of the team SA. Such avenues have also been suggested by work on the measurement of cognitive compatibility (Taylor, 1995), where it is indicated that future systems will need to consider dimensions of social interactions (e.g. goals, shared functions, aiding, advice) as well as the traditional human factors considerations of compatibility (i.e. modality, movement, spatial, conceptual compatibility) to order to achieve cognitive quality in joint cognitive systems.

Advanced HMS ab initio and continuation team training

It has already been discussed that training for a HMS is currently training for the operators only (apologies to maintainers). Usually an engineered system is designed to operate through an interpretation of specifications, often as the designer sees appropriate when there is ambiguity in the specification, and is designed to interface with the human operator(s) in much the same fashion.

In the future ab initio team training for an advanced HMS would consist of :
1) designing the HMS to meet teaming criteria such as outlined in the previous section;
2) training of the human component skills to team with the engineered HMS components.

However, such criteria could not possibly be implemented correctly by design to give the 'right first time' teaming quality required for advanced HMS operations. The design process would need to allow improvement to design through iteration. Moreover, it would only be in the light of actual operating experience, where the tuning of the system is a form of continuation training of both the machine and operator, that the full worth of the teaming concept could ever be realised.

Therefore, to effectively train the machine component of the team would not only rely on a high quality design of the HMS, it would rely on a debrief and tuning of the system after each sortie. This tuning would have to be performed with great care to avoid possible severe degredations in HMS performance. Advice on methods, benefits, and pitfalls inherent in the tuning would have to be included as part of the design process.

Requirements capture and HMS specification

Requirements capture is a difficult enough process for systems engineers considering current advanced HMS systems. Nevertheless the capture of requirements for any system is essential. It can be said that the quality of a system is the design conformance to the system requirements. Requirements are the foundations for system specification.

Future requirements capture for a teamed HMS will be far more difficult than at present and will require a multi disciplinary approach to design. Further, future requirements capture will require that both machine and cognitive functions are captured.

By cognitive functions we refer to the cognitive representations that are required to transform prescribed operator tasks, dictated by proactive mission planning, work procedures and tactics, to the activities that are actually needed, under the auspices of reactive planning and tactics, to adapt job performance in the light of the appreciated work situation and the mission goals (see Boy. 1995, MacLeod et al, 1994).

Current trends are for the development of knowledge elicitation methods relevant to ta (i.e. cognitive task analysis techniques), and also with relation to observable activities (i.e., concurrent verbal protocol analysis). Associated with these current trends are method of assessing operator mental workload, though it is almost certain that what is measured he is different between task and activity. However, beyond the concepts of workload, there is some evidence that the use of dynamic task analytic modelling techniques may aid elicitatic of operator decision processes (Macleod et al, 1993).

Methods of cognitive function capture and analysis are being sought. Without such capture, the cognitive requirements of the machine part of the team cannot be properly considered and specified. It is interesting to consider that cognitive compatibility, currentl; as assessed by the CC-SART Rating Scale (Taylor, 1995), may very well be indicative of th quality of cognitive functions with relation to their application alongside a particular HMS design - a measure of HMS quality for the future?

Final Note

The coverage of this short paper is glib considering the complexity and nascence of the subject. However, the coverage is of a subject area where the approach to technological development must be human-centred if it is to succeed. The aim for an Electronic Crewmember exists outside science fiction. The problem is to sensibly maintain the reality the aim.

References

Boy, G (1995) Knowledge Elicitation for the Design of Software Agents, in *Handbook of Human Computer Interaction*, 2nd edition, Helander, M. & Landauer, T. (Eds), Elsevier Science Pub., North Holland (In press).

Heil, J (1983), *Perception and Cognition*, University of California Press ltd., London

Hollnagel, E. (1995) *Human Reliability Analysis: Context and Control,* Academic Press Lt London.

Hollnagel, E & Woods, D.D. (1983) Cognitive systems engineering: New wine in new bottles, *International journal of Man-Machine Studies,* **18**, pps 583-600.

Novikov, M.A., Bystritskaya, A.F., Eskov, K.N., Vasilyiev, V.K., Vinokhodova, A.G. & Davies, C. (1993) 'HOMEOSTAT -A Bioengineering System' in *Proceedings of 23rd International Conference on Environmental Systems,* Colorado Springs, Colorado, July 12-15, 1993, published by SAE Technical Paper Series, Warrendale, PA.

Poulton, E.C. (1957) ' On the stimulus and response in pursuit tracking', *Journal of Experimental Psychology*, **53**, pps 57-65.

Swezey, R.W. & Salas, E (Eds) (1992), *Teams*, Norwood, NJ: Ablex.

Taylor, R.M. (1995) Experiential Measures: Performance-Based Self Ratings of Situational Awareness, in *Proceedings of International Conference on Experimental Analysis and Measurement of Situational Awareness*, Daytona Beach, 1-3 November. Embry-Riddle Pre: Florida (in press).

Taylor, R.M. & MacLeod, I.S. (1995) Maps for Planning, Situation Assessment, and Missic Control, in *Proceedings of The 1995 International Conference of the Royal Institute of Navigation*, London.

Taylor. R. M. & Reising, J. (Eds), The Human-Electronic Crew: Can We Trust the Team? *Proceedings of the 3rd International Workshop on Human-Computer Teamwork,* Cambridg UK, September 1994, , Report DRA/CHS/HS3/TR95001/01 dated January 1995.

MacLeod, I.S., Taylor, R. M & Davies, C.D. (1995) Perspectives on the Appreciation of Team Situational Awareness, in *Proceedings of the International Conference on Experimental Analysis and Measurement of Situation Awareness*, Embry-Riddle Aeronautic University Press, Florida (In press).

MacLeod. I.S., & Taylor, R.M. (1994) Does Human Cognition Allow Human Factors (HF) Certification of Advanced Aircrew systems? in *Human Factors Certification of Advanced Aviation Technologies,* Embry-Riddle Aeronautical University Press, Florida.

MacLeod, I.S., Biggin, K., Romans, J. & Kirby, K. (1993) Predictive Workload Analysis - RN EH 101 Helicopter, in *Contemporary Ergonomics 1993,* Lovesey, E.J. (Ed), Taylor and Francis, London

Consumer Ergonomics

USER INVOLVED DESIGN OF A PARKING FACILITY FOR BICYCLES

V.B.D. van der Steen, H. Kanis and A.H. Marinissen

School of Industrial Design Engineering
Jaffalaan 9
2628 BX Delft, the Netherlands

This study of a user involved design is primarily concerned with the extent to which usage anticipated by the designers is specified in the design. In a users' trial it is shown that designs which demand specific user behaviour are a source of many and various difficulties in the use of existing bicycle racks. The majority of these difficulties concern required use-actions that were hampered or turned out to be impossible. Difficulties of a cognitive nature were also observed. The findings in the trial were an incentive for a new design for a bicycle stand with a low level of usage-specification, thus allowing it to be used in a variety of ways. In a second trial most of the difficulties observed for existing racks appeared to be solved.

Introduction

In the Netherlands cycling is encouraged by the authorities for several reasons, such as public health, the environmental impact of other forms of transportation and promoting traffic flow in cities. Disregarding weather conditions, in urban environments at least two factors seem to discourage the use of bikes.

Firstly, bicycles are stolen if parked in the street, i.e. at a lamppost, against a wall, or at a bridge railing. Of the 13,000.000 bicycles in use in the Netherlands almost 1,000.000 are stolen every year; a quarter of all these thefts occur in Amsterdam.

Secondly, it seems that bicycle racks, which come in a wide variety of designs, mostly feature poor usability. This discourages the usage of these racks by cyclists who, in consequence, do not benefit from the anti-theft provisions of the racks.

In this paper a user involved design of a parking facility for bicycles, commissioned by the city of Amsterdam, is discussed.

Objective

The general feeling that bicycle racks suffer from poor usability had to be substantiated for the so-called Amsterdam rack, which is currently placed throughout the municipality of Amsterdam (see Figure 1). This figure shows that the Amsterdam rack specifies to a considerable degree the way in which a bicycle should be parked in the racks. One of the wheels must be positioned in the rounded groove, designed to hold the bicycle upright. This positioning may require manoeuvring between adjacent bicycles, while in half of the places constructed in one unit the bicycle must be lifted. Once a bicycle stands in the rack, it is more or less fixed which part of it may be attached to the anti-theft brace. This kind of usage-specification is less urgent for the so-called Sinus-line rack, see Figure 2. The

Figure 1. The Amsterdam rack.

Figure 2. The Sinus-line.

thin tube, that is bent backwards at both ends in order to fit a wheel, does not really clamp a wheel: the Sinus-line is a 'lean-against' rack. In this respect, the positioning of a wheel in the pre-shaped form is not required to attach any part of the bicycle to the rack with a lock. It is noted that the Sinus-line rack can be used from both sides. Obviously, this type of difference in racks affects their usability, and consequently the kind of difficulties users may be confronted with. Thus, for different racks research seeks to address the difficulties users encounter in parking and locking bicycles. In this project, users trialing does not only provide the input for the design process but also the means of evaluating the newly designed bicycle parking facility.

Method

As indicated above, the extent of usage specifying characteristics of racks is seen as an important parameter in the study. There appears to be much variety in available racks requiring specific actions from users. Therefore, besides the Amsterdam-rack in Figure 1, which is the main focus of the study, and the Sinus-line in Figure 2, the so-called U-lock rack shown in Figure 3 was also involved in the users' trial. A U-lock unit comes with two places for a bicycle, the frame-tube of which must be clamped under the saddle. Thus, like the Amsterdam rack, the position of a bicycle put up in the U-lock is completely predetermined. By moving a slide two holes can made to coincide so that a bicycle can be locked up to the rack by means of the pin at the side.

Figure 3. The U-lock.

For participating subjects several characteristics may be deemed relevant as a possible source of difficulties, in particular force exertion in view of the lifting required for the Amsterdam-rack, stature (for the same reason), and clothing which may become entangled. Obviously, the variability in bicycles, including accessories, may also be related to the (non-)emergence of use-difficulties, such as the type of bicycle, i.e. the width of the tyres, the type of brakes (with or without cables), the presence of a basket (in front), of a child's seat or of a pannier (at the back), and the type of lock, e.g. its length and flexibility. Finally, particularly in the case of the Amsterdam rack, difficulties may arise from full racks, especially with the accessories of adjacent bicycles.

Any attempt to account for any or all combinations of these factors in a test-like research approach, even if possible, would be completely premature in view of the speculative character of many suggested relationships. Therefore, this project was focused on the explorative nature of the users' trial. Subjects were observed on site, parking and locking their own bicycle with their own lock in their usual manner. The only selection criteria in the recruitment of participants concerned the heterogeneity in the age of subjects (as a proxy for force exertion), in the stature of subjects, and in the type of bicycle, including accessories. In addition, the observations were carried out in quiet spots in order to record the parking of a bicycle by a subject under different occupation rates, as far as could be arranged in the racks.

All actions of users were video-taped. For each subject the trial was concluded with an open interview about the observed do's and dont's, and about the difficulties that were experienced.

Results

In total, twelve subjects participated. The intention of gathering observations within-subjects throughout the racks almost completely failed due to the large distances between the locations of the racks. Only three subjects were able to park their bikes in both the Amsterdam rack (used by nine subjects) and the Sinus line (used by four subjects). The trial with the U-lock had to be terminated after two subjects due to bad weather conditions. The age of the subjects ranged from 7 to 75 years, while the stature of the eleven adult subjects ranged from 1.50m to 1.95m.

Difficulties in using the racks were various. For the Amsterdam rack 355 difficulties were observed, of all kinds. For the Sinus-line this figure amounts to 37 difficulties, and for the U-lock 14 difficulties. In Table 1 a summary is given of these difficulties. In this table, user activities are confined to parking a bicycle and locking it, since the observations showed that unlocking a bicycle and removing it is less critical, and does not provide new evidence. The charting of difficulties in Table 1 is based on the characteristics of the racks, rather than on user characteristics, in order to enhance the practicability of observed operations for the proposed design activities (cf. Rooden et al., 1996).

An obvious finding is that all kinds of difficulties originate from lack of space between the stand that is used and adjacent stands in the rack. In particular, tall subjects have trouble in squeezing themselves between adjacent bicycles in the Amsterdam rack, both in parking and locking a bicycle. Another finding, that is not much of a surprise, concerns the difficulty caused by the need to lift a bicycle into the upper stands of the Amsterdam rack. This appears to be troublesome, especially for elderly subjects. They explain that the required force is difficult to exert, i.e.

Table 1. Difficulties encountered by users in parking and locking their own bicycle

Functional aspects of the racks	Amsterdam rack (n = 9)	Sinus-line (n = 4)	U-lock (n = 2)
usage anticipated in the design to park a bicycle	no lack of clarity	unclear, also in view of bicycles already parked; use unanticipated in the design may impede the designed parking of 2nd bike	no lack of clarity
space between racks	30cm; pedals, handlebars, clothing greatly impede parking and locking a bicycle*)	90cm; no difficulties observed	67cm; riding a bike in is hampered by adjacent racks, if filled**)
entering the designed fixation for the bike	in trying to position a wheel, often the rounded groove is missed, especially the upper one	impossible with a basket in front	no difficulties observed
stability of bike once in rack	broad tyred bikes like ATB's do not fit in the rounded groove	due to absence of clamp, bicycle is only positioned stably if firmly locked to rack	handlebars and frontwheel of clamped bike easily turn aside, thus causing difficulties in parking a bike in a adjacent rack**)
anti-theft facility	only functional for front part of bikes; difficult to reach due to adjacent bikes *); may be useless due to insufficient locklength	no difficulties observed	designed functionality of slide, to line up two holes, poorly understood; thus, several bikes already parked in the rack were not locked

*) except for both outside places
**) difficulties reduce the further racks are placed apart

beyond control, as there is no room to stand close to the front of the bike.

An intriguing finding concerns the role of rack-characteristics specifying the way of use anticipated in the design. In this respect the Amsterdam rack and the U-lock are more explicit than the Sinus-line. Usage-specifying characteristics, as well as lack of those characteristics, cause difficulties. The Amsterdam rack shows that far-reaching specification of usage goes together with numerous operational difficulties. One of the few problems the Amsterdam rack does not suffer from, i.e. confusion on how to use it, constitutes the main source of difficulties in the case of the Sinus-line. Actually, the usage specifying element of this rack, i.e. the thin tube bent backwards at both ends, is an unfortunate part since it is not understood, and, moreover, it limits the anticipated usage for bicycles with a basket or a child seat in the front. The U-lock seems to be clear as to the way a bicycle should be parked, but appears to be puzzling in the functionality it is named after. Here it is noted that because of the unequal number of participating subjects, which particularly applies to the case of the U-lock, a proper comparison of the functionality of the racks was not possible. This means that for the U-lock, and to a lesser extent also for the Sinus-line, there is a fair chance that observation of more participants would have produced several extra difficulties. However, what has been observed must be seen as relevant, since occurrences found in small samples have a low probability of being rare (Kanis and Vermeeren, 1996).

Design requirements on the basis of the users' trial

The design requirements negotiated in the project did not only involve usability but also positioning in a built-up environment (in view of safety and vandalism considerations), maintenance, aesthetics and the cost, including placement. As to usability, the main requirements for the new parking facility concerned the following topics:
- the minimum distance between two adjacent stands;
- no need to lift;
- the appropriateness for all kinds of bicycles, including those with various accessories;
- ease of usage, i.e. requiring only a few actions from users in parking a bicycle and locking it to the facility;
- the absence of repercussions due to product characteristics, if any, that would be deemed necessary to specify some kind of usage aimed at in the design.

The new design

The findings in the users' trial have been a reason for reflecting on the way in which many bicycles are actually parked in Amsterdam, i.e. leaning against a solid lamp post or a bridge railing. Apparently, both artifacts afford ample possibilities in this respect. Given the design requirements, this consideration is one of the incentives that has inspired the new design shown in Figure 4. Inspiration was also derived from the Amsterdam architecture and the contrast between curved and straight lines, i.e. a simple base completed with some kind of frivolity, as can be seen in the traditional houses along the canals. The stand should feature as a parking facility without dominating the scene. It is usable at both sides. The diameter of the thick tube matches globally the curve of saddles, which is designed as an invitation for use. The bent tube, meant to support the front part of a bike, offers (extra) possibilities for locking up. In this tube a solid bar is placed which will frustrate

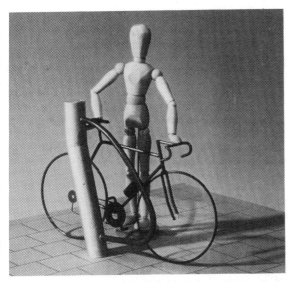

those who try to saw through the tube. Due to the solid, elementary form, vandalism is thought to stand a poor chance.

Figure 4 Model of the new parking facility for bicycles

Assessment of the usability of the new stand in a users' trial

With six prototypes the usability was tested in the way described above. The only location available at the time was at the side of the entry to the city hall, see Figure 5. The drawback of this location is that the stands cannot be approached from both sides. The distance between the stands ranged from 90cm (both outside pairs) to 120cm (in the middle). Five subjects participated. A small number of difficulties were observed, such as pedals being caught, the 'riding on' of a bicycle that is not locked to the stand, and the impossibility of applying small 'brace-locks' for attaching a bicycle to the thick tube. As can be seen on Figure 5, saddles are not always placed against the thick tube. The loosely anticipated way of use in the design, presumably, will be further sapped if stands are approachable from both sides. However, so far there seems to be no reason to expect major functional drawbacks if a bicycle is put up the other way round. As to the distance between the stands, there are no indications that 90cm would be too narrow. Whether the prototypes will be further developed (e.g. with a smaller diameter of the thick tube) and produced is under consideration by the planning authorities of the city of Amsterdam.

Figure 5. The new parking facility in use.

References

Kanis, H. and Vermeeren, A.P.O.S. 1996, Teaching user involved design in the new Delft curriculum, *Contemporary Ergonomics* (this issue).
Rooden, M.J., Bos, A.C., Kanis, H. and Oppedijk van Veen, W.M. 1996, Usability testing of an interface for tele-activities, *Contemporary Ergonomics* (this issue).

DISPLEASURE AND HOW TO AVOID IT

Patrick W. Jordan

Philips Corporate Design
Building OAN, P.O. Box 218
5600 MD Eindhoven, The Netherlands

Telephone +31 40 2733121, Fax: +31 40 2734959
e-mail: c868864@nlccmail.snads.philips.nl

An interview based study investigated the issue of displeasure in product use. Eighteen interviewees were asked to think of a product which they found particularly displeasurable. They were then asked a series of questions about this in order to probe the design properties of displeasurable products and the emotional penalties engendered by their use.

Results indicated that lack of usability, poor performance, lack of reliability and poor aesthetics were the principle factors that were most likely to be associated with displeasurable products. The feelings most commonly associated with displeasurable products were annoyance/irritation, anxiety/insecurity, contempt and exasperation. So whilst usability is an important issue, there are many others which can also affect how pleasurable/displeasurable products are to use.

The implications of the outcomes of this study for the role of human factors in design and evaluation are discussed. It is suggested that — as the representative of the user in the design process — human factors specialists should consider a wider range of issues than those traditionally associated with usability. Similarly, the range of attitudinal measures taken to investigate user responses to product use should be expanded. Usability-based approaches to user-centred design may, then, be insufficient to ensure an optimal experience for product users.

Introduction

A paper by Jordan and Servaes (1995) in last year's 'Contemporary Ergonomics' reported on a study investigating the issue of pleasure in product use. Outcomes of that study indicated that, in order to maximise the experiential aspects of product use, those involved in user-centred design should look both at and beyond usability issues. The results of the study suggested that, whilst usability was an important factor in how pleasurable a product was to use, it was only one of a number of factors influencing pleasurability. Other factors included appropriate functionality, aesthetics, performance and reliability.

This paper reports on another part of the same study — this time looking at displeasure with product use. The research questions associated with this part of the study were:

1. What are the emotions by which displeasure in product use is manifest?,
2. When are these emotions manifest?, 3. What are the properties of a
design that are associated with displeasure in product use?

A 'sub-question' which ran in conjunction with the third research question was
whether or not adequately addressing usability issues would be enough to avoid
displeasure or whether those involved in product design must consider other issues in
order to avoid displeasing the user.

Usability

The International Standards Organisation (ISO) define usability as being: "... the
effectiveness, efficiency and satisfaction with which specified users can achieve specified
goals in particular environments." (ISO DIS 9241-11)... where effectiveness refers to the
level of accuracy and completeness with which users are able to perform tasks, efficiency
refers to the amount of effort required to complete a task and satisfaction refers to the
level of comfort and acceptability associated with product use.

Pleasure and displeasure

'Pleasure' in product use is concerned with the positive feelings and emotions that
are engendered when using a product. Below a definition is offered:

*Pleasure in Product Use: The emotional and hedonic benefits associated
with product use.*

'Displeasure' in product use is concerned with the negative feelings and emotions
engendered when using a product. Again, a definition is offered:

*Displeasure in product use: The emotional and hedonic penalties
associated with product use.*

Method

Participants
A semi-structured interview was conducted with 18 respondents (10 female, 8
male). The vast majority of these were students of less than 25 years. All participants
were residents of Glasgow in Scotland.
Interviewees had been contacted beforehand. It was requested that, before the
interview, they thought of a particularly displeasurable product which they owned or used
(or had previously owned or used). The questions that were asked in the interview
centred around this product.

The Interview
The interview contained a series of open ended questions. Users were asked to say
what their displeasurable product was and to give a general description of it. They were
then asked about the aspects of the product that made it particularly unappealing and
about the feelings engendered by using the product. Finally, they were asked when they
experienced these negative feelings — was it only during product usage, or would, say,
anticipation of using the product induce negative emotions. The interview questions are
listed in table 1.

1. Think of a product that you either do or used to own or use which gives or gave you a great deal of displeasure.
2. Give a general description of this.
3. Which aspects of its design are particularly unappealing?
4. What types of feeling does this product engender?
5. When do you experience these feelings?

Table 1. Interview questions.

Results

Products chosen

The products that participants selected as being particularly displeasurable were as follows: Alarm clock (3 participants), computer (2), stereo equipment (2), car stereo (1), coffee machine (1), cooker (1), electric shower (1), electric toothbrush (1), food mixer (1), kettle (1), personal stereo (1), software packages (1), TV set (1), washing machine (1).

The second interview question was asked to see if users would mention anything about design issues or the emotions associated with the product without being prompted. However, little of interest arose from this question — usually straight factual descriptions of products were given.

Emotions associated with displeasurable products

The emotions that respondents associated with their displeasurable products are summarised in figure 1.

EMOTION

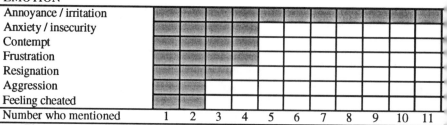

	1	2	3	4	5	6	7	8	9	10	11
Annoyance / irritation											
Anxiety / insecurity											
Contempt											
Frustration											
Resignation											
Aggression											
Feeling cheated											
Number who mentioned	1	2	3	4	5	6	7	8	9	10	11

Figure 1. Emotions associated with displeasurable products (n = 18). (Based on responses to question 4).

Annoyance/irritation was the emotion chiefly associated with displeasurable products. For example, the user of an alarm clock found the buzz tone irritating and would have preferred a more "gentle" sound. Similarly, the owner of an electric cooker was irritated that it took so long to heat up. A feeling of anxiety/insecurity sometimes arose when users were dependent on a product to complete a task, but felt that the product didn't support them as it should. For example, one respondent had problems using her personal computer, which she regarded as being unpredictable and not usable. Sometimes negative feelings could develop into resignation — for example for the owner of a difficult-to-use car stereo.

Four respondents said they felt contempt towards the product, or the manufacturer of the product, that was displeasurable. One respondent had had to take a kettle back to the shop on two separate occasions as it had been faulty when purchased. This had led her to develop "mild disgust" with the manufacturers. Other respondents felt that they had been cheated by those who manufactured their displeasurable product. For example, one participant felt cheated as he had had a number of problems with his electric shower.

Four respondents also mentioned being frustrated with their displeasurable product. For example, one respondent became frustrated when using computer software packages as he had encountered usability problems. Sometimes, feelings of frustration could develop into aggression — the owner of an unreliable stereo said that she sometimes thumped it, for example, if it had "chewed up" one of her cassette tapes.

When are these emotions experienced?

The vast majority of respondents (13 out of the 18) said that they experienced these negative emotions whilst using the product concerned. However a significant minority also experienced these feelings before and after use of the product. It is interesting to compare the times when these negative emotions were experienced with the times that positive emotions were experienced with pleasurable products. Jordan and Servaes (1995) did not give these figures in relation to use of pleasurable products, so here they are in figure 2, alongside those relating to use of displeasurable products. The most striking difference is that negative emotions appear to be more likely to remain after using a displeasurable product than positive emotions do after using a pleasurable product.

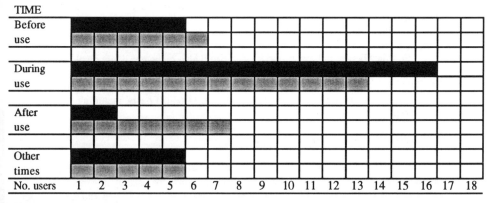

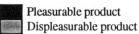 Pleasurable product
Displeasurable product

Figure 2. Times when emotions associated with pleasurable and displeasurable products are experienced (n = 18). (Based on responses to question 5).

Design Properties associated with displeasurable products

Question 3 of the interview asked about the design properties which made a product particularly displeasurable. The design properties associated with displeasurable products are summarised in figure 3. Lack of usability was the factor most commonly mentioned as contributing to a product being particularly displeasurable. For example, the user of a car stereo was "exasperated" as a result of being unable to tune into the stations she wanted to listen to. Similarly, the user of a computer found the machine "daunting" because he perceived it as difficult to use. Unpredictability was also mentioned by two users. For example, the user of a food mixer found it difficult to predict what each of its functions did.

DESIGN PROPERTY

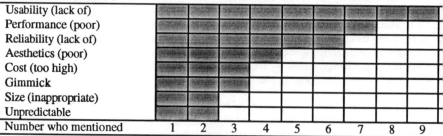

	1	2	3	4	5	6	7	8	9
Usability (lack of)									
Performance (poor)									
Reliability (lack of)									
Aesthetics (poor)									
Cost (too high)									
Gimmick									
Size (inappropriate)									
Unpredictable									
Number who mentioned	1	2	3	4	5	6	7	8	9

Figure 3. Design properties associated with displeasurable products (n = 18). (Based on responses to question 3).

Respondents indicated displeasure with products that didn't perform to a high standard with respect to their central functions. For example, a TV user felt contempt towards his TV set because the sound was distorted. Gimmicks were also not appreciated. One respondent described an electric toothbrush as a gimmick, saying that the whole concept of electric toothbrushes was "ridiculous".

Lack of reliability was mentioned by six of the respondents. This could leave the user feeling cheated, particularly when the product had been expensive to buy. This was the case with one respondent who had paid a lot of money for a personal stereo which proved unreliable.

Poor aesthetics and inappropriate size could also make a product displeasurable. For example, a respondent described the radio-alarm clock that she used as being "ugly". Similarly, the owner of a stereo was displeased that it took up so much space in her room. Another nine properties were mentioned on a one-off basis, including 'inflexibility' and 'complexity'.

Limitations of the study

As reported in Jordan and Servaes (1995), there are a number of limitations on the degree to which these results can be generalised. Because the participants represent a fairly narrow sample, there is no guarantee that these results would generalise to a wider population. The method of data analysis also introduces possible limitations. The approach taken by the analyst was to categorise responses based on his own 'view of the world'. The analyst (the author) is a human factors specialist based in a design department. It might be hoped, then, that the classification used for the design properties is fairly satisfactory. However, the classification in terms of emotions may be less accurate. This classification proved more difficult as there appears to have been little systematic study of emotions on which to draw (although a good overview of what has been done can be found in Plutchik 1994). That the questions in the interview were open ended represents another limitation. This is because it may be that some design properties and emotions are more easy to articulate than others. Similarly, it may be that some responses come to mind more easily than others when questioned on these issues. For example, it could be that the reason that, say, poor performance was mentioned more than unpredictability in the context of design properties connected with displeasurable products, is that issues to do with performance may have come to mind more easily and may have been easier to articulate. It may not, then, be valid to treat the number of mentions that each emotion and design property received as being an index of their comparative importance.

Because of these limitations, it would be prudent to treat the results as giving a 'first pass' at tackling the issue of displeasure with product use, rather than as definitive indicators about the influence of particular design properties or about the emotions associated with displeasure.

Implications for the role of human factors in product creation

Design

The outcomes of the study suggest that designing for usability is central to avoiding feelings of displeasure in product use. Indeed, design issues connected with usability were mentioned more than any others — nine of the eighteen mentioned these (see figure 3). However, there are many other aspects of a product's design that must also be right if displeasure is to be avoided. For example, if the product does not perform its central function adequately, is unreliable, or is not aesthetically pleasing, then users may experience use of that product as being displeasurable.

It may, at first, appear obvious that the above factors are likely to cause displeasure to users. However, often — indeed usually — these issues are perceived as falling outside the domain of human factors professionals. This is reflected both in the training of human factors specialists and their role within manufacturing organisations. For example, few, if any, human factors courses tackle issues such as aesthetics. Similarly, it is not usual for human factors specialists to be involved in specifying the level of reliability that a product must achieve (this is usually seen as a financial/technical issue). However, if the view is taken that it is the human factors specialist's responsibility to ensure that users needs and wants are taken into account in the product creation process, then — judging by these results and those reported by Jordan and Servaes (1995) — he or she should become involved in issues that extend beyond those traditionally associated with usability.

Evaluation

Similarly, users' emotional responses to displeasurable products suggest that, when evaluating attitudinal responses to products, the human factors specialist should look beyond 'satisfaction'. Satisfaction — the level of comfort and acceptability associated with product use — is usually considered as representing the attitudinal component of usability. However, these results (and those reported by Jordan and Servaes) indicate that people can experience pleasure/displeasure when using a product for reasons that go beyond satisfaction/lack of satisfaction.

The most commonly mentioned emotional responses to displeasurable products were annoyance/irritation and anxiety/insecurity. These responses to product use are possibly ones which most in human factors practice would recognise and, indeed check for when investigating attitudinal responses to product use. However, feelings such as aggression, contempt or a feeling of being cheated rarely seem to be investigated by human factors specialists (certainly, this author cannot recall reading a paper within the human factors domain where such an investigation was reported).

References

ISO DIS 9241-11. Ergonomic requirements for office work with visual display terminals (VDTs):- Part 11: Guidance on Usability.
Jordan, P.W. and Servaes, M., 1995. Pleasure in product use: beyond usability. In S. Roberston (ed.), Contemporary Ergonomics 1995. (London: Taylor and Francis).
Plutchik, R. (1994). The Psychology and Biology of Emotions. London: Harper-Collins.

REGIONAL DIFFERENCES IN CAPACITIES OF THE ELDERLY, RELEVANT TO MAN-PRODUCT INTERACTION

Steenbekkers, L.P.A., Dirken, J.M., Houtkamp, J.J.,
Molenbroek, J.F.M., Voorbij, A.I.M.

Delft University of Technology
Faculty of Industrial Design Engineering
Department of Product and Systems Ergonomics
Jaffalaan 9, 2628 BX Delft
The Netherlands

Designers of daily-life products need data on capacities and characteristics of possible future users. Elderly users tend to be neglected in product development. The level and change in capacities with increasing age, are hardly known. A study was started in which this data is gathered on relevant physical, sensory and cognitive capacities of persons of 50 years and over, as well as of the younger part of the population (20-30 years of age) as reference group. In order to decide whether the data should be collected on a national sample or that a regional sample would do, a pilot study was done to gain insight into regional differences in design- and user-relevant capacities and characteristics.

Introduction

This research is part of the Delft 'Gerontechnology Project; redesign guidelines for products also for the elderly'. The project consists of several paths, mainly carried out in parallel.
First an investigation has been made into the state of the art of design of durable, daily-life products (also) for the elderly (Freudenthal, 1993). This resulted in an overview of shortcomings of products and problems the elderly have with products. From here, lacking and urgent fields of design relevant data have been established:
- The first path consists of measurements of physical, sensory and cognitive capacities of users (N=750) to establish a data bank and norms (Steenbekkers et al., 1994).
- The second path is on cognitive ageing and the implications for the design of 'smart' products.
- In the third path, with help of Masters students, product development projects are carried out in industry to test the theoretical insights and preliminary guidelines.
This paper concerns the first path of the project.

Aim

During the design of daily-life products designers need ergonomic data on possible future users in order to be able to generate concepts of new useful products and to adapt products to the characteristics and capacities of the users. Concerning the 'healthy', independently living elderly these data appeared to be hardly available. Where no data are available and data of younger user-groups are used, the design often results in products that cannot or are difficult to be used by many or most of the elderly members of the target group of users.

In the main study of the project we will collect ergonomic data on Dutch user-groups of 50 years and over, in order to gain insight into the level and variance of capacities of users at certain ages. These data will be a basis for a design-relevant databank on user characteristics and design-guidelines will be set up. In order to be able to decide whether the data in the main study should be collected on a national sample or that a regional sample would do, a pilot-study was carried out to gain insight into regional differences in these design- and use-relevant characteristics of the Dutch population.

Variables

In our study several groups of capacities of the elderly, relevant to man-product interaction, are studied. Physical, sensory and cognitive capacities of the independently living and 'healthy' elderly (and, for comparison, of the younger part of the population) are being assessed in a large sample (N=750; 150 per age group: 75 males, 75 females). A representative part of these variables is assessed in a smaller (N=118) pilot-study.

Variables measured in the pilot-study are:
- static anthropometry: body mass, stature, shoulder height seated, buttock-knee length, popliteal height, shoulder breadth, hip breadth seated
- functional anthropometry: step length, walking velocity normal and fast
- force exertion: gripping force
- fine motor skills: peg-board test
- range of movement of joints: wrist (extension, flexion, ulnar deviation, radial deviation) and forefinger (flexion)
- balance when standing (eyes open, eyes closed)
- visual variables: acuity and contrast sensitivity (reading test with variables: contrast, light intensity, letter size)
- tactile capacities: texture sensitivity (operationalised in a test on sorting pieces of sandpaper)
- single and choice reaction time on visual and auditive stimuli
- preferences for colour

and
- ADL functioning (assessed by means of a questionnaire, to compare capacities with actual product use in daily life).

Where possible, the measurements were according to existing measurement methods, otherwise a method and equipment were developed in our laboratory. The latter was the case for the following groups of variables: step length, walking velocity, tactile capacities, preferences for colour and ADL-functioning. During the measurements we tried to imitate the daily-life situation as much as possible. This implies that, for example, the persons wore their own glasses, which might have not been corrected to optimal vision.

Pilot-study

The goal of this pilot-study was to determine whether significant differences exist among elderly, living in different regions and in different degrees of urbanisation. The subjects in the study had to be born and bred in the municipality where the measurements took place.

We expected differences to exist for the physical variables, because it is known that, generally speaking, the Dutch in the Northern part of the country are taller than the Dutch in the Southern part (Roede and van Wieringen, 1985). So for the anthropometric variables differences are expected, as well as for measurements which might have a relationship with anthropometrics, like for instance step length (related to leg length) or gripping force (related to muscle mass).

For the sensory and cognitive variables in the pilot we had hardly any idea whether or not differences could be expected and, if so, in what direction.

Subjects

In the pilot-study the aforementioned variables were measured on a sample of 118 persons, between 70 and 75 years of age, living in one of three regions of the Netherlands (West, North-East, South-East). In each region a village and a town were selected, which are comparable with regard to number of inhabitants and degree of urbanisation between regions. In each municipality one or more family doctors were contacted and asked to select a number of their healthy patients, according to our requirements. These persons received a letter from Delft University of Technology, together with a letter from their family doctor in which the study was explained and they were asked to participate in this study. The response to this written request was 36%.

Results of the pilot

Analysis of the variables on level of education, occupation and living situation revealed hardly any statistically significant differences between the groups of participants. No differences were found between the regions, only some between degrees of urbanisation. Level of education of the men living in the towns appeared to be higher than that of men in villages. This might be of influence on the results of the cognitive variables.

For each variable mean values and standard deviations per degree of urbanisation and per sexe are presented in table 1. The results per region are presented in table 2 for males and in table 3 for females.

The differences between the mean values for males and females, and between the degrees of urbanisation are tested with a t-test for independent samples. The differences between regions are tested with a one-way ANOVA. The results show that, as expected, for many variables statistically significant differences are present between the mean values for males and females. Differences between regions and between degrees of urbanisation are less obvious. A summary of these test results is given in table 4 and table 5.

Table 1. Results per variable according to degree of urbanisation and sex.

	Males (n=58)				Females (n=60)			
	Town (n=28)		Village (n=30)		Town (n=32)		Village (n=28)	
	Mean	SD	Mean	SD	Mean	SD	Mean	SD
dy mass [kg]	84.1	12.6	80.0	9.8	72.2	11.2	75.7	1.9
ture [cm]	173.6	7.6	173.3	6.2	161.5	5.9	160.3	5.4
ing height [cm]	91.0	3.7	90.2	3.5	84.2	2.6	83.6	3.0
ttock-knee length [cm]	61.8	3.1	61.7	2.7	59.9	3.0	60.6	2.9
oliteal height [cm]	47.9	3.0	48.3	2.5	43.2	2.6	43.6	2.0
oulder breadth [cm]	45.6	3.6	45.7	2.6	42.3	2.3	43.2	3.1
breadth seated [cm]	39.2	2.5	37.9	2.3	38.4	2.6	39.3	2.6
ance time, eyes open [s]	25.9	22.8	27.9	25.8	25.3	23.6	10.0	12.5
ance time, eyes closed [s]	3.6	3.8	3.4	4.1	2.3	1.3	1.9	1.4
p length [m]	1.35	0.17	1.45	0.15	1.18	0.19	1.13	0.14
lking velocity, normal [s]	8.78	1.40	8.46	1.40	9.20	1.39	9.49	1.44
lking velocity, fast [s]	6.47	1.04	6.49	0.97	7.24	1.07	7.44	1.05
gboard time [s]	102.2	17.7	107.6	26.3	99.7	43.7	102.8	20.0
sual acuity	0.75	0.23	0.77	0.22	0.67	0.21	0.70	0.18
Reading test:								
ey on white, 10 Lux [pt]	11.3	1.8	11.7	1.5	12.6	.	12.6	.
ey on white, 100 Lux [pt]	11.7	1.5	12.0	1.3	11.7	1.3	11.3	1.3
ey on white, 1000 Lux [pt]	9.7	1.9	10.5	1.6	10.5	1.8	10.4	1.8
ey on black, 10 Lux [pt]	11.5	1.5	10.8	1.6	11.4	1.6	10.9	1.6
ey on black, 100 Lux [pt]	8.9	1.4	8.5	1.6	9.1	1.4	8.9	1.7
ey on black, 1000 Lux [pt]	7.0	1.5	6.9	1.3	6.8	1.4	7.0	1.4
ite on black, 10 Lux [pt]	9.5	2.0	9.0	2.0	9.5	1.8	8.9	2.1
ite on black, 100 Lux [pt]	6.7	2.0	6.7	1.6	6.6	1.4	6.7	1.4
ite on black, 1000 Lux [pt]	4.9	1.3	5.1	1.2	5.0	1.1	5.3	1.1
ick on white, 10 Lux [pt]	7.9	2.0	7.8	1.6	7.9	1.7	7.9	1.9
ick on white, 100 Lux [pt]	5.6	1.5	5.6	1.4	5.6	1.1	5.6	1.3
ick on white, 1000 Lux [pt]	4.4	0.9	4.2	0.6	4.7	1.1	4.5	1.0
action time, visual single [s]	0.35	0.06	0.34	0.08	0.39	0.13	0.35	0.06
ovement time, visual single [s]	0.29	0.10	0.27	0.07	0.34	0.11	0.33	0.06
action time, visual choice [s]	0.49	0.11	0.48	0.08	0.59	0.13	0.59	0.22
ovement time, visual choice [s]	0.33	0.10	0.34	0.10	0.42	0.14	0.52	0.33
action time, auditive single [s]	0.35	0.09	0.41	0.17	0.53	0.51	0.38	0.11
ovement time, auditive single [s]	0.30	0.14	0.29	0.06	0.37	0.13	0.35	0.07
action time, auditive choice [s]	0.70	0.19	0.69	0.22	0.87	0.27	0.82	0.22
ovement time, auditive choice [s]	0.60	0.42	0.54	0.27	0.77	0.30	0.75	0.42
ipping force right [N]	386.9	70.7	413.6	76.1	253.9	56.1	241.3	56.2
ipping force left [N]	384.6	78.0	380.3	80.4	236.6	47.2	224.5	47.8
nd flexion [degree]	57.5	12.4	57.1	11.6	59.5	8.4	54.5	9.6
nd extension [degree]	52.9	13.2	53.4	12.7	57.9	9.6	57.9	9.4
nar deviation [degree]	40.1	11.2	40.6	9.0	39.0	7.0	40.4	9.1
dial deviation [degree]	17.6	5.5	20.3	6.3	22.2	7.5	22.0	7.9
refinger flexion [degree]	52.4	13.0	55.6	9.7	52.5	10.5	50.5	10.2
ctile sensitivity score	2.1	2.1	2.6	2.4	2.4	1.9	3.1	2.2

Table 2. Results per variable according to region, for males.

Males	Region					
	West (n=21)		North (n=18)		South (n=19)	
	Mean	SD	Mean	SD	Mean	SD
Body mass [kg]	80.6	13.4	84.6	11.5	81.2	8.6
Stature [cm]	172.8	6.8	176.6	7.2	171.1	5.6
Sitting height [cm]	90.7	3.9	91.0	3.3	90.0	3.6
Buttock-knee length [cm]	61.1	2.6	63.5	2.9	60.9	2.4
Popliteal height [cm]	47.6	2.6	49.6	3.1	47.3	1.9
Shoulder breadth [cm]	45.3	2.4	46.5	3.2	45.4	3.7
Hip breadth seated [cm]	38.1	2.5	38.5	2.6	38.9	2.5
Balance time, eyes open [s]	22.5	21.8	34.8	25.9	24.4	24.6
Balance time, eyes closed [s]	4.4	5.1	2.5	1.2	3.5	4.2
Step length [m]	1.35	0.17	1.49	0.14	1.38	0.15
Walking velocity, normal [s]	8.94	1.68	8.37	1.08	8.49	1.32
Walking velocity, fast [s]	6.73	1.04	6.24	0.66	6.43	1.18
Pegboard time [s]	103.0	24.8	110.0	16.6	102.2	24.8
Visual acuity	0.80	0.25	0.79	0.14	0.68	0.24
Reaction time, visual single	0.34	0.05	0.36	0.09	0.35	0.07
Movement time, visual single	0.31	0.09	0.25	0.05	0.28	0.10
Reaction time, visual choice	0.47	0.08	0.49	0.09	0.50	0.11
Movement time, visual choice	0.35	0.09	0.34	0.13	0.31	0.06
Reaction time, auditive single	0.38	0.20	0.39	0.12	0.38	0.10
Movement time, auditive single	0.30	0.07	0.27	0.07	0.30	0.16
Reaction time, auditive choice	0.65	0.14	0.79	0.24	0.65	0.19
Movement time, auditive choice	0.57	0.38	0.61	0.31	0.53	0.37
Gripping force right [N]	382.6	80.0	384.5	68.8	436.2	62.2
Gripping force left [N]	355.5	73.4	393.0	86.1	402.0	72.2
Hand flexion	53.6	11.3	57.9	10.5	60.8	13.2
Hand extension	52.2	11.3	52.4	15.4	54.9	12.3
Ulnar deviation	37.6	9.6	39.8	8.1	44.0	11.6
Radial deviation	17.9	5.9	20.8	5.4	18.5	6.6
Forefinger flexion	53.1	12.1	50.7	12.4	58.2	8.6
Tactile sensitivity score	1.9	1.7	2.4	2.7	2.7	1.9

Table 3. Results per variable according to region, for females.

Females	Region					
	West (n=20)		North (n=19)		South (n=21)	
	Mean	SD	Mean	SD	Mean	SD
Body mass [kg]	77.9	11.2	75.1	13.3	68.9	8.4
Stature [cm]	162.0	4.5	163.0	5.5	158.1	5.8
Sitting height [cm]	84.8	2.5	84.7	2.4	82.3	2.7
Buttock-knee length [cm]	61.3	2.9	60.8	3.2	58.7	2.3
Popliteal height [cm]	43.2	2.1	44.3	1.9	42.7	2.6
Shoulder breadth [cm]	43.2	2.7	42.8	2.8	42.3	2.6
Hip breadth seated [cm]	39.6	2.4	39.0	3.3	37.8	1.8
Balance time, eyes open [s]	23.7	23.6	16.3	18.9	14.8	18.9
Balance time, eyes closed [s]	1.8	1.3	2.1	1.0	2.4	1.6
Step length [m]	1.17	0.16	1.20	0.22	1.10	0.10
Walking velocity, normal [s]	9.12	1.54	9.15	1.52	9.71	1.15
Walking velocity, fast [s]	7.02	0.97	7.11	0.91	7.82	1.11
Pegboard time [s]	97.8	19.9	95.5	18.0	109.3	52.2
Visual acuity	0.75	0.15	0.67	0.16	0.63	0.25
Reaction time, visual single	0.35	0.09	0.37	0.06	0.40	0.14
Movement time, visual single	0.33	0.07	0.34	0.12	0.34	0.07
Reaction time, visual choice	0.58	0.22	0.57	0.13	0.61	0.18
Movement time, visual choice	0.45	0.32	0.43	0.15	0.52	0.25
Reaction time, auditive single	0.51	0.65	0.44	0.17	0.43	0.13
Movement time, auditive single	0.38	0.16	0.34	0.08	0.37	0.06
Reaction time, auditive choice	0.83	0.24	0.85	0.20	0.86	0.31
Movement time, auditive choice	0.66	0.27	0.73	0.49	0.90	0.42
Gripping force right [N]	240.7	63.8	262.3	58.1	241.3	46.3
Gripping force left [N]	223.6	55.1	243.4	39.7	226.1	46.4
Hand flexion	56.7	7.7	58.9	10.2	56.1	10.0
Hand extension	58.5	8.0	58.5	10.3	56.8	10.1
Ulnar deviation	38.0	9.8	40.8	8.3	40.2	5.6
Radial deviation	22.4	6.8	23.4	8.4	20.6	7.8
Forefinger flexion	54.0	8.1	48.9	10.2	51.7	12.1
Tactile sensitivity score	2.3	1.8	3.3	2.1	2.7	2.2

Table 4. Statistically significant results of the one-way ANOVA with region as factor, according to sex.

Males				Females			
buttock-knee length	F=5.63	p=0.006	North >South; North >West	buttock-knee length	F=4.66	p=0.013	West>South
popliteal height	F=4.53	p=0.015	North >South	body mass	F=3.51	p=0.037	West>South
stature	F=3.43	p=0.040	North >South	stature	F=4.93	p=0.011	North>South
gripping force right	F=3.55	p=0.037	South>North; South>West	sitting height	F=6.35	p=0.003	West>South; North>South
step length	F=4.11	p=0.022	North >West	walking velocity	F=3.92	p=0.025	South>West

Table 5. Statistically significant results of the t-tests for independent samples, according to sex

Males				Females			
hip breadth seated	t=2.05	p=0.045	town>village	balance, eyes open	t=3.25	p=0.002	town>village

Conclusions

Results show that some statistically significant differences exist, but not for all groups of variables systematically and not between the different regions. The differences concern, as expected, a few of the physical variables in the study, mainly the body dimensions.

This suggests that it will not be necessary to perform the measurements for the main part of the study in municipalities all over the country. For regional differences in body dimensions corrections can fairly easily be applied to the data.

Main study

In the main part of the study the variables will be assessed on a sample of 750 persons, divided among five different age groups: 20-30 years of age, 50-60, 60-70, 70-80, 80 years and over and residing in one region of the country. The results will be: a design-relevant data-bank, a basis for new design-guidelines and standards, a survey of products and components to be improved and an enlarged theoretical framework of man-product interaction.

References

Freudenthal, A. 1993, *Gerontechnologisch Produktontwerpen* (Gerontechnological Product Design), Report Faculty Industrial Design Engineering, Delft University of Technology, Delft.

Roede, M.J. and J.C. van Wieringen 1985, Growth Diagrams 1980, Netherlands third nation-wide survey. *Tijdschrift voor Sociale Gezondheidszorg*, supplement 1985, p 1-34.

Steenbekkers, L.P.A., J.M. Dirken, J.J. Houtkamp, J.F.M. Molenbroek, A.I.M. Voorbij 1994, *Capaciteiten van Ouderen* (Capacities of the elderly), Report Faculty Industrial Design Engineering, Delft University of Technology, Delft.

Hands and Holding

The Measurement of Skin Friction

W.P. Mossel

Faculty of Industrial Design Engineering,
Dept. of Product and Systems Ergonomics
Delft University of Technology
Jaffalaan 9, 2628 BX Delft
The Netherlands

A new measurement instrument is described to measure static skin friction. The principle of this measuring instrument uses an equilibrium of forces. An angle is measured at the moment of breaking the equilibrium. This angle is a measure for the maximum frictional force. This principle is different from the existing systems for friction measurements. The accuracy of the system is discussed and the derivation of the expected inaccuracy is given. The results of three measurement series are given and compared with measurements made with other equipment. The values of the measurements are put in both the linear and the logarithmic friction model.

Introduction

If a designer develops something, in which skin friction plays a role, he or she needs at least a reasonable estimation of the value of the frictional force that the user can perform. For the estimation of frictional force usually the law of Amontons (1699) is used. This law describes the maximum frictional force F_f for non-lubricated surfaces as a function of the normal load F_n and the coefficient of friction μ:

$$F_f = \mu\, F_n \dots\dots\dots\dots\dots\dots\dots\dots\dots\dots\dots\dots\dots\dots\dots\dots\dots\dots\dots\dots(1)$$

It is the 'classical friction law' and for many cases still valid. It is further called the linear model. The independence of the frictional force from the size of the area of contact is explicit in this equation. For soft surfaces and especially for human skin the linear model generally does not apply. Bobjer et al. (1993) show this clearly. Bobjer concludes for the frictional characteristics of human palmar skin:

- It is not directly proportional to the normal force (load) F_n. The coefficient of palmar friction is negatively affected by the load, regardless whether the skin is in a normal condition, sweaty or exposed to oil or lard and regardless of the texture of the surface.
- It is dependent on the size of the surface areas in contact with the skin.

Is it possible to formulate an equation for the skin friction that gives the designer a tool for estimating the friction? The equation described by Mossel & Roosen (1994), the so called logarithmic model, coincides with the conclusions of Bobjer (1993):

$$F_f = M \, c_p \, A_t^{q-1} \, F_n^q \dots \dots \dots \dots \dots \dots \dots \dots \dots \dots \dots \dots \dots \dots (2)$$

in which A_t stands for the contact area and q for some dimensionless exponent smaller than 1. M is a proportional factor. It has a dimension of $[(MPa)^{1-q}]$ in case the forces are expressed in [N] and the area in [mm²]. The factor c_p is the pressure distribution factor, dimensionless and dependent on q. It expresses the number of times the pressure p in a point is bigger or smaller than the average pressure p_m on the contact surface. This formula can be seen as a refinement of the equation of Comaish and Bottoms (1971). It can be derived from the equations of Hertz (1895) and the supposition that the frictional stress τ is proportional to the pressure p according to $\tau = M p^q$.

The question is to what extent M and q are useful constants for a designer, in other words are M and q constants for a wide population or 'constants' only for each single subject. For instance, it remains to be seen whether M is 'constant' because in it the modulus of elasticity of the skin is hidden. The modulus of elasticity of the skin is, amongst others, dependant on the humidity of the air and the age of the subject.

To get an insight in the values of M and q, and to determine if the logarithmic model gives a better description of the skin friction, measurements are needed. Usually the skin friction is measured dynamically, (Comaish & Bottoms (1971), Bobjer et al (1993), Gerrard (1987)). The measurements of Mossel & Roosen (1994) can be considered as static, because of the very low speed applied. For the designer, often static friction values are more valuable. In the next paragraph a new instrument for the measurement of static friction is described.

The measurement instrument

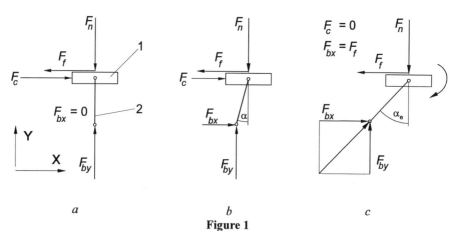

a   b   c

Figure 1
Principle of the measuring instrument

The principle of the measurement instrument is given in figure 1. A plane (1) is supported by pivoting bars. In the figure this is simplified by one bar (2). The subject brings, with the tip of his/her right index finger, a normal force F_n to bear on the plane and the maximum frictional force F_f. The forces on the plane (1) are given in figure 1a. The frictional force keeps an electrical contact closed. The contact force F_c balances the frictional force F_f and the force F_{by} balances the normal force F_n. See figure 1b. When the bar (2) is rotated over an angle α a force F_{bx} has to be exerted on the bar to achieve equilibrium of the bar. Owing to the balance of the horizontal forces of the total system the contact force F_c will decrease and F_{bx} will increase. The situation of figure 1c will arise by the maximum size of $\alpha = \alpha_e$. Then $F_c = 0$ and $F_{bx} = F_f$. The plane (1) is now in a labile equilibrium and will flip in the indicated way by a minimal increment of α. The angle α_e is measured when the plane (1) flips and the contact between the plane and the frame is broken. The normal force F_n is continuously measured. The value of the frictional force F_f can be now directly calculated from the values of F_n and α_e

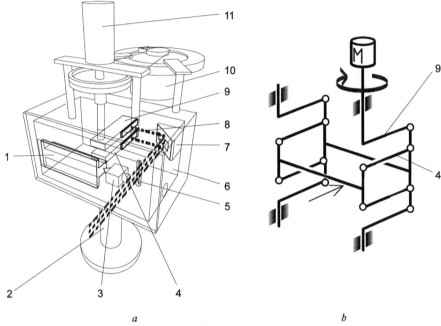

a   b

Figure 2
The measuring instrument

A measurement instrument has been built based on this principle, see figure 2a. The plane (1) is placed vertically to avoid the influence of the weight of the bars (4 and 9). The bar (4) is coupled with the plane (1) and the bar (9) coupled with the frame. There are four double bar combinations, in figure 2a one one set is drawn, in figure 2b the complete set. The double bar combinations are interconnected as can be seen in figure 2b. With the double bar system the plane does not move when the bars are rotated with the motor (11). Via a 2:1 snare transmission the angle of rotation is measured with a pulse generator (10).

To measure the correct angle α_e, the bars have to be perpendicular to the plane (1) in the start situation. The plane (1) is parallel to the backplane (6) of the frame. On the backplane a prism (7) is fastened. It has two planes under 45° and one parallel to the backplane. On one of the four double bar combinations two flat mirrors (8) are fastened. If the images of the hair (5) via the mirrors coincide with the hair itself, the bars are perpendicular to the plane (1). The last fine tuning can be done with the adjustable contact (3). The contact (3) is the above-mentioned contact kept closed by the force F_c.

The normal force F_n is measured twenty times per second with a 3-component piezoelectric force transducer. The force transducer (not drawn) is placed between the pole (2) and the frame. No forces other than the normal force on the plane (1) should be exerted on the chassis. Otherwise, they will be interpreted as normal force.

A control system controls the speed of the motor with the pulses fed by the pulse generator. During the measurement the motor runs with a speed of 2 rpm and turns back to its start position with the double speed after the breaking of the contact. The pulse generator delivers 20.000 pulses per revolution to the control system and the pulses are also fed to the computer. From these pulses the angular rotation of the motor is calculated.

The way of measurement

When a measurement starts, first the bars are set perpendicular to the plane and the pulse generator is reset to zero. The contact between (1) and (3) is in the closed position. The desired normal force is adjusted at the amplifier for the piezo element. The subject places his/her right forefinger on the plane and pushes by means of the friction between the finger and the plane with the maximum frictional force to keep the contact closed. (The stainless steel plane surface and the finger have been cleaned with a tissue saturated with 50% alcohol and wiped dry). On a computer screen the subject sees a circle that moves up or down with more or less normal force. The subject has to try to keep the circle between two preset lines and keep the contact closed. After a short time the motor starts running. During this running the normal force and the angular rotation of the bars are fed into the computer. When the plane flips, the contact is broken and the motor stops. The momentaneous normal force and the angular rotation are stored in a file.

Accuracy

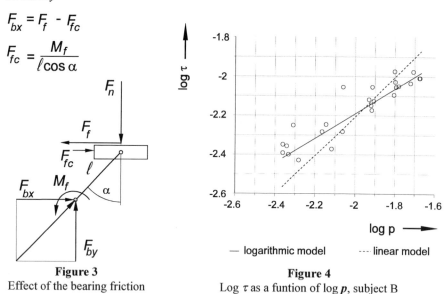

$$F_{bx} = F_f - F_{fc}$$

$$F_{fc} = \frac{M_f}{\ell \cos \alpha}$$

— logarithmic model   --- linear model

Figure 3
Effect of the bearing friction

Figure 4
Log τ as a funtion of log p, subject B

In figure 1 the friction of the bearings is neglected. The friction of the bearings prevents the plane from flipping. In the same way as in figure 1c the situation is given in figure 3 but with a total friction moment M_f of the bearings added. The compensation of the friction moment results in a force F_{fc}, which equals $M_f / (\ell \cos \alpha)$. ℓ stands for the length of the bar. So at the moment the resultant of F_{bx} and F_{by} stands in the direction of the bar the frictional force F_f measured is not $F_n \tan a$ but bigger viz. $F_n \tan \alpha + F_{fc}$.

For an estimation of the influence of the error the following values are used. The length l of the bar is 50 mm. Thirty-six deep groove ball bearings play a role in the movement. Suppose a normal force of 5 N. (The measurements are done with a maximum normal force of 4 N). According to the bearing manufacturer the friction moment can be calculated as $5*10^{-3}$ Nmm per bearing, thus resulting in a total friction moment of 0,18 Nmm. The extreme skin friction coefficients measured by Mossel & Roosen (1993) give a $\tan\alpha = 0,35$ and $\tan\alpha = 1,19$. The result is that the value of F_f will be resp. 0,2% to 0,09% too high. This inaccuracy is neglected.

Measurements

As described in Mossel & Roosen (1994) both the equations of the linear and logarithmic model are logarithmically transformed to achieve straight lines in the graph of the frictional force against the normal force. The equation (1) as well as equation (2) has, after the transformation, the general form $Y = aX + b$. By means of the method of the least squares a and b are calculated for best fitting line $Y = aX + b$ through the measuring points. The proposition is that the linear model is true. Then a should be 1. If not, the logarithmic model is true. Then it should be tested if $a \leq 1$ (Comaish & Bottoms, 1971, Bowden & Tabor, 1974). In every series of measurements the rejection of the linear model is checked under the assumption of the logarithmic model. An error probability of 0,01 is chosen so that the result of the statement has a certainty of 99%. The rejection (or not) of the linear model is done on the basis of the Students-t table.

In the above-mentioned equations stands Y for log (τ) and X for log (p). Because $p = F_n / A_t$ and $\tau = F_f / A_t$, the contact surface A_t has to be determined. The surface is depending on the force F_n and the subject. To determine the area, the plane (1) in figure 2a is blocked and a piece of paper is placed on the plane. The finger of the subject is inked with an inkpad. The subject is asked to push with the inked finger on the piece of paper, in the same way as is done in the way as described in the foregoing paragraph, but now starting with a lower force than indicated and rising until the circle is between the lines and then removing the finger. The force F_n is fed in the computer. This is repeated for the nominal forces 0,5 N, 1 N, 2 N, 3 N and 4 N. The areas of the finger prints are measured and plotted in a graph with logarithmic scales against the corresponding normal force. With the method of the least squares a straight line is calculated by which the areas belonging to the measured normal forces in the measurements can be specified.

A measurement is repeated five times with the same nominal normal forces as mentioned above. So one set of measurements consists of twenty-five points in the graph. The factor c_p is taken equally to 1. This assumption will be further elaborated in the next paragraph.

Results and Discussion

Table 1 gives results of three sets of measurements (A, B, C1) made with the new friction meter. In the measurements with subjects A and B the linear model can be rejected with the specified certainty of 99%. In figure 4 the measurement with subject B is represented as an example. The graph shows that the logarithmic model fits better than the linear model. All the measurements within the subject indication 'C' are with the same subject. The values of C(2), C(3) and C(4) are from the measurements of Mossel & Roosen (1994). The q in the measurement with subject C(1) is bigger than 1. This means that neither the linear nor the logarithmic model fits. Very striking is the big difference in the value of μ compared with the values measured before. A reason could be the incorrect execution of the instructions given to the subject. When the subject is distracted too much by trying to bring the circle of the normal force between the lines, he/she possibly forgets to keep the contact closed by means of the frictional force. The problem of not exerting the maximum frictional force by the subject is a point of further research. However, it seems that this can be achieved with good instructions.

Table 1
Values for μ, M and q found in n measurements

			subject			
	A	B	C(1)	C(2)	C(3)	C(4)
n	25	25	25	25	43	35
μ	0,32	0,63	0,38	1,19	1,03	1,16
M	0,031	0,119	1,627	0,195	0,086	0,755
q	0,521	0,632	1,314	0,686	0,564	0,500

In practice it is too complicated to calculate a value for c_p. The value of c_p lies between 1 and 0,844 according to Roosen (1993). These values are derived from theoretical pressure distributions. Probably, the pressure distribution in practice will be somewhere between homogeneous and parabolic. This means that c_p can be expected to be about 0,95. Further research is needed. On the first sight the influence of c_p not equal to 1 is negligible for most skin friction problems in designs.

More research is needed to give a better basis to the logarithmic model.

Acknowledgements

Thanks are due to Michel Eijgermans and Edwin Kallemein for the development of the computer software for the measuring equipment.

References

Amontons, G. 1699, De la résistance causée dans les machines, Mémoires de l'Académie Royale de Science, 206.

Bobjer, O., Johansson, S. and Piguet, S. 1993, Friction between hand and handle. Effects of oil and lard on textured and non-textured surfaces; perception of discomfort, *Applied Ergonomics*, **24** (3), 190-202.

Comaish, S. and Bottoms, E. 1971, The skin and friction: Deviations from Amontons' laws and the effects of hydration and lubrication, British Journal of Dermatology, 84, 37-43.

Gerrard, W.A. 1987, Friction and other measurements of the skin surface, *Bioengineering and the Skin*, 3, 123-139.

Mossel, W.P. and Roosen, C.P.G. 1994, Friction and the skin, *Proceedings of the Ergonomics Society's 1994 Annual Conference*, (Taylor & Francis, London) 353-358.

Roosen, C.P.G. 1993, *Wrijving en de huid*, masters project, (T.U. Delft Industrieel Ontwerpen, Delft)

A contribution to the understanding of the role of digital pulp in hand grip performance.

George E. Torrens

Department of Design and Technology,
Loughborough University,
Loughborough,
Leicestershire. LE 11 3TU

The paper will discuss current understanding of the role of digit pulp and offer additional information about its role in providing an effective grip upon an object during task performance. The characteristics of digit pulp will be defined based on its anatomy, biomechanical properties and physiology during task performance. Analogies will be described to demonstrate how digit pulp may interact with the underlying, bone, connective tissues and skin to provide an effective counter to slippage. Assessment of the analogies using models and non-invasive testing of human subjects will be described. The results of the assessment will be used to discuss the constraints within which digit pulp may be effective in opposing slippage.

Introduction

The information given is directed towards ergonomists and designers who wish to have a better understanding of the interactions that take place at the hand and handle interface. The main use of a hand is to provide a stable connection between a body and a task object. From a study of literature this field and previous work (Brown, Torrens and Wright 1992) a division of the complex interaction is proposed. The interaction between hand and task object may be divided into three categories:

- Micro interaction with the surface of the task object: involving skin tissues.
- Intermediate interaction with the texture and features of the task object.
- Gross interaction with the overall features and shape of the task object: where the whole hand is used to produce a grip pattern.

Intermediate interaction: the role of Digit Pulp.

Studies of the functional anatomy of digit pulp and connective tissues of the hand have been documented by a number of authors. Stack (1973) has cited most of the key

functions and references. From his list the following functions for digit pulp can be
derived:

- The pulp protects the underlying structures, tendons muscles and nerves from
  pressure during gripping.
- The connective tissues surrounding the digit pulp stabilise skin and pulp during
  gripping
- The connective tissues also maintain the shape of the digit
- There may also be some assistance to blood flow due to the intermittent compression
  of the pulp area

Models of interaction

As a subject grasps an object the skin and underlying tissues of the digit are
compressed until a predetermined load has been applied to the task object. The subject
decides on the applied load from experience of grasping similar task objects. The load is
likely to be an over-estimation of the load required to maintain grip. Studies by Edin,
Westling and Johanson (1992) support this claim. In a static load the subject will
gradually reduce the grip force until it is just above the point of slip (MacKenzie and
Iberall 1994). In a dynamic task situation the forces are likely to be higher to maintain grip
while countering minor perturbations. If slippage does occur the force applied by the digits
increases until slippage has stopped.

Digit pulp has two distinct areas, proximal ungual pulp and distal ungual pulp.
Proximal ungual pulp is encased by a number of septae and ligaments that constrain the
movement of the pulp (Shrewsbury M. and Johnson R.K., 1975). The fatty tissue encased
within a fiborous septae that make up the ungual pulp has been described by Wood Jones
(1941) as granular in appearance. Wood Jones also likened the subcutaneous tissue of the
palm and sole of the foot to a rubber-like pad. The granular appearance can be seen in the
majority of dissection photographs and diagrams used by Landsmeer (1976) and Foucher
(1991). From the photographs and drawings shown in the above references the granules
scale to approximately millimetre diameter.

Contact with the task object deforms the subjects digit pulp area. In commonly used
grip patterns the load is applied through the proximal digit pulp. Where the task involves
fine manipulation or the gripping of small objects the distal pulp is used. The more
compliant area of the proximal pulp may inhibit manipulation by enveloping the object.
The proximal pulp may have too much free movement for the required fine adjustments to
position. Approximately 3-4 millimetre of movement can be seen in the skin and digit pulp
when a force is applied. The amount of skin and connective tissue movement may depend
on the depth of proximal digit pulp.

From the given descriptions of digit pulp the following analogies may be applied to
help identify the most appropriate measurement techniques.

A simple analogy to digit pulp is that of a modern car tyre. Although this structure
acts at much higher forces than the digit pulp the basic function is the same, to provide
grip and absorb minor perturbations in the road surface. If the tyre is too hard the
available grip area is limited. If the tyre is too soft the outer walls are too lax in retaining

the tyres shape and slippage occurs more easily. This suggests there may be a high and low limit of pressurisation of the ungual pulp pouch at which grip is viable of specific objects and tasks. Simpson (1970) has described a demonstration of the digit pulp performance by using a two leather pouches attached to flat plates on a modified pair of pliers. He found using the padded pliers objects required much less force to be picked up than when using a plain metal equivalent.

An explanation for the rigidity of the normally compliant proximal ungual digit pulp is that the fatty tissue within the granules may be viewed as a three-dimensional oil-filled chain mesh linking the stable, flat backing plate of the ungual pouch to the non-regular shape of the task object. The amount of movement within the mesh is limited by the length of granule chain in the opposing direction of movement. The oil may be considered incompressible which indicates the change in shape is accommodated by the elasticity in the granule casing. Collagen has a tensile strain value of 8 to 10 percent and an elastic modulus of 8 x 10 7 Nm2 (Williams and Lissener 1992).

The granules limit the movement of the fatty material within the ungual pouch. This avoids excessive deformation and reduces the time taken for the digit to regain shape. The vascular arcade that is interwoven within the granules helps to re-inflate the previously compressed areas. When applied to the proximal ungual digit pulp conventional measurement of hysteresis should include consideration of the subjects blood pressure involved during the process of re-inflation.

The effectiveness of the digit and object interlocking may also be undermined through lubrication by sweat. In the case of high shear forces being applied to the skin surface sweat lubrication may be seen as a fail safe mechanism to avoid delamination of the skin from the underlying fasica, or de-gloving of the skin and underlying tissues from the phalanx. From the documented information regarding digit pulp characteristics the following aspects would seem to have the main influences over digit pulp performance during grip:
- The depth of proximal ungual pulp on each finger will indicate the capacity of the digit to accommodate a specific surface features on the task object.
- The quality of the encapsulating ligamenture, septae and the tissue surrounding the proximal ungual pulp granules will affect the digits ability to mould around a specified task object surface.

The performance of a small digit compared with a larger digit is likely to be quite different. A larger digit may have difficulty in manipulating a small object. A small digit may have difficulty interlocking with a surface made up of widely spaced surface contours. A relationship is likely to exist between the dimensions of the proximal pulp of a digit and the surface features with which it may interlock. The surface requirements of digit pulp will be different to the surface requirements of the skin.

How can the characteristics of digit pulp be measured?

The following pilot studies may assist in the choice what and how to measure the characteristics of digit pulp. A simple method of recording digit anthropometrics and predicting the performance of digit pulp involves photographing the distal part of the digit before and during the digit applying force to a flat plane. The deformation that occurs can be used as a predictor of grip performance. The deformation indicates a combination of volume of digit pulp and its compliance when deformed. The result will be a ratio of the original depth of the digit to that of the digit when deformed. Viewing anatomical studies of the distal phalanx the pulp occupies approximately one third of the depth of the digit indicating the percentage of change in depth will be small. A pilot trial using the equipment found the results to be variable, mainly due to unevenness of pressure applied by the subjects. It was found a large number of subjects could be processed within a short time. Development of the test is continuing.

The measurement of grip performance has mainly been concerned with the skin surface (Pheasant ,1975 and Buschholz, 1988) and Bobjer, Johansson and Piguet (1993). The textures used by Bobjer et al (1993) were square 0.5 millimetres wide grooves cut 0.5 millimetres deep into a flat surface. The textured surfaces performed better when contaminants were present, but not as well as a dry flat surface. Bobjer et al also reported subject discomfort when using the textured surfaces. If the underlying tissues of the proximal ungual pulp are configured in spherical granules of 1millimetre diameter as suggested the textures used by Bobjer et al would not optimise the pulp characteristics. In order to assess the performance of digit pulp more effectively an appropriate surface feature is required. Based on earlier descriptions of the scale anatomy of skin and underlying connective tissues A surface feature was produced that was considered to be complementary to the glaborous surface of the digit. The contour of the surface feature followed a vee pattern of 45 degrees with a rounded crest of 1millimetre radius. The pitch between each crest was 8 millimetres and the vertical depth to the trough was 4 millimetres.

Digit performance assessment

Test equipment (See Figure 1) consisted of a beam with a carriage attached at one end. The downward force was measured by strain gauges attached to the mild steel beam. Shear forces were measured through strain gauges attached to the strip of mild steel attaching the carriage to the beam. The gauges were normalised to mild steel and the test equipment calibrated using weights and a pulley mechanism. The equipment was kept at an ambient stable temperature throughout the test. The gauges had been calibrated using weights of 0.1Kg to 3Kg and a spring balance.

Figure 1.
Digit pulp performance test rig showing the subject's hand and the 3 artificial pulp pads.

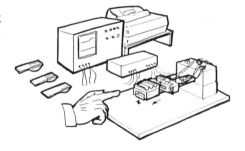

A Caucasian adult male (Technician) and female (administration) were assessed. The subjects were asked to sit down, place their arm level with the test surface, to avoid use of body weight to apply pressure, and press downwards until they reached a marked shown on the oscilloscope screen and then drag their index finger (digit 2) proximally across the textured surface until slippage occurred. There were two marks; Low (1 Kg) and High (2Kg). Slippage was indicated through subject feedback and oscilloscope trace changes. The monitoring rate of the oscilloscope was set at 500 milliseconds. The voltage readings were noted at the point of slippage. The subjects were asked questions about their work while resting for 15 minutes. The pull force was calculated as a percentage of the downward force to obtain a more clear comparison between tests and subjects. The test was repeated using downward force with the surface features lubricated. The lubrication was in the form of low melting point general purpose machine grease. The comparative test of dry to lubricated surface was to try and isolate the effect of skin friction from the pulp performance during gripping. The test was also repeated with artificial finger pads. The purpose of this experiment was to see if the artificial pads had any relationship to the results gained from human digital pads. The downward force generated by the test subjects was replicated when applying the artificial pads to the two surface features assessed. Three finger pads were made each using the same mould to maintain relative volumes. The pad dimensions were 25 millimetres long, 20 millimetres wide and 10 millimetres in depth. The artificial digits consisted of a steel strip backing plate, three inner covers made up of layers of proprietary food grade cling film and an outer cover of poly-cotton glued to the steel strip using an epoxy adhesive. The outer layer of cling film was also bonded to the poly-cotton using a contact adhesive. The three pads contained:

- A proprietary gelatine mix
- An open cell structure foam impregnated with gelatine mix
- Gelatine mix with a filler suspension of partially expanded polystyrene beads of approximately 1millimetre diameter

Results

The following results were obtained:

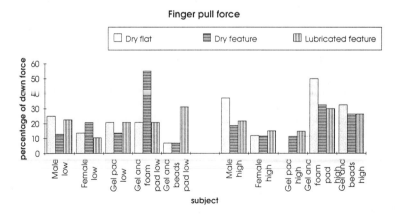

Conclusions

The initial results of the digit performance test indicate that at low downward forces (1Kg) digit pulp is less effective. As the forces involved move above 2 Kg the digit pulp was seen to perform more effectively. From the results of fingers and artificial pads the more fibrous and rigid the pad the more effective it may be at higher forces. The female subject had long thin fingers compared to the large swollen fingers of the male. The volume of digit pulp would also seem to have an effect on performance. There was a notable increase in the performance of the subjects digits at higher forces when the featured surface was lubricated. This observation may have been due to the lubrication softening the skin surface, however the results indicates the pulp interaction has a role within grip beyond increasing skin surface area. During the assessment it was noted that there was a short duration when the subject was reducing the downward force but the pull force was remaining constant. This may be due hysteresis of the pulp. This study has indicated further investigation into aspects of the interaction between the proximal digit pulp and surface features is required.

References

Bojer O., Johansson S.E., Piguet S., 1993, Friction between hand and handle. Effects of oil and lard on textured and non-textured surfaces; perception of discomfort, *Applied ergonomics*, **24**, 190-202

Brown F.R., Torrens G.E., Wright D.K., 1992, Research into optimising hand and body function for tasks in everyday living: the development of a range of "easy use" saucepan handles, In Bracale M. and Denoth F. (ed), *Medicon '92, Proceedings of the VI Mediterranean conference on medical and biological engineering*, (Associazione Italiana di Medica e Biologica, Napoli), **1**, 549-553

Buchholz B., Frederick L.J. and Armstrong T.J., 1988, An investigation of human palmar skin friction and the effects of materials, pinch force and moisture, *Ergonomics*, **31**, 317-325.

Edin B., Westling G., and Johansson R.S., 1992, Independent control of human finger-tip forces at individual digits during precision lifting, *Journal of physiology*, **450**, 547-564.

Foucher G.(ed.), 1991, Fingertip and nailbed injuries, (Churchill livingstone, Edinburgh) 2-3.

Landsmeer J.M.F., 1976, *Atlas of anatomy of the hand*, (Churchill livingstone, Edinburgh) 284-288.

MacKenzie C.L. and Iberall T., Stelmach G.E. and Vroon P.A. (ed.), 1994 , *The grasping hand; Advances in psychology*, **104** (North-Holland, Amsterdam) 222-235.

Pheasant S., O'Neill D., 1975, Performance in gripping and turning - A study in hand/handle effectiveness, *Applied Ergonomics*, **6**, 205-208.

Shrewsbury M. and Johnson R.K., 1975, *The journal of bone and joint surgery*, **57-A**, 784-788

Simpson D.C., 1971, Gripping surfaces for artificial hands, *The hand*, **3**, 12-14.

Stack H.G., 1973, *The palmar fascia*, (Churchill Livingstone, Edinburgh) 3 and 69.

Williams M. and Lissener H.R., LeVeau B.F.(ed.), 1991, *Williams and Lissener's biomechanics of human motion*, 3rd edn, (Saunders, Philadelphia) 39-41.

Wood Jones F., 1941, *The principles of anatomy as seen in the hand*, 2nd edn, (Bailliere, Tindal and Cox, London) 166-167.

Marine Ergonomics

POSITIONING ELECTRONIC AIDS IN SMALL FISHING VESSELS

Stella Mills

Department of Information Technology,
Cheltenham & Gloucester College of Higher Education,
P O Box 220, Cheltenham, GL50 2QF, UK

This paper uses criteria from ergonomic theory to evaluate and compare the positioning of the electronic fishing aids in two 'double-ticket' fishing vessels registered in the UK and with overall length of between 30 and 40 metres. Priority is given to the positioning of the echo-sounders, horizontal sonars and netszondes in relation to the navigational chart together with the radar visual display units. The main conclusion is that there is still room for improving the positioning and visual output from the electronic aids in order to improve ergonomic acceptability.

Introduction

In the UK, the small fishing vessel of between 30 metres and 40 metres registered length requires two Deck Officers, one holding a valid Class 1 (Fishing Vessels) Certificate and the other holding a valid Class 2 (Fishing Vessels) Certificate (Olsen, 1995). Consequently, such vessels are often referred to as 'double ticket', although only one of the officers will be in charge of the vessel at any one time. The electronic equipment carried in the wheehouse of 'double ticket' vessels generally consists of full radio communications often with an Imarsat-C system, a colour video track plotter together with an interfacing Global Positioning System (GPS) possibly digital, radar systems of different frequencies, two or more echosounders, one or two (horizontal) sonars and possibly a netszonde (an acoustical system of transducers for monitoring the position of, and fish entering, the net). For pelagic (mid-water) trawlers and seiners the latter is essential for efficient fishing. Frequently, systems are duplicated to provide additional information (for example, two echosounders of differing frequencies), the second system also acting as a backup. From previous work (Mills, 1993), the skipper's tasks in the wheelhouse while fishing require the use of the video plotter/navigation system, the echosounders, the horizontal sonar and the netszonde often more or less simultaneously in order to position the vessel and the net for optimum fish capture. Thus, in this paper, it is to these systems that criteria from ergonomic theory will be applied in order to ascertain the best possible positioning. Two wheelhouses of 'double ticket' vessels will be used to illustrate the validity of the criteria.

Ergonomic Principles

Although there is a good ergonomic literature generating general design principles for bridges of large merchant and military naval vessels, there has been little study of smaller fishing vessels' wheelhouses. Obviously, the most general principles of bridge design can be applied to all types of naval vessel but because of the specialised nature of the tasks that the fishing skipper must perform, more specific criteria must be developed to accomodate the number and variety of electronic aids 'double ticket' vessels carry.

Witty (Witty, 1984) writing in 1984, suggested that the 'sophisticated array of fish-detection and navigational instruments' which many fishing vessels carry, should be 'grouped together in functional areas'. He correctly classified the echosounder as a navigational aid since it is used to avoid wrecks and an uneven seabed as well as discriminating between different types of seabed. Thus the echosounder must 'often fulfil three functions - to determine the depth of water below the ship (and, in mid-water trawling, below the net), to detect and discriminate fish both in mid-water and close to the seabed, and thirdly to determine the nature of the seabed' (Witty, 1984). Consequently, it may be necessary to prioritise these functions in order to optimise the positioning of the echosounder. Even so, we can formulate the following design principle:
Principle 1
Functional grouping of equipment should be optimised, if necessary through functional prioritising.

Stoop (Stoop, 1990a) suggested that 'principles of informational ergonomics need to be applied to instruments...' particularly through 'integration of information on one display, such as a VDU [Visual Display Unit], instead of presentation on different displays...'. This approach of integrating information to prevent unnecessary physical movement between displays and to help to alleviate cognitive overload is supported by Froese (Froese, 1981) and Harre (Harre, 1987) when considering the more general case of large vessel bridge design. Thus we have:
Principle 2
Information from different displays should be integrated whenever possible.

Continuing his theme of reducing instrumentation, Stoop (Stoop, 1990a) advocated 'linkage between instruments' and 'automation of transfer of information from one instrument to another...'. This is reiterated in another paper (Stoop, 1990b) where Stoop suggested that the 'radar and autopilot' could be combined 'to detect and correct bias errors'. In terms of electronic fishing aids this automated transfer of information could create redundancy of an acoustical display through integration. Consequently, we can formulate a principle which supports Principle 2:
Principle 3
Automation of transfer of information between instruments should be optimised.

These three principles are fundamental to ergonomic positioning of equipment in a wheelhouse and will be applied below to two 'double ticket' fishing vessels.

Practical Examples

Example 1
As can be seen from Plan 1, as far as fishing aids are concerned, this vessel has two radars, one video plotter, two echosounders and one sonar with hard copy output only. There is no netszonde since the vessel is not used for pelagic trawling or seining. Applying

Principle 1 to the vessel's fishing aids, it can be seen that the video plotter is about 1.5 metres away from the echosounder, which, in turn, is about one metre away from the sonar output. Clearly, functional grouping would produce a design where these three equipments are grouped in close proximity. Recalling that the skipper will need engine control while shooting, hauling and positioning the net (Mills, 1993), a more optimum solution would be to place the video plotter beside the primary echosounder and radar and move the sonar output so that these aids could be viewed together when required. The secondary radar

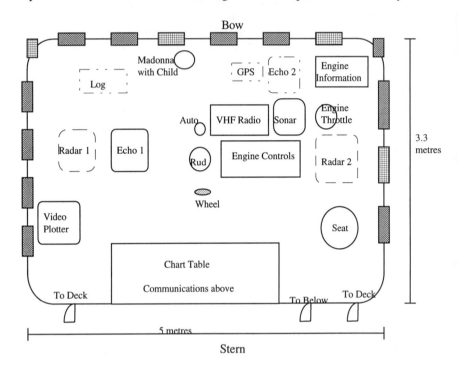

Plan 1: Position of Fishing Aids in the Wheelhouse of Example 1 (not to scale)

Key to Plans 1 - 4

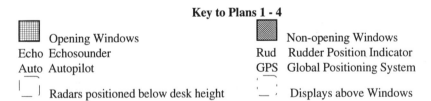

and echosounder should also be moved so that they are readily available for use as a backup but positioned in reverse order so as to avoid confusion. This would also enhance the ability to satisfy Principle 2, since the visual integration of information from the separate outputs used for fishing would be easier from displays positioned in close proximity. At the present time (1995), the transfer of information from one acoustical display to another is rarely achieved; however, integrated trawl systems are available and becoming increasingly used by smaller vessels. Consequently, in order for the fishing aids to satisfy Principle 3, more sophisticated systems need to be purchased. A suggested solution for this vessel with its present aids, giving a more ergonomically acceptable design, is shown in Plan 2.

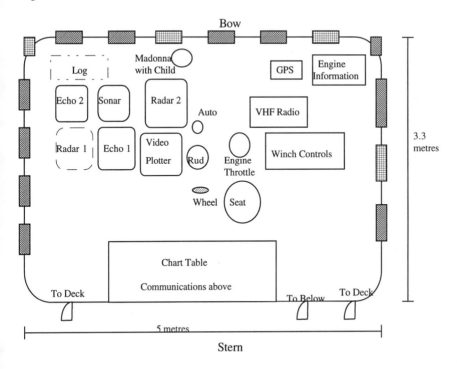

Plan 2: Modified Wheelhouse of Example 1 (not to scale)

Example 2

Plan 3 shows the present positions of the fishing aids in the wheelhouse of the second vessel studied. There is no (horizontal) sonar nor netszonde since the vessel is only used for demersal (seabed) trawling. Two echosounders and radars are carried, the second of each being used as a backup for the primary systems which are those placed to port. If we consider only the primary echosounder, radar and the video plotter in this design, it can be seen that Principle 1 is satisfied. However, the video plotter is too distant from the secondary echosounder and radar for these to be used as backups. Consequently, it is suggested that these secondary aids be positioned above the primary aids to facilitate reading in the case of backup usage but in reversed positions so as to reduce confusion with the use of the primary aids. This positioning would also allow for much easier comparison of data in normal usage and would also facilitate visual integration of information, thus giving better compliance with Principle 2. In order to satisfy Principle 3, more sophisticated equipment would need to be purchased as was seen in Example 1. It is interesting that neither vessels considered here satisfy Principle 3 for even navigational aids, for example by integrating the radar and video plotter using an Automatic Radar Plotting Aid (ARPA). Plan 4 gives a revised design of the wheelhouse showing the suggested modifications.

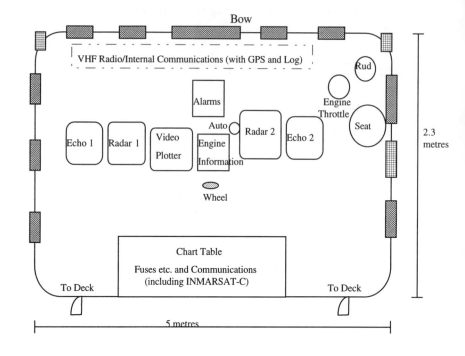

Plan 3: Position of Fishing Aids in the Wheelhouse of Example 2 (not to scale)

Conclusion

Three general and fundamental principles of ergonomics have been formalised and used to evaluate two wheelhouses of 'double ticket' British fishing vessels. The principle of functional grouping (Principle 1) has shown that there is a need to consider the positioning of the aids from a collective functional viewpoint rather than just each individual equipment's usage. With the suggested modifications above, the integration of information (Principle 2), at least visually, can be enhanced but for complete information integration allowing an interchange of data between different aids (Principle 3), more sophisticated equipment is required. In 1995, this is available, for example, in the form of a netszonde which uses object-coding rather than colour-coding for data representation and which also interfaces with the sonar and echosounder, but it is too expensive for the skipper of small fishing vessels. With this integrated display, the three principles are satisfied and, consequently, a more ergonomically acceptable design is achieved.

Acknowledgement

The author would like to express her gratitude to Mr David Lambert of Deep Sea Fish Ltd, Falmouth, Cornwall, UK for arranging the visits to the vessels used in this study.

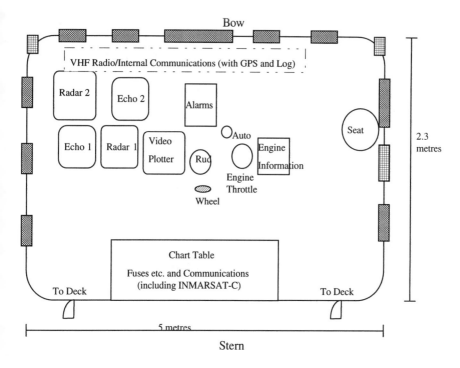

Plan 4: Modified Wheelhouse of Example 2 (not to scale)

References

Froese, J, 1981, Requirements for the Bridge of the Ship of the Future, in *ERGOSEA '81, Second International Conference on Human Factors at Sea, Plymouth, October 5-8, 1981,* 192-216, London: The Nautical Institute.

Harre, I, 1987, Functional Area Intergration on Merchant and Research Vessels, in Data Dissemination and Display - Electronics in Navigation, Paper 27, *Proceedings of the 1987 Conference of the Royal Institute of Navigation,* London, UK, 1987, Royal Institute of Navigation, London, UK.

Mills, S., 1993, A Preliminary Analysis of a Skipper's Interaction with Electronic Fishing Aids, Paper 47, *Proceedings of the First International Maritime Conference on The Impact of New Technologies on the Marine Industries,* Warsash, Southampton, UK, 1993, Southampton Institute of Higher Education, Southampton.

Olsen, 1995, *The Fisherman's Nautical Almanac,* E T W Dennis & Sons Ltd, Scarborough.

Stoop, J, 1990a, Redesign of Bridge Layout and Equipment for Fishing Vessels, *Journal of Navigation,* **43** (2), 215-228.

Stoop, J, 1990b, Redesign of bridge layout and equipment for fishing vessels, Chapter 17, in Work design in practice (ed C M Haslegrave, J R Wilson, E N Corbett and I Manenica), pp. 131-141, *Proceedings of the Third International Occupational Ergonomics Symposium,* Zadar, Yugoslavia, 1989, Taylor & Francis, London.

Witty, J H, 1990, Fishing Vessel Navigation, *Journal of Navigation,* **37** (2), 279-285.

OPTIMISATION OF THE FUZZY HELMSMAN MODEL USING A GENETIC ALGORITHM

Robert Sutton   Gareth D Marsden

Marine Dynamics Research Group
Institute of Marine Studies
University of Plymouth
Plymouth PL4 8AA

In this paper the development of a fuzzy mathematical model which describes the control characteristics of a helmsman whilst manoeuvring a ship is discussed. The helmsman model being optimised using a genetic algorithm. Verification of the model is made by comparing actual helmsman responses with those obtained from digital computer simulations.

Introduction

Human performance modelling techniques have continued to be developed to meet the challenges of designing human-machine control systems. This is particularly so at the skill-based level where humans undertake manual control tasks. Despite the advanced state of this available technology, it would appear not to be fully integrated into the design process. The reason for this being that many of the models which have been created are based on control engineering theory. Thus they are perceived to be complex in nature and therefore have had limited appeal in the human factors community. Hence to alleviate this problem, human performance models founded on control theoretic approaches should possess the simplest structure as is possible.

The work reported herein is concerned with the development of a elementary fuzzy helmsman model which was evolved using a genetic algorithm (GA).

Basic Genetic Algorithm Operators

Based on the Darwinian theory of genetic evolution in which nature tends to favour the survival of the stronger and fitter members of a population rather than the weaker ones, the GA is a global optimisation technique founded upon this natural selection principle. In this optimisation approach, populations of possible solutions are generated by an algorithm with the aid of a number of genetic operators.

Some of the terminology and operators which are used in this approach are described as follows:

The Chromosome

Potential solutions to a given problem will consist of a number of parameters. These parameters are converted into a series of 0s ad 1s which are termed chromosomes. An individual chromosome is, therefore, a member of a population of chromosomes in which a GA will perform its search. The term "chromosome" is a feature of natural systems, and the expression "string" is often used in design studies. A typical chromosome may be described by the following binary code:

$$1\ 0\ 0\ 0\ 0\ 1\ 0\ 0\ 1\ 1\ 1\ 0$$

Whilst operating upon a chromosome, the GA employs schema theory in its search for an acceptable outcome.

Fitness

Fitness is a numerical objective measure of the goodness of the solution to a particular problem. Hence, it is entirely specific to the given problem. The GA attempts to maximise the fitness of an entire population in order to arrive at a global best solution.

Selection

The process by which members of the current population are allowed to progress to the next generation is known as selection. It is a mechanistic procedure which is analogous to reproduction in biological systems. During the genetic evolution it is assumed the population size will remain the same during the process. As a consequence, two mating parents are allowed to create two offspring, however, they themselves must be eliminated. The reproduction process itself is biased towards searching for the more fit members of the population.

To select members from an initial population to participate in reproduction it is necessary to assign to each member a probability of selection on the basis of its fitness measure using stochastic sampling techniques. Hence, a new population is created which has a higher average fitness value.

Crossover

Crossover occurs by selecting pairs of individuals (parents) and choosing a random point is their strings to exchange bits within segments of those strings.

Consider two parents, A and B, in a population and a random point x as follows:

$$
\begin{array}{lll}
A & = & 1\ 0\ 0\ 0\ 0\ |\ 1\ 0\ 0\ 1\ 1\ 1\ 0 \\
B & = & 1\ 1\ 0\ 0\ 0\ |\ 0\ 0\ 1\ 1\ 0\ 0\ 1\ 1
\end{array}
$$

$$\uparrow x$$

Using the crossover operator produces the two offspring, A* and B*, given below:

$$A^* = 1\,0\,0\,0\,0\,\big|\,0\,0\,1\,1\,0\,0\,1\,1$$
$$B^* = 1\,1\,0\,0\,0\,\big|\,1\,0\,0\,1\,1\,1\,1\,0$$

x

Mutation

Mutation manifests itself by randomly inverting one or more bits within the chromosome string, eg:

$$A = 1\,0\,0\,0\,0\,1\,0\,0\;\big|\;1\;\big|\;1\,1\,1\,0$$

After mutation A becomes A* where:

$$A^* = 1\,0\,0\,0\,0\,1\,0\,0\;\big|\;0\;\big|\;1\,1\,1\,0$$

The reason for introducing mutation within the process is to increase the search area of possible solutions and to prevent a GA becoming locked into a local optimum.

The Fuzzy Helmsman Model

The proposed fuzzy model is based on the assumption that the helmsman divides his/her workload into three parts. It is assumed the helmsman initially assesses all input information in an analogue mode. Once the information has been evaluated, a decision is made and appropriate control action is instigated via a neuromuscular dynamic model through to the ship's wheel. The neuromuscular dynamic model is described by a transportation lag. Clearly, the central feature of the model is the decision-making submodel which is formulated using fuzzy logic.

From previous work (Sutton and Towill 1988), the following fixed rule based fuzzy algorithm is presumed to represent the decision-making submodel:

If ε is PB and $\dot{\psi}$ is ANY then δ is PB else
If ε is PS and $\dot{\psi}$ is NS or NB then δ is PB else
If ε is PS and $\dot{\psi}$ is PS or AZ then δ is PS else
If ε is PS and $\dot{\psi}$ is PB   then δ is AZ else
If ε is AZ and $\dot{\psi}$ is NB   then δ is PB else
If ε is AZ and $\dot{\psi}$ is NS   then δ is PS else
If ε is AZ and $\dot{\psi}$ is PS   then δ is NS else
If ε is AZ and $\dot{\psi}$ is PB   then δ is NB else
If ε is NS and $\dot{\psi}$ is NB   then δ is AZ else
If ε is NS and $\dot{\psi}$ is NS or AZ then δ is NS else
If ε is NS and $\dot{\psi}$ is PS or PB then δ is NB else
If ε is NB and $\dot{\psi}$ is ANY then δ is NB

Where the variables yaw error (ε), yaw rate ($\dot{\psi}$) and rudder demand (δ) are represented by the fuzzy sets negative big (NB), negative small (NS), zero (AZ), positive small (PS) and positive big (PB). The shape of the fuzzy sets NS, AZ and PS being characterised by Gaussian functions. Whilst appropriate single sided sigmoid functions were used for NB and PB.

It should be noted that the limits for the variables were taken as $\pm 20°$ for the yaw error (ε), $\pm 1°/s$ for the yaw rate ($\dot{\psi}$) and $\pm 30°$ for the rudder demand (δ). To aid with the evolutionary process of the sets by the GA algorithm, the three universes of discourse were normalised to a range of ± 6. Thus gain terms were required to be introduced into the control algorithm.

Subsequently, the GA is used to generate the fuzzy set shapes and their respective positions on the designated universes of discourse.

Results and Discussion

During this phase of the design process over thirty control strategies were evolved. Throughout the study all the GA optimisations and simulation trials were implemented using the software package MATLAB in conjunction with its dynamic systems simulation library function SIMULINK and a GA toolbox. Using this computer environment, each strategy evolution took approximately two days in running time to develop completely. On completion of the design cycle the fuzzy sets shown in Figure 1 evolved.

The ship model used in both the digital and real-time analogue simulations represents the non-linear yaw dynamics of a Royal Navy warship which is being subjected to sea state disturbances. Verification of the helmsman model was made by comparing the analogue results from an experienced helmsman with those obtained digitally. The data from the helmsman-in-the-loop study being collected via a test rig which contained a compensatory display that presented yaw error and yaw rate information.

Figure 2(a) shows a comparison between the control responses of the fuzzy model and the experienced helmsman. From these typical results, the similarity between the wheel demands can be discerned. In Figure 2 (b), the yaw responses for the model and the helmsman are shown and are seen to be approximately the same.

Conclusions

From the study described, it has been shown that a GA can be used to optimise a fuzzy model to describe the control characteristics of a helmsman. The structure of the model is seen to be uncomplicated and, therefore, it could be used as a design tool.

References

Sutton, R. and Towill, D. R.   1988, Modelling the helmsman in a ship steering system using fuzzy sets.   *Proceedings of the IFAC Conference on Man-Machine Systems: Analysis, Design and Evaluation*, (Pergamon Press, Oxford) 157-162.

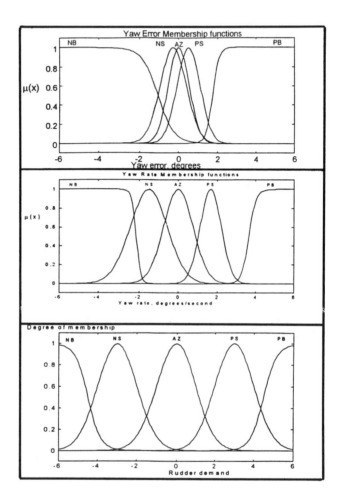

Figure 1. Evolved fuzzy set membership functions for the helmsman model

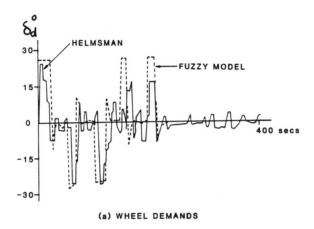

(a) WHEEL DEMANDS

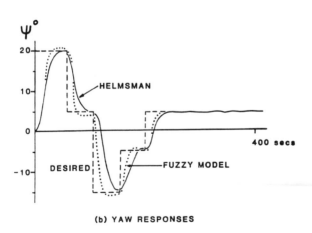

(b) YAW RESPONSES

Figure 2. Comparisons of the wheel demands and yaw responses

Ergonomics Awareness

A EUROPEAN ERGONOMICS NETWORK.

Peter McBride
NORTEL Ltd., Monkstown, N.Ireland

and Alun Batley
NORTEL Ltd., Cwmcarn, Wales

Keywords: Evolution; Devolution; Revolution; Achievement.

This paper will deal with how a team from 3 Regions (12 Nortel/BNR locations in the UK/Ireland) addressed the design, planning, organisation and implementation of an Ergonomics programme for Europe.

Introduction

The programme was planned and organised following NORTEL's Annual Ergonomics Conference in Montreal, Canada, April 1994.
The 12 NORTEL/BNR locations appointed a person (Prime) who would be responsible (on a part-time basis) for implementing and maintaining the Ergonomics programme on their site.

Method

The launch vehicle was MTM's 1 week course in "Human Factors and Ergonomics" and examination for Practitioners. The course was delivered in October '94, to 23 students including Engineers, Facilities Managers, Production Managers, Health Primes and Safety Primes from locations within the UK and Ireland, by a NORTEL Instructor in association with MTM (USA); these 23 students, which included the 12 Site Ergonomics Primes, passed the examination and were certified as MTM "Human Factors and Ergonomics" Practitioners.
(MTM - Methods Time Management Association Ltd.)
Having discussed during the course, and then later via a teleconference, at which all the Site Primes participated, it was decided that they should physically meet together for a given period of time, in order to learn together and to forge a common way ahead.
The first of these meetings was held in December '94, and continued on a monthly basis until May '95. Each meeting was held at a vendor/supplier's premises, and were arranged and minuted by a Site Prime.
At this first meeting, the Primes set out their Strategy Plan for 1995 and QI 1996.
Vendors who were asked and agreed to sponsor the day's meeting were:

Herman Miller ; Ford Motor Co.; Steelcase; GBP/Thorax, and Sven Christiansen. Apart from the Ford Motor Co., the others were either manufacturers or suppliers of office furniture, shopfloor workstations or computer aided products.

As stated, the vendor sponsored the day's meeting with the Nortel Primes getting together and holding their own discussions in the morning; and in the afternoon the vendor talked about their products and invited along "experts" to talk to the group. The Ford Motor Co. was a meeting set up to see at first hand how another major organisation was tackling the issue of workplace design and layout within their industry. Following on from these meetings, it was decided that future meetings would be held every 2 months at either a NORTEL or BNR site. It is planned that invited guests from within the Ergonomics "field" will participate at these meetings.

At the December '94 meeting, the Primes decided that a Steering Committee should be appointed, to drive the whole programme forward.

This team consists of 3 Primes, who because of geographical locations arrange a teleconference every fortnight to discuss and plan the way forward in line with the Strategic Plan.

Following on from this, the Steering Committee felt that the group needed to have more control over their own progress; and approached the EHS Director for Europe with a number of proposals. From these discussions, the group became "empowered" with the full backing of the entire Network to continue to drive the programme forward.

The main reason behind the Primes meeting on a regular basis (as a self-empowered team under the direction of the elected "steering committee") is to provide mutual support, develop knowledge, skills and future strategy.

The 1 week Practitioner's course was repeated in Feb. '95 by the same NORTEL Instructor and again in association with MTM (USA); this was undertaken to increase our resource capacity and broaden it by including employees from our manufacturing locations in France and Turkey and the company's Corporate Design Group. This course had 15 students including 2 full-time Instructors from MTMA (UK);all of whom qualified as Practitioners.

By setting up this second course it permitted MTMA (UK) to initiate the introduction of the "Human Factors & Ergonomics" course into their portfolio.

In March '95, five Nortel practitioners (including two Primes) from the Oct.'94 programme along with MTMA (UK)'s two Instructors attended and successfully completed the Instructor's seminar. These are all now qualified and licenced Instructors /Trainers for the MTMA "Human Factors & Ergonomics" course.

NORTEL (Europe) financed the MTMA (UK) Instructors attendance at the Practitioners course; in exchange MTMA (UK) financed the qualification process fot the 5 NORTEL Trainers.

Having our own Trainers gives us the capacity to cascade Ergonomic knowledge and skills development throughout our region.

The Trainers are responsible for the training of the Response Teams, Health & Safety Competent Persons and Assessors; and selected Engineers, Managers, Supervisors and Purchasing personnel.

Being licenced/qualified as either a Practitioner or a Trainer means that one is permitted to practice or train in MTM's "Human Factors & Ergonomics" for a period of 3 years, after which the licence has to be renewed by resitting an examination.

In June '95, 3 of the NORTEL Trainers along with 1 of MTMA's Instructors delivered their first course, at which all 6 students became qualified Practitioners.

This partnership, between NORTEL (Europe) and MTMA (UK), has been both valuable and worthwhile, and has ensured self-sustainability for our evolving Ergonomics Programme.

The partnership has led to MTMA (UK), with the assistance of the Nortel Trainers, compiling a new student training manual (using the USA version as the base document).

MTMA (UK) have also compiled for first time a Trainer's manual with overhead slides to match.

From July '95 onwards, it was decided that our meetings would be on a two monthly basis and would be held on a nominated Nortel site. At these meetings guest speakers, "experts" or suppliers would be invited along.

In September '95, two Trainers conducted training for a group of NORTEL/BNR's interior designers. The manager of this group of people (9 in total), wanted his staff to have additional knowledge and information concerning ergonomics in relation to their function, but not necessarily to complete the full course and sit the exam. The Trainers put together a "pick & mix" package, ie selecting a number of sections from the MTMA (UK) training manual - suited to this group's needs. The group gained a wealth of knowledge and understanding specific to their requirements; and the Trainers recieved a positive feedback.

This innovation was discussed with and agreed by MTMA (UK). We intend to pursue this option when requested by the different sites and also continue the full 1 week programme.

In October '95, NORTEL's Trainers delivered their second course, at which they used MTMA (UK)'s new student and training manuals, the 7 attending students were all certified as Practitioners.

In November '95, NORTEL's Trainers delivered a further course, at which the 9 students were all certified as Practitioners.

Conclusion

All 12 locations now have a trained and qualified Ergonomics Prime in place, with programmes at various stages of evolution.

Each location Prime has or is in the process of pulling together a trained Ergonomics Response Team; in order to carry forward the programmes, awareness and assessments at their location. The Response Team normally consists of the Ergonomics Prime, the Occupational Health Professional, the Safety Advisor and Engineers.

To date the group have made a number of contacts with "experts" in the Ergonomics field, and with a number of learning institutes, and during 1996 will enhance these contacts with further meetings and debates.

The steering committee is also involved in discussions with other companies and organisations regarding Ergonomics, and how we can help each other; and again during 1996 will have further meetings and discussions.

TEACHING USER INVOLVED DESIGN IN THE DELFT CURRICULUM

H. Kanis and A.P.O.S Vermeeren

*School of Industrial Design Engineering
Jaffalaan 9, 2628 BX Delft
the Netherlands*

The relevance of teaching user-involved design on the basis of empirical research in the new Delft curriculum is discussed. Topics addressed include: the role of presuppositions in research; the necessity to formulate a few clear-cut research questions; different aspects of the practicalities of user trialing, e.g. (un)obtrusiveness of observations, sequential effects and possible consequences of the think aloud technique ; opportunities in the case of small samples; the questioning of the validity of observations and conclusions, and the risk of being misled in the analysis by what is observable. Finally, the identification of re-design requirements on the basis of user trials is dealt with briefly, and user trials are presented as a possible source of inspiration for design solutions.

Introduction

Over the last decade it has been recognised that designers of all kinds of consumer products tend to be poor predictors of the way users behave in practise. For example, Öster et al. (1994) found that users may manipulate a product more efficiently than envisaged by the designers, which may very well mean that design possibilities have been missed. However, what seems to occur much more frequently is that unanticipated usage undermines a designed functionality (Loopik et al., 1994). The issue is that the way users interact with a particular product has been shown to be largely unpredictable, both in the way designed functions are perceived and understood, and in the way people will actually operate a product.

In this paper, the teaching of user-involved design, as implemented in the Delft curriculum, is explored. This implementation involves both the sensitising of students to the central role of users in the ways a new design actually works in practise, and the practicalities of doing empirical research, i.e. exercises in carrying out user trials.

The case of actual usage

Students in industrial design engineering, no less than other people, seem to have difficulty in anticipating the often imaginative ways in which users will manipulate products, and of the consequences of such usage. This applies especially to the versatility of the upper extremities. Human beings seem to consider the operational difficulties of everyday life as self-contained. This tendency of not questioning the effect of external decisions may particularly impede the achievements of designers by limited feedback. This is not to say that unforeseen interpretations of product cues or unanticipated manipulations always produce repercussions, i.e. a poor product performance, unwelcome side-put, or extra effort by the user. The functionality of a design may well be resistant to all kinds of diversity in user activities. However, not much is known about the extent to which this resistance holds, nor in whose perspective, i.e. claimed by designers or actually experienced by users. Ultimately, a users' perspective is decisive, which is something students need to be reminded of over and over again.

The versatility of human beings in understanding and operating a product may be seen not only as the origin of problems, but equally well as a source of remedies, enabling users to by-pass individual difficulties in use. In this respect it would be naive to interpret the absence of expressed use-difficulties as convincing evidence that, by and large, everything is fine with the actual usage of products. There may well be a different explanation for this absence, i.e. the fact that many problems experienced by users are rendered bearable by the adaptive capacities of human beings, and especially by the fact that they seldom accumulate sufficiently to be acknowledged by a user population. There is also the inclination of product users to accuse themselves in case of mishaps (see Weegels, 1996). The growing body of empirical observations, which mainly originate from studies carried out at the Delft School of Industrial Design Engineering, offers the possibility of demonstrating to students the central role of users' 'do's' and 'don'ts' for designed functionalities. This evidence is presented in an obligatory course in the second year of the new curriculum. The central message is that future usage is not to be dealt with speculatively, but must be established on the basis of empirical observations. Ways of conducting these observations are outlined also, in text and talk. However, the competence to carry out user trials can not be learned from paper or by hear-say. Thus, investigative skills are intended to be established through practical work, both in ergonomics exercises and in design teaching.

The teaching of empirical research to industrial design engineering students

In ergonomics exercises and in design teaching different practicalities are outlined in doing research specifically directed at carrying out explorative studies during a design process. The topics subsequently discussed have emerged partly from optional courses in the years preceding the introduction of the new curriculum. The artifact to be studied may be a model, a working prototype or an existing product which has to be redesigned. (See Marinissen (1993) for a review of the various opportunities to anticipate future usage during a design process.)

Presuppositions including use cues
Students must be made familiar with the impossibility of administering an empirical study with an un-conditioned mind, as if a 'tabula rasa'. Presuppositions (rather than 'hypotheses' as, generally, there are no well established theories to refer to) about the realisation of user activities should be explained and made plausible as thoroughly as possible. In practise, the distinction between presuppositions about guiding principles for user activities, and predictions of the actual activities to be observed, appears difficult to make. In this respect the notion of *use cues* proves to be helpful. A use cue is an indication for use derived from product characteristics or from functional aspects. To be effective, use cues must be perceived (visually, audibly, tactiley), and understood. How a perceived cue is interpreted may depend on experience, see Gelderblom and Bremner (1993). But understanding designed cues does not guarantee that use actions like manipulations will proceed as anticipated by the designer(s). Physical conditions such as an impairment may interfere/be dominant (Kanis, 1993). The notion of use cues seems to appeal particularly to students who are asked to articulate their often rather implicit suppositions about actual usage of artifacts, e.g. a prototype, designed by themselves. In the case of existing products with an unknown design history the assignment of use cues is bound to be largely speculative.

Research questions

Students are required to pose a limited number of questions which should be answered by the empirical research. In these research questions the presuppositions should be articulated in such a way that, after the subjects participating in the study have left, the recorded observations include at least:

- the use-actions of the tasked subjects, in terms of *what* they did in operating, or intending to operate, a model or a product, and, if appropriate, also in terms of *how*, for instance types of manipulations;
- which use cues, anticipated or not anticipated in the design, are involved, i.e. perceived and how interpreted;
- the individual characteristics which are thought to play a role, such as experience and physical properties.

As a rule, observations focused on these topics largely circle around difficulties in use encountered by the subjects.

Operationalisation

Given the desirability of lifelike tasks and conditions, students are made aware of the risk of creating artifacts by all kinds of circumstances, e.g. due to carry-over in the course of a task, resulting in sequential effects in case of within-subject research designs, due to obtrusiveness or to biased interviewing. The advantages of unobtrusive observation are stressed, but it is also pointed out that this approach is not compatible with thinking aloud, a technique that is not undisputed but seems indispensable in order to trace perceptions and interpretations. The delicacy of stimulating subjects to verbalise their considerations is emphasised: the role of an interviewer should be distant, for example by ceasing to talk immediately when interrupted by the interviewee, who, in principle, should never be interrupted. Debriefing should be directed towards uncovering the origin of difficulties, rather than 'quantifying' experienced effort, observed performance, etc. by a closed questionnaire, such as a 5-points scale. Therefore, students are required to work with a check-list of functional aspects to be thoroughly addressed by the subject, either spontaneously or, if requested, in hindsight. For example, by the time a subject leaves it must be clear which use cues have (not) been perceived, what interpretations have been made by the subjects, and how the effort involved was experienced in terms of explained individual standards.

The check-list of functional topics should be as comprehensive as possible at the beginning of a user trial, for example on the basis of pilot-studies. However, due to the explorative character of user trials unanticipated functional aspects may emerge in the course of a study, with which only subjects who are still to come can be confronted. Students are taught that the comparability between subjects in a study is least threatened if a new item, which is deemed relevant, is brought up at the very end of a session.

Emphasis is put on establishing experience, particularly in relation to the notion of 'similarity' (cf. Chapanis, 1988), such as in having experience with 'similar products'. Given the users' perspective as the only relevant criterion, students are alerted to the various possible meanings of product similarity, for example similar in appearance to the one tried out, in circumstances of use perhaps, in functional respect, or in the way of operation/ manipulation; This last aspect in particular is usually what matters most.

Sampling of participants

In the literature different approaches can be seen to prevail. In the area of computer interface design the recruitment of experts seems to provide a valuable complement to data derived from user trials, in particular when user trials are limited by a constrained tasking of subjects, e.g. in time and in the range of functions, as compared to those available in current software programmes (cf. Nielsen and Phillips, 1993). However, the opinion of experts must, ultimately, be validated by the observation of users in practise. Also, in many product areas, usage is not confined to people sitting on a chair, striking keys, and looking at a screen: in addition to perceptual/cognitive aspects, physiological elements such as force exertion may also be significant. In these areas there is no expertise to allow for the prediction of future usage, and difficulties in usage, for a specific case on the basis of general insights. In fact, students are urged to be suspicious of expert views. After all, what expertise guarantees the status of the expert? Experts may come in handy for helping to make up the check-list of functional topics (see above), but they should never be used as a panacea for uncertainties about the future usage of a new design.

Basically, students of industrial design engineering should be experts themselves in having learned to be aware of the very limited possibilities in predicting 'do's' and 'don'ts' of future users. And they should also be the experts in carrying out observational studies to generate required insights. In a design process, the number of observations of future users operating a new design(model) is bound to be small for reasons of time and costs. The obvious characterisation of small-scaled trials as 'quick' seems often, for students, to be the equivalent of 'dirty'. The pedantic idea that 'good' research requires large numbers of subjects seems almost ineradicable. Various arguments serve to release students from this narrow-mindedness. User trialing in a design context is not about prediction of averages and dispersion, but about the identification of different types of usage, especially in terms of difficulties in use. In this respect, it is of little concern whether a particular use pattern is observed once or twice in, for instance, some ten subjects. The point is that occurrences found in small samples have a low probability of being rare, see Figure 1. Thus, belittling observations from small samples is turning the world upside down. The disadvantage of small sampling is non-observation of occurrences that may not be rare, rather than rendering observed occurrences irrelevant.

Instead of sampling subjects randomly, students are directed to the possibility of enhancing the efficiency by the selection of subjects on certain constraints, which are thought to be a source of various use-patterns, such as physical impairments (cf. Kanis, 1995) and particular types of experience (Gelderblom and Bremner, op.cit.). For an example of subject selection, see the study by van de Steen et al. (1996) into bicycle racks.

Figure 1. Occurrences which appear in a small sample
have a low probability of being rare.

Validity
Students are offered a glimpse into the tricky area of criteria in empirical research. The attention is largely focused on the issue of validity. Emphasis is placed on questioning the generalisability of observations, made under simulated conditions (e.g. in the lab) to actual practise, or for early models to working prototypes or complete products. In fact, this questioning of the validity of predictions on the basis of empirical research returns to the issue of presuppositions, e.g. an assumed invariance of observations for different conditions such as various degrees of simulation, obtrusiveness etc.

The teaching of analysing observational studies for design

In observational studies, the analysis is often the most time consuming part. This is certainly true in the case of video-observations, which to some extent can be seen as postponed data gathering. Students frequently tend to underestimate the time this phase takes. Major pitfalls are briefly discussed below.

Observable doesn't mean relevant
Students are often in danger of getting lost in all kinds of observable details, which may have no bearing whatsoever on the answering of any of the research questions, e.g. which side of a finger is used how many times. This pursuing of details, once underway, inevitably ends in quantification, which for some students seems to be as scientifically impressive as it is a disillusioning for them to find that they are required to abstain from all kinds of quasi-quantification. Time, for example, is a popular wrong track, presumably because it runs automatically throughout video recordings. Students are pressed to focus on what they see and hear happening over time, always questioning whether recorded observations contribute to the answering of the research questions. And herein lies a problem. While it is considered relevant, in terms of the posed research questions, to describe particular use actions, there seems to be no benchmark for deciding whether a difference in those use actions is generally significant from a design point of view. Equally, for any given distinction it is possible to imagine a particular user for whom it may indeed be relevant. This pleads to some extent for a fine resolution in describing observations from the outset. The resolution of a grid cannot be set, it seems, without arbitrary decisions. The blurring of resolution should require justification by the variability of the performance, the occurrence of side-effects, and the effort of the user for different requirements, assuming that performance, side-effects and effort are always addressed in the research questions in some way or another.

The description of use difficulties
The description of difficulties as demonstrated and voiced by subjects does not self-evidently make sense as incentives for (re)design considerations. As the artifact is the target of design activities, rather than the user, students are urged to detail use difficulties as much as possible vis-a-vis featural and functional aspects of the model, prototype or product under investigation. For an example see the study by Rooden et al. (1996) into an interface for tele-activities.

(Re)design requirements from user trials

Observed usage may reveal an undermined functionality, i.e. a degraded performance compared to the design goals, unwanted side-effects or extra effort by the user. In a redesign the usage which results in this undermined functionality may be dealt with by making the functionality of the new design impervious to this way of use or to make the particular usage impossible. In design practise neither option seems to be realistic. Students are made aware of the limited possibilities designers have in ensuring that users operate in a pre-determined way, and of the risks of introducing new problems with the redesigned characteristics. In particular, the benefits of low use-specificity in a design are stressed, see Van der Steen et al. (op.cit.).

Discussion

Designers are not always the most obvious researchers to charge with the testing of their own designs by playing the Devil's advocate. However, this drawback may be outweighed by the advantage of the direct involvement of the designer(s) concerned. Anyone who is carrying out a user trial should be prepared to be seen as the bearer of bad news. In fact, absence of any cross findings in a trial should raise suspicion. In conclusion, it is noted that user trialing should be seen not only as a means of identifying difficulties users encounter in operating particular artifacts, but also as a possible source of inspiration to explore new design solutions, see the study by Van der Steen et al. (op.cit.)

References

Chapanis, A. 1988, Some Generalisations about Generalization, *Human Factors*, **30**, 253-267.

Gelderblom, G.J. and Bremner, A. 1993, The role of experience in performance on different types of telephone memory retrieval tasks, *Contemporary Ergonomics* (Taylor & Francis, London) 422-427.

Kanis, H. 1993, Operation of Controls on Consumer Products by Physically Impaired Users, *Human Factors*, **35**, 305-324.

Kanis, H. 1995, Manipulation of push buttons and round rotary controls, *Proceedings Human Factors Society 39th Annual Meeting*, 374-378.

Loopik, W.E.C., Kanis, H. and Marinissen, A.H. 1994, The operation of new vacuum cleaners, a user trial, *Contemporary Ergonomics* (Taylor & Francis, London) 34-39.

Marinissen, A.H. 1993, Information on product use in the design process, *Proceedings of the 29th Annual Conference of the Ergonomics Society of Australia*, (Ergonomics Society of Australia, Inc., Downer Act) 78-85.

Nielsen, J., Phillips V.L. 1993, Estimating the relative usability of two interfaces: heuristic, formal, and empirical methods compared, *Proceedings of INTERCHI'93, 1993 (Amsterdam, The Netherlands, 24 April-29 April, 1993)* (ACM, New York) 214-221

Öster, J., Kadefors, R., Wikström L., Dahlman, S., Kilbom, Å. and Sperling, L 1994, An ergonomic study on plate shears, applying physical, physiological and psycho physical methods, *Industrial Ergonomics*, **14**, 349-364.

Steen, van der V.B.D., Kanis, H. and Marinissen, A.H. 1996, User involved design of a parking facility for bicycles, *Contemporary Ergonomics* (Taylor & Francis, London) this issue.

Rooden, M.J., Bos, A.C., Kanis, H. and Oppedijk van Veen, W.M. 1996, Usability testing of an interface for tele-activities, *Contemporary Ergonomics* (Taylor & Francis, London) this issue.

Weegels, M.F. 1996, Accidents involving consumer products, *Dissertation* (Faculty of Industrial Design Engineering, Delft University of Technology, Delft).

Organisational
Ergonomics

COMPARISON OF WORKING SITUATION IN SMALL SCALE ENTERPRISES IN SWEDEN, JAPAN AND KOREA

Kiumars Teymourian and Houshang Shahnavaz
Division of Industrial Ergonomics
Luleå University, Sweden

A comparative study on working condition in small scale enterprises were designed and carried out in three countries namely: Sweden, Japan and Korea. The main tools used in this study were two sets of questionnaires one for employees and the other for managers. In this study six different indices that explained the working situation of companies were developed. These indices are: Physical stress, mental stress, information flow top-down, information flow bottom-up, workers participation and job satisfaction index. The aims of the study were to identify the level of scores for each index in the three investigated countries and to compare these indices' scores between the countries. The results reveal that there are some differences between workers and managers responses regarding experienced working conditions for each index in different countries as well as between the countries.

Introduction

People spend a great deal of their time at the workplace. Therefore the work and the working condition has significant effects on the people's health, satisfaction and perception of stress. The literature of occupational stress show that physical fatigue resulting from poor physical working condition tend to lead to mental fatigue, which can cause stress. Further, it is demonstrated that several aspects of work and work organisation such as lack of individual control over the way work is performed, dissatisfaction with organizational climate and bad interpersonal relationship can contribute to dissatisfaction and feeling of stress, Westlander et al (1995).

Access to information within the workplace create feeling of belonging. Hancok (1986) mentioned that "information is recognized as that which acts to reduce uncertainty". Lawler and Mohrman (1989) point out that , "the managers must regularly and honestly share valid data about the organization and the work unit's performance and future plans. In particular, employees need information about plans for new equipment, new work processes and new schedules".

Moreover, managers can use employees potential to improve the working situations, problem solving and product improvement through participation in the work places. Using participatory method will gives employees a feeling of belonging and ownership, making them more satisfied with their job. An organisation can make the employees satisfied if it creates the situation that raises employees' intrinsic and extrinsic satisfaction, regarding their work and their working condition. Shahnavaz (1994) indicated that, "high quality production starts with employees who have a thorough understanding of their job, knowing the risks involved in the operation, having confidence and motivation to participate in company's day to day affair as well as authority to carry out the necessary changes for improvement."

A comparative study on working condition in small enterprises engaged in different activities and from various branches was designed and carried out in three countries, namely: Sweden, Japan and Korea. Two sets of multi-choice questionnaires were used as a main tools for conducting this study. One was designed for workers, the other for the managers. Different questions or indicators from these two sets of questionnaires that were providing information on specific issues are combined under various indices. One main problem encountered in this study was prevailing of different working conditions in these three countries and consensus on composition of various indices. To solve these difficulties all questions relating to each index were discussed and verified in a common meeting. The indices used were: Physical stress, mental stress, information top-down, information bottom-up, participation and job satisfaction. The objectives of this study were to identify the level of scores for each index and each country and to compare the indices' scores between the countries.

Method and Results

The number of managers and workers who filled up the questionnaire in this study for each participating country were: in Sweden 21 managers and 166 workers, in Japan 67 and 752, and in Korea 110 and 417 respectively.

The questionnaire for employees contained 84 questions, and the one sent to the managers had 128 questions. Since the data were qualitative, the mean proportions of positive answers for the questions in each index that were given by each individual was considered as the score for each index, which varied from 0 to 1. In the cases of indices for physical and mental stress, the scores are indicating the lack of physical and mental stress at work. However, to have a better view for these two indices the score of each index was subtracted from 1 which means the new scores show the existence of physical and mental stress.

The 'H_0 hypothesis' was assumed as: there is no difference between workers and managers in the mean of proportions in group of indices within and between countries. The alternative hypothesis was assumed as: H_a, there is a difference between workers and managers in the mean of proportions in group of indices within and between countries. The Chi-square test revealed that there are statistically significant differences between workers and managers in the mean of proportions in group of indices within and between countries. Figures 1, 2. Shows the level of score for each index for workers and managers for all three countries. The vertical axis is the percentage value. (Proportions which were between 0 and 1 have been multiplied by one hundred).

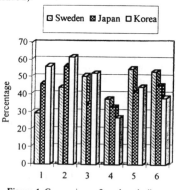

Figure 1. Comparison of workers indices among all three countries.

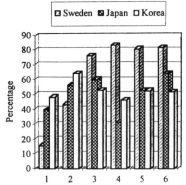

Figure 2. Comparison of managers indices among all three countries.

1. Physical stress. 2. Mental stress. 3. Information flow top-down.
4. Information flow bottom-up. 5. Participation. 6. Job satisfaction.

Discussions and Conclusion

The method used was to identify the score for each index and to compare each index within and between each country among workers and managers. For the physical stress index (table 1), it can be seen that there are high variation between workers responses between the three countries. These responses shows the presence of the physical stress among the workers in each country. Table 1 shows that the majority of Japanese and Korean workers have reported that their working and break time are not adequate. It reveals that the amount of their work is not appropriate. In the case of managers it is almost the same as the workers (table 2). It means the work organization should be changed in a way that both group receive more breaks and improve their working condition.

Table 1. Workers response to each question for the physical stress index.

	Sweden	Japan	Korea
appropriate work amount	0,57	0,43	0,38
no long harmful work	0,7	0,58	0,35
adequate work time and break	0,71	0,41	0,34
no Work tiredness	0,84	0,735	0,677
Index score	0,705	0,539	0,437

Table 2. Managers response to each question for the physical stress index.

	Sweden	Japan	Korea
no long harmful work	0,71	0,627	0,382
appropriate work amount	0,81	0,493	0,409
adequate work time and break	0,9	0,433	0,427
no Work tiredness	0,94	0,851	0,855
Index score	0,843	0,601	0,518

Table 3. Workers response to each question of the mental stress index.

	Sweden	Japan	Korea
no stress related to work	0,12	0,093	0,036
no stress related to personal life	0,35	0,291	0,137
manager listening readily to workers	0,5	0,41	0,35
no monotony work	0,52	0,456	0,35
appropriate work amount	0,57	0,43	0,38
no tension in workplace	0,58	0,44	0,41
not monitoring to managers reaction	0,6	0,43	0,28
distribution of appropriate responsibility	0,63	0,42	0,41
clarity of employees roles	0,64	0,56	0,57
subordinate free to opine	0,77	0,46	0,5
no feeling of loneliness	0,87	0,832	0,8
Index score	0,559	0,438	0,384

Table 3 reveals that in all three countries the majority of workers have reported that they have feeling of stress related to their work and their personal life. The majority of workers reported that their managers are not readily listening to the workers' opinion. In the case of Japanese and Korean workers, the majority of them have reported to have monotonous work, inappropriate amount of work and that there is tension in their work place.

Table 4. Mangers response to each question of the mental stress index.

	Sweden	Japan	Korea
no stress related to work	0,048	0,118	0,018
not monitoring employee alertness on mood	0,048	0,478	0,118
no stress related to personal life	0,381	0,221	0,064
no feeling of loneliness	0,429	0,328	0,282
distribution of appropriate responsibility to workers	0,619	0,582	0,491
Manager free to opine	0,714	0,522	0,636
employee listening readily to managers	0,714	0,701	0,5
appropriate work amount	0,81	0,493	0,409
no tension between manager and employees	0,952	0,343	0,582
no monotony work	0,952	0,582	0,473
Index score	0,567	0,437	0,357

Table 4 shows that almost all managers have reported that they have experienced stress related to their work and their personal life. The majority of them should keep alert on the mood of their employees. The most of Japanese managers reported that there is tension between their employees and themselves.

Table 5. Workers response to the information flow top-down index.

	Sweden	Japan	Korea
no needs for new information for job	0,3	0,079	0,38
information about company plans and policies	0,57	0,49	0,57
receiving necessary information for their work	0,65	0,48	0,62
Index score	0,507	0,35	0,523

In table 5 all three countries gave lowest score to the question that their work does not require constantly new information for their job. The highest score were given by the Swedish and Korean workers for receiving necessary information for their work, while Japanese highest score was for receiving information about company plans and policies.

Table 6. Managers response to information flow top-down index.

	Sweden	Japan	Korea
Manager free to opine	0,714	0,522	0,636
employee listening readily to management opinions	0,714	0,701	0,5
information about company plans and policies to workers	0,857	0,582	0,455
Index score	0,762	0,602	0,53

Table 6. shows that the Swedish managers informing more their employees regarding company plan and policies than their colleagues. The Korean managers give less information regarding the company' plan and policies compared to their colleagues. Japanese managers highest score is for employees listening readily to their opinions and Korean highest score is for the question that managers can express freely their opinions.

Table 7. Workers response to information flow from down to up index.

	Sweden	Japan	Korea
managers seeking employees' advice	0,2	0,25	0,19
managers listening readily to employee	0,55	0,41	0,35
Index score	0,375	0,33	0,27

Table 7 shows that the managers are not seeking actively their employees' advice, and it can be seen in table 8, specially for Japanese managers that they have less direct contact with workers. This means lack of good communication between workers and their managers in Japan and Korea compare to Sweden.

Table 8. Managers response to information flow from down to up index.

	Sweden	Japan	Korea
employee seeking managers' advice	0,714	0,313	0,418
employees free to opine	0,952	0,313	0,509
Index score	0,833	0,313	0,464

Table 9. Workers response to participation index.

	Sweden	Japan	Korea
seeking employees advice by managers	0,2	0,25	0,19
manager listening readily to workers	0,55	0,41	0,35
information about company plans and policies to workers	0,57	0,49	0,57
receiving necessary information for their work	0,65	0,48	0,62
subordinate free to opine	0,77	0,46	0,5
Index score	0,548	0,418	0,446

Table 9 shows the lowest score for "question regarding whether managers actively seeking their employees advice" which is almost the same for all three countries, and the highest score for Swedish workers regarding receiving information about the company plans and policies, which is the same for Japanese workers. Korean workers gave highest score to the question "receiving necessary information for their work".

Table 10. Managers response to participation index.

	Sweden	Japan	Korea
manager free to opine	0,714	0,522	0,636
employee listening readily to managers opinions	0,714	0,701	0,5
employee informed of company plans and policies	0,857	0,582	0,455
employee free to opine	0,952	0,313	0,509
Index score	0,809	0,53	0,525

Table 10 shows that the Swedish managers gave highest score to the question that "employees are free to express their opinions" which is lowest score for Japanese. For Japanese managers' highest score is given to employees listening readily to the managers, while Korean managers highest score is given to the question "managers freely express their opinions to their employees" and Korean lowest score is for employees are informed of company plans and policies.

Table 11 shows all workers have given highest score to "forgetting the time during the work" and the least score for Swedish workers is opportunities for job education. It seems that Japanese and Korean are not satisfied with their current department and

Table 11. Workers response to the job satisfaction index.

	Sweden	Japan	Korea
opportunities for job education	0,28	0,3	0,31
fair promotions of personal	0,32	0,31	0,23
preference to stay in the company	0,41	0,4	0,23
work affection	0,47	0,66	0,5
safety and health concerns by company	0,48	0,35	0,4
no monotony work	0,5	0,45	0,35
managers listening readily to employees	0,54	0,41	0,35
feeling doing something worthwhile	0,55	0,44	0,27
appropriate work amount	0,57	0,43	0,38
work for feeling	0,57	0,52	0,46
employee satisfaction on department and position	0,62	0,23	0,16
clarity of employees roles	0,63	0,56	0,57
adequate working time and break	0,71	0,4	0,34
forgetting time during work	0,81	0,92	0,82
Index score	0,532	,0455	0,383

position. In the case of Korea the score for preference to stay in the company shows that only 23% of Korean workers were willing to stay in their company. From table 13, the highest score for Swedish managers is for the question "feeling that they do something worthwhile", while for the other managers are the question "whether they have a feeling of affection for/interest in their work and /or the things they make at work". The least score for the Swedish managers is for question "if there are obstructions in implementation of their advice at workplace". The Japanese managers least score is for the adequateness of their working time and break time and Korean managers least scores were for two questions: (1) having no long and harmful work and (2) managers idea about their employees that they have affection for /interest in the products they make.

Table 12. Managers response to the job satisfaction index.

	Sweden	Japan	Korea
no obstructions in implementing advice	0,524	0,507	0,527
not having long and harmful work	0,714	0,627	0,382
employees work affection	0,714	0,672	0,382
salary satisfaction	0,714	0,761	0,709
employees work for feeling	0,762	0,582	0,427
appropriate work amount	0,81	0,493	0,409
adequate working time and break	0,905	0,433	0,427
no monotony work	0,952	0,582	0,473
work affection	0,952	0,836	0,709
work for feeling	0,952	0,866	0,755
feeling doing something worthwhile	1	0,701	0,5
Index score	0,818	0,642	0,518

It should be mentioned that the value of each question might be interpreted differently in the different socio-cultural environment, accordingly, not all people may respond equally to the same question. Further, the meaning of question after it is translated to other languages might convey different notion. For example the concept of job satisfaction that is understood in western countries may not have the same meaning and also critical affect on productivity and product quality for other countries. Roland Dore suggests, "low job satisfaction in Japan may imply a restless striving for perfection, an ongoing quest for fulfillment of lofty work values and company goals." Further the size of the companies integrated in the three countries different greatly which could also be a confining factors in the study.

Acknowledgments

The project was supported by grants from the Swedish Work Environment Fund. We are grateful to Prof. M. Kumashiro, department of Ergonomics Univ. of Occupational and Environmental Health in Kitakyushu, Japan, Prof. T. Hasegawa Kindia Univ. and Prof. Y. J. Seo Kyungnam Univ. in Korea. Our thanks to T. Laitila and Prof. U. Zakriksson department of statistic at Umeå Univ. Sweden and finally we thank also all companies in the three participating countries that responded to our questionnaire.

References

Hancok, P.A. 1986, STRESS, INFORMATION-FLOW, AND ADAPTABILITY IN INDIVIDUALS AND COLLECTIVE ORGANIZATIONAL SYSTEM, *Human Factors in Organizational Design and Management*-II, 293-296.
Lawler III, Edward. E. and Mohrman, Susan. A., April, 1989, High- Involvement Management, *Organizational Dynamics*, 27-31. Copyright © by American Management Association, New York.
Dore, Roland. 1973, British Factory, Japanese Factory, *The origins of Diversity in Industrial Relations (Berkeley, CA: University of California Press)*.
Shahnavaz, H. 1994, Managing Human Factors Issues in Future Organizations. *Human Factors in Organizational Design and Management*-IV, 655-661.
Westlander, G., Viitasara, E., Johansson, A. and Shanavaz, H. 1995, Evaluation of an ergonomics intervention programme in VDT workplace, *Applied Ergonomics*, 26, 83-92.

CHANGE PROCESSES IN SMALL WOOD COMPANIES

Kaisa Nolimo and Jörgen Eklund

Division of Industrial Ergonomics and
Centre for Studies of Man, Technology and Organisation
Linköping University
S-581 83 Linköping
SWEDEN

The aim of this study was to describe readiness for change and the
initiation of change projects in small wood companies. Ten
companies were contacted and offered support to plan and start
change projects, including an application for financial aid from the
Swedish Work Life Fund. The results show that prospects of decreased
costs and increased productivity are often driving forces for change
projects in small companies. Quality improvements, demands from
customers, demands from authorities, and to a certain extent
employees´ working environment proposals are also common driving
forces. External influences from other companies, consultants, owners,
government authorities or universities are important for idea
generation and for learning processes leading to decisions of
initiating change projects. However, the time horizon is very long,
sometimes many years. Financial aid from government funds can be a
facilitator in this respect. Small companies often have problems to
spread sufficient information, to obtain a good communication and to
obtain worker participation. Lack of time, bad economy, lack of
competence and previous unsuccessful projects are considered as the
most important obstacles for change.

Introduction

The course in Industrial Ergonomics within the M.Sc. programs in
Engineering at Linköping University, includes a compulsory workplace
investigation. Groups of four students visit companies in the region during one
day in order to study the work conditions at "their" company, the production and
ongoing projects. If needed, basic measurements are also performed to assess the
physical working environment. The results and recommendations to the company
are presented in a report written for the company and the University.
The Swedish Work Life Fund (SWLF) was created in 1990 as a five year
temporary project. The purpose was to support renewal of work organisation,
improvements of working environment and rehabilitation. In total, over 23 000
programs were supported with 11 billion SEK. In order to attract small companies
with less than 50 employees to apply for support to change programs, SWLF
collaborated with the Division of Industrial Ergonomics at Linköping University
in a joint activity. In total, 60 companies, including ten wood companies, were

contacted by telephone and offered a free workplace investigation. The company could use the resulting report as a basis to proceed with an application to SWLF if they wished. Further support to the companies to write the application was also promised. The companies that applied and were given financial aid were obliged to carry through their projects themselves without any particular support.

The aim of this study was to describe readiness for change, initiators of change processes and obstacles and driving forces for change in a sample of small wood companies. Another purpose was to study influences on the companies from so called external actors, such as SWLF and the University.

Methods

Out of the ten wood companies that were contacted 1992, six approved of receiving the workplace investigation. At the follow-up workplace investigation 1994, all ten companies approved of the investigation. However, two companies dropped out since they had no possibilities to receive visits on the day the investigation was scheduled. The student groups spent one day in their company, which included interviews with managers, foremen and most of the employees in the smaller companies. The eight workplace reports formed a base material for this study. As a complement, telephone interviews were made with managers from all ten companies. Also, the students were contacted for additional information. Interviews were also performed with personnel from the SWLF and the supervisors of the students from the University. Some data was collected from documents describing the contacts with the companies and from SWLF official data.

The following questions demonstrate the structure of the interviews. In addition to these questions, other questions, clarification and follow-up questions were asked:

-What reflections and considerations did you make when offered help with the
    application to SWLF?
-What causes an insight and desire to start change projects?
-What obstacles for change projects are there in your company?
-What problems do you consider common in change projects?
-How is the spread of information handled?
-Did the employees participate in the changes?
-How is your company organised?
-What changes in technology are there in your company?
-Do you consider having succeeded with your change projects?
-What kind of external support do you consider your company needs?

The companies had between 4 and almost 50 employees, and represented carpentry and furniture making. They were all situated in the south-east region of Sweden.

Results

All companies had more or less been negatively affected by the recession in the Swedish economy in the beginning of the 1990´s. Several of the companies had recovered at the time of this evaluation. The rates of personnel turnover and sickness absence were low and all companies considered work injuries to be rare.

Readiness for change

The companies were classified in three categories. Two companies were qualified for the first group: "Management consider substantial needs for improvements and consciously try to foresee such needs. Several change projects are in work." Six companies qualified for the second group: "Management has a positive view towards change projects, but are not working particularly active with identification of such needs. Only single projects are in progress." Two companies were classified in the third group: "Management is pleased with the present situation, does not recognise any particular problems and does not see needs of changes or only carries out those changes which are absolutely necessary."

Economic benefits was the most important driving force for change projects. This was particularly so for the two companies who were fighting for survival after having gone bankrupt. Management had a very important role in all companies. Their attitudes to, decisions and actions for obtaining change strongly influenced the employees. Low levels of employee participation caused passivity. As one employee expressed it: "It is meaningless to give proposals, because nothing will happen whatever". Influences from outside the companies were in several cases described as the initiators of change projects. In one case a new share owner introduced new ideas. The contacts with the SWLF and the University in two other cases strengthened two other companies to deal with their problems. In a third company, the discussions with SWLF started a learning process and internal discussions in the management group which was the reason that they became more mature and started change projects several years later. In several cases, proposals for improvements came from employees who were highly involved in their jobs. Recently employed younger workers were more active in that respect than other workers, and education programs were also noted as a facilitator for improvement ideas. Risks for work injuries, musculoskeletal symptoms and deficiencies in the working environment were examples of such factors that evoked proposals from employees.

Obstacles for change

Lack of time was the most important reason mentioned for not working with change projects. In particular, there were few white-collar workers in the companies, and therefore the bottle-neck was often their time to administrate and to be responsible for the change processes. The companies investigated had relatively little paper work and tried to avoid it as much as possible. The demands on formal applications and formal documentation from SWLF was therefore an obstacle for several of the companies, and definitely a time consuming task that they disliked and gave low priority to. Bad economy was also a most important obstacle for change projects, especially for the companies that earlier had gone bankrupt. In a situation with bad economy manning was often reduced to a minimum and investments with a long term pay-back were out of question. It was also lowering the workers' motivation when there was no money to carry out their good proposals. The personnel at the companies investigated had neither methods knowledge, nor sufficient experience from running change projects. Consequently, several of the companies had a dependency on external actors or consultants for running change projects. In a bad economic situation, of course the possibilities of involving external consultants decrease.

Earlier experience of change processes was an important determinator of the readiness for further change. One company avoided further projects because of bad experiences. A few companies had become restrictive when investing time and economical resources in new projects due to the earlier experience that it was difficult to obtain the results aimed at and the expected pay-back. Two companies were clearly inspired by the good results obtained from their change projects, which made them intensify their efforts further. This study also points to the importance of good information and communication between all persons in the company. Conflicts, misunderstandings and problems could often arise in difficult and unusual situations, and especially so if there was not sufficient information available. It was common that information meetings in the companies were held more seldom than planned. The main reason was considered to be lack of time, but the managers in some companies considered this sort of meetings unimportant.

It was common that the employees felt that they did not participate in the change processes. At the same time, management considered with some disappointment that the employees were not interested to participate in new projects. Management often considered the companies so small that all necessary information could take place over the coffee table, but the employees often felt that the information was insufficient or they could feel difficulties to bring up controversial issues to discussion. Especially one company was a good example in this case when the employees participated in the planning, installation and running-in of a new machine with a very good result.

Most of the managers initially felt that SWLF was not the right thing for them. They did not know enough about SWLF and considered an application and a project too large for them. Also many felt a kind of suspicion against governmental funds and interference. Later on, most of the companies became positive to SWLF. A majority of the companies were at the follow-up positive to government support, but they strongly disagreed with tendencies of bureaucracy and demands on extensive documentation and application procedures.

Goals of the change projects
The most common goals with the projects were to decrease the costs or to increase the capacity of the production. Quality improvements were also ranked high. Some companies mentioned better qualifications among the employees or the importance of keeping up with the technology development. The goal was to improve the working environment in a few companies. The level of and ambition for the working environment varied from "only if necessary" to recognition of relationships between work conditions and productivity.

Initiation of change projects
Several of the companies had difficult problems at the time with a strong drop in demand, personnel reductions, a bad economy and time restraints. This situation of uncertainty contributed to unwillingness to take on new obligations. Several of the companies who did not apply for support earlier would do that today. Two of them stated clearly that they were not mature enough earlier, but the discussions in the companies over the years had made them ready for change projects now. They had even initiated projects on their own, and they knew what projects they would carry through if they could get external support.

Discussion

Several of the cases reported here support Kurt Lewins model of change, starting with an unfreezing phase (Lewin 1951). It can be noted that this phase could take several years, until changes started in the second phase. The reason can be the small size of the companies, where management was less familiar with working in change projects and also had fewer discussion partners. This study points to the importance that small companies get opportunities to different types of contacts with other companies, consultants, personnel from Universities and other organisations. Such activities have been shown in other studies of SWLF projects to improve the competence for change (Aronsson et al 1995). The difficulties to initiate change projects in these companies were on the other hand contrasted with the observation that non-complicated practical improvements could be very rapid and common in some companies, although there was no formal system for this, nor any use of terms as Kaizen or continuous improvements (see Imai 1986).

This study shows that worker participation in the changes varied very much between the companies, in spite that they were all small. It also demonstrates that it is difficult to obtain a high level of participation, even in small companies. This finding is consistent with other results (Neumann 1989). It should be noted that small companies are not an exception of this. It seems as if the change methodology used in the companies and the approach of the managers were more important for the results than other factors. Aronsson et al (1995) identified support from management in co-operation with unions and employees as important for obtaining results, which are in agreement with the results from this study.

The results emphasise the importance of performing evaluation studies over long time periods, since the processes that are evaluated take place over long time periods. A limitation in this study is that parts of the assessment of the companies was based on the results from the student groups, one group in each company, and the work was performed in limited time. Interpretations therefore differed between the student groups. Also their experience and frame of reference was limited. However, the study points to several important similarities and differences between these small wood industry companies and larger companies in other branches. Experiences from small companies in other branches indicate that the results from this study to a very large extent can be generalised for small companies.

Conclusions

Driving forces for change projects in small companies are often prospects of decreasing costs and increasing productivity. Quality improvements, demands from customers and demands from authorities are also common as driving forces, and to a certain extent also work environment issues after proposals from the employees. External influence from other companies, consultants, owners, government authorities or universities are important for idea generation and for learning processes leading to decisions of initiating change projects. Government funding can be a facilitator in this respect, even though the time horizon is very long, sometimes many years before projects are started.

Even small companies often have problems to spread sufficient information and to obtain a good communication, and the result can be conflicts or misunderstandings that counteract the progress of the project. A low level of worker participation in the changes is not uncommon. Management plays a very important role for attitudes to change and is decisive of the actions taken.

Lack of time and bad economy are considered as the most important obstacles for change. Previous unsuccessful projects, and lack of experience and competence are also important obstacles for initiating change projects.

Small companies are often, but not always, positive to government support, given minimal bureaucracy and uncomplicated documentation and application procedures.

References

Aronsson, G., Svensson, L., Leksell, K. and Sjögren, A. 1995, *Förändringskompetens,* (Arbetslivsinstitutet, Stockholm) (In Swedish)
Imai, M. 1986, *Kaizen - The key to Japan´s competitive success,* (McGraw Hill, New York)
Lewin, K. 1951, *Field theory in social science*, (Harper, New York)
Neumann, J.E. 1989, Why people don`t participate in organizational change. Research in Organizational Change and Development, **3**, 181-212.

THE SICK BUILDING SYNDROME, STRESS, AND PERCEIVED AND DESIRED PERSONAL CONTROL

Sean M. Bolas and Joanne O. Crawford

Industrial Ergonomics Group,
School of Manufacturing and
Mechanical Engineering,
The University of Birmingham,
Edgbaston,
Birmingham B15 2TT.

This study, based partly upon the job demands-job control model, predicted that low perceived personal control within an office environment, coupled with high work demands would predict a higher prevalence of self-reports of symptoms of the sick building syndrome (SBS). This prediction was not fulfilled. However, the introduction of the personality measure Desire for Control as an additional variable within the regressions resulted in support for the contention that it is the interaction of perceived control with the level preferred control that is the salient and consistent predictor of symptoms of SBS. Support for the job control - job demands model in relation to SBS symptoms, arose only with the inclusion of this trait. Psychological stress and dissatisfaction deriving from various work-related sources significantly correlated with symptom measures.

Introduction

The sick building syndrome (SBS) has in recent years become a universal problem that has the potential to cause great disruption to employee efficiency (Raw, Roys & Leaman 1993). Formally, SBS is defined as a diffuse set of symptoms of non-specific aetiology, the onset of which is correlated with working in certain buildings (Wilson and Hedge 1987). Irritation of the eyes, throat and nasal mucous membranes are typical; but skin irritation and fatigue are also included within the list of symptoms.

There is an overwhelming number of studies concerned with the physical characteristics of problem buildings, which implicate, among many other factors, ventilation, general climatic parameters, volatile organic compounds, dust, ozone, and cigarette smoke (see Raw 1992 for a review). Much of this research is based on physical factors, and generally fails to investigate the importance of the occupant's perceptions and experience of their environment, including the psychosocial environment (Ryan and Morrow 1994). Where perceptions of the psychosocial climate and environment are included, strong associations with symptoms are often found, and within reviews of the literature, psychological factors are afforded much credence (Baker 1989). Occupational stress has repeatedly been associated with SBS (Crawford and Hawkins 1995; Hedge, Erickson and Rubin 1993; Hodgson, Muldoon, Collopy and Olesen 1992; Morris and Hawkins 1987). This has led some researchers to suggest that stress sensitises the individual to poor air quality within the office (Hedge, Erickson and Rubin 1993).

In the wake of Karasek's (1979) job demands - job control model, there has been an emphasis on control at work and its affect on health (Sauter, Hurrell and Cooper 1989). Bain and Baldry (1995) apply these principles directly to SBS, suggesting that it is the systematic reduction in employee control of the environment and within work practices found in so-called 'intelligent buildings' that predisposes workers to symptoms of SBS. It would perhaps be safe to say there has been no investigation of perceived control at work in relation to SBS.

For this study, it was predicted that low perceived personal control within an office environment would be associated with symptoms of SBS. Based upon the theory of Karasek (1979), the second main hypothesis was that perceived control and demands would be interactive in predicting symptoms of SBS. Additionally, the reported symptoms of those with a preference for high levels of control should be disproportionately more severe at low levels of perceived control. This was based on the premise that the deleterious consequences of impoverished levels of perceived personal control would be commensurate with the discrepancy between desired and perceived levels of control.

Finally, it was predicted that reported stress and satisfaction levels stemming from various sources at work would correlate with symptoms. This followed from previous findings showing strong associations between these psychosocial aspects and symptoms of SBS.

Method

Participants

The setting for the study was a regional office of a large international parcel distribution firm in the West Midlands, England. Following extensive complaints from employees regarding environmental factors, management had agreed to an environmental and psychosocial survey of the workforce.

Apparatus

Each employee was given a booklet composed of the following five sections:

i) A checklist of symptoms of SBS based on that of Hedge, Erickson and Rubin (1993). For each of fourteen symptoms the respondent first states its frequency as 'never', '1-3 times per month', '1-3 times per week', or 'every day', and in cases where the symptom was experienced, whether it improves when away from the office. Only those symptoms that are work-related are presumed indicative SBS. Two measures are adopted: the 'person symptom index' (PSI); and the 'weighted person symptom index' (WPSI). The first of these refers simply to the number of symptoms of the fourteen that are experienced. The second acknowledges the importance of the frequency with which these symptoms occur.

ii) An in-house scale of perceived control. Items for this were adapted from Greenberger (1981). The questions pertain to six domains of perceived control, namely: *task control*, regarding the order, choice, methods and quality of performing tasks (4 items); *control of work pacing*, regarding the speed of working and the arrangement of breaks (4 items); *control over work scheduling*, regarding the timetabling of work (4 items); *control over the physical environment*, regarding the orientation, personalisation, and locality of one's workstation (3 items); *control over the office climate*, regarding ventilation, temperature, humidity and lighting (4 items); *social interaction*, regarding privacy (2 items). Also included within this section are five items designed to provide a measure perceived work demands; these are based on Karasek (1979). Three items concerning satisfaction with the office temperature, humidity and lighting conclude this section. Respondents mark their agreement with each item of the control, demands and satisfaction scales using five point increments.

iii) Burger and Cooper (1979) Desire for Control Scale (DC). This is a 20-item personality measure devised to assess the extent of control that is desired by the respondent in various aspects of their lives. It is therefore not specific to the work context. The DC scale has been well validated and applied in diverse contexts (Burger 1992). Respondents' agreement with each item is indicated using scores from 1 to 7; the maximum score being 140.

iv) The Occupational Stress Indicator (OSI) (Cooper, Sloan and Williams 1988). This is an extensively used and validated stress and satisfaction scale designed to elicit respondents' assessments of many psychosocial aspects of work. Only the responses to the following two sections were appropriate: 'Sources of pressure in your job', which is divided into six areas; and 'How you feel about your job', which again is divided into six areas.

v) Finally, a number of items asked for information concerning the following: gender; age; marital status; number of children under five living at home; educational status; smoking habits; alcohol consumption; asthma; hay fever; eczema; hours per day in the office, engaged in VDU work, and engaged in paper work; tenure; years in the present job field.

Procedure

Ninety-six employees were issued with a nine-page booklet (from hereon referred to as the basic questionnaire) containing the symptom checklist, the in-house control scale, the DC scale and items regarding personal characteristics. In addition to this booklet a group of 51 employees, those at every other occupied desk, also received the OSI. Environmental measures were recorded at all workstations; this was concurrent with questionnaire distribution.

Results

The response rate for the OSI was 81% (N=41) and that for the basic questionnaires 77% (N=74). The number of basic questionnaires used in analysis was therefore 74, and of these 41 were accompanied by the OSI. The PSI and WPSI for the total sample were 4.81 (st. dev. = 3.18) and 31.31 (st. dev. = 28.51) respectively. Symptoms characteristic of SBS were greatly in evidence. Lethargy was the most prevalent (77% of the sample), followed by dry throat (61%), headaches (58%), blocked nose (47%), sore throat (43%), itching eyes (42%) and runny nose (32%). The PSI and WPSI were not significantly different across gender.

Hierarchical regression analysis was performed to test Karasek's job demands - job control model for those 74 employees returning the basic questionnaire, employing the PSI and WPSI as outcome variables. For each domain of control, the specific control score, the work demands score, and the interaction of these were entered into the regression analysis. Control over climate was not included as a domain in these analyses since Karasek never addressed this aspect of control. Variables were selected on the simple basis of predictive value using the 0.05 significance level to enter and remove them. Although Karasek prescribed a 'special interaction term', the product term of control and demands is used here since it is believed to be more representative of true interaction.

No support was found for the interaction model, with the PSI and WPSI as outcomes. In the case of the PSI, only two of the five regressions contained significant predictors. These were perceived task control (t = 2.11; p < 0.05) and work pace control (t = 2.65; p < 0.05); low perceived control in these was associated with a greater PSI. Using the WPSI as the outcome variable, only one regression included a significant interaction term. This was in the domain of control of the social environment (t = -2.56; p < 0.05), but its direction was contrary to Karasek's model. In addition, the perceived control of work pace was found to be significant (t = 2.90; p < 0.05). Work demands was the single significant predictor in the three remaining regressions (t = -2.37; p < 0.05).

The role of the Desire for Control was investigated by including this as an additional variable within the regression analyses. In these cases the regression for each domain of control contained three main factors, three two-way interactions and one three-way interaction. It is clear from these further analyses that, with the PSI as the outcome, all of the regressions, except that relating to social interaction, include the interaction term involving DC and the specific control measure. That is, the interaction terms involving DC and perceived task control (t = 2.43; p < 0.05), work pace control (t = 2.68; p < 0.05), work schedule control (t = 2.13; p < 0.05) and control over the physical surrounds (t = 2.50; p < 0.05) were all significant predictors. In each case this was the only significant term. Most importantly, however, all of these interactions were in the predicted direction; high DC coupled with low perceived control was associated with elevated levels of the PSI. The same interaction terms were significant when the outcome variable was the WPSI, and again these were in the predicted direction (t = 2.90, 3.22, 2.58, and 2.45 respectively; p < 0.05). Interestingly the interaction terms involving perceived control over tasks, work schedules, physical surrounds and social interaction with perceived work demands also reached significance (t = -2.53, -2.04, -2.68, and -2.56 respectively; p < 0.05), and their direction agreed with the job control - job demands model.

The DC score correlated 0.22 and 0.27 with ratings of task and work pace control. It cannot therefore be concluded that an employee's Desire for Control score is independent of their perception of the amount of control the environment offers.

Table 1 Correlations and partial correlations of SBS symptoms with the satisfaction and stress scales on the OSI (N = 41). (Significance denoted by * = p<0.05, ** = p<0.01)

Satisfaction deriving from:	WPSI	PSI
Personal relationships	-0.04 (0.02)	-0.27 (0.00)
Achievement, value and growth	-0.22 (0.11)	-0.25 (0.110
Job itself	-0.31 (0.07)	-0.35* (0.00)
Organisational design and structure	-0.40* (-0.16)	-0.43**(-0.16)
Organisational processes	-0.40* (-0.17)	-0.32* (-0.15)
Broad measure of satisfaction	-0.39* (-0.03)	-0.38* (0.02)
Stress deriving from:		
Factors intrinsic to the job	0.38* (0.11)	0.53** (0.19)
The managerial role	0.38* (-0.02)	0.52** (-0.05)
Organisational structure and climate	0.37* (-0.02)	0.52** (-0.05)
Career and achievement	0.32* (0.10)	0.42** (-0.12)
Home / work interface	0.38* (0.05)	0.50** (0.02)
Relationships with others	0.36* (0.04)	0.46** (0.02)

Data from those returning the OSI (N = 41) allow correlations and partial correlations to be computed for the WPSI and PSI with various areas of satisfaction and stress (Table 1, above). Significant correlations exist between the WPSI and the PSI and four of the satisfaction subscales and all of the six stress subscales. In all cases, higher levels of dissatisfaction and stress are associated with elevated reports of symptoms.

All data from the total sample (i.e. excluding data from the OSI) were submitted to hierarchical regression analysis. Variables included were: the six measures of perceived control and their interaction with DC; the demands rating and its interaction with the DC score; the DC score; items relating to satisfaction with the office climate; the personal characteristics (listed earlier); and the environmental measures of air temperature, humidity and lighting. Satisfaction with air temperature in the office accounted for significant variance in symptom reporting for the PSI and WPSI, the figures being 17.57 % (t = 3.92; p < 0.05) and 7.79 % (t = 2.64; p < 0.05) respectively. The only additional significant predictor to emerge in both cases was the interaction term involving perceived control over the work pace and the Desire for Control, the variances explained by this term being 7.62 % (t = 2.69; p < 0.05) and 12.61% (t = 3.22; p < 0.05) respectively.

Discussion

These results, based on office employees reporting high prevalences of various SBS symptoms, substantiate previous findings. Symptoms were highly correlated with nearly all subscales of the satisfaction and stress components of the OSI, although partial correlations for each were negligible. The findings corroborate those of, among others, Hodgson, Muldoon and Olesen (1992) and Hedge, Erickson and Rubin (1993).

In addressing the possible role of personal control and preference for control, the present findings extend previous work on the psychology of SBS. While certain domains of control do seem to predict the incidence and prevalence of symptoms within simple three term regression analyses, the introduction of the personality variable Desire for Control (DC) allows a far greater proportion of variance in the PSI and WPSI to be explained. The interaction term of DC and perceived control over the work pace emerges as a significant factor, along with satisfaction with the air temperature in the office, even when all other personal characteristics and environmental data are included. This gives support to the idea that it is the way the individual perceives his or her environment that may be important in SBS.

Clearly cross-sectional studies such as these do not imply causality. It is not difficult to imagine that the experience of symptoms, whether derived from physical or psychosocial aspects colour the way the environment is perceived. It may be the case too that the propensity to report symptoms increases with stress or dissatisfaction. Indeed, an area that has not been sufficiently addressed concerns the way in which reporting of symptoms reflects an outlet for frustrations with the work situation and the way in which the organisation is managed.

Pennebaker (1994) and Leeshaley and Brown (1992) provide discussions concerning the various factors (although not with specific application to SBS) that may affect self-report of illness - factors that appear irrelevant or innocuous but in reality may confound results of studies that use self-report data elicited by symptom checklists. It would seem an appropriate, if belated, time to study the psychology of self-report of ill-health in relation to SBS, irrespective of whether such a programme would appear to question the solidity of the foundation upon which SBS research rests.

References

Bachmann, M.O. and Myers, J.E. 1995, Influences on the sick building syndrome symptoms in three buildings, Social Science and Medicine, **40**, 245-251.

Bain, P. and Baldry, C. 1995, Sickness and control in the office - the sick building syndrome. Technology, Work and Employment, **10**, 19-31.

Baker, D.B. 1989, Social and organisational factors in office building-related illness. Occupational Medicine: State of the Art Reviews, **4**, 607-624.

Burger, J.M. 1992, *Desire for Control; Personality, Social and Clinical Perspectives*, (Plenum, New York).

Burger, J.M. and Cooper, H.M. 1979, The desirability of control, Motivation and Emotion, **3**, 381-393.

Cooper, C.L., Sloan, S.J. and Williams, S. 1988, *Occupational Stress Management Guide* (NFER-Nelson, Oxford).

Crawford, J.O. and Hawkins, L.H., 1995, Sick Building Syndrome and Occupational Stress. In S.A. Robertson (ed), *Contemporary Ergonomics 1995,*(Taylor and Francis, London) 207-212.

Greenberger- D 1981, *Personal control at work: Its conceptualisation and measurement, Technical Report 1-1-4*, (Univ. of Wisconsin - Madison NR 170-892).

Hedge, A., Erickson, W.A. and Rubin, G. 1993, Why do gender, job stress, job satisfaction, perceived indoor air quality and VDT use influence reports of the sick building syndrome in offices. In H. Luczak, A. Cakir and G. Cakir (eds) *Work with Display Units '92* (Elsevier Science, Amsterdam).

Hodgson, M.J., Muldoon, S., Collopy, P. and Olesen, B. 1992, Sick building symptoms, work stress and environmental measures, *Environments for People. Indoor Air Quality '92, Proceedings of the ASHRAE Conference* (ASHRAE, Atlanta) 47-56.

Karasek, R.A. 1979, Job demands, job decision latitude, and mental strain: Implications for job redesign, Administrative Science Quarterly, **24**, 285-308.

Leeshaley, P.R. and Brown, R.S. 1992, Biases in perception and reporting following a perceived toxic exposure, Perceptual and Motor Skills, **75**, 531-544.

Morris, L. and Hawkins, L. 1987, The role of stress in the sick building syndrome. In B.Siefert, H.Esdon, M.Fischer, H.Ruden, and J.Wegner (eds) *Indoor Air '87, Proceedings of the 4th International Conference on Indoor Air Quality and Climate vol. 2*, (Berlin: Institute for Water, Soil and Air Hygiene) 566-571.

Pennebaker, JW 1994, Psychological bases of symptom reporting- perceptual and emotional aspects of chemical sensitivity, Toxicology and Indust. Health, **10**, 497-511.

Raw, G. 1992, *Sick Building Syndrome: A Review of the Evidence on Causes and Solutions*, (HSE Contract Research Report No 42/1992).

Raw, G., Roys, M.S. and Leaman, A. 1993, Sick building syndrome, productivity and control, Property Journal **August** 17-19.

Ryan, C.M. and Morrow, L.A. 1992, Dysfunctional buildings or dysfunctional people: An examination of the sick building syndrome and allied disorders, Journal of Consulting and Clinical Psychology, **60**, 220-224.

Sauter, S.L., Hurrell, J.J. Jr, and Cooper, C.L. 1989, *Job Control and Worker Health*, (John Wiley, Chichester).

Wilson, S. and Hedge, A. 1987, *A Study of Building Sickness*. (Building Use Studies Ltd, Surrey).

Anthropometry

VALIDITY AND USABILITY OF 3D ANTHROPOMETRIC COMPUTER MODELS IN CONSUMER SAFETY USES

Pyter N. Hoekstra

Delft University of Technology
Faculty of Industrial Design Engineering
Department of Product and Systems Ergonomics
Jaffalaan 9, 2628 BX Delft, the Netherlands

At our Department we are currently engaged in a project for the Dutch Consumer Safety Institute: using 3D-Anthropometric computer models in predicting reachability of product-parts by small children (up to 4 years of age) on behalf of safety standards. This paper deals with the levels of validity in the use of these models and with the issue of usability: in what way can these models (as generalized abstractions of only some human characteristics or capabilities) be used in simulating extremes. It is reasoned that only with the application of data from 3D volumetric scanning techniques can we reach a valid level for functional postures of the computer models. The usability should be restricted to a library of product-related postures illustrating for expert users or ergonomists the level of assessment.

Introduction

At the Faculty of Industrial Design Engineering of the Delft University of Technology ADAPS (Anthropometric Design Assessment Program System) is used as a commonly available workspace assessment tool for design and graduation projects, as a more or less mandatory course in Computer Aided Anthropometric Assessment (CAAA), see e.g. Ruiter (1989) and as a research project in e.g. simulating some aspects of directing a 3D-Anthropometric computer model's line-of-sight and displaying the corresponding field-of-view, Hoekstra (1993).

At this moment we are engaged in an on-going research project for the Dutch Consumer Safety Institute in predicting reachability of product-parts by small children (up to 4 years of age) on behalf of defining safety standards.
Stages in this research are:
- the definition of 'reachability' and the manner in which this could be simulated using the 3D-Anthropometric computer models in the ADAPS package;
- the development of new computer models for the required specific age or motor control categories: 0-3, 3-6, 9-12, 12-18, 18-24, 24-36 and 36-48 months using the

most recent anthropometric survey of Dutch children, see e.g. Steenbekkers (1993);
- assessing the predictive capacities when using the computer models compared with
real-life behaviour of a test-population of children of the same categories.
In this research specific attention is paid to the issue of 'expert use' of assessment
programs like ADAPS in relation to the issue of safety and the definition of safety
standards. Parts of this research will be reported elsewhere, parts will be described and
presented by my colleague, Ruiter (this issue). In what follows in this paper we will
try to focus on two issues:

- validity	- the levels of validity in the use of 3D-Anthropometric computer models based on data from essentially *univariate* anthropometric surveys;
- usability	- in what way can these computer models (as generalized abstractions of only *some* human characteristics or capabilities) be used in simulating extremes, especially in regard to product safety.

Validity

We will use the word 'validity' to stand for: acceptability by force of reasoning
or logic (idealy the result of an algorithmic IF ... THEN ... operation); the property of
possessing confirmation elsewhere. This use is consistent and analogous with the use
by Tainsh (1995) when defining a valid design. As discussed elsewhere, Hoekstra
(1995), we can speak of levels of validity because when tracing the use of CAAA
from its roots onwards we find that each succesive stage (with corresponding validity)
we pass is either hierarchically dependant from or uses derivatives of earlier stages.
In CAAA we find 3D-Anthropometric computer models that can be manipulated into
functional postures, integrated in a display of a workspace. After a number of
inspections the process is ended with a prediction about the mimicked real world
situation. In this process we trace the following five stages.

Population
Do we have the relevant data that is going to be input, e.g. in defining the
anthropometric computer model? When we look at an anthropometric model on a
computer display, we see a transformation of the original data (derived from a human
population and defining the model) using mathematical and computer-graphics
algorithms. To be valid, the model's geometry should possess a (as close as possible)
one to one mapping with the original human population that was measured. Values for
missing or non-existent anthropometric variables can be estimated, cf. Phillips and
Stevenson (1993), but it should be understood that the variables measured in standard
('classical' one dimensional) anthropometry are essentially univariate and should not
in principle be combined into multivariate dimensions. So we can speak of a 5th
percentile Head Length and of a 5th percentile Head Circumference; to speak of a "P5
Head" (or a "P5 anthropometric computer model") is nonsense, cf. Robinette (1993).

Anthropometric Model
In what way does a 3D-anthropometric computer *model* represent selected
human characteristics? Only if human characteristics can be described by single
numbers, can a (in essence univariate) anthropometric model be used to arrive at valid

assessments (and for that specific characteristic only). If we start with human stature and derive an anthropometric model's length, we can use the last to determine or assess e.g. the height of a door; to determine the fit of a helmet we need at least the digitized contours of a human head via 3D-surface anthropometry, see e.g. Robinette (op. cit.) or Brooke-Wavell, Jones and West (1994), to realize a valid model.
In this respect we have to conclude that the use of the anthropometric computer models in ADAPS, as an integration of univariate variables, is, though valuable and useful, essentially not valid in predicting multivariate dimensions.

Functional Postures

What is the relation between data derived from standard anthropometric measurements and the data we chose to describe functional postures of the anthropometric model? As yet we do not know. Standard anthropometric measurements uses standardized postures, it has to. The validity of a model based on those measurements only, is thus restricted to those postures only. If we would like to simulate a functional posture in a cockpit and predict 'Reach to Controls' we need more factors besides Armlength. Among these are Seat Position, Shoulder Hight, Shoulder Width, Restraint System and Motivation, see Rioux and Bruckart (1995), all measured in that same functional posture. Except for the last two factors we could 3D-surface scan a human subject in this posture and transform the data into a valid anthropometric model. There is no valid model however for the interaction of the human body with any restraining system (even when disregarding the influence of Motivation).

Assessment

What are the assumptions in computer algorithms and in human evaluation? Suppose our CAAA-software would contain an algorithm for collision detection between model and workspace. If such a collision occurs positions could be automatically stored. The graphics e.g. via changing colours, could notify the user of such an event. This implies that if, and only if valid human evaluation can be expressed (assuming such an evaluation exists) and translated into a series of algorithmic commands, can we speak of valid Computer Assessment. By definition computer algorithms are open for inspection and its assumptions can in this way be traced and checked. Most users of CAAA-systems however do not have access to the software source-code or even the knowledge that is contained within.

Prediction

Can CAAA-based evaluation provide predictions for the 'real world' where our target population will interact with the realized design? According to Tainsh (1995) we can see the design process as aiming to enhance the validity of the design.
For a design to be valid he lists four categories of validity, each one increasing in assessment value. Basic for all four is the notion that evaluation requirements must have been expressed beforehand and in such a way that a check to ensure that the requirements have been met is made possible. Although our levels of validity are consistent with the notion of possessing increased assessment value, the latter is certainly not the case in CAAA at this moment. Only the user of CAAA-software who is both expert ergonomist and knowledgeable about what's 'behind the computer display' can provide valid predictions for the real world.

Usability

As stated before we are primarily interested here if, and if so the way in which we can use anthropometric computer models (as generalized abstractions of only some human characteristics, capacities or capabilities) to simulate extremes in reachability especially in regard of product safety for small children. If we now again trace the assessment process from its roots upwards we can annotate the previous encountered stages with the following.

Population

For our research question we do have data about the target population, see Steenbekkers (op. cit.). The parts that we can use later to define our anthropometric computer model for use in ADAPS however are the result from classical, essentially univariate anthropometry. Although the dataset contains information about motor control there is no specific data that can be used to describe excursion ranges for the model's joints. For this we have to use older, sometimes non-consistent and again essentially one dimensional datasets (extension for the upper leg e.g. measured in one specific posture only). The usability is thus restricted to 'what is measured' or what can be logically derived.

Anthropometric Model

An anthropometric computer model in ADAPS consists of a set of linear branched chains, containing twentyfive links (bodymembers). Relative to the links surface-points are constructed that define the model's outside geometry. As all ingredients thus used are univariate the model as a 3D-entity is essentially non-valid. The usability is high however in that it integrates 1D-data (lengths, widths etc.) into 3D-information (space requirements), at least if the expert user is aware of the one dimensional nature of the model's basis and is able to translate a 3D-design problem into a series of 1D-solutions.

Functional Postures

A designer or ergonomist is hardly ever interested in the standardized postures for anthropometric measurements. What is encountered in the real world are human beings performing tasks in postures specifically related to those tasks that thus could be described as functional (sitting, reaching, operating controls, looking at the world outside or just resting). Since it is quite easy to manipulate the computer model into functional postures in most CAAA-systems (using reach algorithms or retrieving stored postures from libraries) the usability is high, see e.g. the successful use of CAAA-systems by major automobile and aircraft manufacturing companies, cf. Porter (1995). We do find the same restrictions here as in the paragraph before but the more so for our specific case: in simulating extreme reaches we certainly need an expert user or ergonomist to decide if our extreme posture can be thought of as a simple integration of three 1D joint-angles or that real 3D surface or volumetric data is necessary for that specific posture. We also have to bear in mind that small children, discovering and exploring the world could adopt postures that are for them very functional, where we wouldn't think so for a 'grown-up' situation. They could also show extremes ("what is this thing under my seat that I can just touch") that are certainly not the early precursors of 'Reach to Control'.

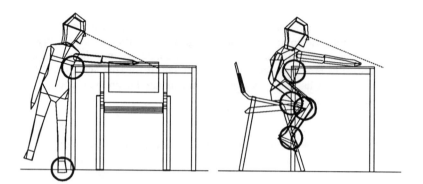

Figure 1. Encircled: areas of interaction between model and workspace.

Assessment

In CAAA assessment means inspection, algorithmically and/or by the human user, of the relationships between the anthropometric model and the simulated workspace. If we take a look at some reach simulations illustrated in Figure 1 we can easily see that the usability of CAAA is very much dependent upon the differences in series of interactions we are confronted with. The child on the left stands on tiptoe, the amount of which influencing the position of the armpit, the interaction with the table and thus its maximum reach. For the situation on the right we have a different and even longer chain of interactions that requires human expertise to assess properly. As discussed by Ruiter (this issue) we already get different results in simulating maximum reaches between expert and novice users of ADAPS. The best we can offer here in usability is a library of product-related postures (defined by experts) illustrating for (again) expert users or ergonomists the level of assessment. These libraries could then form the basis for the definition of safety standards.

Prediction

Since this level is essentially dependent on or derived from earlier stages we have to conclude that its usability as yet still lies entirely outside 'the computer domain' and is restricted to human expertise only.

Conclusions

By retracing the stages we pass when using CAAA we have seen that each successive stage is characterized by an increase in *possible* assessment value but a concurrent decrease in validity. The emergence of 3D-surface anthropometry techniques and the use of volumetric imaging of the human body - using ultra sound, X-ray CT, MRI, nuclear scintigraphy combined with conventional projection radiography - Rioux (1995), would extend the validity of CAAA to the level of Functional Postures. For the last two stages we would still require human beings, either as observable subjects or as expert users of CAAA-programs. Regarding the issue of usability: in our case of predicting reachability of product-parts by small children the best we can offer at this moment is a library of product-related postures (defined by experts) that can illustrate for expert users or ergonomists the possible level of assessment. At the same time these libraries could serve as a possible basis for the definition of safety standards.

Acknowledgement

This research was made possible through funding by the Stichting Consument en Veiligheid (Dutch Consumer Safety Institute). Opinions expressed in this paper are the author's responsibility only.

References

Brooke-Wavell, K., Jones, P.R.M. and West, G.M. 1994, Reliability and repeatability of 3-D body scanner (LASS) measurements compared to anthropometry. *Annals of Human Biology*, **21**-6, 571-577.

Dennison, T.W. and Gawron, V.J. 1995, Tools and Methods for Human Factors Test and Evaluation: Mockups, Physical and Electronic Human Models, and Simulation. *Proceedings of the Human Factors and Ergonomics Society 39th Annual Meeting - 1995*, (HFES, Santa Monica, CA) 1228-1232.

Hoekstra, P.N. 1993, Some Uses of "Active Viewing" in Computer Aided Anthropometric Assessment. *Proceedings of the HFES 37th Annual Meeting - 1993*, (HFES, Santa Monica, CA), **1**, 494-498.

Hoekstra, P.N. 1995, Levels of Validity in Computer Aided Anthropometric Assessment. In A. de Moraes and S. Mariño (eds.), *Proceedings of IEA World Conference 1995, 3rd Latin American Congress, 7th Brazilian Ergonomic Congress*, (ABERGO, Rio de Janeiro) 77-80.

Kohn, L.A.P., Cheverud, J.M., Bhatia, G., Commean, P., Smith, K. and Vannier, M.W. 1995, Anthropometric Optical Surface Imaging System Repeatability, Precision, and Validation. *Annals of Plastic Surgery*, **34**-4, 362-371.

Phillips, S. and Stevenson, M. 1993, An anthropometric data base for combined populations, and the importance of its use on manual handling. In W.S.Marras, W.Karwowski, J.L.Smith and L.Pacholski (eds.) *The Ergonomics of Manual Work*, (Taylor & Francis,London), 211-214.

Porter, J.M. 1995, The Ergonomics Development of the Fiat Punto - European 'Car of the Year 1995'. In A. de Moraes and S. Mariño (eds.), *Proceedings of IEA World Conference 1995, 3rd Latin American Congress, 7th Brazilian Ergonomic Congress*, (ABERGO, Rio de Janeiro) 73-76.

Rioux, M. and Bruckart, J. 1995, Data Collection. In K.M.Robinette, M.W.Vannier, M.Rioux and P.R.M.Jones (eds.) *3-D Surface Anthropometry: Review of Technology*, Draft Report, AGARD, personal communication.

Robinette, K.M. 1993, Fit testing as a helmet development tool. *Proceedings of the 37th Annual Meeting of the Human Factors and Ergonomics Society*, (The Human Factors and Ergonomics Society, Santa Monica CA), **1**, 69-73.

Ruiter, I.A. 1989, Education in Computer Aided Anthropometric Assessment. In E.D. Megaw (ed.), *Proceedings of the 40th Annual Conference of the Ergonomics Society*, (Taylor & Francis, London) 138-143.

Steenbekkers, L.P.A. 1993, *Child development, design implications and accident prevention*, (Delft University Press).

Tainsh, M.A. 1995, Human factors contributions to the acceptance of computer-supported systems. *Ergonomics*, **38**-3, 546-557.

ANTHROPOMETRIC DESIGN OF A SIZE SYSTEM FOR SCHOOLFURNITURE

Johan Molenbroek and Yvonne Ramaekers

*Delft University of Technology,
Faculty Industrial Design Engineering,
Jaffalaan 9,2628 BX, Delft, Netherlands*

This paper is the result of an investigation by the Delft University of Technology for the Dutch Standardcommittee of Schoolfurniture. The purpose of this study was finding a scientific way to determine a size system for chairs and tables for educational institutions. We argue for the popliteal height to be a far better key dimension than the body height for defining module classes. Other design-critical measurements have been derived from literature. Based on these measurements, it is shown how a size system can be optimized.

The system proposed in this paper, is not complete. It is an impulse to a discussion about the way of designing a size system.

Introduction

Figure 1 compares the CEN-proposal (N47R,1992) to Dutch children in the age of 4-20 years. (Steenbekkers, 1993). The purpose of a standard for the dimensions of school furniture should be, that furniture designed corresponding to the dimensions in the standard, is compatible with the body-measurements of the students. The CEN-proposal should fit European children, but has large overlap of the modules. Through this overlap of body height classes it is possible, for example, that a student with a body height of 140 cm can choose between three chairs. This would almost certainly lead to wrong seating heights. Comparatively large toleration factors, as proposed in CEN N47R, can lead to an unfavourable shift in furniture dimensions. Another problem is the body height grouping. Module no. 4, according to the CEN-proposal, is suitable for children with body heights of 120 cm to 165 mm. These children have very different body measurements. We therefore conclude that the CEN-proposal lacks adjustment to the body dimensions of the target group.

Figure 2 compares the present Dutch standard NEN 3531 (1977) with the same group of children. The present CEN-proposal is less suitable for the Dutch children than the NEN

3531. The NEN (although this could be improved after 20 years) is still better adjusted to the dimension of the children, than the current CEN-standard.

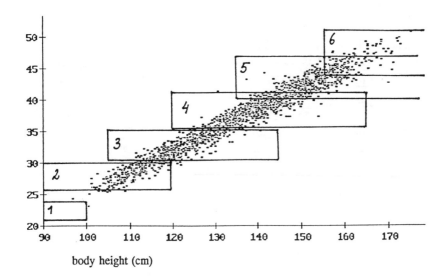

Figure 1 Popliteal height versus body height for Dutch children (age 4-12 from Steenbekkers) in comparison to the CEN size system

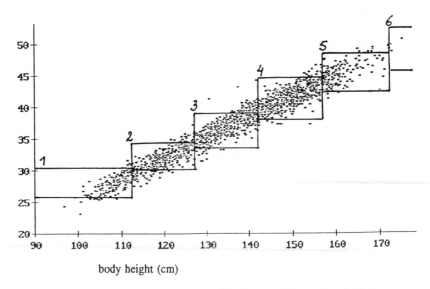

Figure 2 Popliteal height versus body height for Dutch children (age 4-12 from Steenbekkers) in comparison tot the current Dutch standard NEN 3531

Design-critical measurements for school furniture

Table 1 gives an overview of the furniture dimensions that are dependent on body dimensions.The seat height is further investigated in this paper to demonstrate the methodology involved.

Table 1: Critical body dimensions

	product dimensions		body dimensions
chair	h5:	seat height	1: popliteal height
	t4:	seat depth to back support	2: lumbar-popliteal depth
	y:	height frontal point back support	3: waist height while sitting
	h6:	height lowest point back support	4: buttock clearance
	h7:	height highest point back support	5: height lowest point scapula
	b3:	seat breadth	6: hip breadth, sitting
table	h2:	vertical span below table	7: thigh clearance sitting
	t2:	horizontal knee space	8-9: buttock-knee distance minus body depth
	t3:	horizontal clearance below the table	10: buttock-feet depth
	h1:	table height	11: elbow height sitting

Table 2: remaining measurements

	product dimensions
chair	r2: radius of back support
	r1: radius of front of seat
	β: angle between seat and back support
	δ: angle between seat and horizontal
table	b1: breadth
	t1: depth
	α: angle between desk and horizontal

The measurements in table 2 are not di related to static body measurements bu more to functional measurements and t biomechanical criteria

Starting-point for a new standard

In present standards, the body height is usually taken as the measurement to assign a module to a student. This assumes that the body height is a key measurement for the seat height. The body height of children in the Netherlands is normally measured about every year by school doctors/nurses for children aged 4-12 years. Since children in this age group can have a grow velocity about 8 cm a year, the question can be asked during which period this height value is a proper key to the right furniture.

Although the correlation between body height and popliteal height is large, it can vary trough age and when using popliteal height as we suggest, their is a direct hint for the user andt their caregivers to see if the chair fits. See figure 3. We propose therefore that the popliteal height is an easy key variable to be used.

re 3: Measuring the popliteal height.

Measuring the popliteal height as in figure 3 has not to be done in cm or mm but in size marks, which is quite easy, when one leg of a table is prepared according to the standard.

Description of the sources used
It would be perfect if we had a raw dataset from every European country with the relevant variables for schoolfurniture. Then we could make an European dataset an create a size system as we did for the Dutch data. But this data doesn't yet exist
However trough searching in Ergonomic Abstracts, in CHILDDATA (Norris and Wilson,1995) in ERGODATA(Pelsma,1987), in PEOPLESIZE (1993) and in own archives we found several sources. A selection of these fulfilled the foloowing crieria:
1. The investigated population should contain students in the age of 4-20 years.
2. The data should be as recent as possible.

3. The investigated population has to match the European target population: age, sex, (ethnic group) and activities should roughly be the same.
4. The sample should ideally contain at least 40 subjects per age group per sex and per measurement.

For the measurements elaborated in this paper 4 or 5 European countries per age group per measurement have been found. To form a mixed population several formulas where used from Roebuck et al. (1975) and (Molenbroek, 1994), which are left out this paper.The comparison of available 17 European sources is done (Ramaekers et al ,1993). In this paper only a short list is summarized hereafter:
The data from West-Germany (BRD) were collected between 1968 and 1974 by Jürgens, and are also published in the DIN 33402 (1981). The data from East-Germany (DDR) were collected during a long period of time, and were updated to 1982 by the authors Flügel et al.(1986). These data are very well suited to supplement missing values in data in other sources. The British (GB) data consist of processed data from samples between 1960 and 1977, according to Pheasant(1986). The Dutch (NL) data for the 4 to 12 year old children where collected by Steenbekkers; this investigation is representative for this age group in the Netherlands. In the investigation of Kemper, 600 students in the age of 12-17 years where examined during 4 years. These data are not representative for this age group in the Netherlands, but they are an indication of the measurements in this group (Kemper, 1985). The same applies to the investigation by Molenbroek of Dutch university students (18 years and over) (Molenbroek, 1994). Further there were also data found from Belgium, Switzerland, Spain, Italy,Norway Sweden and Turkey, some of these were traced after an expensive search in the online database 'ERGODATA' in Paris.
From comparison of the body heights and the popliteal heights, the Dutch measurements turn out to be the largest values.

Calculation of the module dimensions

We will now show how furniture dimensions are derived from our primary measurement: the popliteal height. In this paper this is done for the corresponding product dimension: seat height:
The seat height should be somewhat smaller than the popliteal height plus the heel height (about 2 cm) to avoid pinching of the veins and nerves in the upper legs. Therefore, the seat height of each module is derived from the corresponding smallest popliteal height without shoes. If the chairs are classified according to seat height, then the other critical chair dimensions (for example seat width) should be adapted to this.
The percentiles of body measurements mentioned below, belong to the extreme European populations. This implies that if, for example, the biggest width of a body measurement is significant for the product measurement, the maximum measurement of the broadest population is taken. In Ramaekers et al. (1993) was presented the P1 (first percentile), P50 and P99 popliteal height, per year and per country, of 4-7 year old European children.. It can be said that, in general, it is better for children to be seated a bit too low than too high. Furthermore, the popliteal height is an indicative measurement for the module and should be accurately recorded.

Therefore, the modules are arranged starting with the smallest values, which is 215 mm. If one assumes a maximum comfortable angle between upper and lower leg of 120° while seated, the other module heights (see table 3) can be derived from this.

Table 3 Module seat heights

no	seat height	popliteal height		step height
		minimum	maximum	
1	215 mm	215 mm	250 mm	35 mm
2	250 mm	250 mm	290 mm	40 mm
3	290 mm	290 mm	335 mm	45 mm
4	335 mm	335 mm	390 mm	55 mm
5	390 mm	390 mm	450 mm	60 mm
6	450 mm	450 mm	520 mm	70 mm
7	520 mm	520 mm	600 mm	

The calculated values are rounded off to 5 mm.

Comparison of calculated product dimensions against actual body dimensions

The seat heights are drawn in figure 4, which also shows the popliteal height plotted against the body height, as a way of comparison of this proposal with the present CEN-proposal (Dutch 4-12 year olds). This new proposal appears to be better fitted, the body height groups are smaller and there is less overlap.
This also shows again that the popliteal height is a better measurement to determine the correct module. If the body height were used here, students could still choose between two modules.

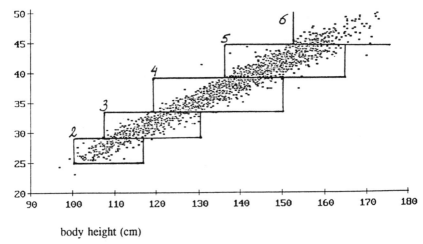

Figure 4 The fit of the new proposal for Dutch children age 4-12

Status
    At the end of 1995 the CEN standard committee has accepted our recomendation to use popliteal height as a key dimension, also that the their is a need for a higher size mark for the tallest student. Unfortunately their is still discussion going about the design of the size system .

References
CEN/TC 207/WG5/TG1 N47R, July 1992, *Chairs and tables for educational institutions,- functional dimensions and their test methods* (Danish Standards Association, Hellerup, Denmark).
DIN 33402, 1986, *Körpermaße des Menschen, Teil2* (Deutsches Institut für Normung)
Kemper, H.C.G. 1985, *Growth, Health and Fitness of Teenagers, Longitudinal research in international perspective,* Vol 20 in: Med. and Sport Science, (Karger, Basel) 66-80.
Flügel,B., H.Greil und K.Sommer, 1986, *Anthropologischer Atlas, Grundlagen und Daten DDR,* (Verlag Tribune,Berlin).
Molenbroek, J.F.M.1994, *Made to Measure, Human body dimensions for designing and evaluating consumer goods,* (Delft University Press, Delft) (in dutch with english summary).
NEN 3531 ,1977, *Schoolfurniture:Tables and chairs* (Dutch Standards Institution, Delft).
Norris,B and J.Wilson, 1995,*Childata, The Handbook of Child Measurements and Capabilities,* (Instute for Occupational Ergonomics,University of Nottingham).
Pheasant,S., 1986, *Bodyspace, Anthropometry, Ergonomics and Design,* (Taylor & Francis, London).
Ramaekers, Y, Molenbroek, J.F.M. and J.M. Dirken ,1993, *Investigation into a system of measures for chairs and tables for educational institutions.Document 56* (CEN/TC207/WG5/TG1).
Roebuck, J.A. K.H.E. Kroemer and WG Thomson,1975, *Engineering Anthropometry Methods* (Wiley,London).
Steenbekkers, L.P.A. ,1993, *Child development, design implications and accident prevention* (Delft University Press, Delft).

USE OF THE ADAPS COMPUTER MAN-MODEL BY EXPERT AND NOVICE USERS - a pilot study

Iemkje A. Ruiter   A.J.M. van der Vaart

Delft University of Technology
Faculty of Industrial Design Engineering
Department of Product and Systems Ergonomics
Jaffalaan 9, 2628 BX Delft, The Netherlands

We studied the way two groups of ADAPS-users (expert and novice users) used the ADAPS child-model to determine the maximum reach distance of a 4-year old child in a given posture.
The experts tend to predict larger reach distances than the novice users. Where the novices use the possible link rotations of the model quite uniformly, the experts show a larger variation.

Introduction

One of the research projects of the Faculty of Industrial Design Engineering of Delft University of Technology is ADAPS (Hoekstra, 1993; Ruiter, 1995a). ADAPS stands for "Anthropometric Design Assessment Program System". The ADAPS computer program enables the user to evaluate the dimensions of 3-dimensional workspaces with the use of computer man-models. These man-models consist of a frame (25 links, connected by joints) and surface points (Figure 1).

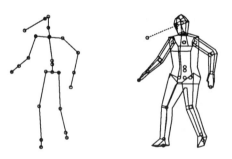

Figure 1. Frame and surface of the ADAPS model.

Recently we were confronted with a question asked by the Dutch Consumer Safety Institute: "Is it possible to use the ADAPS child-models (Ruiter, 1995b; Steenbekkers, 1995) to estimate the reach of children in order to create safety standards?". To answer this question it is necessary to have a definition of 'reach', to know how well the models match real children and to have insight in the way people use the models to predict the reach of children.

This paper deals with the third topic. We compared the way two different groups of ADAPS-users (expert and novice users) use the models. We wanted to find out if these two groups differ in their prediction of maximum reach (in cm). We were especially interested in the way the users translate a described posture (e.g. "sitting on the floor, reaching as far forward as possible") into a posture of the man-model: what use do they make of the possible rotations of the links and to what extent (angle) do they rotate these links.

Experiment

Two groups of ADAPS-users were formed, one group consisting of 6 expert users, the other of 7 novice users. The expert users were (mostly) recently graduated students of our Faculty who had used ADAPS intensively during their graduation project. Before the experiment started they got a short refresh course in ADAPS. The novice users were students who had enlisted for the ADAPS-course. To this group we showed how to manipulate the model. Each user was provided with a scheme of all possible ranges of motion of the links of the model.

Both groups got the same task. They had to imagine a 4-year-old child, sitting on the floor, legs apart, instructed to move an object on the floor (a toy-train) with one hand as far forward as it could manage. The task was to manipulate the child-model into this position in order to determine the model's prediction of the reach distance. To get an impression of the intra-individual variation the experts performed the task three times, with a one week interval.

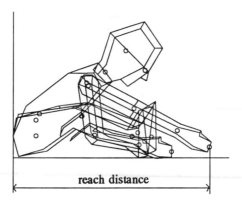

Figure 2. Definition of reach distance.

Table 1. Reach distances of novices	
novice	reach distance (cm)
1	68
2	68
3	69
4	69
5	70
6	72
7	72

Table 2. Reach distances of experts			
expert	reach distance (cm)		
	I	II	III
1	61	62	68
2	67	67	74
3	70	74	74
4	73	76	77
5	72	73	78
6	72	77	79

Results

Maximum reach.

The reach distance is defined as the horizontal distance between the outer points at the back of the buttocks and the marker on the edge of the model's hand (Figure 2). The results of the novice ADAPS-users are presented in Table 1, those of the experts in Table 2.

Use of degrees of freedom.

The reach distance of the model is determined by the orientation of the body as a whole (the orientation of the basepoint), of the trunk (consisting of the pelvic, lumbar and thoracic link), of the shoulder and of the arm. The orientation of the head does not influence the reach distance, neither does the orientation of the legs (this is true for the model, in real children the orientation of the head and legs might influence the orientation of the trunk).

We found that in 24 out of the 25 trials the arm was fully stretched, due to the fact that all users used the reach algorithm (a standard feature of the ADAPS program) for positioning the arm. Because we are interested in the difference in strategies both expert and novice users apply to manipulate the child-model into the described position, the results for the arm will be left out.

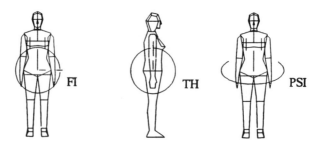

Figure 3. Possible rotations of the basepoint.

Table 3. Ranges of motion used by novice users.

novice	basepoint FI TH PSI			pelvic TH	lumbar		thoracic		shoulder	
	FI	TH	PSI	TH	FI	TH	FI	TH	FI	TH
1				30		30				
2				30		30				
3				30		30				
4				30		30				
5				30		30				
6				30		30				
7				29		29				

Table 4. Ranges of motion used by expert users.

expert		basepoint			pelvic	lumbar		thoracic		shoulder	
		FI	TH	PSI	TH	FI	TH	FI	TH	FI	TH
1	I	5			-20		25		25	30	
	II	-5			-10		30		30	30	
	III	10					20		15	35	
2	I	13							30	35	
	II	23								35	
	III				30				30	35	
3	I	-20			30		30		30	35	-10
	II				30		30		30	35	-10
	III				30		30		30	35	-10
4	I						30		30	35	
	II				30		30			35	
	III				30		30			35	
5	I	-15			15		30	30	30	35	
	II	-15			20		30	35	20	35	15
	III	-5			30		15	35	30	35	15
6	I			5	20		20	20	30		30
	II			10	15		20		30	35	30
	III			10	20		20		30	10	30

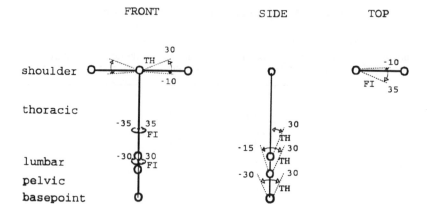

Figure 4. Possible rotations of the links of trunk and shoulder.

Figures 3 and 4 present an overview of the possible rotations (and their maximum ranges of motion) of the basepoint, the links of the trunk and the shoulder link. Table 3 presents the ranges of motion of the basepoint and the pelvic, lumbar, thoracic and shoulder links of the model as used by the novice users, Table 4 presents the same information from the expert users.

Conclusions

We wanted to know if there is a difference in the prediction of maximum reach as determined by expert and novice ADAPS-users. When we compare the maxima (out of the three attempts of each expert) predicted by the experts to the distances predicted by the novices we see that in general the experts tend to predict larger maximum reach distances. The range of the predicted distances is larger for the experts (11 cm) than for the novices (4 cm). The intra-expert ranges differ from 4 cm to 7 cm; they show a tendency to be smaller than the inter-expert range.

When looking at the strategies used to manipulate the model into the given position we see that the novices are suprisingly uniform in their approach. They do not use the basepoint, thoracic link and shoulder link and bend the pelvic and lumbar links to their maximum. The differences in reach distance therefore depend entirely on the position of the arm, which explains the relatively small range of the predicted reach distances.

The experts show much more variation in their strategies. Five out of six experts use a rotation of the basepoint in at least one of their attempts. In 5 cases the body is rotated backwards; they seem to introduce an inhibition of the rotation of the pelvis. Forward rotation of the three links of the trunk is used by almost all experts. Two of them also use a rotation around the length-axis of the thoracic link, probabely to simulate the extension of one side of the trunk when reaching with one hand.

All experts use rotation of the shoulder link. This might explain why the experts tend to predict larger reach distances than the novices who did not use the shoulder link at all. When we compare the three attempts for each expert we see that they tend to keep using the same strategy. They change details but do not use a completely different approach.

It will be clear from the results that there are many possible interpretations of a described posture. The main conclusion is that if you want to use child-models for the creation of safety standards, it is necessary to lay down the postures unambigiously.

Discussion

The results of the novices were thus uniform that we wondered if this could be due to the introduction they got. So we repeated the experiment with a new group of 6 novice users. During the introduction we paid extra attention to the possible shoulder rotations. The result was that this time 3 out of the 6 novices used the shoulder. Surprisingly this did not lead to larger reach distances.

Because we were interested in how far real children are able to reach forward we asked 5 children to perform the reach task. Compared to real children even the maximum of all predicted reach distances (79 cm) was too small; we found that one of the 5 children managed to obtain a reach distance of 87 cm (scaled to the body length we used for the model). Whether this is due to shortcomings in the fitting of the model or to the way the model is used is subject to further research.

References

Hoekstra, P.N. 1993, *ADAPS manual*, (Faculty of Industrial Design Engineering, Technical University Delft, Delft)

Ruiter, I.A. 1995a, Teaching the use of computer man-models. In L.M. Pacholsky (ed.), *Proceedings of the 12th Seminar of Ergonomics Lecturers*, (Poznan University of Technology, Poznan) 177-182.

Ruiter, I.A. 1995b, Development of computer man-models for Dutch children. In S.A. Robertson (ed.), *Proceedings of the Ergonomics Society's 1995 Annual Conference*, (Taylor & Francis, London) 328-333.

Steenbekkers, L.P.A. 1995, Anthropometry of children 2 - 12 years of age in the Netherlands. In S.A. Robertson (ed.), *Proceedings of the Ergonomics Society's 1995 Annual Conference*, (Taylor & Francis, London) 322-327.

Alarms

ALTERNATIVE MEANS OF PERSONAL IDENTIFICATION FOR AN ALARM SYSTEM

Simon Rubens

Siegel & Gale
27 Fitzroy Street
LONDON
W1P 5AF

Jane Dillon & David Coward

Royal Mail Technology Centre
Wheatstone Road
Dorcan
SWINDON SN3 4RD

Simulations of six different means of identification for an alarm system were developed and tested among Royal Mail employees. One means of identification used physical recognition while the other five required some recall or recognition on the part of the user. The findings indicate that users generally prefer the concept of physical recognition as it places no demands on memory or recognition, but only if the system can reliably confirm identification within 1 to 3 seconds. The means of identification with the most potential for immediate implementation is a variation on the conventional PIN system with icons added to the number keys. Users are able to exploit the icons to remember the sequence of key presses.

Introduction

In order to unset the alarm system protecting Royal Mail and Post Office Counters' sites, users must enter a 4 digit PIN (Personal Identification Number) within a fixed period of time (usually about 30 seconds). If they fail to enter the PIN within three attempts, or they exceed the fixed period of time, the alarm activates. The incidence of false alarms is a small but significant problem.

Initial research indicated the common cause of such alarm activations is that some users can become temporarily unable to recall their PIN or they transpose the digits on recall and entry.

In order to explore the possibilities for overcoming the problems experienced by users with the PIN system, several alternative means of personal identification were identified through brainstorming and tested alongside the conventional PIN. In total six means of identification were tested. They are

PIN- the user enters a 4 digit PIN using a conventional PIN key pad

PIN plus Icon - the user enters a 4 digit PIN using a conventional PIN number key pad but with icons added to the number keys (e.g., a house on the '1' key, a steam locomotive on the '5' key and a cat on the '9' key).

Password - the user enters a 6 character password using an alpha keypad with a QWERTY layout.

Personal Information - the user enters two pieces of personal information (place of birth, mother's maiden name) using an alpha keypad with a QWERTY layout.

Image Selection - the user recognises and selects 1 out of 9 related images on each of four successive multi-image displays.

Hand Recognition - the user places his or her hand on a metal plate until detection is confirmed by the system.

The conventional PIN, PIN plus Icon, Password, Personal Information and Image Selection means of identification are essentially active and cognitive while hand recognition is passive and physical.

Method

Apparatus

Simulations of each of the six means of identification were developed on an Apple Macintosh® using Hypercard®. Each simulation initially presented the subject with the image of a front door to a building and the instruction to click on the door handle to enter the building. The simulation then presented the subject with the image of a control panel appropriate to the particular means of identification. For testing on the cognitive systems, the user interacted with the computer by means of a mouse. For testing on the physical system, the user interacted with the computer by means of a mouse and a dummy hand detection plate, which appeared to be wired up to the computer.

Subjects

A total of 50 subjects were recruited from Royal Mail's Technology Centre. A different group of 10 subjects was tested on each of the five cognitive systems. The physical system was tested on 33 of the 50 subjects.

Procedure

The testing was conducted over three sessions at 7 day intervals.

The purpose of the first session was to introduce subjects to the study and get them to select and enter their identification data into the computer.

For PIN, PIN plus Icon and Password, subjects were required to choose their PIN or Password from five alternatives and were then prompted to enter it. For Personal Information, subjects were prompted to enter their place of birth and then their mother's maiden name. For Image Selection, subjects were prompted to select an image from each of the four successive displays of 9 related images. The act of selection also entered the data.

Subjects entered data with a mouse, but were advised that the actual system would require physical contact to be made with the control panel. Subjects clicked where they would expect to touch the panel. After subjects had entered their data, the simulated control panel confirmed that the data had been accepted.

At the second session the simulation presented subjects with the image of the front door of a building and an instruction to click on the door handle. After clicking on the handle, the simulation presented them with the control panel and an instruction to enter their identification data, or to select their choice of images to unset the system, as appropriate to the subject. If they did this correctly within three attempts, the simulated control panel indicated that the system was now unset. The computer recorded the time of each key press.

During the second session, subjects were introduced to Hand Recognition as a means of identification. The computer instructed subjects to place their left hand on the detection plate and click on an 'OK' screen button. The simulated control panel then indicated that the subject's hand characteristics were being read. After 10 seconds it indicated that they had been read and were entered into the system.

At the third session subjects were again tested on their cognitive means of identification and on Hand Recognition. For testing on Hand Recognition, the computer instructed subjects to place their left hand on the detection plate and click on the 'OK' screen button. After 10 seconds the simulated control panel indicated that the hand characteristics had been recognised and the system unset.

After testing in the third session, subjects rated, on 5-point scales, the ease / difficulty with which they were able to recall and enter their cognitive identification data in absolute terms and in comparison with a conventional PIN. Subjects were also asked to indicate preference for the hand recognition system vs. a means of identification that requires recall and entry of information (i.e., any cognitive one).

Results

Absolute success/failure

In session 1, one subject failed on Password and another on Image Selection. In session 2, one subject failed on PIN.

Time to enter data

Table 1. Mean time to initiate and complete entry (session 2)

Means of identification	Initiate (secs)	Complete (secs)
PIN	1.8	5.5
PIN plus Icon	2.6	6.3
Password	2.3	7.8
Personal Identification	2.8	23.7
Image Selection	2.1	9.9

Rating of ease/difficulty

Table 2. Absolute recall and entry (mean rating)

Means of identification	Recall	Entry
PIN	3.9	4.8
PIN plus Icon	4.2	4.4
Password	4.6	4.2
Personal Identification	4.9	4.4
Image Selection	4.0	4.5

1=very difficult, 2=difficult, 3=neither, 4=easy, 5=very easy

Table 3. Recall and entry compared to PIN (mean rating)

Means of identification	Recall	Entry
PIN plus Icon	3.8	3.2
Password	4.1	2.8
Personal Identification	4.6	2.9
Image Selection	4.1	4.1

1=much harder, 2=harder, 3=neither, 4=easier, 5=much easier.

Preference for physical vs. cognitive means of identification
Of the subjects who were tested on Hand Recognition, 76% preferred the concept of passive physical recognition to active cognitive identification.

Discussion

Conventional PIN
Initiation and completion of entry was quickest for the conventional PIN. It was rated as the hardest to recall but the easiest to enter. Only 1 subject failed to correctly enter their PIN at the second session but it is likely that subjects did not experience the same level of stress as in a real life situation. A minority of subjects indicated that they were unsure whether they were entering the correct PIN although in fact they did. Three subjects initially transposed the numbers but then entered the right sequence on their second or third attempt. Several of the subjects who were tested on the other cognitive means of identification commented on the problems that they have in recalling PINs for purposes such as a cash point.

The findings indicate that difficulty in PIN recall is a problem specific to certain people. For those who do not experience difficulty, PIN offers a quick and convenient means of identification.

Users prefer to be able to chose their own PIN and would like to be able to use the same one for all their identification needs.

Subject comments indicated that infrequency of usage and the need to hold several different PINs in memory affect ease of PIN recall.

PIN plus Icon
PIN plus Icon was considered easier to recall than a conventional PIN and almost as easy to enter. After PIN, it was the quickest to enter.

Four subjects found the icons of great help in recalling the key sequence and used them to construct and memorise a 'story'. One subject was dyslexic and found that his story helped him avoid transposing the digits on recall. Other subjects who do not experience difficulty in recalling PINs, chose to ignore the icons on the number keys as they but they were not distracted by them.

Several subjects pointed out that the most effective icons are those that directly relate to the number with which they are associated, e.g., the cat for the '9' key worked better than the steam locomotive for the '5' key.

Password
Subjects indicated that a password is easier to remember than a PIN because associations can more easily be made and because a word is remembered as a single unit rather than as a sequence of letters. Subjects also indicated that a password may be easier to recall because they are unlikely to store more than one in memory. The length of the word is not likely to affect recall but it slightly affects the time required to enter it.

All the subjects were familiar with a QWERTY keyboard. Entry of a password is likely to be harder for subjects who were not regular keyboard users.

Personal Information
Personal Information is considered 'very easy' to recall because the information is already encoded in memory and no conscious effort is associated with extraction of this information. Several subjects commented on the increased possibility of keying error due to the considerable length of the entry.

Although a majority (63%) of subjects tested on Personal Information expressed a preference for this system of identification over a conventional PIN several pointed out that they would prefer to use a PIN if the information was required to be entered on a regular (e.g., daily basis) as entry of personal information involves significantly more keying. Again, entry of a password is likely to be harder for subjects who were not regular keyboard users. This is reflected in the long mean completion time of 23.7 seconds.

Image Selection

It was notable that several subjects could not recall the images they had selected at the first session until they saw and recognised them on screen. As subjects had been able to select their initial choice of images, they could have selected ones that had particular associations for them. If images were allocated to subjects at random, their ability to recognise them might be reduced. The presentation of images by successive screen meant that subjects did not need to memorise a sequence. The size of each image made entry of data easy. The change in screen after each entry provided useful additional feedback that the entry had registered.

Hand Recognition

A significant majority of subjects (76%) expressed a preference for physical identification over any cognitive means, but only on the basis that it would be fast and reliable.

Several subjects were sceptical about system reliability particularly if their hand was cold, bandaged or dirty. The recognition time in the simulation of 10 secs was felt to be far too long. Subjects expressed concern that the system might be found at, some time in the future, to be harmful to health. Although subjects appreciated the lack of effort required on their part, they indicated that they would feel helpless if the system were to deny them entry.

One subject expressed concern over infringement of privacy and another expressed concern that such a system might leave one more vulnerable to physical attack (e.g., your hand could be cut off by criminals and used to gain entry). It is important to note that all the subjects work in a safe environment where the likelihood of attack is perceived as minimal.

Conclusions

The findings highlight the potential for a passive physical recognition means of identification, such as a system that recognises the characteristics of a person's hand, over an active cognitive one that requires accurate recall and entry of data. However, existing technology cannot provide the reliability and speed of recognition that users may demand.

Of the cognitive means of identification that were tested in this study, PIN plus Icon would seem to offer the greatest potential for immediate implementation. The icons are of assistance to those individuals who are prone to forget their PIN or transpose the digits but are unlikely to distract those who prefer to use the PIN in the conventional manner. Importantly, the PIN plus Icon system is almost as fast to use as the conventional PIN. Speed of entry is of particular importance to users if they feel vulnerable to attack. Finally, the cost of modifying control panels to the PIN plus Icon system would be small compared to the other means of identification that were tested

It is important to note that all of the subjects were of above average intelligence and familiar with technology and computers. As a next stage, trials should take place in a live environment with typical users of alarm systems.

WARNING LABEL DESIGN: TRADING OFF COLOUR AGAINST FONT SIZE

Judy Edworthy

Department of Psychology,
University of Plymouth,
Drake Circus, Plymouth,
Devon, PL4 8AA, UK

Austin Adams

Department of Psychology,
University of New South Wal
Sydney, New South Wales 20.
Australia

This study looks at the effects of the four simple design variables of font size, white space, border width and colour on the subjective perceived urgency of a set of warning labels in which the word 'WARNING' is printed in a range of font sizes as well as in two different colours regularly used on warning labels, red and black. The results show that the perceived urgency of the label increases monotonically as font size and border width increases, but that the amount of white space around the word has little effect. The use of red as opposed to black considerably increases the perceived urgency of the label. The effects of font size and colour are compared, and the rule of thumb for the designer is that a warning in black needs to be about twice the font size as one in red to be judged equally as urgent.

Introduction

There has been a considerable amount of interest in the effects of design features of warning labels on both subjectively-derived and objectively-observed dependent variables thought to influence warning use and compliance. Aside from the considerable interest in issues such as layout, comprehension, spatial and temporal proximity as well as many other variables, there has been some interest in relatively simple design variables such as hazard words (such as 'Danger', 'Deadly', 'Attention and so on), colour (whether a warning is presented in monochrome, or perhaps red, t most typical colour, or some other attention-getting colour), as well as the font size of the lettering conveying the warning. These types of variables typically impart relatively little information about the precise risk or hazard in question, but can serve as the reader's first cue to the importance of the risk or hazard, and may even direct their attention towards the hazard when otherwise this might not have been the case. Edworthy and Adams (1996) refer to these features as 'iconic' aspects of the warning A considerable amount is known about the effects of these variables both on subjective measures and directly observed compliance levels. For example, Silver and Braun (1993) found that 10-point print is perceived as being more readable than 8-point print and that a 2-point contrast between the signal word -- the word placed at the top of the warning -- is more effective in readability terms than a 4-point contrast. In the past, work on warning labels has typically considered monochrome warning labels, but mc recent research has shown that the colour of a warning label can have a considerable effect on its alerting qualities. For example, Chapanis (1994) showed that decreasing levels of hazard are associated with the colours red, orange, yellow and white. This h been demonstrated in various guises in a number of studies, most recently by Braun and Silver (1995) who have shown that the colour red is associated with the highest level of hazard, followed by orange, then black, then green, then blue. Braun and Silver's work also shows that actual compliance levels are affected by the colour of the warning. This work, along with a further study (Braun et al, 1994), shows that signal words and colour can interact so that the impact of a signal word can be enhanced or degraded by the colour in which it is shown, so that a specific word presented in red will be rated as more hazardous than that same word in, say, green. The importance of this to designers is that by charting these tradeoffs between colour and signal word, a variety of designs for specific products can be proposed which may be equally alerting, but can accommodate the design limits within which the designer may be working. The designer may be limited by, for example, the colours available for use, or the size of the final warning label. Braun et al's 'isoperformance curves' (1994) for colours and signal words thus have a great deal of use in practical warning design.

In the study presented here, we investigate potential tradeoffs between four important elements in warning label design so that rules-of-thumb may be generated for the designer faced with practical restrictions on the design of the warning label itself. Typically, a decision might need to be made about the colour of a warning label, and most likely this will be whether the label should be in monochrome, or have some of its elements in colour. The most likely colour, if any is used, is red as this has clear hazard and danger associations. Another feature the designer might have to make a decision about is the size of font used. In practice, restrictions will be placed on the font size possible by the information which needs to be conveyed by the label, as well as the product on which the label will eventually be placed as in many cases this product will be quite small -- for example, a pesticide bottle. Both of these variables were investigated in the experiments reported here, together with a third and fourth, the amount of white space around the signal word itself, and the thickness of a border surrounding the signal word, in order to begin to generate these rules-of-thumb. Subjective assessments of the perceived urgency of sets of warning labels were derived using a paradigm used in many studies of auditory warnings (e.g. Edworthy et al., 1991; Hellier et al., 1993).

Experiment One

Method

Seven levels of three variables were manipulated with respect to the signal word 'Warning', which headed a detailed set of warning instructions which might be found on a pesticide label. The instructions themselves were held constant in every way throughout -- it was only the word 'Warning', and its surrounding white space and border, which was varied in the study. The font size used varied from between 8 and 32 points in 4-point intervals; the amount of white space between the signal word and its surrounding border varied between 2 and 32 points in 5-point intervals; and the width of the border sorrounding the word 'Warning' varied from between 1 to 8 points in either 1-point or 1.25-point intervals, depending upon the border widths available from the desk-top publishing package used to produce the warning labels.

A set of 21 stimuli were generated in which the three variables were systematically varied. As a full 7x7x7 design would have produced too many stimuli to be judged, the mid-point of the two other variables was fixed as each of the seven levels of the variable of interest was manipulated. Thus for example the seven font sizes were varied systematically from 8 to 32 points whilst the white space and the border width values were fixed at 17 points and 4.5 points respectively in generating the stimuli representing the 7 font size levels. Font size was held constant at 20 points when either of the other two variables was being manipulated.

Participants, 24 in all, were asked to judge the extent to which the word 'Warning' as shown on the label contributed to the sense of urgency associated with the label by drawing lines to represent the perceived urgency of the label. Short lines represented low-urgency warning labels, whilst long lines represented high-urgency labels. The maximum length allowed was 374mm, which was the size of the page on which judgements were to be drawn. Participants were allowed to look through the labels before they started, and were required to rate each label twice, with a short break between each run of 21 stimuli. This procedure has been used and validated for assessing the perceived urgency of auditory warnings (Hellier et al., 1993; Hellier et al., 1995).

Results and Discussion

The effects of inter-observer variability caused by different choices of moduli wa removed by using a method advocated by Engen (1971). The transformed means vari from 29.5mm to 353.8mm, with a mean of 148.7mm and a standard deviation of 58.6mm. Each of the three variables was subjected to separate repeated-measures analysis, and polynomial contrasts were used to examine the trends for each of the variables. Table 1 shows the linear equations of best fit for each of the three variables For font size, there was only a significant linear effect ($F=120$, df=1, 23, $p<.001$). Fo white space there was also only a signficant linear trend ($F=6.55$, df=1, 23, $p=.02$). For border width there were significant linear ($F=72.18$, df=1, 23, $p<.001$), quadrat ($F=7.74$, df=1, 23, $p=.01$) and cubic ($F=4.72$, df=1, 23, $p=.04$) trends. However, t quadratic and cubic trends accounted for little of the variance.

Table 1. Linear best-fit functions using nominal as well as absolute x values

Variable	Absolute x values	Nominal x values	Variance accounted for
Font size	$y=6.3x+20.6$	$y=25.3x+45.5$	99.2%
White space	$y=.819x+135.4$	$y=4.1x+133.0$	71.1%
Border width	$y=10.7x+101.8$	$y=12.4x+100.3$	93.5%

In the case of all three variables, a large proportion of the variance was accounted for by the best-fit functions. When the three slopes were compared in pairs, it was found that white space differed from border width ($F=12.4$, df=1, 23, $p<.01$), border width differed from font size ($F=23.3$, df=1, 23, $p<.001$) and white space differed from font size ($F=120.$, df=1, 23, $p<.001$).

Table 1 shows the functions for the three variables in two forms. In the first, absolute values are shown. This is where the absolute value of x, in point size, is used in the equation. According to this, border width produces the largest exponent, suggesting that it has the greatest impact on the perceived urgency of the warning label However, the second column shows the regression analysis carried out with nominal, rather than absolute, values. In this analysis, the values 1 to 7 are placed along the x axis instead of the absolute point sizes. This approach is more ergonomically valid, because each of the three variables covered the whole of the range of point sizes that would have been possible given the ultimate size of the warning label (which would be determined, in turn, by the size of the bottle on which it would be placed). Thus the variable having the greatest impact in practice -- given the range over which it might typically vary -- is font size, rather than border width. White space has much less impact on either count.

This experiment therefore shows that font size has the greatest impact on the perceived urgency of a warning label, if account is taken of the range over which it is able to vary, given other constraints. Border width can have a significant impact, but is unable to vary over such a large range in practice. In fact, larger border widths quickly become unpleasing aesthetically (Adams and Edworthy, 1995) and are thus unlikely to be used. White space appears to have little impact on the perceived urgency of warning labels, and so for practical purposes can be largely ignored if the urgency of the label i the main focus of concern.

Experiment 2

It is well-documented that colour affects the perceived urgency, arousal strength and compliance rates associated with warning labels (e.g. Chapanis, 1994; Braun and Silver, 1995). In the first study presented here three features of monochromatic warning labels were explored. In the second study, the effects of a single colour -- red -- was explored in a similar study. Here, the white space variable was ignored because it was shown in Experiment 1 to have little impact on the perceived urgency of the warning labels even though it varied as much as was possible given the total size of the warning label.

Method
The same 7 levels of font size and border width were investigated as in Experiment 1. Two sets of 14 stimuli were generated, one for labels in monochrome, as before, and another where the word 'Warning' was in red throughout. As before, only the middle value of the second variable was used whilst the other was being systematically varied. Therefore the 7 font sizes were each portrayed in a warning with a border width of 4.5 points, and the 7 border widths were each portrayed with a font size of 20 points. A total of 28 stimuli were thus generated.

Participants, 24 in all, were asked to rate the perceived urgency of each of the 28 warning labels by drawing a straight line commensurate with the urgency of the label as before. Each stimulus was presented twice, once in each run, with a rest in between.

Results and Discussion
As the study was completely within-subjects, the data was subjected to a repeated-measures analysis using polynomial trend contrasts for the seven-valued variable. The results showed that the red stimuli were rated as significantly more urgent than the monochrome stimuli ($F=150.8$, $df=1$, 23, $p<.001$). The only significant trends in the data were linear ones. The best-fitting straight lines, and the proportion of variance accounted for, were as follows: Red font size, $y=15.9x+73.6$, $R^2 = .99$; monochrome text size, $y = 12.7x+47.2$, $R^2 = .99$; red border width, $y = 8.9x+100.4$, $R^2=.98$; and monochrome border width, $y=6.4x+70.8$, $R^2=.90$. In each case, x refers to the nominal, rather than the absolute, x values for reasons stated earlier. There was also a significantly greater slope for the red than for the monochrome stimuli ($F=20.0$, $df=1$, 23, $p < .001$).

The best-fitting trends allow the calculation of trade-offs between colour and font size. These two dimensions are, not surprisingly, those which have the greatest impact on the perceived urgency of warning labels in this study. Thus for text size, a black signal word would need to be 17.3 points in size to be equally urgent to a red one in 8-point size. For a red signal word to be as urgent as a black one of 32 points, it would need to be 19.8 points in size. The contrast is even more striking for border width as a red border of only 1.84 ponts in width is of equivalent urgency to one of 8 points. A black border would need to be 6.05 points in width to be equally urgent to a red one of 1 point.

General Discussion

The studies presented here have demonstrated that some design features have a greater impact on the perceived urgency of warning labels than others. Specifically, font size and colour have greater effects that border width and white space. Border width in absolute terms has a considerable impact on perceived urgency, but in practice it can range over such a small range before it becomes aesthetically unpleasing that it has relatively little use in practical design. Because the methodology involved psychophysical techniques, it is possible to calculate precise tradeoffs between design variables and the most useful of these is that a signal word in black needs to be approximately twice as large as a signal word in red in order to be equally as urgent. This is useful for design purposes, for if a designer is able only to design in monochrome then he or she would be well advised to use relatively large print, at lea~ for the signal word(s), if the product is relatively hazardous. By contrast, a designer working on the design of a small warning label (which may, of necessity, be small because of the size of the product on which it is to be placed) may find that the warni will have a greater impact if the hazard word(s) are presented in red rather than in monochrome.

We have also considered the issue of cross-variable comparisons, and have argued that because each of the variables under consideration effectively covered the entire range in each case, then it is more appropriate to compare their effects in relativ rather than in absolute, terms. This argument is developed elsewhere (Edworthy and Adams, 1996). Finally, further dimensions to this study (reported in Adams and Edworthy, 1995) considered both composite stimuli and aesthetic reactions to the warning labels. Responses to the composite stimuli largely replicate the studies repor here, and the aesthetic judgement studies show that designers find designs using mid values of variables generally to be more aesthetically pleasing than those which use th most extreme values. Thus the practical concern of designing labels which are both perceived as urgent, and are well-designed and pleasing, presents the researcher with new and intriguing research questions.

References

Adams, A.S. and Edworthy, J., 1995, Quantifying and predicting the effects of basic text display parameters on the perceived urgency of warning labels: Tradeoffs involvi font size, border weight and colour. Ergonomics **38**, 2221-2237.

Braun, C.C., Sansing, L., Kennedy, R.S. and Silver, N.C. 1994, Signal word and colour specifications for product warnings: An isoperformance application. *Proceedings of the 38th annual conference of the Human Factors and Ergonomics Society*, (Human Factors and Ergonomics Society, Santa Monica) 1104-1108.

Braun, C.C and Silver, N.C. 1995, Interaction of signal word and and colour on warning labels: differences in perceived hazard and behavioural compliance. Ergonomics, **38**, 2207-2220.

Chapanis, A. 1994, Hazards associated with three signal words and four colours on warning signs. Ergonomics, **37**, 265-276.

Edworthy, J. and Adams, A.S. 1996, *Warning Design: A Research Prospective.* (Taylor & Francis. London).

Edworthy, J., Loxley, S.L. and Dennis, I.D. 1991, Improving auditory waring desig Relationship between warning design parameters and perceived urgency. Human Factors, **33**, 205-231.

Engen, T. 1971, Scaling methods. In J Kling and L Riggs (eds) *Experimental Psychology* (Methuen, London).

Hellier, E.J., Edworthy, J., and Dennis, I.D. 1993, Improving auditory warning design: Quantifying and predicting the effects of different warning parameters on perceived urgency. Human Factors, **35**, 693-706.

Hellier, E.J., Edworthy, J. and Dennis, I.D. 1995, A comparison of different techniques for scaling perceived urgency. Ergonomics, **38**, 659-670.

Silver, N.C. and Braun, C.C. 1993, Perceived readability of warning labels with varied font sizes and styles. Safety Science, **16**, 615-625.

THE DESIGN OF PSYCHOLOGICALLY APPROPRIATE AUDITORY INFORMATION SIGNALS

Elizabeth Hellier & Judy Edworthy

Department of Psychology,
City University,
London, EC1V OHB.

Department of Psychology,
University of Plymouth,
Plymouth, Devon.

In previous research, changes in the acoustic parameters that are commonly used in the design of auditory warnings and information signals have been mapped to changes in semantic associations. In this paper, knowledge of how acoustic changes affect listeners interpretations of sounds is used to design a set of trend monitoring sounds (trendsons) for the helicopter flight environment. Four trendsons that are psychologically appropriate for the situations that they monitor are proposed. For example, a trendson for Rotor Overspeed is proposed that increases in speed, fundamental frequency and regularity of rhythm. These acoustic changes have been shown to be associated with meanings that are psychologically appropriate to the condition of rotor overspeed. The potential advantages for the learning and retention of auditory information of designing by reference to psychological appropriateness is discussed.

Introduction

Some previous research has investigated how changes in simple acoustic stimuli invoke changes in the semantic associations which a listener attributes to those sounds (e.g. Loxley 1991, Edworthy, Hellier & Hards 1995). In particular, Edworthy et al 1995 revealed how changes in acoustic parameters commonly used in the design of auditory warning and auditory information signals are mapped to changes in semantic associations. Identical stimuli were constructed which varied dynamically along one of four acoustic parameters, speed, pitch, rhythm or inharmonicity, while all other aspects of the stimuli were held stable. Eight different stimuli, ascending speed, descending speed, ascending pitch, descending pitch, slowing down, speeding up, ascending inharmonicity and descending inharmonicity were presented to participants who were required to indicate how appropriate a range of adjectives were as descriptors of the stimuli. In this way, different levels of each acoustic parameter were mapped to changes in meaning.

This work revealed two important sources of information that can be used to facilitate the design of auditory information signals. Firstly, the meanings most highly associated with each acoustic parameter were revealed. For example, it was shown that the adjectives 'rising', 'controlled' and 'straining' were rated as most appropriate descriptors of an ascending pitch stimulus. Secondly, analysis was also conducted between acoustic parameters to reveal, in instances where an adjective was rated as an appropriate descriptor of more than one stimulus, which stimulus it described best. For example, 'dangerous' was a more appropriate descriptor of ascending speed than of any of the other stimuli.

This data has direct implications for the design of auditory information. Some adjectives were more appropriate descriptors of some stimuli than of others. This shows the semantic reactions likely to be invoked by acoustic changes as well as the relative strengths of these associations. This is useful for design purposes, for example, the results can be applied to the design of auditory information signals so that they convey appropriate semantic cues for the situation that they are indicating. The data also shows which of several stimuli are most associated with a particular adjective, and can guide the design of auditory information signals in instances where a particular semantic reaction is desirable, but might be communicated by two or more parameters. The designer would be able to know which acoustic parameter to select to communicate the desired meaning, that is, which parameter was most strongly associated with the desired meaning. Data of this type thus facilitates the design of psychologically appropriate auditory information signals that invoke semantic reactions appropriate to the sense of the situation that they are indicating. The application of this information to the design of psychologically appropriate trend monitoring sounds is described below as an example.

Trend Monitoring Sounds

Trend monitoring sounds ('trendsons') provide auditory feedback to an operator by sounding when a physical parameter being monitored begins to exceed normal limits. They are precursors to warnings, their primary function is to provide immediate feedback to the operator on the nature of a physical trend; whether it is going up, or down, and ideally the speed with which it is changing. Thus trendsons change as a physical parameter changes, mirroring changes in the physical parameter through the sound itself.

In order that trendsons are psychologically appropriate, and thus more easily and quickly learned, identified and retained, they could be designed to provide an 'auditory metaphor' for the situation being monitored by invoking that situation through an acoustic structure that results in predictable and appropriate connotations for the operator (Edworthy et al 1995). For example, an increase in trendson speed could be used to indicate an increase in helicopter rotor speed. When the rotors are rotating too fast (rotor overspeed), the trendson could be analogous to the situation being monitored because the increasing trendson speed would mimic actual increases in rotor speed; the trendson would also be appropriate to changes in the situation in a more abstract, psychological sense because increasing trendson speed would indicate an increasingly urgent situation as rotor speed increased. It is well established that increases in the speed of an acoustic stimulus increases its urgency (Edworthy et al, 1991; Hellier at al 1993; Momtahan, 1990).

Four different trendsons for the helicopter flight environment have been designed so that they are psychologically appropriate, by reference to the semantic reactions that the acoustic changes used to construct the trendson invoke. These trendsons are described below.

Rotor Underspeed

In the case of rotor underspeed, a trade-off between appropriate and contradictory semantic associations for possible indicating parameters exists. An analogous trendson for rotor underspeed might decrease in pitch and speed to indicate a sense of slowing down and falling. However, decreases in both pitch and speed also communicate decreasing urgency, which is contradictory to the sense of the trendson (which indicates an increasingly urgency situation). This dilemma was resolved by using one of the indicating parameters (pitch) to indicate connotations that are analogous to the situation being indicated (falling, dropping etc.) and the other indicating parameter, speed, to indicate increasing urgency. The indicating parameters for rotor underspeed and their semantic associations are shown in Table 2. It is important to note that without research into the semantic associations of different acoustic stimuli, the potential for contradictory information in the sounds that designers create would remain unacknowledged. Now the effects of potentially contradictory information can be minimised and neutralised.

Table 2. Semantic Associations of Indicating Parameters used to Communicate Rotor Underspeed

Indicating Parameters	Semantic Associations
Increasing Speed	Dangerous, fast, urgent, faltering, high, jerky, straining
Descending Pitch	Dropping, falling, controlled, powerful, steady

Positive G

The indicating parameters that were selected to communicate positive-G were decreasing inharmonicity and slowing rhythm. Table 3 shows the semantic associations shown to be highly associated with those parameter changes. For this trendson all of the connotations invoked by the parameter changes were analogous to the situation of increasing G-force, there were no associations with more psychological attributes of the trendson such as increasing urgency.

Table 3. Semantic Associations of Indicating Parameters used to Communicate Positive-G

Indicating Parameters	Semantic Associations
Decreasing Inharmonicity	Full, heavy, powerful, solid
Slowing Down	Heavy, straining, tight

Negative G

The indicating parameters selected to communicate negative-G were increasing inharmonicity and increasing pitch. Table 4 shows the semantic associations shown to be highly associated with those parameter changes. The indicating parameter changes resulted

Psychologically Appropriate Trendsons

Each trendson was constructed from a short pulse of sound. The pulse was 100 ms. in length, with a regular envelope and eight regular harmonics. This basic pulse of sound was repeated six times to form a short burst of sound. A trendson consisted of five bursts of sound, each burst representing a different level of the trendson. The time course of a trendson mirrors the physical changes in the parameter that it is monitoring by changing from Level 1, when the parameter begins to exceed normal limits, through to Level 5, when the parameter is at the boundary of safe limits and warning is imminent, as the physical parameter changes. For each trendson, different indicating parameters were manipulated to indicate change from Level 1 through to Level 5, while other aspects of the trendson were held stable.

Consultation with pilots and aerospace scientists elucidated connotations that were appropriate to the four trendsons, rotor overspeed, rotor underspeed, positive-G and negative-G. For each trendson, indicating parameters were selected to communicate meanings which were psychologically appropriate to the sense of the situation being indicated on the basis of the research described above. Where possible, parameter changes were selected which invoked connotations that were both analogous to the situation being indicated (for example 'heavy' for positive-G) as well being appropriate in a psychological sense (for example 'urgent' or 'dangerous' for an increasingly hazardous situation). It was not always possible to achieve both. Parameter changes that resulted in some connotations that were psychologically appropriate for the situation being indicated but other connotations that were contradictory were avoided unless the appropriate connotations were demonstrably stronger than the inappropriate connotations.

Rotor Overspeed

The indicating parameters (those aspects of the sound that changed between Levels 1 and 5) which were selected to communicated rotor overspeed were increasing speed, increasing pitch and increasingly regular rhythm. Table 1 shows the semantic associations that have been shown to be highly associated with those parameter changes by previous research. All of the indicating parameter changes result in connotations that are appropriate to the sense of the trendson rotor overspeed. Some of these connotations such as 'fast' and 'jerky' are analogous to the situation itself, while others, such as 'dangerous' and 'urgent', are appropriate to the trendson in a more abstract psychological sense.

Table 1. Semantic Associations of Indicating Parameters used to Communicate Rotor Overspeed

Indicating Parameters	Semantic Associations
Increasing Speed	Dangerous, fast, urgent, faltering, high, jerky, straining
Increasing Pitch	Dangerous, high, rising, straining, urgent
Increasing Regularity	Jerky

in some connotations that were appropriate to the sense of the trendson negative-G. Although the connotations indicated in Table 4 in italics were more contradictory to the sense of the trendson these effects were thought to be outweighed by the more appropriate connotations which were stronger.

Table 4. Semantic Associations of Indicating Parameters used to Communicate Negative-G

Indicating Parameters	Semantic Associations
Increasing Inharmonicity	Empty, light , loose, weak, *controlled, safe, steady*
Increasing Pitch	High, straining, rising, *controlled*, full, jerky, steady

Conclusions

The examples above have focused on trendson design to demonstrate how knowledge of the semantic associations invoked by different sounds can be used to design auditory information signals that are appropriate to the situations that they indicate. It is envisaged that by designing auditory information that is psychologically appropriate to the situation that it is indicating we might be creating more informative auditory information that 'sounds like' the situation it indicates and is thus easier to learn and retain. Furthermore, an awareness of the semantic associations that result from auditory changes which a designer might propose can help to prevent us from building contradictory information into auditory signals. Such contradictory information may present an ambiguous signal to the operator and result in response uncertainty.

References

Edworthy, J., Loxley,S. & Dennis,I. 1991, Improving Auditory Warning Design : Relationships Between Warning Sound Parameters and Perceived Urgency. *Human Factors*, **33(2)**, 205-231.

Edworthy, J., Hellier, E. & Hards, R. 1995, The Semantic Associations of Acoustic Parameters Commonly used in the Design of Auditory Information and Warning Signals. *Ergonomics*, **38 (11)**, 2341-2361.

Hellier, E., Edworthy, J., & Dennis, I.1993, Improving Auditory Warning Design: Quantifying and Predicting the Effects of Different Warning Parameters on Perceived Urgency. *Human Factors*, 35(4), 693-706.

Loxley, S. 1991, *An Investigation of Subjective Interpretations of Auditory Stimuli for the Design of Monitoring Sounds*. Unpublished Masters Thesis, University of Plymouth.

Momtahan, K. 1990, *Mapping of Psychoacoustic Parameters to the Perceived Urgency of Auditory Warnings*. Unpublished MA thesis, Carleton University, Ottawa, Canada.

Risk and Error

A HUMAN RELIABILITY ASSESSMENT SCREENING METHOD FOR A CANADIAN REACTOR UPGRADE

Fiona M Bremner
Carolyn J Alsop

AECL
Chalk River Laboratories
Chalk River
Ontario, Canada, K0J 1J0

The National Research Universal (NRU) nuclear reactor is one of the world's largest and most versatile research reactors. NRU began operating in 1957, and is being upgraded as part of AECL's continuing commitment to operational safety. As part of these upgrades, Probabilistic Safety Assessment studies are being carried out, and it has been recognized that the assignment of Human Error Probabilities is an important part of these studies This paper provides a summary of the Human Reliability Analysis (HRA) method, which has been developed for the identification, quantification, screening and reduction of human errors. This HRA method has been developed from existing methods and revised to take advantage of NRU operational experience. This paper discusses the development of the method, its key features and its testing and evaluation.

Introduction

The National Research Universal (NRU) Reactor is a 130 MW, low-pressure, heavy-water-cooled and -moderated research reactor. The reactor is used for research, both in support of Canada's CANDU development program, and for a wide variety of other research applications. In addition, NRU plays an important part in the production of the world's medical isotopes; e.g., generating 80% of worldwide supplies of molybdenum-99

NRU is owned and operated by Atomic Energy of Canada Ltd. (AECL), and is currently being upgraded. As part of these upgrades, both deterministic and probabilistic safety assessments are being carried out. It was recognized that the assignment of Human Error Probabilities (HEPs) is an important part of the Probabilistic Safety Assessment (PSA) studies, particularly for a facility whose design predates modern ergonomic practices, and which will undergo backfitted modifications.

A simple Human Reliability Assessment (HRA) method was used in the initial safety studies; however, following review of this method within AECL and externally by the regulator, it was judged that benefits could be gained for future error reduction by including additional features.

The project consisted of several stages: needs analysis, literature review and development of method (including testing and evaluation). This paper discusses each of these stages in detail.

Needs Analysis

The NRU Reactor has 38 years of operating experience, but information has not been recorded in a form suitable for HRA use. Therefore, a method was required that would supply generic data from the literature, with a weighting based on the extensive operating experience.

The needs analysis performed at the start of the project provided a number of criteria based on past experience of the existing method and on the future needs of the project. These criteria were defined as follows. The NRU HRA method should:
* be an extension of the original simple method,
* be directed towards incorporating human factors considerations into the assessment of existing systems and the design of upgrades,
* be able to quantify both diagnostic and execution errors,
* make effective use of limited resources,
* be adaptable for use throughout AECL,
* be compatible with other Canadian nuclear industry methods,
* be useful as a coarse screening approach for setting system 'design' targets, and
* be useful as a fine-tuning HRA method, as new system details evolve.

Review of Existing Methods

A number of available HRA methods were examined for their suitability; they included the Darlington Probabilistic Safety Evaluation (DPSE) (King, Raima and Dinnie, 1987), Human Error Assessment and Reduction Technique (HEART) (Williams, 1988), Technique for Human Error Rate Prediction (THERP) (Swain and Guttmann, 1983), and an internal HRA method applied to CANDU reactors.

Each method was rated against the specific NRU HRA acceptability criteria above, as well as more general criteria, derived from Humphreys (1988). The review indicated that no single method was appropriate for the specific needs of the NRU Upgrade Project. However, the elements of best practice were elicited for inclusion in the NRU HRA method. These elements are listed as follows:
* a comprehensive, mutually exclusive, human error classification scheme,
* a database of generic HEPs,
* a broad range of Performance Shaping Factors (PSFs),
* a method of modelling dependency between human errors,
* auditability, and
* identification of practical means of reducing the error likelihood.

The methods reviewed were then re-examined for the best representation of each of these elements. These elements were then incorporated into the NRU HRA method.

Development of the Method

One of the principal features of the NRU HRA method is that errors are classified into pre-accident errors, which encompasses operational and maintenance errors, and post-accident errors made during remedial action in a situation following an incident. Pre-accident errors were judged to be dominated by errors in task execution; however, post-accident errors were subdivided into diagnosis and execution parts.

The major steps in the method were identified as:
- Identification of Errors,
- Assignment of a Basic HEP,
- Assessment of PSFs,
- Assessment of Dependency between Errors,
- Assessment of Recovery Options, and
- Application of Error Reduction Mechanisms.

Identification of Errors

The error classification scheme for execution errors was developed from Swain & Guttmann (1983), and the scheme for diagnosis errors was derived from Reason (1990).

Assignment of a Basic Human Error Probability

In the draft revision of this method, the Basic Human Error Probabilities (BHEPs) for execution errors were taken from HEART (Williams, 1988). In HEART, these BHEPs are then multiplied by negative PSFs (i.e., those which may affect performance adversely) as appropriate. However, following the pilot study and correspondence with other HEART users, it was felt that:
- Particularly for post-accident errors, the resulting HEPs were inconsistent with other methods, (e.g., THERP, DPSE, etc.).
- Inter-user consistency was found to be low when users were simply asked to pick the most suitable task description. This was due in part to several task descriptions being equally appropriate for the selected NRU tasks.
- When experts were using HEART, they tended to only use two of the BHEPs, and this choice was based mainly on the probability value and not the task description.

As a result of these findings, it was concluded that this lack of comparative validity and consistency could present a problem. A brief literature search was then carried out to investigate existing BHEP databases in an effort to identify a simple comprehensive database 'free' of negative PSFs, to which the appropriate PSFs could then be applied. The results of the literature search are shown in Figure 1. Most of the databases in Figure 1 contain negative PSFs, particularly at the more knowledge-based end of the spectrum (e.g., SHARP (Hannaman and Spurgin, 1984). This was taken into account, and the BHEPs selected for use in the NRU method were extrapolated from the data in Figure 1.

A common basis for many of the databases was shown to be the complexity of the task. 'Complexity' can encompass several factors, but in practical terms, this was as valid a scale as could be found in the literature. In the NRU method, in order to choose between levels of complexity, a flowchart to guide the analyst was developed. The BHEPs included in the NRU HRA method are listed in Table 1.

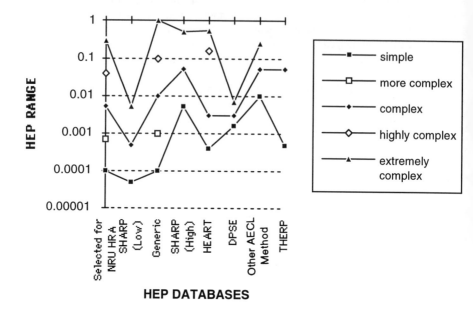

HEP DATABASES

Figure 1: A Comparison of HEP Databases

Table 1: Execution Basic Human Error Probability Database

BHEP Task Description	BHEP
Knowledge-based, extremely complex task	0.4
Rule-based, highly complex task within a long procedure	0.2
Rule-based, highly complex task within a short procedure	0.06
Rule-based, complex task within a long procedure	0.02
Rule-based, complex task within a short procedure	0.008
Skill-based, less complex task within a long procedure	0.003
Skill-based, less complex task within a short procedure	0.0009
Skill-based, simple task within a long procedure	0.0003
Skill-based, simple task within a short procedure	0.0001

Table 2: Diagnosis Basic Human Error Probability Database

Time	Basic Human Error Probability
0-15 minutes	1.0
16-20 minutes	0.1
21-30 minutes	0.01
31-60 minutes	0.001
>61 minutes	0.0001

The BHEP database for diagnosis errors is based on the Time Reliability Correlation (Dougherty and Fragola, 1988). The Nominal Model was used (listed in Table 2), onto which negative PSFs are applied.

Assessment of Performance Shaping Factors (PSFs)

The potential PSFs and the method in which they are applied to modify the BHEP were taken from HEART (Williams, 1988). HEART lists thirty-five 'Error Producing Conditions', but not all were relevant to NRU. This list was reduced to nineteen relevant PSFs. It was considered very important to relate the choice of PSFs to the experience of activities within NRU. Therefore, a survey of staff was carried out in order to identify which task-specific PSFs should be included in the HRA screening method. The PSFs selected for inclusion are listed in Table 3.

Table 3: Summary of PSFs incorporated into NRU HRA Method

Pre-Accident	Post-Accident Diagnosis	Post-Accident Execution
Poor Procedures	Unfamiliarity	Unfamiliarity
Unfamiliarity	Information	Poor Procedures
Lack of Job Aids	Overload	Poor Feedback
Insufficient Checking	Poor Procedures	Design Mismatch*
Poor Feedback	Excess Alarms	
Design Mismatch*		

* A mismatch between the operator's mental model of the plant and the actual design

In order to ensure that users are as consistent as possible, guidance tables were produced regarding the appropriate weighting of each PSF, based on actual NRU experience and human factors experience.

A criticism of the application of PSFs in HEART is that the multiplication of factors can cause HEPs to exceed 1.0. HEART deals with this in a simplistic cut-off manner. For the NRU HRA method, the asymptotic function $HEP = 1 - e^{-p}$ for $p > 0.1$ was used for modelling higher probability errors

Assessment of Dependency between Errors

Dependency was felt to be necessary to allow assessors to take credit in the PSA for changes in the task that would result in decreased potential dependency. THERP's dependency model was one of the few available, and fulfilled the NRU criteria. The relevant parts were selected and incorporated into the NRU method. To increase inter-user consistency, guidance has been given for the assessment of dependence as complete, high, moderate, or zero for pre-accident errors, and complete or zero for post-accident errors.

Assessment of Recovery Options

Recovery steps, (e.g., component surveillance or verification inspection programs that are independent of the maintenance or testing task itself) are modelled separately from the relevant error. They are assigned an HEP of 0.1, a figure commonly given to inspection tasks in Canadian HRA methods and consistent with values in THERP.

Application of Error Reduction Mechanisms

Error Reduction Mechanisms are identified by assessing where the main contribution to the error probability comes from. Error Reduction Mechanisms can not only be applied to the PSFs, but also to the dependency levels, and BHEPs. Specific NRU guidance is given on each of the BHEPs, PSFs, and dependency levels to indicate the most appropriate mechanism to reduce the error probability.

Testing and Evaluation

Once the method was developed, two evaluations were carried out, to ensure that the method was reliably consistent with a range of users and valid in terms of comparisons with other methods. To fulfill the first objective, a pilot study was carried out. Four realistic potential errors were identified and supplied, along with necessary background information, to ten assessors/engineers who were asked to quantify the errors using the revised method. The results identified a number of areas where the model was weak, and a number of changes were incorporated. The second evaluation was then performed by comparing our developed method against a number of well-known HRA techniques, by quantifying the same four errors. The results showed that the NRU method fell within the range of the other methods.

Conclusions

The NRU HRA screening method described in this report has been developed to address a number of criteria, all of which have been met. It is resource-efficient, gives initial direction toward the identification of risk-significant errors, and provides a means of reducing the potential for those errors. The NRU HRA method takes advantage of over 38 years of operating experience whilst benefiting from the use of established methods. The method was implemented in 1995 and will now be used in all PSA studies within the NRU Upgrade Project. The method will continue to be monitored and, if necessary, improved, to ensure that it achieves its targets of quantifying, screening and, most importantly, reducing human errors in the NRU Reactor.

References

Dougherty EM Jr and Fragola JR, 1988, Foundations for a Time Reliability Correlation System To Quantify Human Reliability, *Proceedings of the 1988 IEEE fourth conference on human factors and power plants.*.
Hannaman, GW, and Spurgin, AJ, 1984, *Systematic Human Action Reliability Procedure (SHARP)*, EPRI NP 3583, Palo Alto, California.
Humphreys P, (ed) 1988, *Human Reliability Assessors Guide*, SRD, Culcheth, , England.
King, FK, Raima, VM, and Dinnie, KS, 1987, *Darlington Probabilistic Safety Evaluation - CANDU Risk Assessment*, 8th Annual CNS Conference, June 1987, St. John, New Brunswick, Canada.
Reason, J, 1990, *Human Error*, Cambridge University Press.
Swain AD and Guttmann HE, 1983, *A Handbook Of Human Reliability Analysis With Emphasis On NPP Applications*, USNRC NUREG/CR-1278, Washington DC.
Williams JC, 1988, The HEART Methodology Explained, *Proceedings Of The IEEE Fourth Conference On Human Factors And Power Plants*, Monterey, California.

MAPPING KNOWLEDGE UTILISATION BY NUCLEAR POWER PLANT OPERATORS IN COMPLEX SCENARIOS

Fiona Sturrock and Barry Kirwan

Industrial Ergonomics Group
School of Manufacturing & Mechanical Engineering,
University of Birmingham, Birmingham, B15 2TT. United Kingdom.

This study aimed to map the different types of knowledge real nuclear power plant operators used during diagnosis of three complex emergency event scenarios, in an attempt to gain insights into how operators problem solve, and why in some circumstances diagnosis fails. Six categories of knowledge were used to map the knowledge operators were perceived to activate during their diagnoses. The results revealed individual differences in the types and levels of knowledge activated during diagnosis, and also differences in the success of diagnosis. A better understanding of the types of knowledge used by real operators has implications for ergonomics areas such as operator training and interface design.

Introduction

Any complex or dynamic process is not guaranteed to run trouble-free even with the most sophisticated process control equipment. Thus an essential feature of any central control room is the operator who has the unique ability to diagnose in dynamic abnormal events and emergencies. The way in which an operator uses his/her knowledge to solve such a problem, however, is not clearly understood (Hoc et al., 1995). Research into the area of problem solving has produced only limited understanding of the skills required to detect and diagnose faults within the complex and dynamic environment of a nuclear power plant (NPP). Six main problem solving strategies have been identified from the literature which have been used to solve problems in a variety of situations. These strategies are: symptomatic; topographic (Rasmussen, 1984); hypothesis-driven; data-driven (Yoon and Hammer, 1988); analogical (Gick and Holyoak, 1980); and backward chaining problem solving strategies (Chi et al., 1985). Investigations of the results from three recent diagnosis-related experiments undertaken by the OECD Halden Reactor Project (HRP) in Norway, however, revealed that the identification of problem solving strategies,

utilised by real operators, and using traditional strategy definitions such as topographic and symptomatic, was somewhat unreliable and unstable. The operators appeared to continually shift between strategies, and the observed pattern of problem solving behaviours did not always appear to conform to the strategy definitions. From these data questions therefore arise as to how strategies are used during fault diagnosis, and
how important they are in the diagnosis and system stabilisation process. Additionally, more practical questions arise as to how to go about training operators to become more effective and flexible in their diagnostic approach, and how to design interfaces to support shifting strategies.

It therefore may be potentially more useful to focus on the types of knowledge required for diagnosis. This entails investigating the structure of diagnostic knowledge with the aim being to eventually train operators in the use of such knowledge, and correspondingly design interfaces which better support operator problem solving through better and more pertinent or flexible information presentation. Thus, rather than focusing on strategies, which necessarily are predicated on knowledge, attention was directed to the different sources and types of knowledge utilised by such strategies, irrespective of which strategies were being used at the time.

Types of knowledge

The investigation of the results from the HRP experiments led the authors to distinguish between six types of knowledge (see Figure 1) which were used by the operators to diagnose scenarios. A brief description of each of the six knowledge types is given in Table 1.

Table 1. Description of knowledge types

Knowledge Type	Characteristics
1	Textbook theory; no practical experience/practice; general NPP knowledge, including functional information and goals of the process.
2	More thorough understanding of the process; practical experience; general fault knowledge; understanding of the logical inter-relationships of the plant systems.
3	Detailed mechanical (functional) knowledge; knowledge of cause and effects; normative knowledge (may also be held by designers and maintenance personnel).
4	Maximum integration of systems, i.e. how changes in one system affect another system.
5	Contextual knowledge; detailed plant layout knowledge; useful for common cause failures.
6	Tactical knowledge; knowledge of specific faults and multiple faults; problem solving skills and realisation of prompt and delayed consequences.

Each of the first four types of knowledge appear to be predicated on the previous type, e.g. operators must have type 1 knowledge (textbook theory) before they can proceed to type 2 (logical inter-relationships). This, however, does not appear to be the case for types 5 and 6. The knowledge contained in these types seems to be related to practical experience, e.g. dealing with faults, and length of experience in actually controlling the process. Type 5 is thought to be activated from type 3, because an operator would require normative knowledge, e.g. which valves can be manipulated; normal values for temperature; flow around the plant, etc.; in order to activate type 5. Once activated, type 5 could interact with either type 3 or 4. Type 6 is thought to be initially activated from type 2, because operators should be able to solve certain faults with minimal technical knowledge, e.g. from classroom practice. However, type 6 could then interact with any of the subsequent types. Overall, therefore, there are different 'levels' of knowledge, as shown on the figure by highlighting the direction of increasing 'depth of knowledge'. Whether this is true 'depth of knowledge' in the cognitive psychology meaning (e.g. Eysenck and Keane, 1990) is debatable. If nothing else, however, increasing levels of knowledge do represent an increased training load, and an increasing amount of knowledge and complexity of knowledge structuring.

It is not proposed that each knowledge type is a separate entity, as there will be a degree of overlap between levels, but it is suggested that this (as yet weak) descriptive model may allow knowledge utilisation during diagnosis to be usefully classified. With further development of this approach, a task analysis tool could be developed to model knowledge utilisation and determine recommendations for training programmes and interface design. This is an ultimate and desirable goal. However, the first question is whether the proposed knowledge types can be reliably categorised. A small scale study designed to test the reliability of categorisation is described below.

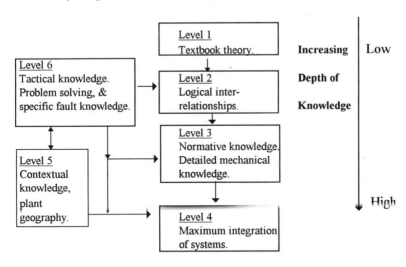

Figure 1. Proposed Type of Knowledge Hierarchy

Evaluating the types of knowledge categorisation system

As a small scale test of the categories, two human factors experts independently 'mapped' the types of knowledge used by operators during three realistic and diagnostically challenging emergency event scenarios. The knowledge time-line maps were created using the raw data from the experimental study, which included eye-movement tracking (EMT) videos, concurrent verbal protocol (VPC) transcripts, debriefing transcripts, and narrative stories (see Kirwan et al., 1995). An example of a knowledge time-line map is shown in Figure 2.

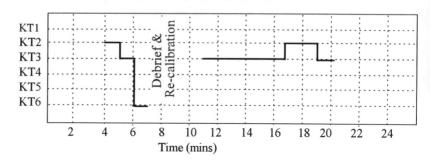

Figure 2. Knowledge time-line map

In Figure 2, scenario time is shown along the x axis and the levels of knowledge along the y axis. The operator's diagnosis, in this particular scenario was incomplete. The scenario consisted of two faults *(busbar failure* and *erroneous closing of a relief valve)* and although the operator was following the correct diagnostic path, a full diagnosis was not carried out in time. The gap between 7-11 minutes in the map was a period of debriefing and recalibration of the eye tracking equipment.

The minimum type of knowledge required to make a successful diagnosis of this scenario was type 4 knowledge for the busbar failure if the busbar alarm was missed. Diagnosis of the second fault required a minimum of type 3 knowledge. From the analysis of the knowledge maps it was noted that some operators did not have, or were unable to activate, the minimum level of knowledge required to diagnosis this, and other scenarios. As a result their diagnosis was unsuccessful, either wholly or partially.

The expert's knowledge maps (for each operator and each scenario) were then compared in two different ways: time-based activity sampling analysis (i.e. single comparison made every minute throughout the scenario time); and sequential comparison analysis (comparison of the actual trajectories throughout the whole scenario). From these analyses the percentage agreement and disagreement between the two experts was calculated. Table 2 is a summary of the time-based activity sampling results showing that the experts independently agreed over 90% of the total scenario time on the knowledge levels they perceived the operators to have used. There was 100% agreement for eight out of the twelve subject-scenario combinations. The results from the sequential comparison analysis (comparing the actual trajectories of the maps over time) found very similar knowledge map results, showing 100% agreement in eight out of the twelve subject-scenario combinations. Less than 10% disagreement occurred between the expert's knowledge maps, and the disagreements which did occur were only one level apart, e.g. type 3 and 4, which seemed to hinge on differences in individual interpretations of single verbal protocol statements. Disagreements were to be expected as the knowledge mapping methodology was somewhat subjective in nature.

Table 2. Expert knowledge map comparison summary table: time-based results.

	Inexperienced ops (mins)	Experienced ops (mins)
Total scenario time	92	91
Total agreement time (%)	84 (91.3%)	82 (90.1%)
Total disagreement time (%)	8 (8.7%)	9 (9.9%)

The experts were able to map both experienced and inexperienced operators knowledge to the same level of accuracy. This preliminary study suggests, therefore, that this methodology can be used to map both experienced and inexperienced operator's knowledge. However, it must be noted that only two experts participated in this study, and further investigation into the validity and reliability of this approach is required.

Insights on interface design based on knowledge type mapping

From the knowledge maps and scenario information it was possible to look at a selection of the VDU format interface displays for each scenario and identify critical information required for diagnosis. Four criteria were used:
- *Information missing*, i.e. the actual context-specific [knowledge type] information required by the operator was not available on the expected or most appropriate format;
- *The level of detail was wrong or inappropriate*, i.e. the information required to diagnose the situation was presented but the type of information presented could only lead to a successful diagnosis if the operator utilised a deeper level of knowledge and reasoned about that information;
- *Information inappropriately presented*, i.e. the information required is presented on the display of interest but due to its presentation the operator is unsure of its meaning;
- *Information is presented peripherally*, i.e. the information is there but it is hidden or buried in other information which makes it difficult to find, i.e. it is not salient. As a result the operator must search for the information.

Results of the interface evaluation revealed there were many areas where operators may have had problems gaining information, required for diagnosis, from the interface. For example, in one scenario a *busbar failure* occurred which caused many pumps to trip. The important alarm (*switch gears under voltage*) is quickly scrolled off the screen due to the large number of alarms for the seemingly random pattern of pump trips. The operator requires deep reasoning (knowledge type 6) in order to diagnose the busbar failure as there are no direct indications that the pump trips are related to the busbar failure. The presentation of information on the VDU display transformed a fairly simple problem into a very difficult and complex one. Taking type of knowledge into account, recommendations can be made for interface redesign. For example, if all the pumps which have a common electricity source were connected by flashing lines (which denoted electrical supply) and which only appeared when, for example, a busbar fails, (i.e. are not visible during normal plant status), then the operators could immediately recognise the existence of an electrical problem and be able to trace the path of the power supply back to the root cause of the problem. This would considerably reduce the level of knowledge required to diagnose the fault (possibly to type 2 or 3), and therefore general diagnoses of this fault would be potentially more successful.

Summary

The results indicate that there is potential reliability and utility in mapping types of knowledge for diagnosis in NPP scenarios. Insights can be gained which are relevant to training needs analysis and interface design (only the latter have been explored in this paper due to the paper length restrictions). The small scale reliability test of the method gave positive results, suggesting that the method of knowledge type mapping can be applied with a degree of robustness. The next stage of research is to apply the approach to a wider set of NPP scenarios. If the results for the approach are still positive, then the next step will be to consider the development of a task analysis tool, based on mapping knowledge types, which can be used prospectively for determining training needs and interface requirements for operator diagnostic support.

References

Chi, M.T., Feltovich, P.J. and Glaser, R. 1985, Categorisation and representation of physics problems by experts and novices, *Cognitive Science*, **15**, 121-152.
Gick, M. and Holyoak, K.J. 1980, Analogical problem solving, *Cognitive Psychology*, **12**, 306-355.
Hoc, J.-M., Amalberti, R. and Boreham, N. 1995, Human operator expertise in diagnosis, decision-making, and time management. In J.-M. Hoc, P.C. Cacciabue, and E. Hollnagel, (Eds.) *Expertise and Technology*. (New Jersey: Lawrence Erlbaum Associates).
Eysenck, M.W. and Keane, M. 1990, *Cognitive Psychology - A Students Handbook*, (Hove (UK): Lawrence Erlbaum Associates).
Kirwan, B., Kaarstad, M., Hauland, G. and Follosoe, K. 1995, See no evil, hear no evil, speak no evil: Verbal protocol analysis, eye movement analysis, and nuclear power plant diagnosis, *Contemporary Ergonomics, Proceedings of the 1995 Annual Conference of the Ergonomics Society*, (Taylor & Francis, London) 249-255.
Rasmussen, J. 1984, Strategies for state identification and diagnosis in supervisory control tasks and design of computer based support systems. In J. Ranta (Ed.) *Advances in Man-Machine Systems Research*, **1**, (Greenwich, JAI Press) 139-193.
Yoon, W.C. and Hammer, J.M. 1988, Aiding the operator during novel fault diagnosis, *IEEE Transactions on Systems, Man and Cybernetics*, **18**, 297-304.

Acknowledgements: The authors would like to thank both the HRP staff working on the Human Error Project, and the operators, for their co-operation and enthusiasm.

DEVELOPING AND VALIDATING ESCAPAID: A COMPUTER MODEL OF EVACUATION FROM COMPLEX SPACES

G.A. Munley*, R.J. Mortimer*, J.O. Oyinloye** , S.V. Worth** and J.C. Williams***

*Department of Psychology and Speech Pathology,
The Manchester Metropolitan University, Manchester M13 0JA

**Electrowatt Engineering Services (UK) Ltd, Birchwood, Warrington
Cheshire WA3 7QS

***Previously with Electrowatt Engineering Services (UK) Ltd
H.M.Nuclear Installations Inspectorate, Bootle, Liverpool L20 3LZ

This paper describes two validation exercises that were conducted to evaluate and direct the development of a software model of evacuation from complex spaces. The computer model, ESCAPAID, is a mathematical flow model used to represent the movement of humans during emergency situations. The validation exercises were conducted on two sites; one provided a sample that had no specific training in emergency procedures whilst the second contained participants who were regularly drilled for such events. The exercises were conducted to evaluate basic movement characteristics and to identify areas where behavioural modelling is required to supplement the mechanistic modelling presently in place within the model. Results indicate that the model is fairly well calibrated in terms of predicting evacuation patterns over time, however, further development is required in order to predict accurately exit choice behaviour.

Introduction

The public inquiry into the Piper Alpha disaster Cullen (1990) made a number of important recommendations for assessing the safety of offshore installations. Amongst these was the recommendation that all offshore installations complete a formal safety assessment and that this formal safety assessment takes the form of a safety case. An important part of the safety case is the demonstration that safe evacuation, escape and rescue (EER) from an installation are possible in the event of an emergency. This requirement to demonstrate safe EER within the safety case and the not insubstantial danger associated with running simulated emergencies from which EER times can be established, has led to the development of a number of proprietary computer models of human behaviour that can be used to predict times for EER activities. Miles (1992) however, has identified the computer modelling of human behaviour in emergency situations as an area of concern for the Health and Safety Executive Offshore Safety Division (HSE-OSD) the regulatory body involved with the evaluation of offshore safety cases. The principal cause for concern is the data on which such models are based and their validation. Searches of the literature reveal very few published validation studies of computer models of human evacuation behaviour in emergency situations. Models that have reported validation studies include BGRAF (Ozel 1988) and Evacnet+ (Kisko and Francis 1985).

In this paper we report the results of two exercises that were conducted in order to validate and aid the development of ESCAPAID a computer model of evacuation from comp' spaces. ESCAPAID has been developed by Electrowatt Engineering Services (UK) Limited, and runs on a PC. The modelling is based on a network representation of the escape routes within the relevant built environment, with the movement of occupants being treated as 'parti flow' along the network. The network is built up as a three dimensional set of arcs and nodes. An arc represents a specific section of an escape route such as a corridor, stairway or door. The data required for each arc include its length, width and speed at which people can move the arc. A node can represent either the initial location of occupants, an end muster point or a junction of two or more arcs. Node flow characteristics are defined in terms of their maximum flow rate and queue behaviour. Queue behaviour is specified as either SIRO (serve in random order), or FCFS (first come first served). Flow along the arcs is calculated by the program using the arc velocity and length characteristics. Other input data items required are the start time, print time steps, number of occupants on each route and parameters (mean and standard deviation values) to define a probabilistic distribution of time taken for the people to respond the alarm. ESCAPAID then tracks the movement of people through the escape routes from their starting locations to an end muster point, noting in particular, the number of people that have arrived at each of the end muster points.

ESCAPAID was used to model two sites and to predict evacuation behaviour from thes sites during simulated emergencies, fire drills in both cases. The sites chosen differed in terms of their size, number and type of occupants making up the populations. The aim of the study was to assess whether ESCAPAID would predict accurately the evacuation behaviour of occupants both in terms of exit choice and evacuation rate.

Method

Evacuation Sites

Two sites were chosen to evaluate ESCAPAID. The first site comprised the building housing the psychology and speech pathology department at the Manchester Metropolitan University. This provided a relatively large two storey environment comprising large and sma teaching areas, staff offices, a general office, a canteen, a student common room, three stairways, a number of corridors and ten exits. The second site comprised office accommodation on a military site in the UK. This provided a smaller single storey site comprising offices, rest rooms, a tea bar, one corridor and four exits. For each site a scale pla of the building was drawn, which was then used as a blueprint for developing the arc and node representation of the site for input into ESCAPAID.

Data Collection

At each site a fire drill was conducted, by means of routine drills that were not specially commissioned for the purposes of this study. The participants in the studies were not aware that the drill would take place. Immediately prior to the activation of the alarm signalling the commencement of the drill, a head count was made of the number of occupants in each of the main areas of each of the sites. At the Manchester Metropolitan University site each of the ten exits was designated as a muster point for the purposes of this study. The egress rate of people from each of the ten exits was then measured in terms of the number of people exiting per ten second time frame. At the military site there was one designated muster point on a hard standing 15m away from the building. Here the rate of people arriving at the muster point per 10 second time frame was recorded. In addition to recording the rate of mustering from each of the buildings each person leaving the building was given a questionnaire to complete. This questionnaire asked for basic demographic information and an indication of the person's role in the building, i.e. staff, student or visitor. The questionnaire also included a scale drawing of the relevant building on which respondents were asked to show their starting point when the alarm sounded and the route they had taken to their chosen exit.

Participants

One hundred and sixty three participants were involved in the exercise conducted at the Manchester Metropolitan University site, of these 72% were female and 28% male. The role of the participants in the building was as follows; 88% were students, 11% staff and 1% visitors. Ages ranged from 12 to 50 years with a mean of 29 years. At this site there is normally one fire drill per year. Evacuation instructions are posted at various points in the building and members of staff are circulated with emergency evacuation procedures. However, there is no formal training in these procedures. For this reason this site has been designated as having a population that is untrained in emergency evacuation procedures.

At the military site data on twenty five participants were collected. Participants were all male members of the armed forces whose ages ranged from 20 to 54 years with a mean of 35 years. The participants at this site were regularly drilled in emergency evacuation procedures and were thus, for the purposes of this study, designated as a trained population.

Modelling of Evacuations

In the current version of ESCAPAID, in order to model the evacuation of persons from a building, the programmer must supply the model with information describing an individual's starting point in the building and provide the expected exit route taken by this person. For the university site used in this study two computer models of the evacuation were built. The first was classified as the predicted model of the evacuation. This model programmed people's evacuation behaviour assuming it would be consistent with instructions contained within emergency evacuation procedures, that is, "on hearing an alarm leave the building by the nearest available exit." Thus, using the data gained from questionnaires and head counts, each person in the drill was assigned an appropriate starting point in the building and was programmed with their ideal evacuation route. The second model of evacuation developed for the university site was termed the actual model of evacuation behaviour. This used information from the questionnaires to assign each person taking part in the exercise their starting point and to program the actual route they took in evacuating the building. At the military site only one model of evacuation was produced as the predicted and actual evacuation behaviour was identical.

Results

University Site

The first stage in analysing the data from the university site was to compare evacuation behaviour in terms of exit choice between the predicted model of evacuation and the actual evacuation behaviour demonstrated. In the predicted model a person's choice of exit is expected to be that exit that is closest to their current position at the point at which the alarm sounds. Data showing the number of persons predicted to evacuate using each exit and the actual numbers using the exits are shown in table 1. A chi-square analysis applied to exit use behaviour revealed that the distribution of exits used was different to that which would be predicted based on an assumption that people will comply properly with emergency procedur $X^2=41.155$, df=5 p<0.001.

Table 1. Predicted and actual exit choices

Exit number	1	2	5	6	7	10
Predicted	44	11	1	73	0	34
Actual	68	0	1	76	8	10

As people did not evacuate the building in a way that was consistent with the original ESCAPAID model of evacuation, there was little point in conducting any further analysis on this model of evacuation. For this reason the second ESCAPAID model for evacuation from the university site was developed. This model programmed each individual's escape route usi the data gleaned from the questionnaire given to participants immediately after they had exite the building. This second computer model was used to make comparisons between the actual exit rate of people from the building and the exit rate predicted by ESCAPAID. In order to make these comparisons a series of Pearson product moment correlation coefficients were computed. These compared the actual egress rate at each exit with the egress rate predicted b ESCAPAID. Egress rates were measured as the number of people leaving the building per 10 second time frame. The correlation coefficient for each exit is shown in table 2.

Table 2. Correlation between predicted and actual egress rate

Exit number	Correlation	df	Significance
1	0.65	19	p<0.01
5	-0.08	13	p>0.05
6	0.82	14	p<0.01
7	-0.07	20	p>0.05
10	0.27	14	p>0.05

From table 2 it can be seen that at two exits, 1 and 6, significant correlation was obtained between the egress rates predicted by ESCAPAID and the actual rate of egress. Interestingly these were the exits that were used most heavily in the evacuation. ESCAPAID, however, was less successful in modelling egress rates from the less heavily used exits.

Military Site

At the military site it was sufficient to produce only one ESCAPAID model of the evacuation as all participants left the building using their predicted exit route. A Pearson product moment correlation coefficient computed to compare the predicted muster rate, at the one muster point for this building, with the actual muster rate was found to be non-significant ($r=$ -0.11, df=13, p>0.05) indicating little agreement, in terms of egress rate, between the ESCAPAID model and the actual evacuation.

Discussion

Results from these preliminary studies have provided a number of important pieces of information for the developers of evacuation models generally, and ESCAPAID, in particular. Firstly, the results from the university site suggest that the early version of ESCAPAID was able to model reasonably accurately, in terms of egress rate, the evacuation of fairly large numbers of people through individual exits. This suggests that the treatment of queuing behaviour within the model by means of a first come, first served rule is appropriate in this particular situation. (It should be noted that all of the evacuation models developed and run within the study used the first come first served model of queuing behaviour.)

Other results from the university site, however, suggest that other aspects of the model need further development. Although ESCAPAID modelled egress through heavily used exits quite well, the results for less heavily used exits showed low agreement between predicted and actual egress rates. This suggests problems with the movement parameters, such as walking speed, time to descend stairs, etc. or with the time delay used to represent the time taken for people to respond to the alarm. None of these explanations, however, present serious problems for the model developers as they can be manipulated within the existing ESCAPAID software.

The area where there is greatest need for development within ESCAPAID, is in the area of modelling exit choice behaviour. The version of ESCAPAID tested had no facility for predicting how people will respond in the emergency situation, hence, the escape route taken by each individual or group of individuals has to be specified by the analyst. The way that this is normally achieved is to predict people's evacuation route behaviour based on emergency evacuation procedures. The results from this study, however, show that when dealing with untrained populations, people do not always conform with such emergency procedures. This finding is consistent with results from research into people's behaviour in a number of major incidents that suggests that familiarity with a route, rather than the availability of an exit, is a major determinant of people's exit choice behaviour, e.g. Bryan (1980), Sime (1985) and Donald and Canter (1990). Even within trained populations, although emergency evacuation procedures may be routinely followed, these may not always be appropriate due to prescribed routes being unusable because of the position of the hazard. Certainly in the Piper Alpha disaster, normal evacuation procedures were not appropriate and those that survived attributed their successful escape from the platform, in part, to their knowledge of its layout (Cullen 1990). Therefore, it is imperative that any computer model of evacuation has the ability to model people's decision making processes during an emergency, rather than rely on preconceived emergency procedures to predict how people will behave in practice.

Although there is a body of literature that identifies some of factors that influence people's evacuation from complex spaces in emergency situations, much of this is based on interviews conducted with survivors from emergencies or from forensic evidence from those who did not survive and is therefore qualitative in nature. What is required in order to aid the development and allow proper validation of computer models of evacuation like ESCAPAID, is a series of experiments that manipulate the factors identified in the literature, i.e. familiarity, role, quality of warning information, etc. in order to determine quantitatively their impact on evacuation. Obviously such studies would need to be based on simulated emergencies and there is the issue of transfer of data from simulated emergencies to actual events to consider. Some authors, e.g. Pauls and Jones (1980), argue and provide evidence that there is sufficient similarity between people's behaviour in drills and real emergencies to make information gleaned from drills useful. Others, for example the Civil Aviation Authority (Muir 1989), have investigated the use of competitive evacuation trials from aircraft in order to generate more accurate data for people's behaviour in emergency situations.

Whichever type of evacuation trial is used, an important issue for future studies of thi type is the quality of the data collected during the evacuation. In this study although attempt were made to be as rigorous as possible during data collection, the technology at our disposa necessitated, to a large extent, reliance on self reports of evacuation behaviour. For this reas the data and analysis reported here are fairly coarse, i.e. the majority of the analysis is conducted on egress rates rather than on egress time for participants. This mode of analysis was necessitated because, although our data collection methods allowed us to identify the st point, egress route and exit point of each individual involved in the exercise, we were unable time individually each participant. Therefore for future studies we intend to investigate the of electronic tracking equipment in order to record an accurate protocol, including timing, of each individual's evacuation. It is only in this way, we believe, that the type of data necessar to properly develop and validate computer models of evacuation can be obtained.

References

Bryan, J.L. 1980, *An Examination and Analysis of the Dynamics of Human Behaviour in t MGM Grand Hotel Fire, Clark County, Nevada. November 21 1980,* (National Fire Protection Association, Quincy, M.A.).

Cullen, W.D. 1990, *The Public Inquiry into the Piper Alpha Disaster,* (HMSO, London).

Donald, I. and Canter, D. 1990, Behavioural aspects of the King's Cross disaster. In Canter, (ed.) *Fires and Human Behaviour,* 2nd edition (Wiley, Chichester) 15-31.

Kisko, T.M. and Francis, R.L. 1985, A computer program to determine optimal building evacuation plans, *Fire Safety Journal,* 9, 211-220.

Miles, R.W. 1992, Human factors research for offshore safety. In *Proceedings of the Practicalities and Realities of Human Factors in Offshore Safety.* Aberdeen, 1992.

Muir, H.C. 1989, *Aircraft Evacuations: the Effect of Passenger Motivation and Cabin Configuration Adjacent to the Exit.* Civil Aviation Paper 89019.

Ozel, F. 1988, Calibration and validation of computer model BGRAF. In Sime, J.D. (ed.) *Safety in the Built Environment,* (E. & F.N. Spon, London) 75-86.

Pauls, J. L. and Jones, B.K. 1980, Building evacuation: research methods and case studies. I Canter, D (ed.) *Fires and Human Behaviour,* 1st edition (Wiley, Chichester) 227-249.

Sime, J.D. 1985, Movement toward the familiar: person and place affiliation in a fire entrapment situation, *Environment and Behaviour,* 17, 697-724.

RISKS TO FORKLIFT TRUCK DRIVERS IN COLD STORES

Nicholas Franzini and Rachel Benedyk

Ergonomics and HCI Unit
University College London
26 Bedford Way
London WC1H 0AP

An ergonomic evaluation of forklift truck driving within a frozen food store was carried out, with emphasis on the possibility of injuries to the musculoskeletal system. As the drivers had to operate in a cold store environment, it was thought that the risk of long-term musculoskeletal injuries might be exacerbated. The approach involved analysis of the driving tasks, questionnaire surveys of the drivers to determine the extent of the risk of injury, and comparisons with similar tasks not performed in cold environments. The results showed a higher incidence of musculoskeletal pre-symptom reporting by the drivers involved in cold store work. Evaluation of the driving tasks identified areas that were considered to provide a potential risk to the musculoskeletal system. The investigation concluded that there was a probable correlation between cold temperatures and potential musculoskeletal injuries for such drivers.

Introduction

The company in question use forklift trucks as part of their everyday storage and distribution operations. A study by the Swedish Transport Research Institute (Tergo Reach Trucks, 1987) has shown that forklift truck driving causes postural problems amongst drivers, and has shown that it is the neck and back that are most at risk of injury through driving. They also concluded that such injuries may occur if the drivers operate the trucks for as little as two hours per day, and they have produced 'limit values' concerning the calculated risk values (determined by their medical staff) in relation to the number and degree of head turns carried out by the driver, based on measurements taken of the load on the sternomastoid and trapezius muscles with skin electrodes. It appears that the major problems are as a result of sideways head movements, which occur throughout the driving period. Bending the neck back during stacking is a lesser problem, although the combination of this bending with simultaneous head turning may also increase the risk of musculoskeletal symptom reporting to the neck and back.

By the nature of the products being handled at the company in question, these forklift operations take place in a freezer store, with the main part of the store being maintained at minus twenty-seven degrees centigrade, and the outer loading bay being maintained at minus eighteen degrees centigrade. Of concern therefore is the possibility of long-term injury as a result of repetitive movements combined with repeated exposure to the low temperatures, and of the exacerbation of any pre-existing injury that the drivers might have by this combination. This final point is plausible, as the repetitive nature of the tasks of forklift truck driving can often result in work-related upper limb disorders to the driver over a long period of time, and it is known that cold temperatures may cause problems in relation to existing muscle strain and soreness (Pheasant, 1991).

From the literature it seems to be the case that those in full sedentary occupations are likely to suffer from problems to the back. It also appears that low temperatures may contribute to back problems. The fact that the drivers are sitting fairly still, not generating much heat, and working in a cold environment suggests that they may be particularly prone to back problems (Williamson, Chrenko and Hamley, 1984 ; Pheasant, 1991).

The literature also suggests that the most common form of work-related neck problem is static muscle loading which comes from potentially damaging working posture (Grieco, 1986), possibly also exacerbated by psychological stress, and that the jobs that tend to result in the highest incidence of neck problems have one thing in common - they require the focusing of the eyes on one point whilst performing a task in a fixed position, and thus the entire posture of the neck, head and upper limbs tends to be fixed throughout the duration of the task (Pheasant, 1991). Although the limbs are not completely fixed, the sight-line aspect of the above statement suggests that forklift driving could be potentially harmful to the neck.

An ergonomic evaluation of forklift truck usage in the cold store was set up. Its first objectives were to record the physical symptoms related to the job and to the type of truck being driven, and a timeline analysis of the tasks performed by the drivers.

The subsequent investigation attempted to show a link between exposure of operators to a cold environment and the incidence of musculoskeletal problems suffered during the performance of their operational tasks. The amount of individual variation in the response to the cold may affect any potential results, and account was taken of time working in the cold store as drivers working for longer might experience different effects, in terms of injury or their subjective reaction to the cold. However, it was expected that individual differences would make it difficult to determine accurately the extent to which low temperatures affect the incidence of injuries (which may or may not be attributable to the cold in the first place).

Method

The field-based method was designed to obtain the maximum amount of information relating to the task of forklift truck driving and the potential risk of injury with a limited number of subjects. Whilst it is acknowledged that the type of protective clothing worn may be a contributory factor to the potential risk, the emphasis of this study was intended to focus on other job design factors which may cause a risk to forklift drivers in cold stores.

Thus the main emphasis of the design was upon a series of questionnaires distributed to the drivers relating to musculoskeletal problems experienced. A task description and postural analysis of the driving tasks were also performed in order to identify task-specific actions thought likely to increase any potential risk of injury.

A variation on the Nordic Musculoskeletal Questionnaire (Kuorinka, Jonsson et al., 1987) was decided upon for distribution to the drivers. This is a fairly detailed questionnaire that may be answered quickly and easily. It also allows a number of symptoms to be reported, and thus avoids the omission of pre-symptoms such as aches and discomfort, which may be thought of by the subjects as otherwise too minor to mention. It is also easily adapted to cover different timespans - in this case the timespans of interest covered a range from twelve months to a few hours. The questionnaire was distributed combined with a diagram to the side of the questions depicting the body areas related to the questions, to try and eliminate any misunderstanding relating to areas of body parts in the responses. A draft copy of the questionnaire was commented on by selected drivers to determine whether they found it easy to understand and relevant to problems that they had experienced in the past.

Each driver was given an initial questionnaire to complete regarding the incidence of ache, pain or discomfort in certain body areas over the past twelve months. This was followed by a pack of twenty questionnaires per driver, to be completed at different times during the day - once directly at the end of the shift and once after a defined period of rest. The experimental period for the questionnaires lasted for two weeks (ten working days) during which the drivers completed two questionnaires per day.

A total of twenty-eight drivers were given the questionnaires. Of these twenty-eight drivers, twelve were involved in permanent cold store work (Location A), nine were involved in part-time cold store work (Location B), and seven performed no cold store work (Location C). Due to the nature of the study, there was no opportunity to match the subjects nor to vary the conditions for individual drivers. The experimental design was intended compare three types of job design in different environments, allowing for a comparative testing of the incidence of symptoms over the study period for the three types.

For each condition, drivers would compare the daily incidence of discomfort at the end of the shift with the daily incidence of discomfort following a period of rest after the shift. It was expected that there would be a higher incidence of positive reporting (i.e. aches, pains and discomfort occurring) on the questionnaires filled in after a period of rest, due to the predominance of myalgic conditions (especially in neck pain and associated conditions) where the pain tends to come on gradually (Pheasant, 1991), and also due to any effects of outside influences on the drivers' musculoskeletal systems (e.g. outside interests such as weightlifting).

The driving tasks were to be observed in order to identify areas of the tasks performed that may give rise to a potential risk of injury due to the repetitive nature of the movement or the duration over which a particular posture must be maintained. The physical movements involved in the task completion were logged for frequency and duration using a timeline analysis. This was combined with a postural analysis to show the frequency and extent of those movements likely to predispose injury, viz. :

- Forward turn of the head from sideways-on driving position *(horizontal plane)*
- Backward turn of the head from sideways-on driving position *(horizontal plane)*
- Backward bending of the neck/head *(vertical plane)*

The movements were recorded over five-minute driving periods, and the extent of the movements was measured by analysis of photographs taken during the performance of the following tasks :

- Driving forwards in a sideways-on driving position
- Driving backwards in a sideways-on driving position
- Loading a pallet onto a shelf at the following heights :
  Shelf height 0m ; Shelf height 1.8m ; Shelf height 3.65m ; Shelf height 5.48m

Results and Discussion

Results of Frequency of Predetermined Movements, Duration of Movement, and Postural Analysis

All movement frequencies are averages taken from 50 five minute observation periods. Angles are mean angles from the 'normal' position, which was the position observed when the driver was sitting in the forklift truck at rest with no task to perform. Duration times are mean times. Results are from Location 1 only, with ten drivers all participating in permanent cold store work.

1) Forward turn of the head *(horizontal plane)* : frequency = **4.2** : angle = 55° : duration = 12 seconds

2) Backward turn of the head *(horizontal plane)* : frequency = **4** : angle = 57° : duration = 12 seconds

3) Backward bending of the neck/head *(vertical plane)* : frequency = **2.1**
3a) Shelf height 1 : angle = 0° from normal on vertical plane : duration = 20 seconds
3b) Shelf height 2 : angle = 0° from normal on vertical plane : duration = 36 seconds
3c) Shelf height 3 : angle = 30° from normal on vertical plane : duration = 65 seconds
3d) Shelf height 4 : angle = 40° from normal on vertical plane : duration = 76 seconds

It can be seen that the tasks most likely to contribute to any musculoskeletal symptom reporting are the forward and backward travel and the stacking tasks that involve the top two shelves (3.65m and 5.48m height) because of the length of time postures are held and the severity of the neck angles. The repetitive movements are not high in frequency, but awkward postures do have to be maintained for long periods of time during the performance of certain tasks, and it is possible that this static loading may be the cause of much of the ache, pain or discomfort reported by the drivers. The drivers also gave verbal confirmation that the stacking tasks involving the top two shelves caused them the most problems in relation to musculoskeletal symptom reporting.

Driver Backgrounds
Location A had the oldest drivers (mean age 36yrs) and Location C had the youngest drivers (mean age 33yrs). The mean length of time spent driving trucks was 8yrs 9mths for Location A, 6yrs 9mths for Location B, and 6yrs for Location C. Of the total of twenty-eight drivers who participated in some or all parts of the study, thirteen participated in a hobby or interest outside the work environment that they thought might cause ache, pain or discomfort to specified body areas, such as the neck (see Table 1).

Table 1. Driver Backgrounds
n = 28

Location	Age (mean)	Length of Service (mean)	No. of Drivers with Hobbies	Incidence of Ache, Pain or Discomfort Over 12 Months
Location A Drivers = 13	36	8yrs 9mths	3	30
Location B Drivers = 8	34	6yrs 9mths	5	24
Location C Drivers = 7	33	6yrs	2	12

Daily Incidence of Discomfort
Across all locations, neck discomfort accounted for 34.2% of the total incidence (after shift and after rest), upper back discomfort accounted for 4.9% of the total incidence (after shift and after rest), and lower back discomfort accounted for 30.8% of the total incidence (after shift and after rest). The remaining 30.1% consisted of discomfort to other body areas (such as knees and wrists). The incidence of reporting was higher for the period directly after the shift than for the time after a period of rest for Location A and Location C. However, for Location B the incidence of reporting was slightly greater after a period of rest (see Table 2). This was due to the fact that here the incidence of lower back discomfort actually increased following a period of rest, suggesting that the lower back experienced the least improvement in symptoms following a period of rest. The incidence of pain in the

unspecified areas (such as the feet and knees) did decrease following a period of rest for all the locations. However, of interest is that reporting was still evident even after a rest, suggesting that some musculoskeletal discomfort is not alleviated between driving shifts.

Table 2. Daily Incidence of Ache, Pain or Discomfort

Location	After Shift *(no. of reports made)*	After Rest *(no. of reports made)*
Location A	100	73
10 drivers		
Location B	59	61
8 drivers		
Location C	24	7
7 drivers		

Different Locations

The incidence of ache, pain or discomfort varied by location, over all time categories (see Table 1 and Table 2).

The total incidence across all body areas for the past twelve months was 30 for Location A (mean of 2.5 incidences per driver), 24 for Location B (mean of 2.6 per driver), and 12 for Location C (mean of 1.7 per driver).

The total incidence across all body areas directly after a shift over a two week period was 100 for Location A (mean of 10 per driver), 59 for Location B (mean of 7.3 per driver), and 24 for Location C (mean of 3.4 per driver).

The total incidence across all body areas after a period of rest over a two week period was 73 for Location A (mean of 7.3 per driver), 61 for Location B (mean of 7.6 per driver), and 7 for Location C (mean of 1 per driver).

The results from this study show that the drivers who are constantly exposed to the cold environment (Location A) suffer from a higher incidence of ache, pain or discomfort directly after the shift than do the drivers at the other locations. Whether this leads to long-term problems is not obviously apparent, and the incidence of ache, pain or discomfort per driver after a period of rest, although much higher in Location A than Location C, is slightly lower than Location B (where the proportion of cold store work is much lower).

The results also showed a high incidence of problems to the same body areas for drivers from all three locations, suggesting that there are specific tasks relating to forklift truck driving which place the driver at risk, particularly in relation to the body areas of the neck, back, and knees. However, it must be considered that particular job design features (such as shifts worked and tasks performed) and workspace design features (such as warehouse layout) may influence any results collected to such an extent that assumptions made about both the equipment and the subjects prove to be invalid.

The environment in which the work takes place is vital in determining if there are any potential risks to injury to the musculoskeletal system, and it may be that the cold temperatures result in changes in behaviour (physiological, psychological and postural) in drivers in cold stores which are not apparent or do not occur in forklift drivers who do not have to operate in the cold. Thus the trends displayed by this investigation are not conclusively linked to any specific aetiology, but provide useful inferences relating to environment, job design, and workspace design. Any may add to the risk of musculoskeletal injury to the drivers.

Summary

The task of driving a forklift truck results in a certain amount of risk of potential musculoskeletal injury to the driver. Whilst this study has not been able to quantify this risk it has identified factors (work/rest schedules, extremes of static posture, repetitive tasks) which, either taken alone or combined with other factors, cause an increased risk to the driver in terms of his experience of ache, pain or discomfort.

It would appear that the incidence of musculoskeletal discomfort is correlated to some extent with the time spent in the cold store, in that the drivers in the two locations which involved the use of forklift trucks in cold stores reported higher incidence of ache, pain or discomfort, particularly after a period of rest.

A number of confounding variables in this study, relating to the drivers' backgrounds and the differences in the tasks that the drivers had to perform (both driving tasks and other tasks), need to be checked in order to ensure the validity of these conclusions.

Recommendations

- Certain aspects of the forklift driving tasks should be altered to try and minimise the incidence of potentially injurious movement. In particular, areas relating to neck movements (both on horizontal and vertical planes) should be examined in order to attempt to reduce the degree of movement required, and the duration that postures involving awkward neck postures have to be held.
- The length of time spent in the cold store should be minimised wherever possible in order to determine any effects on the risk to drivers of potential musculoskeletal problems.
- The length of rest and recovery following forklift truck driving work in the cold should be investigated and optimised.
- For the purposes of further research it would be useful to produce results from a controlled situation, where there is an opportunity to match subjects in terms of workload, type of task, and previous incidence of ache, pain or discomfort.

References

Grieco, A. 1986, Sitting posture : an old problem and a new one, Ergonomics **29**
Kuorinka, I., Jonsson, B., Kilbom, A., Vinterberg, H., Biering-Sorenson, F., Anderson, G. and Jorgensen, K. 1987, Standardized Nordic Questionnaires for the analysis of musculoskeletal symptoms, Applied Ergonomics **18** No.3
Pheasant, S. 1991, *Ergonomics, Work and Health* (MacMillan Press)
Tergo Reach Trucks. 1987, *Promotional Leaflet* produced by Atlet Ltd., Thame, OXON OX
Williamson, D.K., Chrenko, F.A., and Hamley, E.J. 1984, A study of exposure to cold in col-
Applied Ergonomics **15** No.1

RAILWAY SIGNALS PASSED AT DANGER - THE PREVENTION OF HUMAN ERROR

J. May, T. Horberry, A.G. Gale

Applied Vision Research Unit, University of Derby,
Mickleover, Derby, DE3 5GX, UK. Tel\Fax 44 1332 622287
E-mail: AVRU@derby.ac.uk

The potential for disaster with each Railway Signal Passed at Danger (SPaD) is high. This paper forms a literature review of previous research concerning the occurrence of SPaDs in relation to human error and discusses potential causal factors in relation to situational, individual and perceptual/attentional issues. It is shown that the causes of SPaDs are multifactoral and therefore there is no one solution to SPaDs. The de-skilling of the drivers' task is a serious problem and is not always sufficiently addressed by the introduction of vigilance devices.

Introduction

A SPaD occurs when a train passes though a red stop signal instead of stopping before the signal as required. While most SPaDs simply result in near miss incidents each one carries with it a high potential for disaster and loss of life and therefore their reduction is of great importance. Some SPaDs are attributed to signal failures or mechanical faults but the more common cause is human error with 85% of SPaD cases being attributable to human causal factors. While vigilance devices and safety systems such as the AWS (Advanced warning system) and ATP (Automatic train protection) have been introduced to monitor the speed of the train and the signals the driver passes through, and warn him accordingly, it may be many years before such systems are implemented in all cabs and on all lines. The problem of SPaDs must therefore still be addressed. Previous literature which has focused on the human error aspects of SPaD causation is revised here and identified causal factors are discussed.

Situational Factors

Driver Activity

The activity the driver is performing may make them more susceptible to performing certain types of SPaD. Three distinct driving situations where SPaDs occur have been identified in the literature. These arise from different causal factors and therefore need very different solutions. They are:-
- Driving on a main line - Most SPaDs occur because the driver has seen the red signal but made an assumption that the signal will change to a less restrictive

aspect before he passes through it. These assumptions arise through past experience of driving on that line.

- Starting a train after having stopped completely (e.g. moving away from a station)- When starting away from a location the driver must check a signal which tells them that the line ahead is clear, if they forgets to do this then a SPaD may occur. SPaDs can also arise from poor signal placement with respect to the driver's line of sight and mis-communication with other staff as if the driver cannot see the signal themselves they may ask another member of staff to check it for them.
- Driving in railway yards and sidings - SPaDs often occur because of poor communication with other staff resulting in the driver believing that it is safe to proceed through a red light.

Day of the Week

No relationship between SPaD occurrence and the day of the week was found in some studies (e.g. Williams, 1977). Driver's who had more frequent shift changes and were working the first day after a change in shift however were noted to be more likely to experience a SPaD. The important factor therefore may not be a particular day of the week, but the timing of when a SPaD occurs in relation to the driver's shift pattern. Van der Flier and Schoonman, (1988) support this by demonstrating that the distribution of SPaD incidents throughout the days of the week corresponded with the change of the drivers' shifts.

Time of Day

Van der Flier and Schoonman, (1988) found most cases of SPaDs occurred during morning hours (12am to 6am and 8am to 12pm), where as other studies (e.g. Williams, 1977) found no relationship between time of day and SPaD occurrence. It may be possible that this is due to interaction with other factors such as shift rotation and time at work.

Time into Shift

Several studies (e.g. Williams 1977) have demonstrated that SPaD occurrence is greater after the driver has driven for a certain period of time. Disagreement however arises regarding when precisely in the driver's shift most SPaDs occur and Van der Flier and Schoonman (1988) find no significant effects for time into shift. Again this may be masked by the driver's shift pattern. For example some train drivers may have a tendency to drowse whilst driving (Endo and Kogi, 1975) and the time at which this is most frequent varies depending on the night of the shift rotation (Kogi and Ohta, 1975) with the second or third night being the most common.

Number of Drivers in the Train Cab.

Recently the number of train drivers present in the cab has often been reduced from two to one. Job analysis techniques revealed that this should pose no additional health and safety problems (Smith and George 1987). SPaDs however occur more frequently in single man cabs (Williams, 1977), and it may be that the second person confirms the signals as the driver passes through the signal.

Personal Factors

Personality Factors /Previous SPAD Involvement

Verhaegen and Ryckaert (1986) found positive correlations between error frequencies (delayed reactions to yellow lights and speeding) and the personality dimensions of extroversion, neuroticism and emotional instability. Drivers displaying minor psychiatric, and psychosomatic symptoms were also found more likely to experience a SPaD. Drivers who have had previous SPaD involvement, demonstrate worse performance on multiple reaction tasks, and report less job satisfaction than in a matched control group. They are also more likely to be involved in future SPaD incidents (Van Der Flier and Schoonman, 1988) but it is difficult to explain why this is the case and there is little evidence to identify some drivers as "accident prone".

Driver Morale and Motivation

SPaD incidents have been attributed to drivers' low morale and poor job satisfaction (Van der Flier and Schoonman, 1988). Many drivers reported feelings of alienation, isolation and redundancy. They also reported anxiety concerning; vandalism, poor pay, working conditions, organisation of work, and relations with management. Drivers blamed SPaDs on low concentration rather than poor equipment design, expressing concern at the de-skilling of the train driving task and the hours worked in a monotonous environment.

Age/Length of Service/ Track or Rolling Stock Experience

Research regarding age is inconclusive (Williams, 1977). The number of years service which the driver had completed was not important (Van der Flier, 1988) but when driving certain trains for the first time or after a long break the driver was more likely to commit a SPaD. A rise in SPaDs was also associated with introducing new signalling schemes and braking systems, highlighting a need for adequate training.

Perceptual/Attentional Difficulties

Signal design/placement

The frequency of missed signals and the drivers' response time have been associated with the signal intensity in relation to; its background, the intensity of the signal light and the frequency of flashing (Mashour and Devine 1977). There is no evidence to determine why any one signal should present a higher risk factor but a high percentage of SPaDs occurred at identified black spots; the most frequently reported hazard was a signal behind a bend (Van der Flier, Schoonman 1988) and more SPaD incidents occurred at signals which have been installed for less than 6 months (which emphasises the importance of the driver's route knowledge).

Incorrect assessment of track position.

Drivers often use their own route knowledge to anticipate the position of the next signal (Buck 1963). This is important as the signals are often only visible for a short length of time and at varying positions within the visual field. If the driver is unsure of their position then the chances of missing a signal increases. This is particularly important at night or under conditions of poor visibility. In order to determine their position on the track the driver often uses several sensory cues including auditory, visual and vibratory. Newer trains are often insulated against these cues, thus making the drivers' task in this respect, more difficult.

Selecting the wrong signal.

This often occurs when a number of signals are displayed in one location and the driver selects a signal which is inappropriate for their track. This is determined partly by the signal's position in relation to the track the driver is travelling on and also the drivers' route knowledge (Buck 1963).

Vigilance/monotony.
    The drivers' task has become more de-skilled and monotonous over recent years because of longer journey times and the growing automation in the cab causing boredom and monotony. The driver's attention may therefore wander away from the task and be diverted for long enough to miss a signal when it is within their potential visual field. The problem of a general lowering of train driver vigilance in a monotonous work environment is frequently mentioned (e.g. Endo and Kogi, 1975). Reason and Mycielska (1982) argue that errors relating to these factors occur because of misplaced competence rather than incompetence. This is often associated with highly skilled or habitual experiences where the task is largely automatic and attentional demand is low. Errors occur when attention is focused on something else or in familiar environments where the expected is the norm and vigilance is therefore reduced. An example of this is when drivers cannot recall the route they have just been driving. The probability of making such errors increases with task proficiency.

    Vigilance devices have been introduced to reduce such decrements of attention. These often take the form of buzzers or bells which the driver has to cancel. As a vigilance decrement has been shown to occur long before an auditory decrement the effectiveness of such devices at maintaining the driver's attention to the task is debatable. It is also possible that due to the frequency with which the driver has to cancel the sound they may have learnt to respond automatically without thought. In this case the vigilance device would have failed to bring the driver's attention back to the task and they still may pass through a red light.

    Wilde and Stinton (1983) found that certain types of vigilance devices not linked with direct control of the train could in fact, divert the driver's attention away from the task of driving the train to the task of cancelling the warning and thus fail to appropriately focus the driver's attention. They argue that vigilance devices should direct the driver's attention to some specific train driving task such as speed control. This is supported by Buck (1963) who states that certain types of railway vigilance devices are ineffective in relation to attentiveness but are relevant to maintaining wakefulness.

Driver Assumptions.
    A driver may sometimes have such high expectations of what a signal may be that he may fail to look at it. If this inference is incorrect an erroneous response may be made and he may in fact unconsciously pass through a red light. Even if the driver does look at the signal any false expectations they may have might restrict the perception and assimilation of true information (e.g. a signal may be interpreted as orange when it is in fact red). False expectations usually arise from past experience driving on that line and the fact that normally only approximately 2-3% of all signals through which the driver passes are red (Buck 1963).

Sensory error/mis-perception.
    Conditions of fog, sun, rain, etc. can distort brightness, contrast and distance perception thus increasing the possibility of error. Colour signals require a precise definition of hue and colour and any possible driver colour blindness needs to be established. Visual acuity is also important and eyesight should be tested regularly.

Obtaining information from another source
    In some situations, such as in a station, the signal may not be in a position that is easily observed by the driver who may therefore rely on other people to relay the

correct information regarding the signal aspect. If this information is incorrect then it is possible that a SPaD may occur (Buck 1963).

Severity of SPaDs

Different causal factors of SPaDs may affect the subsequent overrun past the signal. SPaDs with longer overruns have the greater potential for damage or collision with another object. A SPaD arising from a misjudgement of the braking distance means that the signal has been seen by the driver and they are aware of the need to stop. A SPaD which occurs because the driver has for some reason disregarded or failed to look for the signal means that the driver is unaware of the need to stop and therefore continues unaware of the potential dangers ahead. SPaD incidents where the driver is unaware of the need to stop therefore have the greatest potential for disaster and should to be addressed first. The importance of the driver's actions in determining the outcome of SPaDs is demonstrated in the diagram below.

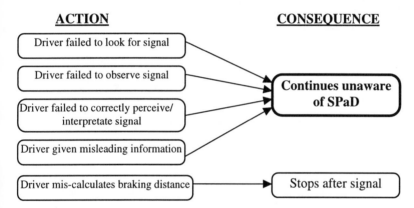

Diagram 1. Driver actions in SPaD causation and the subsequent awareness of the need to stop.

Conclusions

Precise data regarding SPaDs are difficult to determine as the driver group is dynamic and subjected to many external influences (e.g. new traction units, new signalling systems, staff turnover) however several conclusions can be drawn from this review of human factors and SPaD incidents:-

• Many causal and contributory factors are operating in the occurrence of SPaDs and these may have a cumulative effect.
• Different types of SPaDs occur because of different causal factors and different driver activities. Therefore SPaDs cannot be eradicated by one single solution.
• Different SPaDs have different severity outcomes and those with the highest potential for loss of life and disaster must be addressed first.
• Further research is needed on the role of shift work and organisational factors. Some drivers are more likely to be involved in SPaDs than others, although it is not clear why. An inability to cope with shift working patterns may be a cause.
• Lack of vigilance and attention are central issues, both appear to be associated with poor motivation and commitment. It is important that cab design should not de-skill the drivers' task or reduce the motivational levels of the driver who

should be an active and integral part of the driving task.
• Vigilance devices implemented into the cab environment should focus the driver's attention on the task of driving the train and not be distracting. This is important when designing fail-safe equipment such as the ATP system to reduce SPaDs. The sensory cues the driver uses to obtain their track location should not be under estimated and not designed out of the cab completely.
• Proper attention must be given to cab design and internal environmental conditions to ensure compatibility between cab visibility and signal placement and encourage driver alertness and efficiency.
• Drivers should have an active role in the organisational/workplace decision process to encourage feelings of control, commitment and motivation to the job.
• Appropriate training is important regarding the use of new or rarely used equipment. Signals which are passed frequently should be highlighted and addressed even if only by making the driver aware of them during training.
• It is important that good communication systems between the driver and other railway personnel (e.g. the guard, signalman or control room) are maintained. The appropriate personnel should receive training regarding the effective and accurate transmission of information.

The European Commission is seeking to improve the integration and inter-operability of European Transport Networks throughout Europe. Specific projects such as EUROCAB, EUROBALISE and EURORADIO have addressed improvements in the safety and ergonomic aspects of cab interfaces, signalling systems and radio control. Rail signalling is one of the most severe barriers to rail inter-operability and if the project is to be successful the problem of SPaDs will have to be successfully addressed.

References

Buck, L., (1963) Errors in the perception of railway signals, *Ergonomics* **11** (6).
Endo, T., Kogi, K., (1975), Monotony effects of the work of motormen during high speed train operation. *Journal of Human Ergology* **4**, (2), pp129-140.
Kogi, K., Ohta, T., (1975), Incidence of near accidental drowsing in locomotive driving during a period of rotation. *Journal of Human Ergology,* **4**, (1), pp 65-76.
Mashour, M. and Devine B. (1977) Detection performance and its relationship with human capabilities for information processing. *Reports from the Department of Psychology No. 495.* The University of Stockholm.
Reason J., Mycielska K., (1982) *Stress and fatigue in human performance.* J.Wiley.
Smith M.C., George D., (1987) *Health, stress and locomotive engineers.* O.P.R.A. Ltd. New Zealand.
Van Der Flier H., Schoonman W., (1988), Railway Signals Passed at Danger, Situational and Personal Factors. *Applied Ergonomics* **19** (2), pp1235-1241.
Verhaegen P.K., Ryckaert R.W., (1986). Vigilance of train engineers. *Proceedings of the Human Factors Society 30th Annual Meeting.* Human Factors Society, Santa Monica (California).
Wilde, G.J.S. and Stinson, J.F. (1983) - The monitoring of vigilance in locomotive engineers. *International Journal of Accident Analysis and Prevention* **15** (2) 87-93
Williams J., (1977) Railway Signals Passed at Danger. Paper presented to the Annual Conference of the Ergonomics Society.

CONTRIBUTORY FACTORS FOR OUTDOOR FALLS IN POSTAL DELIVERY EMPLOYEES

Tim Bentley and Roger Haslam

Health and Safety Ergonomics Unit,
Department of Human Sciences,
Loughborough University of Technology,
Loughborough, Leics, LE11 3TU

Outdoor falls are the largest cause of accidents and lost time within the delivery function of the Royal Mail. Accident-independent methods of investigation were used to provide data on contributory risk factors for outdoor falls to postal delivery employees. Possible factors identified include the use of unsafe working practices, which appear to be reinforced by rewards associated with reduced task time; use of footwear inappropriate for heavy use and slippery conditions; and the condition of public and private premises.

Introduction

Slips, trips and falls on the level represent a huge cost to industry in terms of lost production, medical and compensation costs, and human suffering. The Health and Safety Executive (HSE) figures for 1992/3 show that slips, trips and falls on the level represent 33 % of all major injuries, and 20 % of all over three day injuries to employees (HSE 1994). For those industries where a significant number of over three day injuries caused by a slip, trip or fall on the same level were reported to the HSE, the highest incidence rate occurred in postal services and telecommunications. Although the grouping of these two industries in HSE figures makes the picture unclear, it appears that postal services compare badly to other organisations for this type of accident. Postal delivery employees work for up to four hours per day, six days per week, exposed to the variable and unpredictable conditions of weather and walking surfaces. Fovergill (1995), in interviews with fall accident victims who had attended an accident and emergency department, found that potentially avoidable environmental factors such as uneven surfaces and inadequate lighting contributed to more than half of all falls occurring in public places. When the effect of adverse weather conditions is added to these environmental risk factors, it is perhaps unsurprising that outdoor falls represent the largest cause of accident and lost time within the delivery function of the Royal Mail, with outdoor falls representing nearly 30 % of all reported accidents and over 36 % of all days lost (Bentley and Haslam, 1996).

The aims of this study are to identify possible contributory risk factors for outdoor falls to postal delivery employees, and make recommendations for their removal or reduction.

Methods

Table 1 provides a breakdown of accident-independent methods used in the study and the information collected. The techniques draw on the experience and knowledge of postal delivery employees and managers, who are considered 'subject matter experts'. The information collected was used to produce a description of possible contributory factors for outdoor falls to delivery employees, and to supplement findings of accident-centred methods of investigation such as statistical analysis of accident data and a detailed accident follow-up survey (Bentley and Haslam, 1996).

Table 1. Accident-independent methods

Method	Information collected
Interviews with personnel responsible for safety (N=17)	Training of new recruits; Safety training; Type and quality of equipment and footwear provided; Management safety activities.
Discussion groups with postal delivery employees (3 groups of 8-10 employees)	Experience of fall accidents and near misses; Behavioural factors related to falls on delivery; Task factors; Situational influences on behaviour ; Experience of use of footwear and equipment; Safety attitudes and activities of management, supervisors and employees; Perception of risks of fall accidents.
Questionnaire survey of postal delivery employees and delivery office managers Response rates: Managers 48 % (25 respondents) Employees 39 % (110 respondents)	Physical environmental factors related to fall accidents; Behavioural factors which increase the risk of fall accidents; Reason unsafe behaviours used on delivery; Organisational factors which employees believe relate to fall accidents; Employees' ideas for reducing the incidence of outdoor fall accidents to delivery employees

Results : contributory factors identified

A summary of possible contributory risk factors for outdoor fall accidents in postal delivery employees, as identified by employees and managers, is presented in figure 1. Factors are grouped into three main sections: 1. Individual characteristics and behaviour; 2. Footwear and equipment; 3. Physical environment.

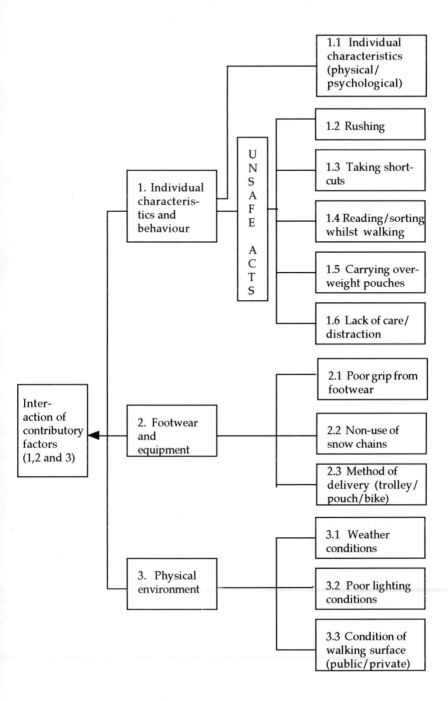

Figure 1. Summary of possible contributory risk factors
for outdoor falls in postal delivery employees

Individual Behaviour : Unsafe Practices and Time-Saving Behaviours

Figure 2 provides a breakdown of unsafe practices identified most frequently by employees and managers as increasing the risk of a fall accident whilst on delivery. A majority of employee discussion group members and questionnaire respondents argued that the use of these practices was necessary to meet Royal Mail's targets for delivery of first post on the busiest days of the week, and because of the increasing quantity of mail they must deliver within the time allocated for each delivery route. The mail load is unevenly distributed across days of the week and across the two delivery rounds. Later days of the week tend to be much heavier than Mondays and Tuesdays, and up to 90 % of mail is taken out on the first of the two deliveries. In addition, employees noted that they prefer to deliver mail at a regular time each day as they are aware that customers often rely on their good time-keeping. This sometimes leads to rushing, particularly when adverse weather or heavy work loads cause them to lose time.

Managers, on the other hand, tended to explain the use of unsafe practices in terms of the business' 'job and finish' policy which allows employees to go home once they have delivered the last letter of their second delivery. Official timings of walks assume safe practice (e.g. walk up and down every drive), therefore any time the employee is able to save by taking short-cuts, rushing or carrying over-weight pouches, contributes to an earlier finish time.

Reading whilst walking was considered by both employees and managers to greatly increase the risk of fall accidents. Employees argued that to check the address for the next delivery point whilst stationary would incur unacceptable time costs which outweighed safety considerations. In addition, employees noted that this practice was used to help avoid 'miss-deliveries' (delivery of mail items to an incorrect address), and to follow the route on an unfamiliar round.

Reasons given by employees for carrying overweight pouches included the unequal distribution of mail between days and rounds, and the bulk and weight of heavy mail items (e.g. magazines, advertising and business mail).

Further organisational factors associated with the use of unsafe practices include the passing on of unsafe habits and time-saving behaviours through 'on-the-job' training methods, and difficulties related to supervisory monitoring of delivery employees, due to their remote location.

In summary, unsafe, time-saving practices appear to be habitually used by a large proportion of delivery employees. It appears the use of unsafe working practices is reinforced by the rewards of reduced task time, and is perceived by a large number of employees as necessary during the busiest periods of work. Any attempt to encourage employees to adopt alternative safe practices may be unsuccessful in the long term while these types of situational influences on behaviour exist.

Footwear and Equipment

Employees, managers and safety personnel noted that the choice of footwear offered by Royal Mail to its delivery employees does not offer a high level of protection from either slipping (the first fall event in over 50 % of all outdoor falls to delivery employees, (Bentley and Haslam, 1996)) or ankle injury (the most commonly injured body part in all outdoor fall accidents to delivery employees). Grip is worn away, particularly at the vital heel area, after only a few months' wear (verified in a series of inspections of employee's shoes, N=38), but footwear is often replaced just once a year.

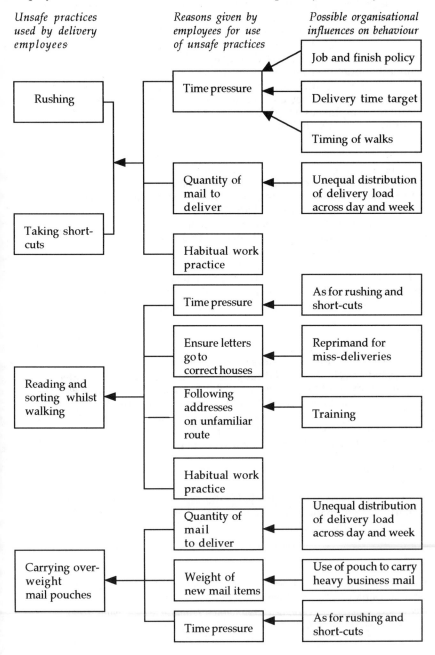

Figure 2. Unsafe practices: reasons for their use and possible organisational influences on behaviour

Trolleys are designed as an alternative to carrying mail pouches manually, but appear to be under used. In discussion groups, employees argued that trolleys slowed them down, could not be used for remote walks or on walks with many steps, and produced a security problem when left unattended. Snow chains are issued to delivery employees as protective equipment for working in snow, but were described by employees in discussion groups as being effective only in fairly deep snow, and subsequently under used. Accident statistics for outdoor falls to delivery employees suggest the incidence of falls increases markedly during extended periods of snow fall (Bentley and Haslam, 1996). For this reason it is important that employees are made aware of the conditions under which they should wear snow chains.

Physical Environment

Unsurprisingly, the weather and walking conditions underfoot were mentioned most often as contributory factors in outdoor falls to delivery employees. Of the measures to reduce the risk of falls in adverse weather conditions, the most promising include the provision of superior occupational footwear, lighter pouch loads (mail broken up into smaller individual pouch loads), and the allowance of additional time to complete deliveries in adverse weather conditions.

Obstacles and contaminants left on home owners' drives represent a major hazard, and are particularly problematic when there is insufficient lighting of the area. Likewise, the unpredictable condition of public pavements and roads was described as a major risk factor.

Conclusion

Possible contributory factors for outdoor falls to postal delivery employees have been identified through the use of accident -independent techniques. Many outdoor falls are avoidable events. Measures to reduce the incidence of these accidents might include: rewarding and reinforcing safe rather than unsafe practices; provision of safe and unobstructed walkways on private premises; and the supply of appropriate equipment and training to employees to help them undertake their work in greater safety.

Acknowledgements

The authors wish to acknowledge the support of the Royal Mail, and particularly wish to thank John Leaviss, Mike Dixon and Ian Cooper for their close interest with this work.

References

Bentley, T., and Haslam R., 1996 , Outdoor falls in postal delivery employees: A systematic analysis of in-house accident data, *Proceedings of the 1st International Conference on Applied Ergonomics, Istanbul, 21-24 May, 1996*.
Fovergill, J., O'Driscoll, D., and Hashemi, K. 1995, The role of environmental factors in causing injury through falls in public places, Ergonomics, 38, 220-223.
Health and Safety Executive 1994, Key fact sheet on slipping and tripping injuries to employees between 1986/87 and 1992/3., Statistical Services Unit, Room 512, Daniel House, Bootle, Merseyside.

Thermoregulation

EFFECTIVENESS OF A NECK COOLING PERSONAL CONDITIONING UNIT AT REDUCING THERMAL STRAIN DURING HEAT STRESS.

Lisa Bouskill and Ken Parsons

Department of Human Sciences
Loughborough University of Technology.
Leics. LE11 3TU

Eight male subjects underwent two 1-hour exposures, in simulated South European conditions (t_a = 39.9 ± 0.2 °C, t_r = 39.6 ± 0.2 °C, rh = 27.1 ± 0.3 %), of which the last 50 minutes involved an exercise protocol aimed at simulating a typical activity level of RAF ground crew. A neck cooling collar supplied by a personal conditioning unit was worn in one exposure. A significant lowering of 4-point mean skin temperature and reduction in subjective whole body discomfort were recorded during cooling. Similarly, cheek and forehead temperatures were also lowered although this was not reflected by head subjective data. Oral and aural temperatures, heart rate and water intake were all highest with neck cooling, these data being attributed to the extra exertion when carrying the personal conditioning unit. If the unit was not so heavy, neck cooling in this way may prove more effective.

Introduction

An increasing number of industrial and aerospace environments pose thermal stress problems. Specific instances include the activities of helicopter and other aircrews, airfield ground support crew, fire-fighters, racing drivers, mine workers and blast furnace operatives. There are several possible avenues by which it may be possible to overcome such heat stress, including reducing exposure times, using alternative clothing, reducing metabolic heat load, shielding from any radiative heat source or reducing environmental temperature. These methods on the main are often either impractical or too expensive to realistically be considered. However, instead of concentrating on changing a person's general environment it has previously been suggested that altering their microclimate, the thermal environment next to their skin, produces better results. Several studies have shown that heat removal by circulating cool water through pipes in close proximity to the skin is an effective method for the alleviation of heat strain (Gold and Zornitzer, 1968; Shvartz, 1971; Shvartz and Benor, 1971; Webb and Annis, 1968; Webb, Annis and Troutman. 1972). Similarly, air-ventilated clothing may also be used for microclimate control (Allen,

Belyavin, Flick and Higgenbottam, 1981; Griffiths and Boyce, 1971), but as reported by Nunneley and Maldonado (1983) cooling by a liquid-conditioned garment (LCG) has many advantages. Full and part coverage LCG's have been used for body cooling in the U.S.space program and in certain industrial settings. Where several layers of clothing are worn in addition to such suits limb movement may be impaired. By using a neck cooling collar as opposed to full suits or even torso cooling vests such problems should be overcome.

Efforts to remove metabolic heat by partial or differential body cooling have been undertaken for many body regions (Gold and Zornitzer, 1968; Nunneley, 1970; Billingham, 1959; Burton, 1966; Webb and Annis, 1967). From these studies it appears that three main factors determine the efficiency and effectiveness of regional body cooling. These being: the amount of vascularization and the proximity of large blood vessels to the skin in the region being cooled, the surface area / mass of the cooled region and the size of the surface area being cooled.

Considering these points and given the surface area of the neck, it's cooling is surprisingly effective at reducing heat strain, decreasing strain by 53% (Shvartz, Aldjem, Ben-Mordechai and Shapiro. 1974).

Under South European (flying) conditions personal conditioning units have proved efficient at improving thermal comfort when used with liquid conditioned vests (Cohen, Allan and Sowood. 1988). However, they have not been tested in the same way with neck cooling. Similarly, neck cooling has been investigated (together with liquid conditioned vests) under simulated South European conditions but with thermocirculators providing the cooling medium. This study examines the effectiveness of neck cooling when the cooling medium is provided by a personal conditioning unit which is carried around the waist.

Method

Eight healthy, physically active males, aged 19-23 years, participated in this study. They were fully informed of the objectives, procedures and possible hazards of the experiment. Experiments were conducted in February when the subjects were naturally non-acclimatised to heat.

The test protocol consisted of two exposures to a simulated South European Environment ($t_a = 39.9 \pm 0.2$ °C, $t_r = 39.6 \pm 0.2$ °C, rh $= 27.1 \pm 0.3$ %). Each exposure lasted 60 minutes, of which the first 10 minutes were spent sat at rest and the remaining 50 minutes were spent exercising. The exercise for this protocol involved a step test with a Reebok™ step set to a height of 250mm. Subjects were required to step in time with a metronome set at 60 beats per minute. With one movement per second this translated to 15 full steps per minute. From Wadsworth's equation (1985), using the mean subject weight (73.35 kg) this gave a work load, as represented by metabolic rate, of 295 Wm^{-2}. The extra weight of the personal conditioning unit (3.5 kg) increased this figure by 14 Wm^{-2}.

Condition A (no neck cooling) was utilised to obtain baseline strain data. In Condition B neck cooling was provided for the 50 minutes of exercise. During neck cooling the inlet and outlet temperatures of the neck collar were maintained at 18.7 ± 0.2 °C and 19.4 ± 0.7 °C. Flow through the collar was constant at 0.8 lh^{-1}. The order of testing was balanced to reduce possible effects of training and acclimatisation. A given subject was always studied at the same time of day, to avoid biorhythm and circadian influences, and with several days separating experiments to prevent acclimatisation.

Subjects reported to the laboratory at least 1-hour after the preceding meal and having been asked to refrain from alcohol, caffeine and tobacco for 12-hours prior to testing. Before instrumentation and following exposure subjects were weighed semi-nude. During the study aural, forehead, cheek, neck (larynx), neck (ear) and Ramanathan 4-point mean skin temperatures were recorded at 1-minute intervals, all thermistors having previously been calibrated in a water bath against a mercury-in-glass thermometer. Sublingual temperature was recorded using a mercury-in-glass thermometer prior to exposure, at the initiation of testing, and after 15 minutes and 45 minutes of exercise. Heart rate was monitored using a cardiac sports tester and recorded at 1-minute intervals. Samples of expired air were collected in Douglas bags prior to exposure, and after 20 minutes and 50 minutes of exercise. Throughout the exposure subjective assessments for whole body and head perceptions, using the ASHRAE 4 point discomfort (1 = not uncomfortable, 2 = slightly uncomfortable, 3 = uncomfortable, 4 = very uncomfortable) and 8 point sensation (1 = cold, 2 = cool, 3 = slightly cool, 4 = neutral, 5 = slightly warm, 6 = warm, 7 = hot, 8 = very hot) scales were requested of the subjects at 10-minute intervals. In both exposures, subjects wore shorts, T-shirts, socks and tennis shoes. In Condition B they also wore a webbing belt for attachment of the personal conditioning unit. Subjects were allowed drinks of water on request, but to avoid water temperature (~15 °C) influencing oral temperature readings drinking was discouraged immediately prior to such readings being taken. Without neck cooling subjects drank a mean of $0.5 \, 1 \pm 0.2 \, 1$ compared with a mean of $0.5 \, 1 \pm 0.1 \, 1$ with neck cooling.

Results

The absolute mean end values for oral temperature, left aural temperature, right aural temperature and heart rate for conditions A and B are shown in Table 1. The gains of these parameters throughout exposure are shown in Table 2.

Table 1. Absolute mean end values.

	No Cooling	Neck Cooling	Difference
Heart Rate (bpm)	140	146	7
L. Aural Temp (°C)	37.8	38.0	0.2
R. Aural Temp (°C)	37.8	38.0	0.2
Oral Temp (°C)	37.5	37.6	0.1

Table 2. Gains throughout exposure.

	No Cooling	Neck Cooling	Difference
Heart Rate (bpm)	60	68	8
L. Aural Temp (°C)	1.3	1.6	0.3
R. Aural Temp (°C)	0.7	0.9	0.2
Oral Temp (°C)	1.1	1.4	0.3

Throughout both exposures heart rate steadily rose. Wilcoxon's matched-pairs signed ranks test did not identify any point of significance comparing these data between exposures. Left aural temperature again rose consistently throughout both exposures and was highest with neck cooling ($P = 0.05$). At the end of exposure left aural temperature gain was significantly higher with neck cooling (P < 0.05). Likewise, right aural temperature rose throughout both exposures and was highest with neck cooling (P = 0.05), but there was no significant difference with respect to end value gain between conditions. Oral temperature also rose gradually throughout both exposures. It's gain between beginning and end values was significantly higher with neck cooling (P = 0.05).

4-point mean skin temperatures were statistically lower with neck cooling (P < 0.05). Neck 1 (larynx) temperatures were consistently and statistically lower with neck cooling (P = 0.05). Neck 2 (ear) temperatures remained fairly constant throughout both exposures and there was no statistical significance between exposures. Cheek temperatures were consistently higher with no neck cooling (P < 0.05) and similarly, forehead temperatures were also consistently higher with no neck cooling (P < 0.05).

Metabolic rate, as calculated from Douglas bag samples, was usually highest with neck cooling, but not to the point of achieving statistical significance. Values of metabolic rates during Conditions A and B are given in Table 3.

Table 3. Metabolic rates.

	No Cooling	Neck Cooling	Difference
Rest (Wm^{-2})	76	75	1
20 mins (Wm^{-2})	218	236	18
50 mins (Wm^{-2})	260	268	8

With the exception of 2 subjects weight loss was consistently higher with neck cooling, although not to the point of reaching statistical significance (P > 0.05).

Subjective ratings of sensation and discomfort for head and whole body at the end of exposure are given in Table 4. Only one point of statistical significance was found comparing subjective data responses throughout both exposures. Whole body sensation throughout both exposures was consistently highest in without neck cooling (P = 0.05).

Table 4. Subjective data at end of exposures.

	No Cooling	Neck Cooling	Difference
Sensation			
Body	7.1	6.8	0.3
Head	7.1	7.3	0.2
Discomfort			
Body	3	3	0
Head	3	3	0

Discussion

The neck collar used in this study has been shown to significantly cool the neck (P = 0.05 for Neck 1 data) throughout the period of exposure. Neck 1 skin temperatures were seen to rise gradually towards the end of the neck cooling exposure thereby suggesting that the collar became less effective with prolonged exposure. However, at the end of each cooling exposure the ice pack was never fully melted indicating that there was a capacity for a longer cooling period.

Since the thermistor at the Neck 2 (ear) site was located above the level of the neck cooling collar it is understandable that the skin temperature at this site was not seen to change significantly during exposure.

Although not increased to a statistically significant level, heart rates were consistently highest during the cooling exposure. It appears reasonable that heart rate was not lowered with neck cooling, as would be expected with an alleviation of heat stress, because of the extra physical exertion involved in carrying the personal conditioning unit. Similar reasons can be attributed to the fact that sweat loss was not significantly reduced during the cooling exposures. However, it is a point of contention whether a reduction in sweating is beneficial. Although a reduction in facial sweating can ease the problem of blepharospasms, sweating is the root of heat loss by evaporation. By reducing heat loss by evaporation excessive heat storage is more likely.

A slight increase in water intake was also observed with neck cooling, again possibly because of the extra physical exertion when carrying the personal conditioning unit. Provision of this extra water in an 'in-the-field' scenario may not be possible. Subsequently, on this basis, neck cooling may not be a suitable option in some instances.

The statistical significance ($P < 0.05$) observed comparing 4-point mean skin temperature between exposures correlates with the fact that neck cooling is thought to cool venous blood returning to the heart via the jugular vein. This cool blood is then pumped around the body thereby cooling other areas of the body and as such resulting in a lower 4-point mean skin temperature.

Cheek and forehead temperatures were seen to be significantly lowered with neck cooling (both $P < 0.05$). This suggests that cooling the neck also resulted in direct cooling of the facial arteries. However, these lowering of temperatures was not largely reflected in the subjective responses. Other than for whole body sensation ($P = 0.05$) no other subjective data responses showed a statistical improvement with neck cooling.

Conclusions and Recommendations

4-point mean skin temperature was lowered significantly with cooling, this was also reflected in a significant improvement in subjective assessment of whole body discomfort, but not in sensation. Cheek and forehead temperatures were lowered significantly with cooling, but not to the point of significantly improving thermal sensation or discomfort ratings of the head. Sweat loss was less with neck cooling, but not significantly so. Water intake was highest with cooling, but not to the point of reaching statistical significance. Heart rates, oral temperatures and aural temperatures were all higher with neck cooling, thereby contradicting previous studies' findings. The extra weight carried, in the form of the personal conditioning unit, during neck cooling may explain these data.

The detrimental effects including increases in core temperature, heart rate and water intake preclude active cooling with a neck collar and personal conditioning unit since they far out weigh the positive effects of reduced facial temperatures and 4-point mean skin temperatures and improved whole body sensation.

This study has indicated that the main drawback with using a personal conditioning unit and neck collar is the weight of the cooling unit. A practical way of assessing the level of this extra stress would be to repeat this study with a third exposure in which the neck cooling collar and personal conditioning unit are worn but without the cooling unit being switched on. A further possible extension to this study would be to conduct more work looking at the effectiveness of this system with different work loads. It may be that the system works adequately with less strenuous work loads. Another possible line of research from this study is that of the effect on evaporation levels of neck cooling. Finally, given that the main problem highlighted by this study is that the weight of the cooling unit is its biggest drawback then it would probably be useful to look at ways of making the system lighter.

Acknowledgements.

The authors would like to thank the subjects for their time and patience and acknowledge the help (and patience) of Mr T Cole who supervised the thermal chamber.

References.

Allan, J.R. Belyavin, A.J. Flick, C.A. and Higgenbottam, C. 1981, Detection of visual signals during induced cycles of core temperatures. RAF Institute of Aviation Medicine, Farnborough (Hants), England. Report No. 592.

Billingham, J. 1959, Heat exchange between man and his environment on the surface of the moon. British Interplanetary Society. **17**, 297-300.

Burton, D.R.1966, Performance of water conditioned suits. Aerospace Medicine. .**37**, 500-504.

Cohen, J.B. Allan, J.R. and Sowood, P.J. 1988, The effectiveness of head or neck cooling when used with a liquid conditioned vest during simulated sorties in the European fighter aircraft. RAF Institute of Aviation Medicine, Farnborough, (Hants), England. Report No. 555.

Gold, A.J. and Zornitzer, A. 1968, Effect of partial body cooling on man exercising in a hot environment. Aerospace Medicine. **39**, 944-946.

Griffiths, I.D. and Boyce, P.R. 1971, Performance and thermal comfort. Ergonomics. **14**, 457-468.

Nunneley, S.A. 1970, Water cooled garments: a review. Space Life Sciences. **2**, 335-360.

Nunneley, S.A. and Maldonado, R.J. 1983, Head and / or neck cooling during simulated cockpit heat stress. Aviation Space and Environmental Medicine. **54**(6), 496-499.

Shvartz, E. 1971, Effect of a cooling hood on physiological responses to work in a hot environment. Applied Physiology. **29**, 36-39.

Shvartz, E. Aldjem, M. Ben-Mordechai, J. and Shapiro, Y. 1974, Objective approach to a design of a whole-body, water-cooled suit. Aerospace Medicine. **45**(7), 711-715.

Shvatz, E. and Benor, D. 1971, Total body cooling in warm environments. Applied Physiology. **31**, 24-27.

Wadsworth, P.M. 1985. Laboratory evaluation of ISO/DIS 7933 (1983) analytical determination of heat stress. BSc (hons) project. Human Sciences Department, Loughborough University of Technology, Loughborough, LE11 3TU.

Webb, P. and Annis, J.F. 1967, Biothermal responses to varied work programs in men kept thermally neutral by water cooled clothing. NASA. Cr-739.

Webb, P. and Annis, J.F. 1968, Cooling required to suppress sweating during work. Applied Physiology. **25**, 489-493.

Webb, P. Annis, J.F. and Troutman Jr, S.J. 1972, Human calorimetry with a water cooled garment. Applied Physiology. **32**, 412-418.

REDUCING RUNNER'S RISK IN SUDDEN HOT/HUMID WEATHER: WILSON'S HOT WEATHER HEURISTIC

Tay Wilson

Psychology Department
Laurentian University
Sudbury, Ontario, Canada

In the current more variable weather conditions, runners are more frequently being exposed to sudden, unaccustomed bouts of hot/humid weather. There is a clear need for an "on the move" heuristic protective behavioural criterion which runners might actually use. What is here called the "Wilson hot weather heuristic" based upon the runner's ability to "effortlessly" return to the earlier pace within a few minutes of being forced to drastically reduce pace because of heat stress is explored here in terms of hidromeiosis (sweat reduction associated with wetted skin) and some mathematical modelling of parameters involved in the solution of the relevant heat balance equation.

In recent years the weather appears to have become more variable forcing distance runners to face sudden bouts of high temperature/high humidity training condition to which their bodies have had little time to adapt. The effects of heat stress can be rapid and catastrophic. Leithead and Lind (1964) noted that an electrician collapsed in convulsions and coma with a body temperature of 42^0 C one hour after entering a cramped space aboard ship. In the 1995 world masters track and field championships at Buffalo, U.S., events one afternoon were cancelled because ambulances attending to heat overcome runners became overloaded.

Exercise in the heat (see Parsons, 1993), leads to initial sympathetic vasoconstriction to divert blood flow to active muscles, increased cutaneous blood flow (vasodilation), decreased central nervous system blood volume, decreased effective circulatory volume due to sweat loss, decreased stroke volume and increased heart rate to maintain cardiac output. Evaporation of sweat is the dominant method for maintaining a stabilized core temperature. Sweating stimulates vasodilation to increase the blood supply and sweat gland fluid. Local skin temperature rises can stimulate some inactive sweat glands and increase overall sweat production. Sweat rates of one litre per hour are common (1litre = 675 W of heat loss, NIOSH, 1986). For review see Kerslake (1972) and Edholm and Weiner (1981).

Distance runners tend to be an independent, self-driven lot whose major mistake is to over-do work-outs to the point of exhaustion/injury. E.g. British former 10 km world record holder Brendon Foster, on his own admission, would have jeopardized his gold medal Commonwealth race by doing a "four minute" mile in a workout consisting of running the straights and jogging the curves if he had not been stopped by his coach. In the danger posed by sudden hot/humid weather, simple admonitions from self or others to "ease-up" are likely to be ineffectual. What is desirable in these circumstances is some sort of "during the run" self monitoring technique which is acceptable to athletes and which might keep them from self-destructing. The goal in this study was to search for a useful self-monitoring technique to enable runners to train safely but hard during sudden hot/humid weather.

Method

The running course consisted of repetitions of a four km, hilly, three lobe, wind protected, route through a lightly wooded area in Sudbury, Canada with nine reasonable size hills on each loop which allowed ample opportunity for sudden heart rate increases. The nature of the course allowed several easy exits to civilization and water. Because, the goal was to develop a workable self-monitoring technique for running in sudden heat stress in the absence of equipment, it was considered important to have, as subject, an experienced runner with good knowledge of the course willing to introspect quite closely about his body activity; moreover, because of the risk involved, the subject had to be the experimenter. The subject then was 54 years old with 40 years experience in competitive running at various distances who had over ten years run more than a two thousand timed kilometres over the chosen course. This experience provided a solid phenomenological base upon which to evaluate subjective effort over sections of the course. During a sudden hot/humid spell, immediately after having spent a month in much cooler/drier conditions, the subject, clothed in t-shirt and shorts, endeavoured to run as hard and as often as possible over the course as if he were peaking for a major race while, at the same time, noting what he was experiencing during and immediately after the runs. As some degree of protection against injury, the runner had a one hour nap within an hour of completing workouts. During this time his average weekly running distance was kept at least 20% above his weekly average for the previous year.

Results

A summary of the runners training versus weather appears in table 1. The overall level of activity was rated subjectively by the runner as extremely stressful. E.g., at the completion of almost every workout, he had to hold on to the hand rail of the staircase leading to the change rail in order to avoid strong leg cramps. Leg cramps in bed at night or during the run were experienced twice. Unaccustomly, twice after runs, the runner felt his t-shirted back to be virtually burning in the sun. Despite the interest in the experiment, on several days the runner felt compelled to swim rather than run. Finally, the running times in the hot, humid conditions were about .3 ms^{-1} (two standard deviations) slower than the comparable three year average (viz. in table 1, on the cool Aug 21 the running distance increased by 50% and on the cool Aug 24th the running speed was .3 ms^{-1} above the monthly average). The goal of subjecting the runner to strong heat stress appears to have been fulfilled. Were any useful behavioural self-monitoring cues discovered? Yes. Twice (day 4 and 14) the runner noticed that within a few minutes of being forced to drastically reduce pace by about 30 s km^{-1} because of heat stress, he had made an seemingly effortless return to a "floating" run at the original pace. Moreover, on those days, the runner appeared to recover somewhat better and was able to perform reasonably well the next day given the conditions. It must be emphasized that this experience was as novel as it was dramatic. In previous runs over about 2000 km representing more than 500 laps in moderate or cool weather, the subject had experienced only two outcomes, either a maintainance of relatively constant pace throughout the run; or a gradual deceleration in speed, unless, of course, the first lap was deliberately run slowly.

Discussion

What explanation can be evinced for these results? The major way people acclimatize to heat is by "training" their sweat glands. Over a number of days, maximum sweat production is greatly increased, distribution is changed and a given submaximal rate is achieved at lower skin or core temperature. Some acclimatization also occurs by increasing blood volume and a fall in NACL content of sweat and urine (see Parsons, 1993, fig 8.4, p175). Over five days acclimatization rectal temperatures can fall 1^0C, 'max' pulse rates 30 beats per minute and sweat loss can increase 0.15 kg\70 kg body weight\hr. Acclimatization is proportional size and duration of body temperature increase; acclimatization may not require exercise.

Two major problems occur with sweating at higher body core temperatures: dripping which is ineffective for heat loss while depleting valuable body water and hidromeiosis.   Hidromeiosis is a local reduction of sweat in hot, humid conditions associated with wetting of the skin (Kerslake, 1972). Edholm and Weiner (1981) suggest, as cause, that continually wetted skin reabsorbs water leading to a hydrated and swollen epidermus with poral closure. (Note so-called sweat gland 'fatigue' probably does not occur.) The decline in sweating is exponential and tends toward zero; however, recovery is rapid. The decrease in sweating promotes a further, often rapid, increase in 'core' temperature to beyond 38-39^0 C and heat stroke may occur with mental and behavioral degradation, sweating and thermoregulation failure and death.

The best explanation of the "effortlesss" return to pace reported above appears to be in terms of hidromeiosis. That is, some time during the heat stress run the skin reaches maximum wettedness, body heat continues to build, the epidermis swells and sweat production is rapidly (exponentially) curtailed. Slowing down for a short time allows epidermal swelling to reduce, sweating and concomitant cooling to occur and the runner experiences an effortless return to the former running pace.

A mathematical check on this interpretation can be made by calculating the sweat rate and degree of skin wettedness required to handle the generated metabolic heat of a runner in such conditions; e.g., by substituting appropriate values into the practical heat balance equation of ASHRAE 1989 - see Parsons, 1993, p11). Since the calculation is somewhat involved, a listing of key terms, defining equations and some key values used in this calculation appears in table 2. For air, radiant and skin temperatures of 22, 25 and 35 ^{0}C respectively, relative humidity of 50%, body weight and height of 178 kg, 1.83 m, air velocity of .5 ms^{-1} and Morrissey et al (1984) calculated metabolic rate of 500 W m^{-2} for an overall running speed of 3 m s^{-1}, an overall skin evaporation requirement to maintain thermal equilibrium was calculated as 414 W m^{-2} while the actual sweat evaporation rate was calculated as 156 W m^{-2} . Taking into account sweat efficiency (loss due to drippage) it was calculated that 1.86 litres m^{-2} of sweat would be required to reach equilibrium.   This value is beyond normal sweating capability and core temperature elevation would be expected with concomitant running speed degradation. Repeating these calculations for a, perhaps more realistic, metabolic rate of 250 W m^{-2} yielded an overall skin evaporation

day	rel hu	dry blb	wet blb	dew pt	wnd m/s	run km train	run spd m/s
						Table 1. Training regime during Aug 95 hot weather adaptation (sp = speed, sw = swim, st = stairs, w = weights	
1	96	15.7	15.3	15	4.4	0	
2	64	18.4	14.2	11.5	1.7	6 w	2.90
3	79	21.9	19.3	18	5.0	8	2.77
4	96	19.2	18.2	17.7	4.4	8 w	3.04
5	81	17.7	15.7	14.4	2.2	8	3.03
6	65	22.7	18.3	15.7	5.6	cramp 0	
7	66	21.7	17.5	15	3.3	cramp 6	NT
8	64	24	19	16.8	4.4	4 w	3.27
9	56	24.8	19	15.5	5.0	8	2.95
10	74	23.5	20.1	18.5	2.8	8	NT
11	59	28	21.8	19.2	7.8	s w	
12	49	22.7	15.9	11.6	3.9	s	
13	72	19.3	16.3	14.2	3.9	s	
14	82	21.5	19.3	18.3	3.3	8	3.02
15	43	25.9	17.1	12.3	5.6	10	2.92
16	53	27	19.6	16.7	2.2	8	3.03
17	52	25.5	18.5	14.8	5.0	4 sp	
18	50	28.8	20.8	17.5	2.2	4 st	
19	50	27.8	19.8	16.4	6.7	8 st sw	
20	34	26.2	16	9.2	2.2	8 st sw	
21	34	25.8	15.4	8.7	8.3	4 st sw	
22	55	18	12.6	8.9	5.0	12 w	3.06
23	64	22.9	17.9	15.7	6.7	8	2.74
24	47	16.7	10.7	5.3	7.8	8 w	3.28
25	37	18.9	11.1	3.9	1.7	8	2.79
26	99	16.4	16.2	16.3	1.1	4sp	NT
27	34	23.5	19.9	18.2	5.6	4	NT
28	56	21.9	16	12.7	7.8	4 w sp	NT
29	14	19.7	13.1	7.6	5.6	8	3:12
30	52	19.6	14	9.5	5.6	2	NT
31	93	19.8	19.2	18.6	3.3	4 s w	2.98
M	**60.3**	**22.1**	**17.0**	**14.0**	**4.5**		**2.99**
S	**20.5**	**3.7**	**2.8**	**4.2**	**2.0**		**0.16**

Table 2. ASHRAE (1989) heat equation formulae, units and values used (values in bold refer to values obtained with metabolic rate M = 500, those in italics refer to values obtained with M = 250, values in normal type are applicable to both M's)

Term, symbol (unit)	Defining equation
metabolic power M and external work W (W m^{-2}) Morrissey and Lion (1984)	M=-75.14+3.11w+v^2(2.72L+87.75)+13.36(w+L)(L/w)^2 where load L = 1, body weight w=178 and wind speed v = 0.5: **M = 500, W = 0**
temp of skin t_{sk}, core t_c radiation t_r, air t_a, wind vel v_a and rel hum ϕ	basic parameters = 35, 37, 25, 22, 0.5 and 0.5 respectively
skin convection ht C (W m^{-2})	C = $h_c f_{cl}(t_{cl} - t_a)$ = 41; h_c = 8.3v$^{0.6}$; cloth area factor f_{cl} =1.03;
skin radiation ht R (W m^{-2})	R = $h_r f_{cl}(t_{cl} - t_r)$ = 19
respiratory conv C_{res} (W m^{-2})	C_{res}= 0.0014M(34 - t_a)/A$_D$ = **4.2**, *2.1*
respiratory evap E_{res} (Wm^{-2})	E_{res}= 0.0173M(5.87 - P$_a$)/A$_D$ = **19.7**, *9.8*
req equil skin evap E_{req} (Wm^{-2})	E_{req} = M - W - C_{res} - E_{res} - C - R = **414**, *176*
req equil sweat S_{req} (W m^{-2} or litres l)	S_{req} = E_{req}/r_{req} = **717 or 1.86**, *305 or 0.795*
sweat evap rate E_{sw} (Wm^{-2})	E_{sw} = M_{sw} x Lv = 156
swt secr rate M_{sw}(kg s^{-1}m^{-2} or l where .26 litre hr^{-1}=100 Wm^{-2})	+4.7 x10^{-5} (wsigb)exp(wsigsk/10.7) = **6.9 x 10^{-5}** or *0.248*
body (skin) ht signal wsigb	t_b - $t_{b,n}$ where $t_{b,n}$ = 36.8; t_{sk}- $t_{sk,n}$ where $t_{sk,n}$ = 33.7
max evap rate achievable with completely wet skin E_{max} (W m^{-2}); R_{ecl} = clothing plus air evap resistance = 0.015 m^{2}kPa W	$E_{max} = \dfrac{(w=1)\ (P_{sks}-P_a)}{R_{ecl}+\dfrac{1}{f_{cl}h_e}}$ =167
skin wettedness w (ND)	w = E_{sw}/E_{max} = 0.92
evap efficiency at required sweat rate r_{req} (W m^{-2})	r_{req} = 1- (w^2/2) =.58
saturated vapour pressure at air temperature p_{sa} (kPa)	$P_{a,s}$ = exp(18.956 - (4030.18/(t_a + 235))) = 2.63
saturated vapour pressure at skin temperature p_{sa} (kPa)	$P_{sk,s}$ = exp(18.956 - (4030.18/(t_{sk} + 235))) = 5.62
air partial vap pres p_a (kPa)	p_a = ϕ p$_{sa}$ = 1.32
radiative heat transfer constant h_r (W m^{-2} K^{-1}) t_{cl} = 28.5, h_r = 4.8 (iterative calculation)	$h_r = 4\sigma\epsilon_{sk}\dfrac{A_r}{A_{Du}}\,[273+\dfrac{t_{cl}+t_r}{2}]^3$

requirement to maintain thermal equilibrium of 176 W m^{-2} or 0.795 litres hr^{-1}, well within normal capabilities.

Conclusion

Subject to further research, if runners persist in training hard in hot condtions despite warnings to the contrary, they might try the Wilson hot weather heuristic, namely, to continue running only if, within a few minutes after a sudden drop in pace caused by heat stress, the runner experiences an "effortless" return to near the original pace. In this case hidromeiosis may have stopped and normal sweating capability may be returning. Better still, don't run.

References

ASHRAE, 1989 Physiological Principles, Comfort and Health, in *Fundamentals Handbook*, Atlanta.

Edholm, O. G. and Weiner, J. S., 1981, *The principles and practice of Human Physiology*, London: Academic Press.

Kerslake, D. M., 1972, *The Stress of Hot Environment*, Cambridge: Cambridge University Press.

Leithead, C. S. and Lind, A. R., 1964, *Heat Stress and Heat Disorders*, London: Cassell.

NIOSH, 1986, *Occupational exposure to hot environments*, National Institute for Occupational Safety and Health, DHHS (NIOSH) Publication No. 86-113, Washington DC, USA.

Morrisey, S. J. and Lion, Y. H., 1984, Metabolic cost of load carriage with different container sizes, *Ergonomics*, 27(8), 847-853.

Parsons, K. C., 1993, Human Thermal Environments, Taylor & Francis, London.

HEAT STRESS IN NIGHT-CLUBS

Marc Mcneill and Ken Parsons

Department of Human Sciences
Loughborough University of Technology
United Kingdom

An Internet survey of behaviour, attitudes and opinions of regular club-goers found that night-clubs were considered to be hot or very hot places where many respondents experienced heat related illnesses. The thermal conditions of a night-club were measured (maximum 29°C air temperature, 90% relative humidity) and simulated in a thermal chamber. Four male and four female subjects danced for one hour. The results showed a rise in core temperature (mean=1.8°C, sd=0.26) and skin temperature (mean=1.34°C, sd=0.48) and a sweat rate of almost 1l/h. Subjects generally felt hot and sticky, preferring to be cooler. The physiological responses compared well with predictions from ISO 7933 and the 2-node model of human thermoregulation (Nishi & Gagge, 1977). The predicted effects of continuous dancing for four hours gave a core body temperature increase to 39.1°C, well above the WHO limit of 38°C in occupational settings. Using ISO 7933 appropriate work-rest schedules for dancing and water requirements were suggested.

Introduction

Every weekend an estimated half a million people in the UK go to raves (all night dance parties) and night-clubs (Jones, 1994). They go to dance, often for long periods of time. The ambient thermal conditions of the night-clubs they dance in are often hot and humid. This can put considerable heat stress on those dancing.

Since 1988 there have been approximately 16 fatalities in UK night-clubs (Henry, 1992; Arlidge 1995). Whilst drugs were often implicated, in most instances heat-stoke was the actual cause of death. It is likely that there are also many less serious heat related problems.

The principle aims of this investigation were to assess the thermal conditions in night-clubs and to quantify the behavioural, physiological and subjective responses of people dancing in them. The study comprised of 4 parts, a survey of behaviour in, and opinions on the thermal conditions in night-clubs; an assessment of the thermal conditions of a night-club, measurement of subjective and physiological responses of dancers in simulated night-club thermal conditions and an evaluation of the accuracy of predictive thermal models in order predict the physiological responses to night-club thermal conditions over a period of time.

Survey of behaviour, attitudes and subjective responses to thermal conditions in night-clubs

The population was identified to be predominantly young people (Jones 1994), covering a geographically scattered area. A questionnaire was distributed to subscribers on the UK-Dance discussion group on the Internet and to people outside night-clubs. In total 54 subjects responded to the survey, 65% male, 35% female.

Night-clubs were considered to be hot or very hot places where 61% of people would prefer to be cooler. Respondents generally preferred to drink soft drinks rather than alcohol, many (76%) used drugs such as 3,4-methylenedioxymethamphetamine (ecstasy) for their stimulation. High priced bottled water and disconnected water supplies were given as reasons for low consumption of liquids. In such environments with increased metabolic activity from dancing dehydration was likely, indeed 88% of respondents had experienced heat related illnesses.

Thermal Audit in a Night Club

Night clubs vary in architecture and interior design and each is unique. The auditorium investigated represented a large, high ceiling type, typical of many institutionalised auditoria. Using a Grant Squirrel data logger fixed in the lighting rig, air temperature, radiant temperature and relative humidity were recorded every five minutes over a period of three nights.

A maximum air temperature of $27^\circ C$ and 82% relative humidity were recorded in the auditorium during each night. These measurements were made at approximately 2m above head height. Using a hand held Solex humidity/temperature meter the maximum air temperature and relative humidity amongst those dancing were $29^\circ C$ and 90% respectively. Air velocity in the empty auditorium was 0.175m/s. With a total of 180 lamps rated between 150-750 watts a significant radiant heat load was expected, however the positioning of the Squirrel data logger prevented the black globe being placed under any direct radiant load, hence the true extent of this thermal load was not seen in the results.

Investigation into the Physiological Responses in a Night-Club

Method
The thermal conditions of the auditorium were simulated in the thermal chamber. A pilot study was conducted to evaluate and improve the experimental methods. Aural, oral and Ramanathan's four point mean skin temperature (Parsons 1993), metabolic rate, heart rate and amount of sweat loss were all measured. Eight subjects were exposed to the

experimental conditions over two sessions. In each session there were four subjects, two males and two females with a mean age of 21.75 years and wearing clothing of an estimated value of 0.7 clo. They were weighed semi-nude then thermistors were securely attached. Subjects danced for 30 minutes with their metabolic rates being taken using the Douglas bag method for 2 minutes after 25 minutes. A five minute break allowed them to rest whilst the music was changed. They then continued to dance for the remaining time, their metabolic rates again being taken after 55 minutes.

Finally they were weighed semi-nude with their clothes being weighed separately in a plastic box. All measurements were repeated to ensure accuracy. After weighing they were given soft drinks, offered a shower and discussed the investigation.

Results

The mean aural temperature rose gradually for 15 minutes before flattening. The inadequacies of the measuring techniques for this application were identified. The ear thermistors for three subjects lost contact and were therefore unreliable. When these results were removed, the mean rose to a maximum of $38.2^{\circ}C$, sd=0.25 (Fig 1). The mean 4 point mean skin temperature rose at a similar rate from $36.9^{\circ}C$-$38.2^{\circ}C$ sd=0.5. The five minute break was sufficient to elicit a decrease in skin temperature of $0.56^{\circ}C$.

The anticipated difference in metabolic rate between the two different styles of music was not found. The discrepancy between the actual and expected results may have been due to more athletic dancing in response to preferred music being heard.

Heart rate was sustained at a mean of 140 bpm. The heart rates of the females (who were considered to be fitter) were lower than those of the males. The heart rate was correlated with the measured metabolic rates in order to estimate the mean metabolic rate for dancing to be 238 W/m^2.

The results suggested that night-clubs present stressful conditions to those dancing. It was not possible to assess the effects of a prolonged exposure in the simulated conditions. The 2-node model of human thermoregulation (Nishi & Gagge, 1977) and ISO 7933 were therefore used to predict how the human thermoregulatory system responds to night club conditions over an extended period of time. The thermal conditions found in the night-club were used in the models, air and radiant temperature being $29^{\circ}C$, air velocity 0.175m/s and relative humidity being either 70% or 90%, the later being the maximum humidity recorded. A metabolic rate of 238W/m^2 was used. The models were run on PC's, the predictions generated being compared with the results from the laboratory investigation to evaluate their accuracy.

The 2-node model over estimated the 4 point skin temperature. It was more accurate with core temperature, (Fig. 1). The predicted core temperature did not account for the five minute rest that was observed in the actual core temperature. Towards the end of the experiment the actual temperature exceeded the predicted temperature. This may have been due to the increase in activity by subjects towards the end of the investigation. ISO 7933 also accurately predicted trends in the rise of core body temperature. The effects of a four hour exposure were then predicted, this being the mean time that subjects

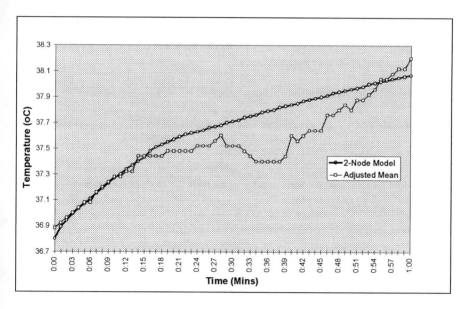

Fig 1. *Predicted and actual (adjusted) core temperatures*

Table 1. *Predicted physiological responses to dancing in a night-club from 2-Node model.*

Exposure (Ta=Tr=29°C)	Final Body Temp (°C)
1 hour (rh=70%)	38.07
1 hour (rh=90%)	40.71

Table 2. *Predicted maximum exposure times and sweat required to maintain heat balance from ISO 7933 (Ta=Tr=29°C)*

Exposure	Alarm Criteria Mins.	Danger Criteria Mins.	SW$_{req}$ (g)
1 hour (rh=70%)	43❶	52❶	997.2
1 hour (rh=90%)	30❶	36❶	1007.4
4 hour exposure, 45 mins. dancing, 15 mins. rest	124❶	149❶	756
4 hour exposure. 40 mins. dancing, 20 mins. rest	312❷	390❷	192.3

❶ Rise in core body temp.
❷ Excessive water loss

in the survey danced for. The 2-node model predicted a core body temperature increase to 39.07^0C (table 1). This is well above the WHO limit of 38^0C in occupational settings. Using ISO 7933 appropriate work-rest for dancing and water requirements were suggested. This can be seen in table 2.

Conclusions and Recommendations

1. Night-clubs operate at stressful temperatures and humidity. This can predispose those dancing in them to heat strain. Suitable measures such as increased air velocity should be taken to reduce the thermal stress.

2. "Chillout" rooms at lower temperatures to the main dance floor should be provided for rest and cooling of body temperature.

3. Frequent rests should be taken between periods of dancing; after 40 minutes of dancing, a 20 minute rest should be taken in a "chillout" room.

4. Adequate amounts of water should be consumed by those in night-clubs to prevent dehydration. It is suggested that this should be 1 litre/hour for active and prolonged dancing, however over consumption of fluids should also be avoided and advice should be provided.

5. Provision of free, cool, drinking water should be made compulsory.

6. The 2-Node predictive model for human response to thermal environments proved to be a fairly good representation of the actual environment observed. With care this can be used to make further predictions.

7. ISO 7933 accurately predicted the times of exposure before alarm limits were reached and over-estimated times before danger limits were reached.

8. Club goers should be educated as to the risks of heat strain illustrated in this report. This could be made possible with a simple wet bulb globe thermometer (WGBT), displaying the thermal conditions in clubs and the likely effects of dancing in the environment.

References

Arlidge J (1995) *Ecstasy drug condemned as 'dance with death'* The Independent Thursday 16 February

ISO/DIN 7933 (1989) *Hot Environments- analytical determination and interpretation of thermal stress using calculation of required sweat rate* International Standard Organisation Geneva

Jones D (ed.)(1994) *Equinox November 1994: Rave New World, programme transcript* Channel 4 London

Henry (1992) *Toxicity and deaths from 3,4-methylendioxymethamphetamine ("ecstasy")* The Lancet, Vol. 340 384-387

NIOSH (1986) *Occupational Exposure to Hot Environments* National Institute for Occupational Safety and Health DHSS (NIOSH) Publication No. 86-113 Washington DC

Nishi Y and Gagge A P (1977) *Effective temperature scale useful for hypo- and hyperbaric environments* Aviation Space and Environmental Medicine 48 97-107

Parsons (1993) *Human thermal environments* Taylor and Francis London

WHO (1969) *Health Factors Involved in Working Under conditions of Heat Stress* Technical Report Series No. 412. Geneva

Acknowledgements

The authors would like to thank Trevor Cole for technical support during the experimental study. In addition, acknowledgement must be made to all the subjects who gave up their afternoons to dance and to Mark Camp who provided the music.

Musculoskeletal Disorders

POTENTIAL MUSCULOSKELETAL RISK FACTORS IN ELECTRICITY DISTRIBUTION LINESMEN TASKS

R J Graves*, A De Cristofano****, E Wright***, M Watt *, and R White**

*Department of Environmental & Occupational Medicine
**Department of Biomedical Physics
University of Aberdeen
***OMS, Aberdeen
****Weaver Associates, London

Medical referrals of manual workers from an electric generation and distribution company showed that 41% reported musculoskeletal disorders. These were found especially in those men working with distribution poles (overhead linesmen). An ergonomic study was undertaken to determine whether there could be a link between these findings and work practices. This involved amending the Nordic questionnaire, taking video recordings of typical tasks, evaluating postures using the Ovako Working Posture Analysing System and biomechanical analyses, and a postal questionnaire survey of musculoskeletal and task risk factors. Problems were reported in lower backs, shoulders, necks and knees. These could be due to forces being applied in awkward postures. Improvements could result from better tool design, and reducing forces applied via awkward postures.

Introduction

Electricity supplies are generated from power plants and transported via transmission towers to underground power lines and lower voltage distribution poles to the customer. The efficiency of distribution depends to an important extent on overhead linesmen being able to maintain these appliances effectively.

Over an eight month period, medical referrals of 87 manual workers from an electric generation and distribution company were examined by the company medical adviser (CMA). This showed that 41% of these reported musculoskeletal disorders. The disorders affected the upper and lower limbs, and the lower back. An important proportion of the referrals (27.8%), were concerned with distribution, especially overhead linesmen working with distribution poles. The linesmen had musculoskeletal disorders including sciatica, cervical spondylosis, disc prolapse, arthritis and tenosynovitis. The CMA identified the referrals as likely to be the most severe cases. Other employees may have had milder symptoms which could have resulted in little or no absence from work. From an audit of these referrals, musculoskeletal disorders were shown to be a likely cause of long-term sickness absence and staff wastage in terms of redeployment and premature retiral. The company was concerned about possible causes and encouraged an ergonomic study to be undertaken to identify whether there could be a link between these findings and work practices.

An informal two day visit was made on-site in order to develop terms of reference. From this it was seen that work activities undertaken by linesmen were both varied and physically demanding. It was hypothesised that activities such as climbing poles and using apparently heavy tools to maintain power lines, may be associated with the reports of musculoskeletal disorders. A series of research questions needed to be answered in order to obtain a clearer picture of potential risk;

    a) Are the linesmen experiencing musculoskeletal aches, pains or discomfort at work, and to what degree?

    b) If they are experiencing discomfort, what body regions are affected?

    c) Has the discomfort prevented normal work activity, and, if so for how long?

    d) Are there any ergonomic factors which may lead to an increase in risk?

It was clear that there was a need to carry out a survey of linesmen to answer the first three questions. In addition, an ergonomic analysis of their tasks was needed to determine whether there were risk factors which could lead to musculoskeletal disorders.

In selecting appropriate methods for conducting a survey, studies were reviewed which had investigated the prevalence of musculoskeletal disorders and analysed tasks in a similar context. It was found that such studies (see Ekberg et al, 1994) had used the Nordic questionnaire successfully, and it appeared to be appropriate for the present study. The Nordic Questionnaire (Kuorinka et al, 1987) was developed to identify which body regions experienced ache, pain or discomfort. The reliability of the Nordic questionnaire, has been shown in studies on its test and re-test reliability. Evaluation of its test and re-test reliability has shown that identical answers have been generated for 80% to 100% of the questions (Deakin et al, 1994). Additionally, the Nordic questionnaire has been used on over 50,000 subjects since it was introduced in 1985 (Kuorinka, op cit.). This questionnaire was thought to be useful because it would allow data to be collected from large sample sizes at a relatively low cost. Further, the method would allow data on discomfort to be related to specific body regions and the length of time it had been experienced.

There was a need to analyse postures and to determine biomechanical issues that would arise from the linesmen tasks to answer the final question. The OWAS (Ovako Working Posture Analysis System) was developed by a steel company, to identify harmful postures in steel workers. A scale was developed to categorise the severity and potential risk of postures observed during working (see Kant et al, 1990). This method was seen as a useful means of allowing an initial simple analysis of postures to provide priorities for more detailed analyses.

It was felt that a more detailed analysis would be needed to gain better information on the forces being exerted to complement the results of the initial process. Biomechanical analysis of postures provides a means of achieving this. This can be undertaken by examining static examples of postures adopted at work and calculating the biomechanical stresses around joints. An example of this approach has been shown by an off-site posture and biomechanical evaluation of luggage handlers undertaken by Stalhammar et al (1986), based on video recordings of postures during work. The use of video as a means of recording postures and analysing it later was seen as being a more practical means of obtaining data than from direct observation on site.

Method

The ergonomic study was undertaken in three stages. The first involved a pilot study using an amended version of the Nordic questionnaire to assess its applicability. In addition, video recordings were taken of postures adopted during typical tasks to provide supplementary information for the questionnaire design and material for postural analysis. The second stage involved analysing the postures using the OWAS to determine any risks. In addition, a biomechanical analysis of the postures ranked as having risk factors was conducted to identify which aspects of the work activities could be contributing to risk. The third stage involved a postal questionnaire survey of all 292 linesmen to obtain information on musculoskeletal discomfort, whether work had been affected, and which tasks were felt to be the causal factors.

Pilot Study

A sample of nine men were selected from company records of all 292 linesmen. The modified form of the Nordic questionnaire was used to identify musculoskeletal discomfort in specific body regions and activities felt to be related to discomfort. Results from the pilot survey showed that 44% had experienced knee discomfort, so it was decided to include questions about knee discomfort in the main survey.

As the main survey was to be postal, a small scale evaluation of the questionnaire's comprehension was undertaken using a subjective rating approach. A rating scale was added after each page to determine the linesman's ease of use of the questionnaire. After completion, the questionnaires were checked to ensure that all the answers were filled in and that the subjective rating scale had been completed.

Main Survey

The finalised questionnaire was sent out to all 292 linesmen with a covering letter. After two weeks, a reminder letter was sent out and team leaders prompted individuals to return the questionnaire. The data were analysed using a statistical package.

Postural and Biomechanical Analysis

The Extreme Posture Checklist (Parker and Imbus, 1992) in conjunction with the video recording was used initially to identify any potentially harmful tasks. For example, a task such as using a crimping tool which involved elevated shoulders and extended elbows with angles greater than 120° between the upper arm and forearm, was identified from the checklist as having extreme postures. Tasks with any extreme postures were analysed further using the OWAS. Such postures were processed using a video copying system, allowing still colour frames to be reproduced from the video recording. Posture analysis from the still frames was performed using the OWAS to categorise the postures into one of the following four Action Categories, (AC):

AC 1 Normal posture: no action required

AC 2 The load of the posture is slightly harmful: actions to change the posture should be taken in the near future

AC 3 The load of the posture is distinctly harmful: actions to change the posture should be taken as soon as possible

AC 4 The load of the posture is extremely harmful: actions to change the posture should be taken immediately.

In addition, the still frames were examined and traced onto paper to provide the basis for free body (or space) diagrams, so that the various forces required to support body regions and weights held by muscles could be calculated (Le Veau, 1992). By examining the application of a force at a distance from a pivot, moments of force can be obtained. The obtained forces were then compared to the voluntary joint strengths to determine moments of force exceeding criteria.

Results and Discussion

Pilot and Main Survey

The majority of the nine linesmen in the pilot study rated most of the questionnaire easy to use. One page appeared to be less easy to understand, indicating that this page required improvement. Other feedback indicated that some numbering did not contain sub-sections, so these were added to help with the questionnaire's flow and comprehension.

There was a response rate of 143 out of the 292 linesmen, (49%). An overview of the results is provided below. The respondents' average height, weight, age, hours worked per week and years in the company were 1.78m, 78.9 kg, 38.5 years, 39.03 and 19.45 years respectively.

Periods within the previous 12 months were used to determine the prevalence of musculoskeletal problems, as periods over a year have been considered to provide vague recollections and inaccurate data (Agius et al, 1994). The data showed that during the last 12 months, 63.6% of respondents had experienced low back discomfort, 33.6% neck, 33.6% knee trouble, 31.5% shoulder discomfort and 18.9% each for hips/thighs and ankles/feet discomfort.

Clearly, a high percentage of respondents appear to have problems with their backs and, to a lesser degree, with their necks, shoulders and knees, and some problems to their hips/thighs and ankles/feet. If these results provide an accurate reflection of discomfort generally experienced by linesmen, then there appears to be a high number of linesmen at risk of musculoskeletal problems. In addition, the problem appears to be continuing, because, in the last 7 days, 11.9% of respondents had experienced low back pain, 8.4% neck trouble, 6.3% knee trouble, and 5.6% wrist/hand and ankle/feet discomfort.

The potential impact of musculoskeletal discomfort over time can be seen from 4.2% of respondents reporting discomfort from low back trouble every day, 25.2% for up to a week, 18.9% for up to a month, and 9.8% for over a month. In addition, 1.4% of respondents had experienced neck discomfort every day, 18.9% for up to a week, 6.3% for up to a month, and 7% for over a month. Further, 0.7% experienced knee discomfort every day, 7.7% had discomfort for up to a week and similar number for up to a month, and 5.6% for more than a month, but not every day.

The possible impact of these complaints on efficiency is shown by the number of respondents reporting that they were prevented from doing normal work during the last 12 months. Back trouble at 19.6% had the highest single effect, with 9.8% affected by neck discomfort and 7.7% with hip/thigh trouble. In addition, 7% of respondents had to change jobs or duties because of low back trouble, 2.8% with neck trouble, and 1.4% with shoulder and knee trouble.

Some indication of respondents' perception of causes over the last 12 months is shown by 7.7% of the respondents identifying looking up at the tops of poles as being related to their neck trouble, 14.7% felt that stretching with the crimping tool was a contributor to shoulder discomfort, while 18.2% identified lifting and bending, and 10.5% stretching out with the crimping tool as related to their low back discomfort. Sixteen percent said pole climbing was a contributory factor to their knee trouble.

Postural and Biomechanical Analysis

A number of tasks with risk factors were identified using the OWAS. Tasks which appeared to exceed Action Categories 1 included lifting up tools via pulley rope, lifting pole platform, hammering nails into poles, circuit testing with operating rod, attaching circuit breakers with operating rod, pole climbing, digging up ground and pavements, using pick-axe to loosen concrete, lifting wooden blocks/ heavy tools, crimping tool work, drilling new poles before putting them up, changing insulators - pole work, changing insulators - using wrench, jointing, swinging breakers on rod up to linesman, tightening cross-arm nuts with wrench, and tightening bolts on insulators. These tasks showed elements which included extreme postures of the neck, back, shoulders, upper and lower limbs, and task risk factors such as repeated/ sustained and forceful exertions away from the body.

The degree of risk can be clarified by using biomechanical analysis. For example, in crimping tool work, the linesman was stretched out horizontally from the pole with the weight and length of the crimping tool away from the body and above the shoulders in order to reach cables overhead. Static biomechanical analysis of the shoulder joint loading can be carried out. Even a simple analysis of the posture described by a hand in an outstretched position with a weight similar to the crimping tool, indicates shoulder joint forces exceeding three times body weight (Le Veau, op cit.).

In addition, operating forces ranged from 10 Newtons, at the start of pumping the hydraulic tool, to 200 Newtons when the tool jerks as it is released back to the start position. The way in which the tool has to be operated means the user could be under considerable strain in pumping the tool and then receiving a sudden jerk as the tool releases. The effect of this is accentuated by the forces being applied at the end of his reach. Detailed biomechanical analyses are on-going to determine the sources of risk in relation to each of the task elements in order to develop solutions.

Ekberg et al (1994) stated that neck, shoulder and upper limb disorders were aggravated by organisational and ergonomic work conditions such as repetitive tasks, excessive forces, static loads and extreme joint positions. Studies (see Hagberg and Wegman, 1987) suggest that repetitive shoulder muscle contractions, static contractions and work at shoulder level, as seen in tasks such as changing insulators, tightening bolts on insulators and crimping tool work, could lead to musculoskeletal risks.

Conclusions

In conclusion, over a quarter of distribution personnel had been found to have musculoskeletal disorders over an eight month period. However, the findings of the survey suggest that this rate of referral may be the tip of the iceberg. This is supported by over sixty percent of survey respondents reporting low back discomfort, a third each neck and knee trouble, almost a third shoulder discomfort and nearly a fifth each for hips/thighs and ankles/feet discomfort over the last 12 months.

Ergonomic analyses of the tasks indicated that there were musculoskeletal risk factors which could lead to this degree of discomfort and possibly the occurrence of disorders. The main risk factors appear to be the application of forces while in awkward postures, especially while working on poles and operating tools at or above head height.

Further work is needed to provide practical solutions. There may be scope for improving task design to reduce the effects of the described postures. For example, where an important contribution to the force is from the weight of a tool, it may be possible to reduce this by providing some degree of support to the tool.

References

Agius, R. M. Lloyd, M. H. Campbell, S. Hutchison, P. Seaton, A. Soutar, C. A. 1994, Questionnaire for the identification of back pain for epidemiological purposes, Occupational & Environmental Medicine, 51, 756-760.

Deakin, J. M. Stevenson, J. M. Vail, G. R. Nelson, J. M. 1994, The use of the Nordic questionnaire in an industrial setting: A case study, Applied Ergonomics, 25, 3, 182-185.

Ekberg, K. Bjorkqvist, B. Malm, P. Bjerre-Keily, B. Karlsson, M. Axelson, O. 1994, Case-control study of risk factors for disease in the neck and shoulder area, Occupational & Environmental Medicine, 51, 262-266.

Hagberg, M. Wegman, D. H. 1987, Prevalence rates and odds ratios of shoulder-neck diseases in different occupational groups, British Journal of Industrial Medicine, 44, 602-10.

Kant, I. Notermans, J. H. V. Borm, P. J. A. 1990, Observations of the postures in garages using the Ovako Working Posture Analysing System (OWAS) and consequent workload reduction recommendations, Ergonomics, 33, 2, 209-220.

Kuorinka, I. Jonsson, B. Kilbom, A. Vinterberg, H. Biering-Sorenen, F. Andersson, G. Jorgensen, K,. 1987, Standardised Nordic questionnaires for the analysis of musculoskeletal symptoms, Applied Ergonomics, 18, 3, 233-237.

Le Veau, B. 1992, *Williams and Lissner's biomechanics of human motion*, (W. B. Saunders Company, 3rd edition, London).

Parker, K. G. and Imbus, H. R. 1992, *Cumulative trauma disorders: Current issues and ergonomic solutions: A systems approach*, (Lewis Publishers, USA).

Stalhammar, A. R. Leskinen, T. P .J. Kuorinka, I. A. A. Gautreau, M. H. J. Troup, J. D. G. 1986, Postural, epidemiological and biomechanical analysis of luggage handling in an aircraft luggage compartment. Applied Ergonomics, 17, 3, 177-183.

MUSCULOSKELETAL DISORDERS ARISING FROM THE HARVESTING OF PALM OIL FRUIT IN MALAYSIA

M F Zainuddin and R A Haslam

Health and Safety Ergonomics Unit,
Department of Human Sciences,
Loughborough University of Technology,
Leicestershire, LE11 3TU

Harvesting palm oil fruit bunches and pruning underlying leaves from tall palms (e.g. > 3 m) are still predominately manual tasks. Although several attempts have been made to introduce mechanisation into the cutting operation, very limited success has been achieved. Thus, the use of a sickle mounted on a flexible telescopic aluminium tube is still the most common harvesting method. This, however, exposes plantation workers to the risk of musculoskeletal discomfort and injury. This paper examines the problems by assessing the task, posture and tools, and their implications for operators, through the use of questionnaire survey and video analysis techniques. The results suggest there a significant problem with musculoskeletal disorders in the industry.

Introduction

Palm oil is the second largest vegetable oil in terms of world production after soya, with Malaysia the largest producer and exporter. In 1994, Malaysian export of palm oil was at 6.65 million tonnes compared to 3.1 million tonnes ten years ago (Porla, 1995). The contribution of the industry to Malaysia's GDP was 7.4% earning £2.2 billion in export revenue (Ahmad, K., 1995).

A palm tree reaches maturity in three years and has an economic life of approximately 25 years. It produces fruit continuously throughout the year, with an average of one bunch per month.

The most crucial and strenuous job in the palm oil plantation is harvesting, an activity carried out everyday. At present, harvesting is done manually using a big chisel with a wooden or iron pipe handle for palm trees less than 3 m in height (figure 1) and using a long telescopic aluminium pole (handle) attached to a sickle for trees more than 3 m (figure 2).

The activity of harvesting comprises the following task elements: walking in the estate to look for ripe fruit bunches, positioning the tool, cutting subtanding fronds, cutting the ripe bunches, cutting each frond into two and stacking along the stacking path, cutting excess stalk from the cut bunch, and finally laying off the bunch on collecting stations at the end of the harvesting path.

Figure 1. Using a chisel for short palm.

Figure 2. Using a sickle for tall palm.

Cutting fruit bunches contributes to 25% of harvesting time and normally each worker has to cover an area of between three to four hectares per day. In the cutting operation, positioning the tool to cut fruit bunches from short palms is relatively straightforward. However considerable effort and time are needed to deal with tall palms more than 6 m from the ground. Manoeuvring a tool which has a 6 m or longer handle exposes the worker to the possibility of musculoskeletal discomfort and injury.

The purpose of this paper is to consider risk factors and symptoms of musculoskeletal disorders arising from the harvesting task. Data collected from a questionnaire survey and video analysis are presented and discussed.

Methods

Questionnaire Survey

A questionnaire survey was performed among harvesters from various estates, mainly in the West Midland region of the Malay Peninsula. The questionnaire was administered as a structured interview, comprising 48 questions. A variety of question formats were used but mainly they either required a yes/no or a force-choice selection response. The questions dealt mainly with incidence of musculoskeletal pain in the upper extremity region, and how discomfort experienced by workers influences their job performance and productivity.

The questions were structured to correspond to potential risk factors, as outlined by Putz-Anderson (1988); Armstrong (1986) and Keyserling et al. (1993). Question content also drew upon existing questionnaires related to musculoskeletal pain and injuries (Kuorinka et al., 1987 and McAtamney and Corlett, 1992).

Posture and Force Analysis

Video tape was used to assess posture and movement. The involvement of posture in musculoskeletal disorders has been well documented. Postural angles of the upper extremity when performing manual work have profound influence on musculoskeletal load and muscle fatigue (Kuorinka et al., 1995). Elements of the cutting task were recorded and analysed to determine the position of the shoulders, arms and hand/wrists using the method given by Armstrong et al. (1982).

Results

Questionnaire Survey

One hundred and fifty workers participated voluntarily in the survey (table 1). The majority of the respondents (87%) harvest fruit bunches from tall palms, with 69% gathering fruits from trees more than 13 m high. The prevalence of fatigue and discomfort among workers was high in all three parts of the body specified in the questionnaire: neck, forearm and hand/wrist regions. However, causative factors and degree of discomfort varied. In the neck region, 98% of subjects reported experiencing discomfort arising from their work. There was a high proportion of discomfort reported in the arms during the cutting cycle.

Table 1. Characteristics of Survey Respondents (n=150).

	Mean	Standard deviation
Age	32 yr	7.8
Height	1.6 m	11.5
Weight	57.2 kg	7.7

In the hand/wrist region, all respondents reported discomfort whilst gripping the pole. The scale of discomfort subjects experienced showed 63% suffered moderate discomfort, 35% experienced noticeable discomfort and the remaining 2% suffered from intolerable discomfort. Furthermore, the results indicated 72% of workers experienced ' awakening ' during their night time sleep due to musculoskeletal pain. With respect to the cutting tool, 48% of the harvesters claimed the use of the existing tool was not suitable, whilst 25% of them were satisfied (27% were uncertain).

Posture and Force Analysis

Postural problems identified through video anaylsis were:

i.   Excessive extension and flexion of both arms whilst holding the pole during cutting (figure 3).
ii.   Ulnar deviation of both hand/wrists whilst pushing and pulling a pole when manipulating the cutting action.
iii. Radial and ulnar deviation and palmar base pressure whilst loosing and tightening the telescopic bracket screw with a spanner (figure 4).

Figure 3. Excessive extension and flexion of both arms.

Figure 4. Radial and ulnar deviation and palmer base pressure.

iv. Bending the head and neck backwards to look for ripe fruit bunches (figure 5).
v.   Supination and pronation of forearms whilst performing the push and pull motion during cutting.

There was also considerable force exerted during the following task elements:

i.   When holding/gripping the pole and maintaining that position.
ii. Forceful exertion of the forearms during the push and pull motion.
iii. Moving the tool between trees (figure 6).

Figure 5. Bending the head and neck backwards.

Figure 6. Moving the tool between trees.

Forces exerted across the various work elements eventually lead to the development of muscular fatigue in the wrist, forearm and shoulder regions.

Conclusions

Results from the questionnaire survey and postural and force analysis demonstrate workers involved in the harvesting of palm oil fruit bunches report a range of musculoskeletal symptoms. The prevalence of discomfort involving the upper extremity region was particularly high. Awkward work posture arising from tool orientation, high force exertion and repetitive work seem likely to contribute to the occurence of musculoskeletal disorders. It may be possible that redesign of the tool handle could improve working posture and minimise force exertion during the cutting operation. Modification of task cycle might also help optimise task performance, reducing unnecessary work load and time per task cycle. Further research into the design of the cutting tool and task is currently in progress.

Acknowledgements

This study was carried out in co-operation with the Palm Oil Research Institute Malaysia (PORIM), Rubber Industry Small holders Development Authority (RISDA), Sime Darby Plantation (M) Bhd, Guthrie Plantation (M) Bhd., Golden Hope Plantation (M) Bhd. The research was funded by the MARA Institute of Technology, Malaysia.

References

Ahmad, K., 1995. Jentera moden atasi masalah buruh. Berita Harian, **30**, 25.

Armstrong, T.J., 1982. Development of a biomechanical hand model for study of manual activities. In: R. Easterby, K. H. E. Kroemer, and D. Chaffin (Eds), *Anthropometry and Biomechanics: Theory and application* (Plenum, New York), 183-192.

Armstrong, T.J., Radwin, R.C., Hansen, D.J., and Kennedy, K.W., 1986. Repetitive trauma disorders: job evaluation and design. Human Factors, **27**, 325-336.

Keyserling, W.M., Stetson, D.S., Silverstein, B.A. and Brouwer, M.L., 1993. A Checklist for evaluating ergonomics risk factors associated with upper extremity cumulative trauma disorders. Ergonomics, **36**, 807-831.

Kuorinka, I., Jonsson, B., Kilbom, A., Vinterberg, H., Biering-Sorensen, F., Andersson, G. and Jorgensen, K., 1987. Standardised Nordic questionnaires for the analysis of musculoskeletal symptoms. Applied Ergonomics, **18**, 233-237.

Kuorinka, I. and Forcier, L. (Eds), 1995. *Work related musculoskeletal disorders (WMSDs), A Reference Book for Prevention* (Taylor & Francis, London).

McAtamney, L. and Corlett, E. N., 1992. Reducing the risk of work related upper limb disorders: A guide and method. Institute of Occupational Ergonomics, University of Nottingham.

PORLA, 1995. Information Leaflet (Palm Oil Registration and Licensing Authority, Ministry of Primary Industries, Malaysia).

Putz-Anderson, V. (Ed), 1988. *Cumulative trauma disorders, A manual for musculoskeletal diseases of the upper limbs* (Taylor & Francis, London).

CONFLICTS AND HARASSMENT AS RISK FACTOR FOR MUSCULOSKELETAL SHOULDER PROBLEMS

Jörgen Eklund

Divison of Industrial Ergonomics
Linköping University of Technology
S-581 83 Linköping
SWEDEN

The aim of this study was to identify relationships between psychosocial factors and indicators of musculoskeletal problems, and to identify how risk factors can be connected to different work conditions. A questionnaire, EMG measurements, ratings of strain and workplace evaluations were used for the assessment of 326 and 15 vehicle assembly workers respectively. Those with the most repetitive and physically demanding tasks had a higher prevalence rate of musculoskeletal problems, compared to the other assembly workers. The victims of conflicts and harassment had a significantly higher level of both mean and static muscle activity and significantly shorter rest pauses in the shoulder muscle (trapezius), compared to the other assembly workers. This relationship was interpreted as causal. One conclusion is that conflicts and harassment constitutes a risk factor for shoulder problems.

Introduction

Research on risks for musculoskeletal problems has over the years identified a large number of risk factors that in one or another way are related to mechanical loads (Hagberg 1987). During the last years, research has to a larger extent focused on psychosocial factors, in addition to the traditional mechanical factors, and several studies have found significant associations between psychosocial factors and musculoskeletal problems (Bongers et al 1993, Johansson 1994). Stress or mental work load have also been found to generate increased EMG activity in the shoulder muscles (Waersted et al 1991).One reason for this broader research focus is that it has become commonly recognised that the causes of musculoskeletal disorders are multifactorial. Another reason is that the explanatory value of mechanical factors seems not sufficient. Also, preventive measures against musculoskeletal injuries, focusing on mechanical load, have not had a particularly good effect.

In other fields of research, several models for stress and health have been developed. They have focused control, freedom of action, work demands and social support at work as important factors for the risk of cardiovascular disease. These factors have also been shown to be important for general health and well-being (Alfredsson et al 1985, Karasek 1979). Assembly line work is often characterised by machine pacing, monotony, short work cycles and mental under stimulation, and has been shown to cause relatively strong stress reactions among the workers (Frankenhaeuser and Gardell 1976 , Lundberg et al 1989). There is also a lack of knowledge about the differences among males and females in this respect, even though the frequency of musculoskeletal disorders in general is higher for females. The reasons for this are, however, not clear. The aim of this study was to identify relationships between psychosocial factors and indicators of musculoskeletal problems, and to identify relationships between psychosocial risk factors and different work conditions.

Methods

The study reported here was a part of a more extensive project, concerning work environment, stress and health impairments in traditional assembly work. The methods used for this part of the study were a questionnaire (326 participants), EMG measurements, ratings of strain and workplace evaluations (15 participants).

The questionnaire included the standardised Nordic questionnaire for musculoskeletal symptoms (Kuorinka et al 1987) and questions on psychosocial work conditions (Wikman 1991). The degree of conflicts and harassment a person was subjected to, was based on the total severity of the answers given to three questions. All participants were vehicle assembly workers from one plant in southern Sweden, with at least 0.5 years of work experience. The response rate was 70%, and 54 of these 326 were women.

EMG measurements were performed on the right trapezius muscle and the left erector spinae muscle on the L3 level with surface electrodes. The muscle activity was recorded during 15-20 minutes, which normally included three complete work cycles for each person in his or her ordinary work task. In the subsequent analysis, static, mean and peak levels and measures of the temporal pattern were determined. For a comprehensive description of the methods used, see Lindberg et al (1993) and Linderhed (1993). Borg's category-ratio scale was used for ratings of perceived strain during work (Borg 1990) and ratings of the person's actual work pace in relation to maximum possible. The workplace evaluations were performed by structured observation, including back flexion and rotation, shoulder abduction/flexion and presence of lifting tasks. Out of 18 participants that were randomly chosen, half of which were women, 15 fulfilled all obligations. 8 of these were women. Regression analyses were used to test the interaction between independent (psychosocial factors) and dependent variables (EMG, work observations, ratings and symptoms).

Results

According to the questionnaire, the prevalence of musculoskeletal symptoms was high, especially in the shoulders and back. In comparison with a Swedish reference group (FSF Metod 1993), significantly higher prevalence rates were seen for neck, shoulders, wrist/hands, upper back, lower back and ankles/feet. As can be seen in Table 1, the females had higher prevalence rates for all body parts except for the knees, compared to the men. This was especially so for the upper extremities such as neck, shoulders and wrists/hands. However, the females were over represented in pre assembly, which was considered less heavy than final assembly.

Table 1. Prevalence (12 months) of musculoskeletal symptoms in %. Significant differences, tested between males and females are marked (Chi-2, significance level 5%, two-sided interval).

Body part	All respondents (n=326)	Females (n=54)
Neck	43	62**
Shoulders	45	62**
Elbows	17	26
Wrists/hands	43	61**
Upper back	35	44
Lower back	59	64
Hips	10	14
Knees	32	24
Ankles/feet	22	37**

The assembly workers with the most repetitive assembly work perceived higher levels of stress and had a higher prevalence rate of musculoskeletal symptoms, compared to the other workers. Those involved in pre assembly had particularly more symptoms in the shoulders and elbows, compared to those in final assembly and those with miscellaneous work tasks. These pre assembly tasks were normally performed standing at a work bench with high demands on upper limb repetitiveness. The workers involved in final assembly on the assembly line had particularly more symptoms from their knees, compared to the others. Final assembly included kneeling postures inside the vehicle and also substantially more trunk movements. They had significantly more trunk rotation and trunk flexion, compared to the others. Work task was in many cases more important than sex as a predictor of symptoms.

The pre assembly workers had a significantly higher static muscle activity and shorter rest pauses for the trapezius muscle compared to the other workers. There was also a tendency that the number of pauses was fewer and that the variation of muscle activity was lower.

From the regression analysis, a significant relationship between higher work pace (in relation to the individual maximum) and symptoms from the trapezius muscles was identified. In the material, there was no significant relationship between EMG activity measures and musculoskeletal symptoms.

The regression analyses identified that victims of conflicts and harassment had significantly higher levels of both median, static and peak muscle activity in the shoulder (trapezius). In addition, they had significantly shorter rest pauses and lower variation of muscle activity. No such relationship between conflicts/harassment and muscle activity from the lumbar back (erector spinae) could be identified, see Table 2. Further, the victims of conflicts/harassment rated higher body strain than the others, although there were no more conflicts/harassment among those who rated higher physical demands in their work or among those who had more physically demanding work tasks.

Table 2. Regression analysis using conflicts and harassment as independent variable and EMG parameters and ratings as dependent variables.

Dependent variable	p-value
Static activity level, trapezius	0.04
Median activity level, trapezius	0.002
Peak activity level, trapezius	0.01
Pause length, trapezius	0.03
Activity variation, trapezius	0.006
Static activity level, erector spinae	0.79
Median activity level, erector spinae	0.81
Peak activity level, erector spinae	0.77
Pause length, erector spinae	0.80
Activity variation, erector spinae	0.98

Those who perceived little social support followed the same pattern as those who were victims of conflicts and harassment, but the strength of the associations were weaker and normally not significant.

Discussion

An important issue of this paper is whether the associations identified in this study can be interpreted as causal relationships or not.

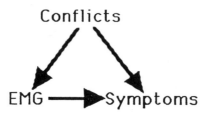

Figure 1. A conceptual model of possible influences.

This study has identified significant relationships between conflicts and EMG activity in the shoulders, and between conflicts and ratings of strain (rather than symptoms). However, no significant relationships between EMG and symptoms have been found, but the study was not primarily designed for that. Many other authors have found support for such a relationship, even though the aetiology behind is not known (Hagberg 1987, Kvarnström 1983).

Dimberg et al (1989) and Linton (1990) found relationships between social support and musculoskeletal disorders, and there is also anecdotal indications that conflicts can evoke shoulder pain. Conflicts can cause strong biopsychophysiological reactions, for example excretion of stress hormones and stress symptoms (Eklund et al 1995). There are obvious similarities between the concepts lack of social support and conflicts. This study shows similar relationships for lack of social support and for conflicts, which supports the findings.

Arguments whether the relationship between conflicts and EMG is causal or not can be systematised as follows. According to the Hill Criteria (Hill 1965), the relationships found in this study must be regarded strong (i), consistent (ii) and specific (iii). There is not sufficient data to evaluate the temporal relationship (iv), nor the biological gradient (v). The biological plausibility (vi) and coherence of evidence (vii) are judged acceptable. There is no experimental evidence (ix), but the analogy (x) with lack of social support supports the findings.

It is more likely that conflicts influence EMG than the opposite. Since there were significant relationships with several characteristics of the EMG signal from trapezius but not from erector spinae, this supports a causal relationship. There was no relationship between conflicts and more physically demanding work tasks, which otherwise would have been an explanation. It can be noted that one of the participants in this study in an interview one and a half years later stated that the neck and shoulder pain that had started in a surrounding of conflicts, disappeared after changing place of work and fellow workers, but with similar work tasks.

Of course, the possibilities that individual dispositions or personalities and selection mechanisms of certain individuals can not be eliminated, even though that kind of explanation seems less likely.

To conclude the discussion, a causal relationship between conflicts and increased EMG activity with elevated static load in the shoulders seems probable. The changes of EMG activity found in the shoulders are all of the nature that they constitute risk factors for musculoskeletal symptoms. This finding does not, however, eliminate the possibility that musculoskeletal symptoms can arise due to conflicts via other mechanisms than EMG changes.

EMG showed weak or no relationships with many of the exposure factors used in this study, which raises doubts about the usefulness of EMG as a general risk indicator for musculoskeletal disorders.

Conclusions

The main result of this study is that conflicts and harassment can cause increased muscle activity, increased static muscle load and shorter rest pauses in the shoulder muscles. These results together with other findings support that conflicts and harassment constitutes a risk factor for musculoskeletal shoulder problems.

References

Alfredson, L., Spetz, C.-L. and Theorell, T. 1985, Type of occupation and near future hospitalisation for myocardial infarction and some other diagnoses, International Journal of Epidemiology, **14**, 378-388.
Bongers, P., deVinter, C., Kompier, M. and Hildebrandt, V. 1993, Psychosocial factors at work and musculoskeletal disease, Scandinavian Journal of Work Environment and Health, **19**, 297-312.

Borg, G. 1990, Psychophysical scaling with applications in physical work and the perception of exertion, Scandinavian Journal of Work Environment & Health, **16**, Suppl.1, 55-58.

Dimberg, L., Olafsson, A., Stefansson, E. et al. 1989, Sickness absenteeism in an engineering industry - an analysis with special reference to absence for neck and upper extremity symptoms, Scandinavian Journal of Social Medicine, **17**, 77-84.

Eklund, J., Chambert, C., Gabrielsson, Å., Lindberg, M., Linderhed, H., Lindqvist, P., Lundberg, U., Nilsson, P. and Strömbeck, B. 1995, Arbetsförhållanden i monteringsarbete, Report LiTH-IKP-R-835, Linköping University (In Swedish).

Frankenhaeuser, M. and Gardell, B. 1976, Underload and overload in working life: outline of a multidiciplinary approach, Journal of Human Stress, **2**, 35-46.

FSF Metod 1994, Referensdata för Nordiska ministerrådsenkäten, Örebro (In Swedish).

Hagberg, M. 1987, Occupational shoulder and neck disorders, The Swedish Work Environment Fund, Stockholm.

Hill, A.B. 1965, The environment and disease: association or causation? Proceedings of the Royal Society of Medicine, **58**, 295-300.

Johansson, J.Å. 1994, Psychosocial factors at work and their relation to musculoskeletal symptoms, Ph.D. thesis, Department of Psychology, Göteborg University.

Karasek, R. 1979, Job demands, job decision latitude and mental strain: implications for job redesign, Administrative Science Quaterly, **24**, 285-309.

Kuorinka, I., Jonsson, B., Kilbom, Å. et al. 1987, Standardised Nordic Questionnaires for the analysis of musculoskeletal symptoms, Applied Ergonomics, **18**, 233-237.

Kvarnström, S. 1983, Förekomst av muskel- och skelettsjukdomar i en verkstadsindustri med särskild uppmärksamhet på arbetsbetingade skulderbesvär, Arbete och Hälsa, 1983:38 (In Swedish).

Lindberg, M., Frisk-Kempe, K., Linderhed, H. and Eklund, J. 1993, Musculoskeletal disorders, posture and EMG temporal pattern in fabric-seaming tasks, International Journal of Industrial Ergonomics, **11**, 267-276.

Linderhed, H. 1993, A new dimension to amplitude analysis of EMG, International Journal of Industrial Ergonomics, **11**, 243-247.

Linton, S.J. 1990, Risk factors for neck and back pain in a working population in Sweden, Work & Stress, **4**, 41-49.

Lundberg, U., Granqvist, M., Hansson, T. Magnusson, M. and Wallin, L. 1989, Psychological and physiological stress responses during repetitive work at an assembly line, Work and Stress, **3**, 143-153.

Waersted, K.B., Björklund, R. and Westgaard, R.H. 1991, Shoulder muscle tension induced by two VDU based tasks of different complexity, Ergonomics, **34**, 137-150.

Wikman, A. 1991, Att utveckla sociala indikatorer. Urval Nr 21, Statistiska Centralbyrån, Stockholm (In Swedish).

The Ergonomics of
Training

INDIVIDUAL DIFFERENCES IN FAULT FINDING SKILL CAN BE RELIABLY PREDICTED AFTER ALL

David L. Morrison
and
Alan LeMap

Department of Psychology,
University of Western Australia
Nedlands, Western Australian 6907

Previous research has failed to consistently link the diagnosis of novel system failures to measures of intelligence. In this paper two studies are reported which examine the development of diagnostic skill from the perspective of Ackerman's (1988) Theory of the Ability Determinants of Skill Acquisition. The studies show that the pattern of correlates between fault finding skill and measures of cognitive performance are predictable providing level of practice, task complexity and difficulty are taken into account.

Introduction

The increasing use of technology in the work place has dramatically altered the role of many employees. Mental agility rather than physical dexterity is frequently required as operators are called upon to diagnose system failures. Often, failures are familiar and there is a prescribed algorithm for finding the solution to the problem. As such, training operators to deal with this class of failure is reasonably straightforward. When problems are unfamiliar, however, the outcomes and durability of training are less certain and it is important, therefore, that those who are most likely to benefit from training are identified.

To date it has proved difficult to find reliable predictors of novel fault finding skill, either before or after training, using standard measures of intelligence (Dale, 1958, Henneman and Rouse, 1984; Elliot 1965, 1967). Most of the measures used in previous research have been measures of intelligence. Contemporary thinking regarding measures of this sort is that they reflect underlying differences in processing capacity or pool of available processing resources. Either way, performance is said to reflect the application of mental capacity to the task at hand. Contemporary research on the development of cognitive skill has, however, revealed that the development of cognitive skill and indeed expertise itself may comprise different modes of cognitive function some of which may not be conscious at all. Contemporary theories of learning (e.g., Anderson, 1983) suggest that the mode of cognitive activity that is brought to bear on any particular task is a function of level of skill and the properties of the task to hand. Tasks on which individuals are inexperienced require relatively more resources but, with experience the reliance on these resources seems to wain. With practice, the knowledge is transformed into a procedural form and environmental (or internal mental states) act as triggers which activate mental routines that are not necessarily resource dependent. If the initial level of resources is limited by some inherent limitation of the individual, then the effort required to sustain performance is greater than when the task gets close to, or exceeds, the resources that are available (Norman and Bobrow, 1975).

The theory of ability determinants of skilled performance (TADSA)(Ackerman, 1988) has been used to account for substantial proportions of individual variability in initial and practiced task performance on complex cognitive tasks. Ackerman argues that as level of skill changes the type of task that should be used to predict performance must also change, In the early stages of skill acquisition, performance is slow and effortful, requiring the expenditure of much mental effort (ie., allocation of controlled attentional resources). After practice automatic processes are said to develop that don't require constant cognitive monitoring. Thus, according to Ackerman (1988), any subject's performance on a complex task such as fault-finding may be represented as a joint function of available cognitive resources and the proportion of the subject's total capability actually devoted to the task. Ackerman (1989) has provided empirical evidence of these effects for simple cognitive tasks and shown how those with higher general intellectual ability (which is equated with resource availability) demonstrate higher levels of performance early in practice. Moreover, as practice progresses, measures of intellectual ability become less related to task performance; a finding which is consistent with the development of resource independence. Interestingly, Ackerman has also been able to correlate performance limitations on criterion tasks after moderate and extensive practice to other measures of ability not dependent on conscious processing activity. At moderate levels of practice, measures of perceptual speed (eg. identification of letters on a page), rather than availability of processing resources, increasingly correlate with task performance. The facility and speed of compilation of production systems that determine performance efficiency are the essence of perceptual speed ability. During this stage of skill acquisition the conditions under which particular mental operations must be applied are learned. Once the compilation processes are complete and performance is fully automated then the relationship between performance and perceptual speed becomes attenuated and performance is limited by non-cognitive psychomotor abilities (Ackerman, 1989). In simple terms, such abilities are reaction times which reflect the delay between a decision having been made and a response being executed. The limiting conditions with regard to the transition from controlled to automatic processing have also been defined by Ackerman (1988, 1989) as being influenced by task complexity and task consistency. Performance on tasks that are comprised of sub-components which make inconsistent cognitive demands do not progress to automatic processing even after extensive practice. With complex tasks the progression from controlled to automatic processing is said to be retarded. In this paper two studies are presented which seek to test the central propositions of Ackerman's theory that: (i) complex task performance is predictable according to level of practice and (ii) the relationship between measures which tap different modes of cognitive control will vary as a function of task characteristics.

It is hypothesised that measures of intelligence which assess availability of general cognitive resources will predict initial, but not later, levels of fault-finding skill. As practice progresses, it is expected that this relationship will diminish and measures of perceptual speed will be the primary determinant of performance level. Following extensive practice, psychomotor speed will best predict performance. While it is expected that these relationships will generally hold true, we expect them to be moderated by task difficulty. Tasks which are consistent (i.e. which contain a limited set of features to be learned) will afford the opportunity for more rapid skill development and hence the moderation of relationships between measures of intelligence will be apparent earlier in practice when compared to tasks that are less consistent.

EXPERIMENT 1

Subjects.
Forty two volunteer undergraduate students from the University of Western Australia (age ranges 17 to 39, median age 20). All of the subjects were female. Subjects were screened to ensure that they were not familiar with fault diagnosis.

Ability Measures
The three different modes of cognitive control were examined in the following ways: (i) . The Advanced Progressive Matrices (Ravens, Court & Raven, 1983) was used as a test of general ability; (ii) Perceptual speed was measured by the Maze Tracing, Number Comparisons, and Identical Pictures tests from the set of Factor Referenced Cognitive Tests described by Ekstrom, French, Harman & Dermen (1976); (iii) psychomotor speed was assesssed by a two-choice reaction time task

The Fault-Finding Task.
The Fault finding task used here is a simplified version of the logic network task developed by Duncan and his colleagues (e.g., Brooke, Cook and Duncan, 1983). The task was simplified in that subjects were not asked to iteratively test the working status of network components in order to identify a system failure. Instead, subjects were presented with a network in which potential failures consistent with a set of network output symptoms were displayed. On some occasions (decided pseudo-randomly) the set of potentially failed components either omitted or included units from those that could be inferred from the pattern of output symptoms. Subjects were asked to make a binary decision. Either the network, as presented, was either a true representation of the faulty set or not. Decision time was the sole dependent variable that was recorded for later analysis. Only those trials on which the subject gave a correct response were included in the analysis of the data presented later. In this experiment the problem structure was repeated every 85 trials such that the number of connections between components varied randomly.

Procedure.
Experimentation was divided into two parts. In part 1 subjects completed the tests of cognitive ability. In part 2, 600 trials on the fault finding task were completed. The testing sessions were spread over two days with all 600 trials on the fault finding task completed in a single session.

Results
The reaction times across all 600 trials (correct trials only) were aggregates into 6 blocks of one hundred trials and a series of multiple regression analyses were undertaken in which the three classes of cognitive performance measure were simultaneously used as predictors of correct solution times. Figures 1a-c show the standardised regression weights for each class of measures plotted across blocks of 100 trials. For each measure the best fitting function, using a least squares criterion, was overlain on the data as shown.

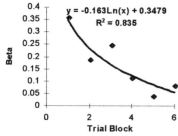

Figure 1a: Cognitive Resources

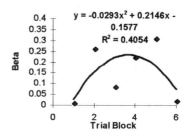

Figure 1b. Perceptual Speed

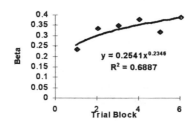

Figure 1c. Psychomotor Speed

The results from Experiment 1 provide a *prima facie* case for the argument that fault finding skills are predictable providing the level of task expertise and mode of cognitive control are simultaneously taken into account. The study just described was limited in that the study is essentially correlational and the curve fitting procedures used to identify the best fitting functions may be capitalising on chance associations. Second, the measurement model of the predictor variables was not as carefully defined as it could have been and there is a chance that the results obtained are partly due to the independent variables being highly inter-correlated. Experiment 2 sought to overcome these limitations by a more careful analysis of the measurement model, replication and manipulation of a key independent variable. A straight forward prediction that can be made from manipulating task difficulty is that increases in task difficulty will change the rate of learning which in turn will affect the relationship between fault-finding skill and measures of perceptual and psychomotor speed.

EXPERIMENT 2

Subjects.

Forty undergraduate students form the University of Western Australia were recruited for this experiment. The age range was 17 to 35, with a median age of 19.

The Fault Finding Task.

In Experiment 2 fault finding task difficulty was manipulated in two ways: (i) *Task Consistency* which involved varying the number of repetitions of network structure. In the high consistency condition the network structure was repeated every 10 trials and after 85 trials for the high consistency condition; (ii) *Task complexity* was manipulated by changing the number of connections between network components.

Procedure and Design.

As in Experiment 1 the data from the cognitive performance measures was collected first. In this session subjects completed the same tasks as in Experiment 1 but with the addition of several new tasks that would be used to define the measurement model. In a second and third session subjects completed 1000 fault finding tasks (500 in each sitting). The design of the study was a 2(complexity) x 2(difficulty) x trials(10 blocks each containing 100 trials). Complexity and difficulty were manipulated across subjects and subjects were randomly allocated to one of four groups.

Results and Discussion

The reaction time data for correct trials were initially analysed via a series of analyses of variance which included an ability and trial block factor in addition to those of complexity and difficulty. Analysis of variance revealed that difficulty and complexity retarded learning rate and there was a main effect of ability for the cognitive resource and perceptual speed factors but not psycho-motor speed. In addition to these effects, the trial block factor interacted with both the cognitive resource and perceptual speed factors. Those with more cognitive resources and those high on the perceptual speed factor improved their performance at a faster rate relative to those who were low on these factors. Finally, a single three way interaction between the cognitive resources factor, tasks consistency and trial block was evident indicating that the most difficult tasks (i.e., the low inconsistency condition) further retarded the progress of the subjects with fewer cognitive resources. This effect is consistent with what would be predicted by the TADSA model.

As in Experiment 1, the data were submitted to a series of multiple regression analyses. Figures 2 and 3 show the beta weights, derived from these analyses plotted across trial blocks. In each case the best fitting function is shown for the experimental conditions expected to show the greatest contrast for each of the cognitive performance measures. Figures 2a-c are the results for the low complexity and high consistency condition. The results shown in Figures 3a-c come from the condition where both complexity was high and consistency low. The figures reveal a striking consistency across conditions in that the perceptual speed and psycho motor factors appear to be reciprocally related in their capacity to predict fault finding skill. When perceptual speed is important, psychomotor speed is less important and vice versa. The results for the "easy" condition are largely in line with the predictions made by the TADSA model. That is, early in practice the cognitive resource factor is the best predictor of performance. As practice proceeds perceptual speed assumes a greater prominence as a predictor. Finally, psychomotor speed is seen as the limiting factor as skill development moves into its final stages for the easy condition. The major differences between the easy and hard conditions is to be seen clearly for the data for the cognitive resources factor (see Figures 2a&3a). For the difficult condition there is a steady decline in the influence of this factor as a predictor of fault finding skill where as for the easy condition there is evidence of a decline in influence followed by an increase. Another difference between the conditions is where the shifts in predictive influence occurs across factors. The influence of the perceptual speed factor in the difficult condition is evident both early and later in practice. The importance of perceptual speed early in practice is unexpected but the later impact of this factor is predictable from within the TADSA framework. According to TADSA the second stage of skill development will be delayed in the more difficult condition as seen here. The early influence of perceptual speed as a predictor of fault finding skill may reflect the difficulties that subjects have with complex displays (ie. those with a relatively high number of inter-component connections), and this effect appears to be sort lived.

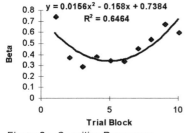

Figure 2a. Cognitive Resources

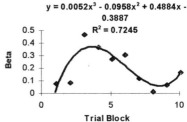

Figure 2b Perceptual Speed

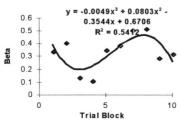

Figure 2c. Psychomotor Speed

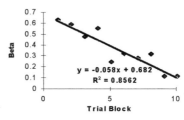

Figure 3a. Cognitive Resources

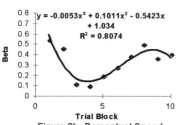

Figure 3b. Perceptual Speed,

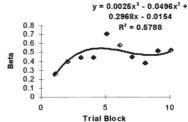

Figure 3c. Psychomotor speed

Summary and Conclusions

The results from the Experiments 1 and 2 suggest that fault finding, at least one aspect of it, is readily predictable providing both task and personal characteristics are taken into account. Ackerman's (1988) Theory of the Ability Determinants of Skill Acquisition provides a fruitful framework from which to understand why previous research has failed to consistently correlate fault finding skill with measures of ability.

Key References

Ackerman, P.L., 1992, Predicting individual differences in complex skill acquistition. *Journal of Applied Psychology, 77,* 598-614.

Ackerman, P.L.,1988, Determinants of individual differences during skill acquisition: Cognitive abilities and information processing. *Journal of Experimental Psychology: General, 117, 288-318.*

Brooke. J.B., Cook, J.C., and Duncan, K.D., 1983, Effects of computer aiding and pre-training on fault location. *Ergonomics, 26,* 669-686.

Dale, H.C.A.,1958, Fault-finding in electronic equipment, *Ergonomics, 1,* 356-358.

TRAINING FOR STRESSFUL ENVIRONMENTS: IMPLICATIONS FOR THE ROYAL NAVY

Mel Forster

Psychology Department
Centre for Human Sciences
Defence Research Agency, Portsdown
Portsmouth, Hants PO6 4AA

mforster@dra.hmg.gb

This study was concerned with identifying stressors encountered by Naval personnel at sea, their effects on operational performance and the potential training solutions for minimising these effects. The methods used for data collection were interviews with Naval personnel to identify sources of stress in individuals and teams; critical incident analyses to identify whether stress has lead to significant error; and interviews with course training instructors to investigate how stress is dealt with in the Royal Naval (RN) training system.

Introduction

Naval personnel are exposed to a variety of stressors at sea that may impact their effectiveness to carry out critical tasks. The potential for serious error that stressful situations engender is well known; risky decisions are made, skilled performance declines, and crucial information is ignored (Driskell, J. E., Hughes, S. C., Guy, W., Willis, R. C., Cannon-Bowers, J. A., & Salas, E., 1991). It is therefore important to understand the effects of stress on task performance, and to determine how training interventions can be used to help teams perform effectively under a range of stressful conditions.

One crucial component in maintaining personnel performance in a stress environment is to provide practice and the exercise of critical tasks under operational conditions similar to those encountered in the real environment. There is however, little empirical guidance available on how to effectively simulate a stress environment for training purposes, or how to design training systems for specific stressor environments. A key issue in this context is stressor fidelity. That is, should stressors that are characteristic of the criterion situation be present with a high degree of fidelity in the training scenario? Furthermore, is it necessary for stressors used in the training scenario to be characteristic of the criterion situation?

This paper will present firstly, an overview of the concept of stress and its effects on trainee performance, secondly a discussion of the design of training scenarios for stressful situations, and finally a presentation of example stressors in Naval operations and possibilities for empirical study.

Stress and the trainee

Performance under stress distinguishes the expert from the novice operator, stress will quickly rob the novice of mental resources and performance deteriorates. Skills that are proceduralised and skills that are automated do not deteriorate as much under stress as skills less well practised. The more complicated a process an individual has to perform, the more likely and potentially more serious is the performance decrement under stress. A trainee's ability to perform cognitive and psychomotor skills learnt under low stress conditions, is known to deteriorate when the level of stress is raised or when the trainee is faced with real responsibility for his actions. An individual's response to stress is seen as an adaptive process, whereby exposure enables the trainee to adopt particular strategies to improve the ability to cope with stress. However, individuals have a limited processing capacity and therefore, the mental resources expended on task performance and strategies adopted to cope with demanding situations must fall within this capacity.

Effects of Stress

The Yerkes-Dodson law (Yerkes & Dodson, 1908) proposes that stress has an inverted U function. The rule states that at low stress levels, attention and motivation are minimal which leads to poor performance. As stress increases, performance increases until a critical level is reached, at which performance starts to decline. With ensuing anxiety and eventually panic, performance deteriorates rapidly. The Yerkes-Dodson law has implications for training. Organisations should not strive to eradicate stress from training, moreover, training should seek to identify stress levels that will arouse individuals sufficiently to act as an effective motivation in the learning of tasks, and should identify which stressors are necessary to elicit the correct level of response.

The classic behavioural response patterns (fight, flight, fright) to stress are well documented. The fight response results in a increase in adrenalin production. Over stimulation of adrenalin decreases the ability of the brain to carry out simple mental tasks, particularly arithmetic ones. In addition, the human has a limited mental processing capacity of 7 simultaneous channels (Miller, 1956). When new problems arise, the brain has to establish priorities for load shedding or time sharing the solution of problems. When subjected to stress, this ability to prioritise is reduced; the cognitive system becomes overloaded which may result in an excessive preoccupation with trivia. This is seen either as tunnel vision, where the individual concentrates on one isolated aspect of the problem, (often the easiest or most familiar rather than the most important), or as the butterfly syndrome where the individual flits from one aspect to another without method, solution or priority (Chalton, 1991).

The fright or freeze response often results in a denial of the existence of a problem or its magnitude, also known as the Cognitive Mediation of Stress and can have disastrous consequences.

Information overload arouses the Autonomic Nervous System which impairs the performance of difficult tasks and causes feelings of inadequacy, frustration pressure etc. The effects of quantitative (more work than time allows) or qualitative (skill/knowledge exceeds individuals ability) overload can quickly lead to reduced performance if the individual has not had sufficient training or experience of the situation. With insufficient training, stress can often result in a regression to a fundamental point in the individuals knowledge or training which is likely to be inappropriate in a complex tactical situation.

The ability to make a correct judgement under pressure, can only be gained through practice and the thorough knowledge of one's own limitations. Within the RN, the Submarine Commanding Officers Qualifying Course and Fleet Air Arms training are courses which candidates may only take once. These courses are specifically designed to push students to the limit in dealing with stress. The philosophy behind these courses is to enable the student to realise his own limitations. A Submarine Commander will not experience such highly demanding situations in real operations, but the training is designed to a) test trainees to determine if they pass the course by subjecting them to the stress levels representative of the real situation, and b) to push the student further than the pass limit to ensure that he is able to realise his own limitations and to be able to recognise signs that he is reaching his upper limit. An important skill of a serving Commanding Officer (CO) is to allow trainee Officers on ship to carry out tasks even when making errors, until the very last moment at which point the CO will bail them out. If the trainee does not see the effect of his actions, he will not learn. This is a very fine line on which the CO has to tread. On these courses it is important to know what type of response pattern an individual will elicit before he takes up a Command appointment.

The philosophy of the training system is to generate stress. Personnel must be aware that certain situations will stress them and they should be able to recognise stress. However, they must also be aware that the activities cannot stop just because they are stressed. In a team, people must be taught to a) recognise and b) expect stress to manifest itself. It is very difficult to teach people to cope with theoretical stress (i.e. stress which is known to occur but has not been experienced), but stress is very difficult to simulate.

Stressor Fidelity

The difficulties in the training of individuals and teams for task performance under stress centres around the problem of stressor fidelity. That is, does the achievement of proficient performance under stress require that stressors that are characteristic of the operational situation be present with a high degree of fidelity in the training for that task, (Friedland & Keinan, 1986)? The argument for high fidelity suggests that the accurate representation of operational stressors in training may allow trainees to experience their reactions to realistic levels of stress, replace anxiety with confidence, and facilitate stimulus generalisations and the transfer of learning. However, the proponents of low fidelity maintain that the presence of stress during training may interfere with task acquisition. Studies have shown that not only does stress reduce efficiency but it can produce interfering responses which impair both performance and learning of tasks (Eysenck, 1976). A compromise theory would suggest that what may be needed is a phased approach where stress intensity is increased progressively with task training, or where separate training sessions are devoted to task and stress learning, (Friedland & Keinan, 1986).

To be effective a simulation should be designed to ensure that the individuals/teams cognitively perceive the situation as stressful, to enable them to experience realistic reactions to stress. The simulation must avoid cues which are artificial or provide extra support than is representative of the real situation.

Stressors in the RN

The military environment exposes personnel to a variety of stressors induced by exposure to continuous hostile environmental demands, coupled with isolation from sources of support in situations where behaviour is highly constrained. In these situations, a number of potential cognitive and emotional stressors exist. However, not all of these fulfil the potential in terms of inducing stress. Many of the variables serve to cause frustration, diminishments in enthusiasm, low morale and ill feeling. It is important to identify variables which actually cause stress per se, and to determine how each effects performance. Few researchers have addressed the question of the comparative effects of different stressors on human performance.

This study was concerned with identifying stressors encountered by Naval personnel at sea, their effects on operational performance and the potential training solutions for minimising these effects. The methods used for data collection were = interviews with Naval personnel to identify sources of stress in individuals and teams; critical incident analyses to identify whether stress has lead to significant error; and interviews with course training instructors to investigate how stress is dealt with in the Royal Naval (RN) training system. A selection of the results are presented below.

Stress in Naval team operations may arise from many different causes:

a. Self doubt. If one believes the task is not one that has been covered in training or if one is aware of a weakness in ones make up that is likely to be exposed.

b. Peer pressure. Others may radiate confidence and that confidence may arise from proven competence or an inability to recognise the problem, either way, the hesitant will be stressed in 'keeping his end up'.

c. Fear. Situations arise at sea which will frighten those experiencing them for the first time, e.g. gunnery work on the upper deck in a force 10 gale as a junior officer.

d. Team pressure. The fear of being identified as a weak link in a mutually dependent team will generate great stress and may prompt error.

e. Keeping up. Different individuals learn/assimilate new information at different rates. Failure at late stages of training is not unusual as a result of failing to grasp fully, a fundamental yet basic principle early on, and "hoping that it won't matter".

f. Fatigue. People can perform continuously and satisfactorily for different lengths of time. In many organisations, those doing the non essential and probably, non complex tasks are the least likely to be spared for a 'breather'. Knowing that the Command has confidence in a team member will increase the pressure to prove that confidence well placed.

This list is certainly not exhaustive but it illustrates the wide range of stressors effecting performance. It would not be possible, however, to simulate all of these variables in a training scenario. Thus it is important to identify whether stressors have to be representative of the operational situation or whether they can be substituted by more easily simulated variables. A number of variables can be manipulated in a Naval training scenario to induce stress. A selection of possible variables is given below:

a. Instructor response. This can be passive and non committal, aggressive or supportive, which can be varied to differ from trainee expectations.

b. Additional tasks. The imposition of additional tasks or considerations, particularly when planning or execution are advanced, can be disruptive, thus increasing the perception of quantitative/qualitative overload.

c. Disruption. The disruption of an ordered plan through a change of task priorities.

d. Navigational constraints. Making the scenario more hazardous by necessitating operations to be carried out in restricted, dangerous positions such as the safety and navigational accuracy of inshore operations, and a reduction in the navigational aids, sensors etc. which will deny important and reliable information.

e. Time constraints. Ordering a specific time for task completion builds stress as deadline increases, giving a time window leaves the student an element of control to carry out the task.

f. Presence of other units. Operations within close vicinity of other units whilst assuming a covert posture in the case of a submarine. In essence, increasing the number of potentially conflicting inputs and enabling the student to establish the correct priorities. An important training aspect is to overload key individuals with tasks so that the trainee learns how to prioritise and delegate.

g. Fatigue. Forcing training/operations when tired.

Conclusions

Should we provide specific stress training which combines practice of coping skills with gradual exposure to stressors, in order to improve performance under stress? Or, by contrast, provide trainees with practice in stressful environments, to allow them to make mistakes, identify their own stress signs and limitations and adopt successful strategies. With sufficient training and high competence it is likely that individuals will be able to cope with stress. If the task is highly learned it may be more resilient to the effects of stress. If the task is not highly learned, specific training interventions may be needed to enhance the performance of critical skills under stress. However, it is necessary to determine how much training is needed to achieve this level of competence. Training in full mission simulators is expensive imposing an upper limit on the amount of training that can be provided. There is likely to be a trade off between the amount of task training and stress training required. Specific strategy training may be able to make up the shortfall of task training that cannot be undertaken. It is suggested that this should be tested empirically.

In addition, research should concentrate on identifying the comparative effects of different stressors on performance. In this way, guidance could be provided on stress variables that could be incorporated into training scenarios to produce specific stress responses.

Any views expressed in this paper are those of the author and do not necessarily represent those of the Defence Research Agency.

© British Crown Copyright 1996/DRA

Published with the permission of the Controller of Her Britannic Majesty's Stationery Office

References

Chalton, Cdr. D. R, 1991, Stress and the submarine command course. The manipulation of stress as an aid to training. *Review of Naval Engineering*, **45(2)**, 25-31.

Driskell, J. E., Hughes, S. C., Guy, W., Willis, R. C., Cannon-Bowers, J. A., & Salas, E. 1991, *Stress, stressors, and decision making* (NTSC technical report). Orlando, FL: Naval Training Systems Centre.

Eysenck, M. W. 1976, Arousal, learning and memory, *Psychological Bulletin*, **83**, 389-404.

Friedland, N. & Keinan, G. 1986, Stressors and Tasks: How and when should stressors be introduced during training for task performance in stressful situation? *Journal of Human Stress*, **12,** 71-76.

Miller, G. A., 1956, The magic number seven plus or minus two-some limits on our capacity for processing information. *Psychological Review*, **63**, 81-97.

Yerkes, R. M. & Dodson, J. D. 1908, The relation of strength of stimulus to rapidity of habit formation. *Journal of Comparative Neurology and Psychology*, **18**, 459-482.

EXPERTISE AT X-RAY FILM READING: THE ROLE OF ATTENTION, AND IDENTIFICATION TRAINING

Paul Sowden, Ian Davies, Margaret Lawler,
Bridget Curry and Sean Hammond

*Department of Psychology,
University of Surrey,
Guildford, Surrey, GU2 5XH*

The success of the national breast cancer screening programme relies on the skill of X-ray film readers in detecting visual signs of abnormalities. Film reading is a complex perceptual skill and even readers with considerable expertise miss signs of abnormality. This paper reports two experiments that investigate the role of attention in learning to detect features in X-rays and the effect of identification training on performance. In both experiments observers searched X-rays of perspex 'phantoms' for low contrast features superimposed on an unstructured amorphous background. Results indicated that learning may occur as a result of simple exposure to target features without even requiring attention. The implications of these findings for the design of film reader training programmes are discussed.

Introduction

Training schedules in screening mammography have generally been designed on a somewhat ad hoc basis taking little account of research findings indicating the nature of the learning processes involved. The formal training period is relatively brief (two weeks to three months) and typically a 'tutorial' approach is adopted. In such an approach trainees are shown examples of and are taught about mammographic features, practice reading screening 'rollers' of X-rays, and also engage in home study of teaching materials such as Tabar's teaching atlas of mammography.

Although this type of training schedule has some face validity and makes a certain amount of intuitive sense, it might be expected that with more systematically designed training programmes, a higher level of screening performance could be achieved. With this aim in mind we have been investigating some of the factors influencing observers learning to read X-rays. In the present paper we report two such studies which partially overlap in that one group of subjects is common to both experiments.

Experiment one

Some of our previous research has shown that improvements in performance are in part due to 'low-level' perceptual learning (Davies, Sowden, Hammond and Ansell, 1994). Such learning is primarily a function of the number of targets a subject is exposed to, but it is not clear from the perceptual learning literature whether attention to the target is required, or whether mere exposure is sufficient. Experiment one addressed this question. One group of observers ('explicit learners') practised a task on two consecutive days (200 trials per day) in which they were required to search X-ray images for small low contrast dots. A second group of observers ('incidental learners') were required to examine the same images for their 'streakiness' on day one, but they were not told anything about the dots. On day two they also practised the 'dot-detection' task. Thus, the incidental learners had the same amount of exposure to the dots, but only half the amount of attentive experience. By comparing the two groups performance after training, it is possible to determine the importance, for learning, of attention to the target features during training. If the incidental groups performance on day two is better than the explicit learners on day one, then this would suggest that learning can occur without attention.

Method

Sample, apparatus and stimuli

Thirty-two subjects were randomly allocated to one of two groups, 8 were male and 24 were female and their ages ranged from 18 to 28. Subjects were paid £5 for participating and in addition, they received a small performance related payment as detailed below.

The stimuli were radiographs of perspex blocks made up of grids of 25 squares. The sides of the squares measured 25mm and there was a 0.35mm diameter hole in the centre of each square. The depth of the holes was either 0.10mm, 0.25mm, 0.50mm or 0.75mm. In addition there was a second hole midway along one of the four hemi-diagonals, of the same diameter and depth as the hole in the centre. Within a grid the four possible locations were used equally often for each hole depth. The holes showed up on the X-ray as small dots whose contrast varied as a function of hole depth: the deeper the hole the greater the contrast of the dot to the background. Perceptibility of the dots varied from below to above typical subject thresholds.

Design and procedure

The explicit learners practised a dot-detection task on each of two consecutive days. They were required to examine each square in a grid and to indicate in which of the four possible locations a dot was positioned. Subjects were instructed to refer to the dot in the centre of each square for an example of what the dot they were looking for in that particular square would look like. Subjects were paid 1p for every dot they correctly detected above chance performance. Subjects completed eight grids on each day (200 decisions per day) and were given feedback on the number of dots they had correctly detected after completion of each grid. In addition the time taken to complete each grid was recorded.

The incidental learners performed a 'cover task' — rating of streakiness — on day one. They were presented with the same grids as the explicit learners, but were not told about the presence of the dots. Instead subjects were instructed to examine each of the four quarters of a square and to decide which was the streakiest. Subjects were given 'bogus' feedback on their performance after completion of each grid such that the feedback received by subjects in this group overall closely matched the real feedback given to the explicit learners on day one of the experiment. They were paid 1p for each 'correct' decision above chance performance. Subjects completed eight grids on this task. On day two subjects took part in the dot-detection task exactly as described for the explicit learners. Questioning of the incidental learners after completion of the experiment indicated that none of them had been aware of the presence of the dots in the grids they were shown on day one.

Results

The results reported below are based on the findings of mixed design analyses of variance with repeated measures on day of the experiment and/or training block (i.e. grids one through eight on each training day).

Do the explicit learners learn?

The number of dots that the explicit learners correctly detected increased between day one and day two ($F(1,15)=9.47$ $p=0.008$; mean day one = 10.49, mean day two = 11.53), but there was no significant increase in detection accuracy within sessions. However, there was a significant change in the explicit learners detection speed within sessions ($F(7,105)=3.35$ $p=0.003$). The relevant mean scores are shown in table 1.

Table 1. Mean grid completion times (seconds).

	Block 1	Block 2	Block 3	Block 4	Block 5	Block 6	Block 7	Block 8
Mean completion time	272	269	260	254	250	249	242	236

Curve fitting revealed a significant linear fit to the data ($r^2=0.978$ $F(6)=268.59$ $p<0.0005$), indicating a trend for detection speed to decrease linearly within sessions.

Do the incidental learners learn without attention?

The incidental learners were more accurate at the dot-detection task on day two than the explicit learners were on day one ($F(1,30)=6.31$ $p=0.018$; mean for group one subjects = 10.49, mean for group two subjects = 12.08), but there were no significant differences between the two groups grid completion times.

Finally, there were no significant differences between the incidental learners and the explicit learners on day two of the experiment in terms of either their detection accuracy or their detection speed.

Discussion

Experiment one examined whether attention to target features was required in order for improvements in detection performance for those features to occur. Results indicated that the explicit learners detection accuracy improved, and at no cost to their detection speed. However, the incidental learners were as good at the detection task on day two as the explicit learners, and were better than the latter group were on day one — it appears that learning can occur without attention. These findings are in contrast to some previous research on low-level perceptual learning phenomena which has indicated that attention is necessary for learning to occur (e.g. Shiu and Pashler, 1992).

Experiment two

The dot-detection task of experiment one could be considered to have at least two separate components: identification of what constitutes a target, and then search for a target. Consequently, the learning observed in experiment one may have resulted from refinement of subjects' 'target templates', or an improvement in the way in which subjects searched the images for features, or both. If learning is based on improvements in both components of the task, then it might be expected that a group of subjects who practice just one component, for instance identifying the dots in a known location, but not searching for them, would not learn as much. However, it might be argued that such a procedure could lead to the formation of more refined target templates and hence equal, but different, learning. If this were the case then subjects who practice both identification and dot-detection might be expected to show the greatest learning. Experiment two investigated whether identification training led to as much learning as shown by the explicit learners in experiment one. On the first of two consecutive days, one group of subjects ('identifiers') were required to practice identifying the dots, presented in the centre of each square in the grids used in experiment one, according to their contrast. A second group of subjects ('identifiers+explicit learners') were required to identify the dots according to their contrast, and then to detect the dot midway along one of the hemi-diagonals in each square. On day two, both groups completed the dot-detection task exactly as described in experiment one. Analysis then compared the learning achieved by the explicit learners of experiment one, with that achieved by the identifiers, and with that achieved by the identifiers+explicit learners.

Method

Sample, apparatus and stimuli

Thirty-two subjects were randomly allocated to one of the two groups, 12 were male and 20 were female, and their ages ranged from 18 to 27. Subjects were paid £5 for participating and in addition, they received a small performance related payment as detailed below.

The stimuli were identical to those used in experiment one.

Design and procedure

The identifiers practised identifying the centre dot in each square on the first day. They were required to examine each dot and to identify it as one of four possible contrast levels ('very dark', 'dark', 'light', 'very light'). Subjects were paid 1p for every dot they correctly identified above chance performance. Subjects completed eight grids (200 decisions per day) and were given feedback on the number of dots they had correctly identified after completion of each grid. In addition the time taken to complete each grid was recorded.

The identifiers+explicit learners practised both identifying and detecting the dots in each square on the first of two consecutive days. For each square they were first required to identify the centre dot in the same way as subjects in the first group and then to indicate which of the four corner locations they thought contained a dot. Subjects were given feedback on their identification performance and on the number of dots correctly detected after completing each grid, and were paid ½p for each correct identification or detection decision made. Subjects completed eight grids on this task.

On day two both groups of subjects completed the dot-detection task exactly as described for day two in experiment one.

Results

As in experiment one the results reported below are based on the findings of mixed design analyses of variance with repeated measures on day of the experiment and/or training block (i.e. grids one through eight on each training day).

Does identification performance improve?
Neither the identifiers or the identifiers+explicit learners significantly improved their identification performance on day one of the experiment in terms of accuracy or speed.

Do the identifiers+explicit learners improve their dot-detection performance?
The identifiers+explicit learners dot-detection accuracy increased significantly over the course of the experiment ($F(1,15)=8.63$ $p=0.010$; mean for day one = 10.1, mean for day two = 11.43), but not within experimental sessions.

Do the identifiers learn as much as the explicit learners?
There were no significant differences between the identifiers and the explicit learners on day two in terms of either their detection accuracy or their detection speed.

Do the identifiers+explicit learners learn more than the explicit learners?
There were also no significant differences between the identifiers+explicit learners and the explicit learners on day two in terms of either their detection accuracy or their detection speed.

Discussion

Experiment two examined whether identification training led to as much learning for a dot-detection task as training on the dot-detection task itself. Further, the experiment examined whether a combined identification plus dot-detection training procedure had an additive effect on subsequent detection performance. Results indicated that the identifiers+explicit learners dot-detection accuracy improved as a result of practice. However, there were no significant differences in dot-detection accuracy between the three groups on day two of the experiment. These findings suggest that practice on the single component of dot identification leads to as much learning for subsequent dot-detection performance as practice on the dual component (identification & search) dot-detection task itself. The fact that the identifiers+explicit learners did not exhibit a greater amount of learning would suggest that there is no additive effect of completing both types of training. One possible explanation for these results is that even in identification training subjects are exposed to the targets, and that it is simply this exposure which induces learning to occur.

General Discussion

Both experiments provided evidence that practice on a dot-detection task leads to improvements in detection accuracy across days, and the results from experiment one suggest that this is not at the cost of a reduction in detection speed. In other words there does not appear to be a speed-accuracy trade-off. However, it is noteworthy that no improvement in accuracy occurs for either the dot-detection or the identification tasks within sessions. It would seem likely that this finding can be accounted for as resulting from subject fatigue masking any within session improvements.

The present experiments are in agreement with our previous findings that detection performance increases as a function of exposure (Davies, Sowden, Hammond and Ansell, 1994; Davies, Roling, Sowden and Hammond, in press). This finding suggests that in order to most efficiently induce improvements in detection sensitivity, in addition to the tutorial approach adopted in many mammography training programmes, it would be beneficial for trainees to engage in some form of massed practice training procedure.

Further, the lack of difference in performance between any of the groups on their dot-detection accuracy on day two of the experiment is also consistent with the proposition that learning is facilitated through simple exposure to target features. This finding is of particular interest when considering the training of X-ray film readers. One of the problems facing mammography screening services is recruitment of trained radiologists in sufficient numbers. An alternative is to train non-medical personnel to carry out routine screening. The present results suggest that radiographers, whose normal job is to produce the X-ray images, may have learned something of the perceptual skill involved in film reading, and thus could start training as film readers at a higher level of skill then those without such incidental exposure. As such radiographers may represent an advantageous group of individuals to train in screening mammography.

References

Davies, I.R.L., Sowden, P.T., Hammond, S.M. and Ansell, J. (1994) Expertise in categorising mammograms: a perceptual or conceptual skill? *Proceedings of the International Society for Optical Engineering*, **2166**, 86-94.

Davies, I.R.L., Roling, P., Sowden, P.T. and Hammond, S.M. (in press) Design of training schedules for medical image readers: the effects of variability and difficulty levels of the training stimuli. *Proceedings of the International Society for Optical Engineering*.

Shiu, L. and Pashler, H. (1992) Improvement in line orientation discrimination is retinally local but dependent on cognitive set. *Perception and Psychophysics*, **52**, 582-588.

HELPING LEARNERS GET STARTED: DESIGNING A MINIMAL MANUAL FOR A TELECOMS TERMINAL

Kevin C. Kerr and Lorna M. Love

Philips Business Communication Systems
Victoria Place
Strathclyde ML6 9BL
Scotland

This paper reports the design and evaluation of a prototype minimal manual for a telecommunication terminal. Input for the design was gathered from forty learners, customer support experts, help desk data and survey material. Design issues surrounding manual content and presentation are discussed within the paper. The prototype manual and the conventional manual were evaluated using think aloud protocols with eight learners. Data from these not only indicated the superior performance of the minimal manual over the conventional manual, but also provided justification for design changes based on learners' input.

Introduction

Although increasing effort is being paid to improving the design of a product's interface there will always be a requirement for supplemental materials that aid users in both paper and on-line form. For 'Hi-End' telecommunication terminals documentation has become an expected component of most systems. Difficulty occurs, when attempting to anticipate what problems learners will have when interacting with telecommunication terminals and what type of documentation would provide the best assistance. Output from a previous study (Jordan and Kerr 1993) highlighted: difficulties for learners in getting started with a telecommunication terminal; and the need for an abridged instruction manual that specifically focused on helping learners get up to speed with the procedures for basic call handling (e.g., transferring, diverting calls). The second author (from here on referred to as the designer) selected 'The Minimal Manual' proposed by Carroll, Smith-Kerker, Ford and Mazur to serve as a foundation for the protoype development of the terminal manual.

Carroll's 'Minimal Manual' was designed to address difficulties people have with state-of-the-art self-instruction manuals in learning to use powerful computing devices. It is briefer, it helps learners co-ordinate their attention between the system and the manual, it specifically trains error recognition and recovery, and it better supports reference use after training (Carroll et al, 1987-88). Minimalist design gambles on the expectation that if you give the learner less (less to read, less overhead, less to get tangled in), the learner will achieve more (Carroll 1984).

The main principles taken by the designer from (Carroll et al, 1987-88, Carroll 1984) as a basis for design were:

Table 1

Principle	Principle Reference
Focus on realistic tasks	P1
Slash the verbiage	P2
Support error recognition and recovery	P3
Guide exploration	P4

Approach

Firstly, content and presentation design issues were tackled by the designer. Then, a first cut prototype was constructed and made ready for usability evaluation by potential learners.

Since an aim of the study was to investigate the benefits of adopting the minimalist approach, to that of the conventional, evaluation of the conventional manual was also undertaken.

Minimal Manual Design

Content

Fundamental to the minimal manual's success would be the detailing of functionality that formed a best match with the needs of learners when first interacting with the telecommunication terminal. Three main sources were used to steer the selection of the manuals functionality viz:

- Survey Material: Particularly (ISDN survey 1992) which detailed the most frequent and commonly used functionality on ISDN terminals.

- Expert Input: Experts within customer support were interviewed for their recommendations.

- Help Desk: Help desk database reports were analysed. Particularly those that detailed guidance on operation of functionality.

The above approach was undertaken to follow principle P1 (c.f., Table 1).

To match principle P2, the following approach was used. When detailing the procedural steps for each function: task information was mentioned only once and a balance was struck between brevity in writing style and providing enough information for the learner to complete the task successfully.

For principle P3, system walkthroughs were used by the designer to formulate predictions on learner errors and possible recovery mechanisms.

For principle P4, each task within the minimal manual would have a dedicated section that encouraged the exploration of related system functions.

Presentation

Issues surrounding the presentation of the minimal manual (e.g. overall size [A4, A5, A6], orientation [portrait, square, landscape], layout [loose pack of cards, joined cards, small booklet, concertina guide]) generated numerous prototypes. These prototypes were discussed with forty learners. Learners considered each prototype in turn, and were encouraged to give their opinion in terms of presentation. These opinions were then used to converge the early designs. The layout of the minimal manual is illustrated in Figure 1 below:

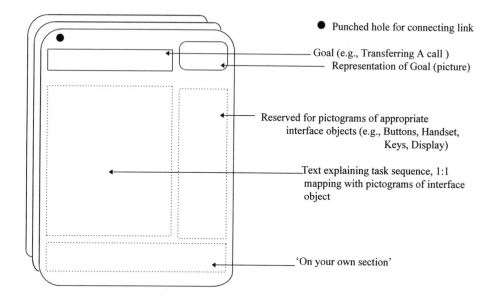

● Punched hole for connecting link

Goal (e.g., Transferring A call)
Representation of Goal (picture)

Reserved for pictograms of appropriate
interface objects (e.g., Buttons, Handset,
Keys, Display)

Text explaining task sequence, 1:1
mapping with pictograms of interface
object

'On your own section'

Fig 1: Layout Of Prototype Minimal Manual

The manual was made out of A6 cards and connected by way of a semi-permanent link. It was divided into three areas. The top area was reserved for highlighting the functional goal that could be achieved by using that specific card (where possible a representation was used to help the learner visualise the task). The middle area was reserved for sections of textual description (procedural steps) on how to carry out the task. A section of text, where appropriate, would be aligned with a pictogram of the terminals interface object that required to be activated (e.g., button press) or referred to (e.g., display). The bottom area was reserved for encouraging exploration of related functionality and, where appropriate, the 'undoing' of the function that had just been completed.

Now that the manual's content and presentation design issues had been addressed, to a degree, the prototype was made ready for usability evaluation.

Usability Evaluation

The purpose of the evaluation was two-fold. Firstly, to gather data about the usability of the prototype and, secondly, to compare the prototype in terms of subject performance against that of the conventional manual. The evaluation was rather typical of the sort of "Quick and Dirty" evaluation carried out in industry (Kerr and Jordan 1995). Eight subjects were involved and limited performance measures (i.e., time and errors) recorded. Notwithstanding, Q&D evaluations "..provide an approximate result.." (Bonner and Cadogan 1991) which was considered satisfactory for the first iteration of the minimal manual's design.

All subjects consented to be filmed carrying out eleven tasks on the terminal with the aid of the two manuals: four subjects used the minimal manual and four used the conventional manual. All subjects:

- were asked to think aloud when carrying out each task; and
- were not familiar with the terminal or the manuals.

The following table details the tasks set:

Task Number	Task Description
T1	Making a call to an external number
T2	Using last number redial
T3	Transferring an incoming call (to free extension)
T4	Putting a caller on hold
T5	Programming a memory key
T6	Diverting/Undiverting a call
T7	Examine system status
T8	Establish a 3-Party call
T9	Answering a call handsfree
T10	Switching between Loudspeaking and Softspeaking
T11	Transferring an incoming call (to busy extension)

Results of Evaluation

The charts below detail a comparison between the minimal manual and the conventional manual by plotting the mean subject performance on each task . Chart 1 details, for each task (T1-T11), the time taken in seconds for successful task completion (note: subjects completed all tasks). Chart 2 details the quality of interaction using a subject error scale: on a per task basis, points were deducted (from the top mark 5: no errors) in relation to the severity (major - minor) of the error made during interaction.

Time On Task   **Quality Of Interaction**

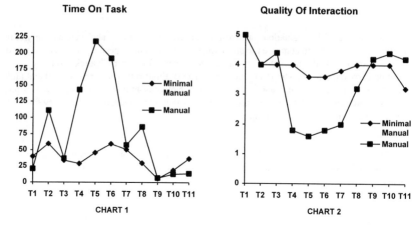

CHART 1   CHART 2

Both Chart 1 and Chart 2 indicate that the minimal manual had a better overall mean subject performance.

Discussion

Our technique here was not just to use Carroll's principles as a tool to prune or minimalise an existing conventional manual, but to design a manual from scratch utilising only minimalistic principles. This approach has placed us in a better light for judging the number of skills (e.g., language, graphical design, usability testing..) that are required to contribute to the production of a successful minimal manual and of the development costs in terms of time and effort. One major development component that appears to be missing form the literature that we have covered during this study is a practical methodology for practitioners in this area. The existence of a methodology would have allowed the designer in this study to systematically implant minimalist properties into the telecom manual rather than running with intuition alone.

That said, the results indicate that subject performance, as a whole, was improved when using the designer's minimal manual in measures of 'time on task' and 'quality of interaction' within this context. Comments noted from the think-aloud data suggest a positive reaction to the minimal manual from subjects: with straightforward usage, ease of reading and ability to scan quickly being suggested as important to the success of such a manual.

Further examination of the think aloud data indicates that the 1:1 mapping of text to interface objects (c.f., Fig 1) by way of pictograms contributed significantly to the minimal manual's superior performance. At first, learners would read a section of text followed by a visual check which matched the pictogram adjacent to the text with the interface object to be attended to. However, in subsequent tasks learners would, at times, neglect the text in favour of considering the pictograms only. This allowed the learners to quickly activate a number of interface objects in sequence and check for the commensurate output on the display at intervals. The effect of this was efficient task completion times in comparison to the conventional manual.

Furthermore, this study reports on the first iteration of the minimal manual's design. Subsequent iterations (using think-aloud data to remove reported design flaws) may make the design more polished and increase its out performance of the conventional manual.

Since the evaluation of both manuals was "Quick and Dirty" with a small sample size it would be clearly rash to try and generalise on the basis of this study. However, despite these limitations we anticipate that this type of approach will produce a manual which will be better designed to meet the requirements of learners in this context.

Conclusion

As a first pass (for the designer) at this problem, recourse to Carroll's work on minimalism served as a good basis in designing a prototype minimal manual for a telecommunication product. Opinions from forty potential learners on manual presentation helped in focusing early designs. Recommendations from customer support experts, analysis of help desk data and survey material all can be considered as valid input when designing a manual from scratch. The prototype minimal manual was shown to out perform the conventional manual in measures of 'time on task' and 'quality of interaction'. Further work is required in the empirical section of this study to increase the confidence in the result as the sample size was small.

Acknowledgement

We thank our colleagues in Philips for advice and comments on this topic.

Note: that the views expressed within this paper are the authors' and not necessarily those of the company they represent.

References

Acceptance of ISDN Terminals, February 1992. Prepared for Philips Komunikations Industrie AG Nurnberg.

Bonner JVH, Cadigan P, 1991, Important issues in successfully using "Quick and Dirty" methods in ergonomics design consultancy. Contemporary Ergonomics 1991, (ED) EJ Lovesey, (Taylor & Francis: London)

Carroll J. M., Smith-Kerker P. L., Ford J. R., and Mazur-Rimetz, S. A., 1987-1988, The Minimal Manual. Human Computer Interaction, volume 3, pp.123-153, Lawrence Erlbaum associates. Inc.

Carroll J. M., 1984, Minimalist Design for Active Users. In proceedings of IFIP INTERACT' 84: Human Computer Interaction (pp. 39-44).

Jordan P.W., Kerr K.C. 1993, Usability testing in the "Real World": Evaluating a multi-function phone system. Contemporary Ergonomics, edited by E.J. Lovesey, (Taylor & Francis: London).

Kerr K. C., Jordan P. W. 1995, An Investigation Of The Validity And Usefulness Of A "Quick And Dirty" Usability Evaluation. Contemporary Ergonomics, edited by E.J. Lovesey, (Taylor & Francis: London).

Meij, H. v. d., and Lazonder, A. W. (1993). Assessment of the Minimalist Approach to Computer User Documentation. Interacting with Computers, 5(4) 355-370.

Task Analysis

THE PRACTICAL INTEGRATION OF TASK ANALYSIS INTO SYSTEM DEVELOPMENT

Roger S Harvey[1] and Michael K Goom[2]

[1]R(LS)HFI
Ministry of Defence
Main Building Rm 2230
Whitehall
LONDON SW1A 2HB

[2]Systems Concepts Dept.,
BAe Dynamics Ltd
PO Box 5
Filton
BRISTOL BS12 7QW

This paper provides a summary of the findings of a two-day workshop, held in September 1995 under the auspices of the UK MOD/Industry MANPRINT Consultative Working Group (a forum for the exchange of information between UK MOD and Industry concerning the use and application of the MANPRINT/Human Factors Integration programme in defence equipment procurement). Some sixty experts in a variety of the disciplines that contribute to system development analysed ten topics concerning Task Analysis and its relationship to the system development processes within defence equipment procurement.

Introduction

Task Analysis (TA) can be considered to be one of the fundamental methods underpinning the successful application of Human Factors to system design and development. Recognising this, the UK MOD/Industry MANPRINT Consultative Working Group (a forum for the exchange of information between UK MOD and Industry concerning the use and application of the MANPRINT/Human Factors Integration programme in defence equipment procurement) organised a two-day Workshop "Task Analysis and MANPRINT". This meeting brought together experts in a variety of the disciplines that contribute to system development in order to discuss and analyse a number of key topics concerning TA and its relationship to the system design and development processes within defence equipment procurement.

In order to set the context for the Workshop the design process was taken to mean the whole process of product development and procurement from concept design to equipment disposal. The Workshop aimed to

disseminate a shared understanding of what TA can contribute to system design and evolution, examine the required level of detail for TA,

determine how best to utilise information derived from TA,

examine what common core of TA techniques might be used,

examine the potential for sharing and re-use of TA data.

Workshop delegates were selected from a broad spectrum of Government, Industry and Academic staff contributing expertise covering areas including Human Factors Engineering, Training, Integrated Logistic Support (ILS), Safety, Technical Publications, Procurement and Contracts and the Military User.

Workshop Topics

In structured discussion sessions ten topic areas were covered during the two-day Workshop:

1.   To define the essential characteristics of TA that are needed to contribute to Human Factors Integration.

2.   To identify the basic information requirements for TA to answer Human Factors Integration questions.

3.   To identify the most appropriate methods of TA to answer Human Factors Integration/system questions.

4.   Specify the level of detail to which TA should be accomplished for each development phase.

5.   For each group within an organisation that uses TA, what are its specific inputs and outputs (by phase) and how are TA outputs used to improve development of systems.

6.   Identify what is needed to ensure that the system design takes into account the results of the TA and whether this should be a contractual requirement.

7.   Identify the key contractual requirements of TA at each development phase. Determine who should "own" the TA at/between these phases, and the mechanisms for transfer between these phases.

8.   Identify the benefits and problems of TA for the Customer and Supplier.

9.   Define the relationship between HFI and ILS in respect of their individual responsibilities for Task Analysis, in order to reduce data gathering and maximise information.

10.   Identify the opportunities and mechanisms for the re-use of TA data within system procurement.

Summary of Workshop Findings

Given the diverse range of background and specialist areas of the delegates there was a surprising degree of consensus in the discussion periods. The main findings are:

The TA process provides a wealth of data which can serve a number of purposes. It is essential that TA is started as early as possible, and is traceable through the total design process,

There is a need for contracting the implementation of TA, at the generic level, but without mandating a specific methodology,

Key issues were agreed to be timeliness, flexibility, transparency and neutrality, all within the context of TA being fit for purpose, cost effective and capable of validation and auditable.

The communication and sharing of TA information, that is a framework for the control and management of TA throughout the design process, is a basis for ensuring that the appropriate decisions can be made at the right time. However it was felt that the current MOD procurement policy mitigated against this.

The interaction between HFI and other key specialisations such as ILS, Safety etc, needed to be better defined.

The TA activities needed to be fully supported by an MOD assessment capability, both in terms of the methods to be employed and the end results.

TA underpins the understanding of the human contributions to system success, which has been undervalued in the past. MOD must avoid seeing TA as a "tick-in-the-box" activity with the consequent less than serious consideration of TA.

There is a need for education and dissemination with regard to the benefits of TA, both to ensure greater awareness of TA and to establish its credibility.

MOD must provide a structured and high level description of user tasks as part of system requirements specifications.

The Workshop provided an intensely practical view of the role of TA in the design process. It was recognised by delegates that to adopt any or all of the recommended courses of action would require a change in the way that the different phases of the design process are currently handled. It was also acknowledged that there was a need for a radical re-appraisal of the way in which different organisations interact with one other if the concept of a data process passed from phase to phase was to be achieved.

The Workshop has provided an opportunity to beneficially affect the understanding of the human element in system design, through the collection and analysis of all task related data in a form usable throughout the design process.

Acknowledgements

The authors gratefully acknowledge the substantial contributions made by the Workshop Topic Leaders, and Organizing Committee, to the success of the event, a majority of whom are members of the UK MOD/Industry MANPRINT Consultative Working Group.

The views expressed are those of the authors and do not necessarily represent Company policy of BAe Dynamics Ltd, or Official policy of the UK Ministry of Defence.

RECENT DEVELOPMENTS IN HIERARCHICAL TASK ANALYSIS

John Annett

Department of Psychology,
University of Warwick,
Coventry CV4 7AL

Hierarchical Task Analysis (HTA) was developed in the 1960's and grounded in the concepts current cognitive psychology. Used initially for determining training needs of individual operators its scope has been widened to encompass a variety of applications. This paper describes the adaptation of the basic principles to the analysis of team skills with particular reference to Naval Command and Control teams. A model of team skills is outlined which identifies observable team processes and their inferred knowledge bases. The methods used to analyse team skills and to prescribe effective training procedures are briefly outlined.

Introduction

Hierarchical task analysis (HTA) originated in the early days of cognitive ergonomics and reflected some of the dominant ideas of the time. First, it was an attempt, but by no means the first, to represent significant information processing features of human work in very general terms which did not beg too many questions about specific cognitive mechanisms. Second, HTA was perhaps the first analytical methodology to recognise the importance of feedback in the control of behaviour. Even simple motor acts are not just 'emitted' but depend on the arrival of sensory feedback within 2-300 milliseconds if the action is to continue smoothly and so the description of behaviour had to include not only a characterisation of the action itself and the conditions under which it should occur but also the sensory feedback which assured the operator that the intended action is accomplished. This turned out to be particularly important in process control tasks. Thirdly, the idea that work could be described in terms of a goal hierarchy turned out to be not only a practical necessity but also matched the seminal ideas of Miller, Galanter and Pribram (1960) on the ways in which complex behaviour is organised. Since the publication of the definitive account of HTA in Annett, Duncan, Stammers and Gray (1971) there have been a number of significant developments but in this paper I want to outline a recent adaptation of HTA principles to the analysis of team tasks.

Team-work

Most published examples of HTA describe the tasks of individual operators yet many of these tasks, including some of those first analysed in the petrochemical industry, are carried out by teams of operators. I became acutely aware of this deficiency when invited to investigate the training of Naval Command and Control (C^2) teams for the DRA. I shall not describe the analyses in detail but will show how the basic principle of hierarchical decomposition can be applied team tasks, but first I shall outline a process model of the structure and organisation of C^2

A Process Model

In a powerful theoretical analysis of research on groups and teams Steiner (1972)pointed out that the productivity of the group, and the variables which determine it, are strongly influenced by structure and organisation. In some cases group performance depends heavily on the ability of the weakest member, in others on the ability of the strongest. Adopting Steiner's approach the C^2 team can be seen as comprising a hierarchically structured set of sub-teams and individuals. For example in a typical warship there are sub-teams concerned with collecting data on the air surface and underwater environments which contribute to a tactical picture for the Principal Warfare Officer who must decide on the deployment of weapons and other assets. Each of these sub-teams is defined by its *common purpose,* that is the variables it attempts to optimise.C^2 tasks are, in Steiner's terms, mostly *divisible* tasks, that is to say the overall objectives can be broken down into different sub-objectives which individuals and sub-teams strive to attain. By contrast a *unitary* task is one in which all members contribute to the overall effort in the same way - a tug-of-war team has a unitary task. The C^2 team's task is an *optimising* task, that is the aim is to produce some optimal outcome, an accurate tactical picture (rather than, say, exert maximum force). C^2 tasks are to some extent *discretionary,* that is to say although individual roles or functions are largely prescribed by the formal structure there is room for some degree of negotiation. The efficiency of teams of this sort would appear to depend very heavily on (a) the competence of individual operators and (b) the effectiveness of communication between members and with other teams.

A key feature of C^2 tasks is that their common purpose relates to the achievement of certain values of the *product* variables by the team processing information concerning the values of other variables representing changing states of the world, the ship and its systems. The work of such teams has been successfully analysed in terms of *control hierarchies*, that is by spelling out the relationships between overall goals and sub-goals of the system. The product of a C^2 team, say successful defence against air attack, is achieved by controlling a number of sub-goals. *The analysis of team skills must take into account the processes by which the team achieves its various goals and sub-goals.*

Team Processes.

The processes by which teams achieve their goals are *communication* and *coordination*. Figure1 illustrates how these two umbrella terms comprise a number of observable behaviours. Communication includes transmitting and receiving messages whilst coordination includes collaboration, or working together, and synchronization. The direction of the arrows indicates the main routes by which the components of the model exert their influence.

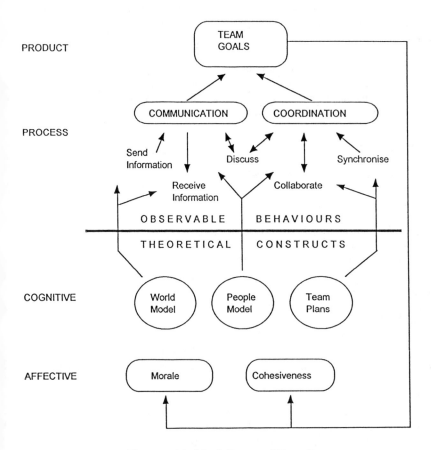

Figure 1 Model of Command Team Processes.

Mental Models.

Figure 1 also recognises a second level of analysis concerned with *cognitive* and *affective processes.* The important cognitive processes are the construction and maintenance of mental models and three types are distinguished in Figure 1. First is a model of the state of part of the world that is relevant to the team task including knowledge of the present disposition of the ship in relation to other friendly and hostile units. Second is knowledge of the other members of the team, what they are doing, or about to do, and even how well they are doing it. This 'people' model will be important in choosing which information to transmit and when to do so and whether or not to offer help. The third represents knowledge of team goals and plans, that is to say what goals the team is aiming to achieve and what strategies and techniques are available and are normally employed. These models are inferred from behaviour. *When any skill is being acquired task-relevant mental models are being learned* and it is therefore important to bear this in mind in the design of training and assessment procedures.

Figure 1 also includes some *affective constructs* namely *group morale* and *cohesiveness.* Arguably his level of description has not, so far, yielded measures powerful enough to justify including in a practical model of team performance. Note the arrow directions suggest that these are influenced by the attainment of team goals. They may well be the *result* of success rather than a contributory cause.

The Analysis of Team Tasks

Both individual and team tasks can be analysed by hierarchical decomposition. A task *process* is the means by which the performer (an individual or a team) achieves the task product or goal. For an individual this means that at a given signal, or under a given set of conditions, one or more actions is executed in order to achieve a new set of conditions - the goal. For a team the process by which a goal is achieved is by joint or complementary action, which involves *communication and coordination.*

In individual HTA the basic processes underlying performance are specified as 'input', 'action' and 'feedback' and it has been found useful to add to these fundamental descriptive categories the *plan* which indicates *how sub-operations are linked together in order to constitute a superordinate operation.* Extending this principle to the analysis of team skills some *fundamental categories of interaction* are (i) *communication* - information is passed between team members and (ii) *coordination* - team members carry out joint activities. These general processes can be expanded to identify features of the task which are likely to require special attention in training.

Communication

Sent messages may vary in respect of the *accuracy* of the information in the message, the *clarity* with which they are transmitted and their *timeliness.* Timeliness implies that the sender knows enough about the problems of the receiver to be able to judge that the message is required or can be dealt with at that time. Training for this feature is thus bound up with an understanding of the wider picture referred to in the literature as 'situational awareness' and with knowledge of the goals and tasks of other team members. *Receiving* messages depends at least in part on controlled attention to significant sources and this in turn is probably dependent on 'situational awareness', that is an understanding of what is going on in the external world and what other members of the team are doing. Again 'mental models' of the world, of other team members and of team plans could be critical in certain situations. *Discussion* can be related either to the clarification of acts of communication, ensuring that information has been correctly transmitted and received, or it could have as its main purpose arranging some joint activity or changing a plan of action. A properly conducted discussion is likely to result in a modification to one or more of the mental models referred to in Figure 1. Joint decision making is normally based on sending and receiving information but goes beyond simple information exchange to include some *discussion of alternative courses of action*. Team training might well be designed to take this factor into account in assessing team performance. Practice, with appropriate feedback, at reaching genuinely joint decisions might be used to modify the behaviour of team members.

Coordination

This is subdivided into *collaboration* and *synchronization*. Collaboration, literally working together, also depends on awareness of the current state of the world, but collaborators must also be aware of the current activities and needs of their fellow team members. The *re-allocation of duties* between team members is likely to arise in special circumstances, such as information overload, or even injury, affecting one member of the team to such an extent that team output will be compromised unless another member takes on part or whole of the duties normally carried out by that individual. One way of developing a model of the collaborators' activities and state of knowledge is to *swap roles* during training so that individuals learn something about the tasks of their colleagues. (b) *Synchronization* means working independently according to a common plan. In the absence of direct communication this depends on the individual's knowledge of *a plan which* may be built up formally by training or may be acquired informally as team members acquire an habitual pattern of actions which may not have been articulated but is known to team members. To the extent that plans are formalised team members need to learn and practise them, but informal plans may only emerge if the team reflects on its own performance in post-exercise debriefing.

The Stop Rule

HTA proceeds by 'unpacking' each 'operation' into a set of sub-operations and applying a rule which states that *if the product of the probability of failure and the cost of failure is unacceptable to the system then either propose a solution such that the criterion is met or proceed to a more detailed level of analysis by decomposing the operation into its component sub-operations*. In a training analysis a proposed solution will normally be a training prescription. In the case of *team-work* operations of the kind described in the preceding sections suitable training prescriptions will include learning communication protocols and disciplines and developing situational awareness by exercises designed to develop the mental models referred Figure 1.

Stress and Environmental Factors.

The circumstances under which any given operation might be attempted may be unfavourable and may have a significant effect on the probability of failure over and above that attributable to the skill level of the operator. If conditions such as environmental or psychological stress due to workload, danger and so on, are anticipated these must be taken into account in applying the p x c rule and, of course, in the training prescription. For example, in the stress of an engagement there might be a higher probability of communication failures or under workload stress individuals may be less likely to monitor the activities of colleagues with whom they should be collaborating. Conventional Naval task analysis specifies duties and tasks without reference to these factors whilst the HTA procedure *requires* they be considered and suitable training be prescribed to deal with them.

Analytical Procedures

HTA is a methodology broadly based on principles rather than a rigidly prescribed technique. The analysis of Naval Command Team skills, which is still in progress, is based on interviews with experts, including senior naval officers and experienced ratings, and direct observation of exercises on shore-based simulators and at sea. Although the duties of individual team members are formally specified and well-documented, informal accounts of what happens during a typical exercise give a much clearer impression of the kinds of team processes which are involved and those which are critical. Formal documentation relating to Command and Control teams and the Action Information Organisation specify what should happen, not what does happen nor how things can go wrong. HTA encourages the analyst to pursue just these questions using a variation of Critical Incident Technique and to attempt to identify objective indices of goal success and failure which are currently subjective and impressionistic. Finally, HTA is intended to lead directly to training prescriptions rather than being just a preliminary stage in specifying training techniques and procedures. Since this kind of team training is extremely expensive the identification of critical team skills should lead to the optimal use of training resources. Whether significant savings can be achieved remains to be seen.

Acknowledgements

This work was supported by a grant from the DRA Centre for Human Sciences, Portsdown. The author wishes to acknowledge the help of David Cunningham and the expertise of the School of Maritime Operations.

References

Annett, J. Duncan, K.D., Stammers R.B. & Gray, M.J. (1971) *Task Analysis.(HMSO,* London)
Miller, G.A., Galanter, E. & Pribram, K.H. (1960) *Plans and the Structure of Behavior.* (Henry Holt, New York)
ISteiner, I.D. (1972) *Group Processes and Productivity.* (Academic Press, New York)

CONCURRENT ENGINEERING AND THE SUPPLY CHAIN, FROM AN ERGONOMICS PERSPECTIVE

M.A. Sinclair, C.E. Siemieniuch, K.A. Cooper, and N. Waddell

HUSAT Research Institute
Elms Grove
Loughborough, Leics. LE11 1RG

To remain competitive, manufacturing companies must make use of the expertise and knowledge that is available from outside the company. This paper discusses some ergonomics issues involved in Concurrent Engineering within supply chains. It discusses some concepts of federated control systems as applied to supply chains. The implications of these for information flows and the maintenance of trust are discussed. All this depends on human knowledge; hence, the structure of supply chain teams is then discussed. Finally, some of the support requirements for such teams are listed, linking to current ergonomics knowledge.

Introduction

'Concurrent engineering' (CE) is a shibboleth in manufacturing industry. Various definitions exist; one such is: "[It] attempts to optimise the design of the product and manufacturing process to achieve reduced lead times and improved quality and cost by the integration of design and manufacturing activities, and by maximising parallelism in working practices." (Broughton 1990). The implications of successful application of this philosophy are shown in Fig. 1 (Clark and Fujimoto 1987b). Clearly, there are large market advantages to be gained. Furthermore, the trend towards 'virtual engineering', where nothing is tangible until the first product rolls off the production lines, emphasises the importance of time shrinkage in design. Coupled with the quality requirement that companies only do what they do best, and buy in everything else, it is evident that concurrent product engineering must extend to the supply chain., as illustrated in Fig 2 (PA Consultants 1989).

From a business perspective, the critical issues are firstly, the outsourcing decisions, and secondly the security and continuity of supply decisions. We ignore the second class of decisions for this paper, and concentrate on the first class.

Outsourcing parts of the product is a critical decision; it can make or break a company, as is discussed by (Davis 1992) . If a firm outsources its key expertise, it may become squeezed out of the market by its erstwhile supplier. Equally, if a company goes it alone, in effect including some non-world-class components, it could lose market share.

We concentrate on the engineering aspects. From this point of view, the problem is one of co-ordinating the efforts of designers and engineers spread over several companies, at different sites. The systems issues of importance are those of control, communication, compatibilities and culture. We discuss these briefly below.

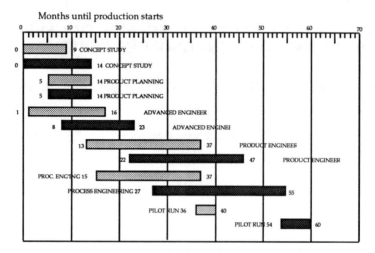

Fig. 1   Comparison of automobile product development times in European (dark bars) and Japanese companies (light bars). From Clark & Fujimoto, 1987.

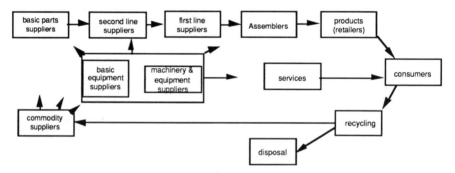

Fig 2.   A generic supply chain, showing the flow of components.

Control issues

In a company, whether paternalistic or 'empowered', there is an hierarchical control system. At the level of the supply chain this is not the case, and one must consider it as a federated control system. The essence of this is that there are firms (or sites) which are under independent control , and which will not be brought under higher level control. In principle, any firm can, if it wishes, restructure its sites or disconnect itself from other firms as it wishes. Consequently, co-operation occurs only by negotiation and acceptance of common policies (e.g. (Martin and Dobson 1991). Until legal contracts are signed, there is no power of enforcement.

The prime requirements in a federated chain are communications (to enable agreements to be reached) and the need for trust (and also its antithesis, mutual suspicion), between sites (see Fig. 3). One must be suspicious of one's partners because each independent firm in the chain has its own goals, which are not necessarily common. Secondly, one's suppliers may supply competitors as well. Equally, one must trust one's partners and reveal sensitive information to them, to gain the benefits of time shrinkage. This raises issues of security of information, and the need for companies to be perceived as secure. This applies to the company's IT systems and procedures, and, in an engineering environment, especially to its people. All three of these are important aspects for ergonomics, and we return to these later.

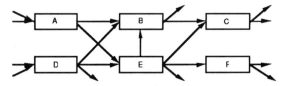

Fig. 3 Part of a supply chain. A - F represent firms, linked by logistics channels .
Two interlinked supply chains are shown, to company C and to company F.

Communications

Given some companies within a manufacturing domain, it is now necessary to enable them to form supply chains. Negotiations accomplish this, using the communication channels available. We discuss firstly some classes of communications, and secondly some of the generic characteristics of supply chains. It is possible to define at least four classes, as a series of levels:

- transactions - information about designs must be communicated;
- operations - co-ordination and control of transactions (who meets, when, why);
- policy execution - negotiate targets, develop operational procedures, etc.;
- strategy - define role and level of participation in supply chain, discuss market information, and settle other policy issues.

Secondly, the following characteristics of supply chains are apparent.

Imperfect information. If company C in Fig. 3 is a chainstore, its information regarding customers is better than anyone else in the supply chain. This is a characteristic of all the companies; their local information is better than their distant information, and is more up-to-date. Consequently, there are inequalities of information within any supply chain. This creates problems for a Concurrent Engineering context, since common information is a key requirement for this approach to work. It points to the importance of communications and tools to allow people to communicate easily (even for co-location of people at one site).

Excluded information. In Fig. 3 company B should not inform company D about company A's new product ideas. Equally, company D should not inform company B about its rival, company E. In general it means that the situation of non-common information is maintained, even enhanced. It implies the distribution of tasks in the product introduction process such that information is eused only where it is permissible. Of course, this is not possible in reality, and leakage of information will occur; it leads to a new definition of security - 'to control leakage, not to stop it'.

Importance of information utilisation. With imperfect information flow, unexpected events will always occur. Hence, companies must maximise their utilisation of the information available to them by the use of appropriate tools, methods and knowledge. Early communication, and rapid, joint decision-making are critical attributes. This can be assisted by agreements among the companies in the supply chain to extend the 'information window' between them (i.e. to provide companies in the supply chain with longer-term sensitive information).

Capabilities

There are two issues; engineering capability (which is a core competence issue), and communication capabilities. We assume that companies are competent, and discuss the second issue. The importance of this is driving many standardisation efforts, such as the Computer-aided Acquisition and Lifecycle Support programme (CALS) and the developing STEP standard (ISO 10303), both now global initiatives. These, however, are not enough; the interoperability of IT systems (and in particular computer-supported co-operative working - CSCW) must be supported, together with the provision of trained people to utilise both the tools and the information.

Culture

Most of the cultural issues stem from the importance of trust in supply chains. Where one must operate with incomplete information, one must have trust in one's partners. For example, in Fig. 3, if company F discusses costs of supplies with companies D and E, in principle company F could squeeze company E very hard to reduce prices, if it wished. In federated systems such as these, the building and maintenance of trust is crucial to success. This relies on a number of things:

Common language and understanding of terms. This problem is exacerbated in transnational supply chains, where both language and cultural differences occur (Kanoi 1991; Mackay, Siemieniuch et al. 1992). If one standardises terms then the knowledge structures that use them will require some standardisation as well; in turn, this implies some commonality of training in concepts and levels of attainment.

Common goals and shared benefits. If one company has hidden goals, this may cause what seems to be aberrant behaviour *vis-a-vis* another company. The inability of the second company to understand this behaviour will destroy trust rather than build it. Hence, there must be a commitment to open discussion of goals, including long-term ones, between companies. This implies that communication between companies in the supply chain must occur at many levels; not just Director level, for example. Companies entering supply chains will find pressures to ensure that personnel at different levels have some degree of common understanding and motivation towards the goals of the participation. Strongly centralised companies with hierarchical structures based on fractionated processes will find this difficult.

Integrity in relationships. This is crucial at all levels; at personal, transactional, operational , policy execution, and strategic levels. Inter-personal relationships are obviously important; but these are not enough. It is also necessary to ensure that operational goals, procedures and organisational structures and rewards will enable integrity to be demonstrated. Finally, a clear commitment to ethical behaviour is required, in mission statements, policies, goals and performance.

Knowledge configurations and decision support

Given an information infrastructure, the resource required to achieve the company's goals is knowledge. A wide range of knowledge classes are required; interpersonal skills, negotiation and communication skills, team building skills, characteristics of competitors and suppliers, understanding of global markets, technical knowledge of products and their manufacturing processes, logistics, organisation procedures, human resources, organisational policies, politics and ethics, etc. The economical and efficient configuration of this knowledge within the organisation is a critical issue (e.g. Prahalad and Hamel 1990; Nonaka 1991; Siemieniuch and Sinclair 1993). This knowledge will be held mainly in human heads, indicating the importance of correct job design (and its corollaries, training provision, decision support and organisational design).

From this discussion, a number of needs emerge; (a) guidelines for the configuration of knowledge across teams, for some optimum level of efficiency to be achieved; (b) continuity of personnel in the engineering teams, (c) importance of full documentation and traceability of designs across the supply chain, to recycle design knowledge into new designs , (d) team design to provide the responsibilities, authority, resources and support to exercise and to develop knowledge structures.

There is also a requirement for decision support tools, as outlined below.

Integration. This is critical with legacy systems, where information 'hidden' in heterogeneous systems. These are issues widely addressed at present; the effort that is being put into Reference Architectures (e.g. (Williams, Bernus et al. 1994; Sinclair and Siemieniuch 1995)) and commercial software development is testimony to this.

Visualisation/Interpretation. Computer screens represent a small porthole into huge datasets of complex information about the produc tlife cycle,. There is a need for usable tools to allow users to present data in novel ways tto extract meaning.

Simulation/Speculation. There is always a need to answer 'what if...' questions. This is a key area for integrated decision support tools.

Dissemination. Tools for this are essential. The requirements form a long list of technical, organisational and end-user-based needs. Some are listed below:

- Conformance of IT systems and applications to global standards .
- Common data structures and formats supporting the standards above.
- Interoperability of software systems.
- Reference architectures and modular software, to allow easy upgrading.
- Appropriate, unobtrusive security, based on the 'ownership' of information, and with information bound tightly to its appropriate applications.
- Comprehensive directory facilities linked to security systems;
- Version management systems and protocols as part of security systems;
- Workflow software, to assist in overcoming synchronisation problems.
- Multi-user applications (and multi-user shells for existing single-user applications)
- Interactive, informal sketching and discussion tools;
- Real-time, broadband multimedia conferencing systems, including point-to-point and group-to-group, able to cope with heterogeneous equipment;
- Browsing systems able handle multimedia databases in real-time operation;
- Usable, consistent, robust, and user-configurable interfaces.
- On-line suppliers' databases ("product libraries")

Support for CE teams in supply chains.

One must consider relationships with suppliers from the perspective of supply teams, not just individuals, because of the several levels of negotiation and the changing purposes of the supply chain. Companies must present a coherent face to each supplier in the supply chains, and maintain internal cohesion. This requires co-ordination within the company. Thus, control will be required within companies to maintain their interfaces with the supply chains.

Finally, CE teams are indeed teams; depending on the nature of the negotiations, the problems, and the state of the business environment, different people within the teams may take leading roles at different times. Again, this emphasises the need for organisational structures to support teams as well as individuals, and the need for commonality of viewpoint among the team members.

Consequently, there is a need for organisational design to ensure clear lines of authority from Board level to the teams. This should provide each team with:

- Access to appropriate information on suitable displays at the appropriate time;
- Appropriate common models of the technical system and its operations,;
- Cognitive skills to manipulate information and models to reach decisions;
- Skills for communicating with other team members and outside personnel;
- The means by which to execute the decision and to evaluate its consequences;
- Responsibility and authority to entrain resources to carry out the actions; and

• Adequate motivation to carry out the tasks.

These are all relevant issues for manufacturing ergonomics, and a wide body of knowledge exists (e.g. Majchrzak 1988c; Clegg and Ulich 1989; Wilson 1991; Wortmann 1991). However, they are usually addressed from the perspective of teams operating in isolation; a key issue in this paper is the need to consider the wider context of supply chains and the altered emphases that result. Two threads always arise from research in this area: firstly, the ability to generate, manage, and transfer relevant information along the supply chain as required; and secondly the ability to create and support teams with the knowledge, support tools, authority, responsibilities and resources, and who can interact with other teams in a continuously evolving manner. The close linkages between these two threads show that ergonomics has a major role to play to ensure that companies and their people can secure their future.

Acknowledgements

This paper is based on research in project P.6599 EAGLE (European Advanced Global Logistics for Enterprises) in ESPRIT-CIME, and project R.2112 SMAC (Suppliers and Manufacturers in Automotive Collaboration) in RACE , for the Commission of the European Union, and project SIMPLOFI (Simultaneous Engineering through People, Organisations and Functional Integration), funded by the Engineering and Physical Sciences Research Council.

References

Broughton, T. (1990). *Simultaneous engineering in aero gas turbine design and manufacture.* Proc. 1st Int. Conf. on Simultaneous Engineering, London,.
Clark, K. B. and T. Fujimoto (1987b). Product development in the world auto industry: strategy, organisation and performance. Brookings Institute Microeconomics Conference, Brookings Institute.
Clegg, C. and E. Ulich (1989). Job design, SAPU, University of Sheffield.
Davis, E. W. (1992). "Global outsourcing: have US managers thrown the baby out with the bathwater?" Business Horizons 35(4): 58-65.
Kanoi, N. (1991). Manufacturing modernisation - Sony's approach, 1st International Manufacturing Lecture, Inst. of Manufacturing Engineers.
Mackay, R., C. E. Siemieniuch, et al. (1992). A view of human factors in integrated manufacturing from the perspective of the ESPRIT-CIME programme. Proc. 2nd Information Technology and People Conf. (ITaP '93), Moscow.
Majchrzak, A. (1988c). The human side of factory automation. San Francisco, Jossey-Bass.
Martin, M. and J. E. Dobson (1991). Enterprise modelling and security policies. Database security, IV: status and prospects., Elsevier North-Holland.
Nonaka, I. (1991). "The knowledge-creating company." HBR 69(6): 96-104.
PA Consultants (1989). Manufacturing into the late 1990s, PA Consulting Group, Bowater House East, 68 Knightsbridge, London SW1X 7LJ.
Prahalad, C. K. and G. Hamel (1990). "The core competence of the corporation." Harvard Business Review 68(3): 79-91.
Siemieniuch, C. E. and M. A. Sinclair (1993). "Implications of Concurrent Engineering for organisational knowledge and structure - a European, ergonomics perspective." Journal of Design and Manufacturing 3(3): 189-200.
Sinclair, M. A. and C. E. Siemieniuch (1995). Reference architectures for manufacturing organisations - a discussion of the ergonomics issues. ECAME '95, Louvain-la-Neuve, Austrian Research Centre, Siebersdorf, Vienna.
Williams, T. J., P. Bernus, et al. (1994). "Architectures for integrating manufacturing activities and enterprises." Computers in Industry 24(2-3): 111-139.
Wilson, J. (1991). "Participation - a framework and a foundation for ergonomics." Journal of Occupational Psychology 64: 67-80.
Wortmann, J. C. (1991). Factory of the future: towards an integrated theory for one-of-a-kind production. ESPRIT-91, Brussels.

DEVELOPMENT OF A HUMAN FACTORS TASK DATABASE FOR HUMAN PERFORMANCE EVALUATION [©]

Brian Buck and Gretchen Greatorex

GEC-Marconi Research Centre,
West Hanningfield Road, Great Baddow,
Chelmsford, Essex CM2 8HN

Task Analysis (TA) can take on many different forms depending on the problems being addressed and the scope of the study. This can often lead to a fragmented and inefficient approach, where data is difficult to share and duplication of effort results. These problems can be overcome if a common framework for TA data can be used to promote the system level sharing and dissemination of data within design teams. Such a framework needs to be generic and flexible if it is to support data collection, retrieval and analysis for all uses of TA data. This paper describes how these issues have been addressed in the development of a Human Factors Task Database (HFTD). A common framework is also shown to have potential to support the re-use of TA data within and between projects, and to provide the TA data necessary to populate human performance models.

Introduction

Task Analysis (TA) involves the study of what an operator (or team of operators) is required to do to achieve a system goal. When used in conjunction with techniques such as Workload and Timeline Analysis, TA can aid identification of incompatibilities between the demands being made on the operator and his/her capabilities. TA is therefore essential to a human-centred design approach.

TA is used by many aspects of the design and development process. These include design, human factors, training needs analysis, safety hazard analysis, user manual development, design for maintainability, manning level analysis and usability testing. Currently, TAs take on many different forms depending on which group is doing the analysis, the problems being addressed and the scope of the study. This can often leads to a fragmented and inefficient approach where data is difficult to share and duplication of effort results. A common framework is therefore needed that supports the collection, retrieval and analysis of TA data relevant to these design efforts. A common framework can also yields other benefits. These include the provision of a generic basis for assessing design from the standpoint of fitness for purpose, the potential for international TA standards, the potential for improved systems integration and the ability to re-use TA data within and between programs.

This paper introduces the Human Factors Task Database (HFTD) as an example of a common framework for TA. HFTD was developed specifically to increase the cost effectiveness of TA within GEC- Marconi. The paper falls into four major sections. The first section deals with the issues relevant to developing a common framework using HFTD as a context. Next, it is shown how a common framework can provide the additional benefit of facilitating the re-use of TA data within and between projects. This is followed by a brief overview of HFTD itself. The last section of the paper summarises ongoing work to extrapolate TA data to new situations and to populate human performance models directly from HFTD.

Building a Common Framework

This section describes the issues that need to be addressed in designing a common TA framework and how these issues have been addressed in HFTD.

Accommodating multiple uses

A common framework must support the user by making it easy to find and extract information relevant to a specific analysis quickly and effectively. This alleviates a common problem that occurs today where a document the size of a large telephone book must be sifted through in order to find relevant data. HFTD allows its users to search through the database using keywords such as domains (*e.g.,* air, land, sea), typical functions (*e.g.,* command, planning, movement), equipment (*e.g.,* binoculars, radio), and task verbs (*e.g.,* march, listen, observe) in order to access quickly the information of interest. Similarly, HFTD allows users to extract categories of data relevant to specific task sequences or operational scenarios. The key words used in these searches are selected from a standard list in the database to ensure consistency.

The user must be able to use TA data to answer specific questions. For instance, if the user is interested in Training Needs or Manning Level Analysis he/she might want to retrieve only information on the skill requirements of tasks. Alternatively, if the user is interested in identifying the high risk tasks in a scenario for prototyping, he/she might want to extract only information on task frequency and importance. HFTD allows the user to print any subset of task information in a pro forma style report. For instance, in the training example above he/she could print the 'Skill Requirements' field, whilst in the prototyping example he/she could print the task 'Frequency' and 'Importance' fields. HFTD also supports the two major techniques of Timeline and Workload Analysis with in-built tools. Timeline and Workload Analysis are important in revealing incompatibilities between the demands made on the operator and his/her capabilities.

Enabling use in multiple domains

A common framework must also support the exhaustive collection of all TA data in all domains. In describing this wide variety of tasks, it must be flexible enough to be able to characterise tasks on several aspects, from the highly physical efforts of a soldier digging a trench to the fine psychomotor requirements of a pilot flying a plane. In developing HFTD, care has been taken to ensure that the framework supports both flexible free text entry to allow description of these various attributes, as well as structured techniques such as task verbs and a broad range of rating scales for capturing the conditions under which the task is performed. These facilities support the definition of a wide variety of tasks across the domains of air, land and sea.

Supporting analysis at various levels of detail

The level of task detail required by potential users depends on the development phase and the type of analyses being carried out. High level TA data is sufficient for concept and feasibility studies, whilst increasingly more detailed TA data is required as the system design progresses. HFTD therefore follows the principles of Hierarchical Task Analysis (Kirwan and Ainsworth (1992)) to allows tasks to be broken down into component subtasks to a level appropriate for the analysis. Hierarchical structures are defined in HFTD by specifying the parent and children associated with each task.

Ensuring consistency

Representations of TA data must be consistent if data from different TA sources is to be combined in generating a TA. HFTD imposes a structure on the collection of TA data that requires the user to enter data in a common format. Tasks names are specified in the standard 'verb-noun' format, using a standard verb list developed from that devised by McCracken and Aldrich (1984). In addition, on-line rating scales are used wherever possible (*e.g.,* to define task importance, keywords and workload values.)

Re-using TA data

A common framework can potentially allow data from existing TAs to be re-worked to form the basis of new TAs. Instead of constructing a TA from 'scratch', data from existing TAs could be retrieved and tailored to provide the building blocks for a new analysis. This has potential to reduce significantly the resource demand of TA. A common framework must address the issues of recording the data collection conditions and data reliability if it is to support the re-use of TA data. These issues are discussed below.

Recording conditions at data collection

TAs are system specific. This is especially true of performance metrics such as time, error and workload. TA data is therefore only directly re-usable if there is a match between the set of conditions for the system under development and those that prevailed during the collection of TA data. The conditions under which TA data was collected must therefore be rigorously recorded if TA data is to be considered for re-use. This exhaustive data set could potentially be used to extrapolate from existing data to new conditions if models for such extrapolations can be developed. (Extrapolation of performance data is an example of the re-use of TA data.) Whilst the recording of condition data imposes a burden on the user, this extra work will be more than compensated for by reducing the data collection necessary in generating new TAs.

HFTD records the conditions that prevailed during data collection by recording information on Performance Shaping Factors (PSFs), where PSFs include such factors as time of day, training, noise and fatigue. HFTD employs a comprehensive set of on-line scales to collect data on PSF levels. This information could also support the extrapolation of performance data to new situations.

Recording data reliability

In addition to information on the conditions under which TA data was collected, the reliability of the stored data should be recorded. This is necessary to enable the database user to judge how much weight to give to stored data. HFTD supports this by recording information on the data source and a rating for the data reliability. The reliability rating scale ranges from 1 (speculative) to 10 (real operational data).

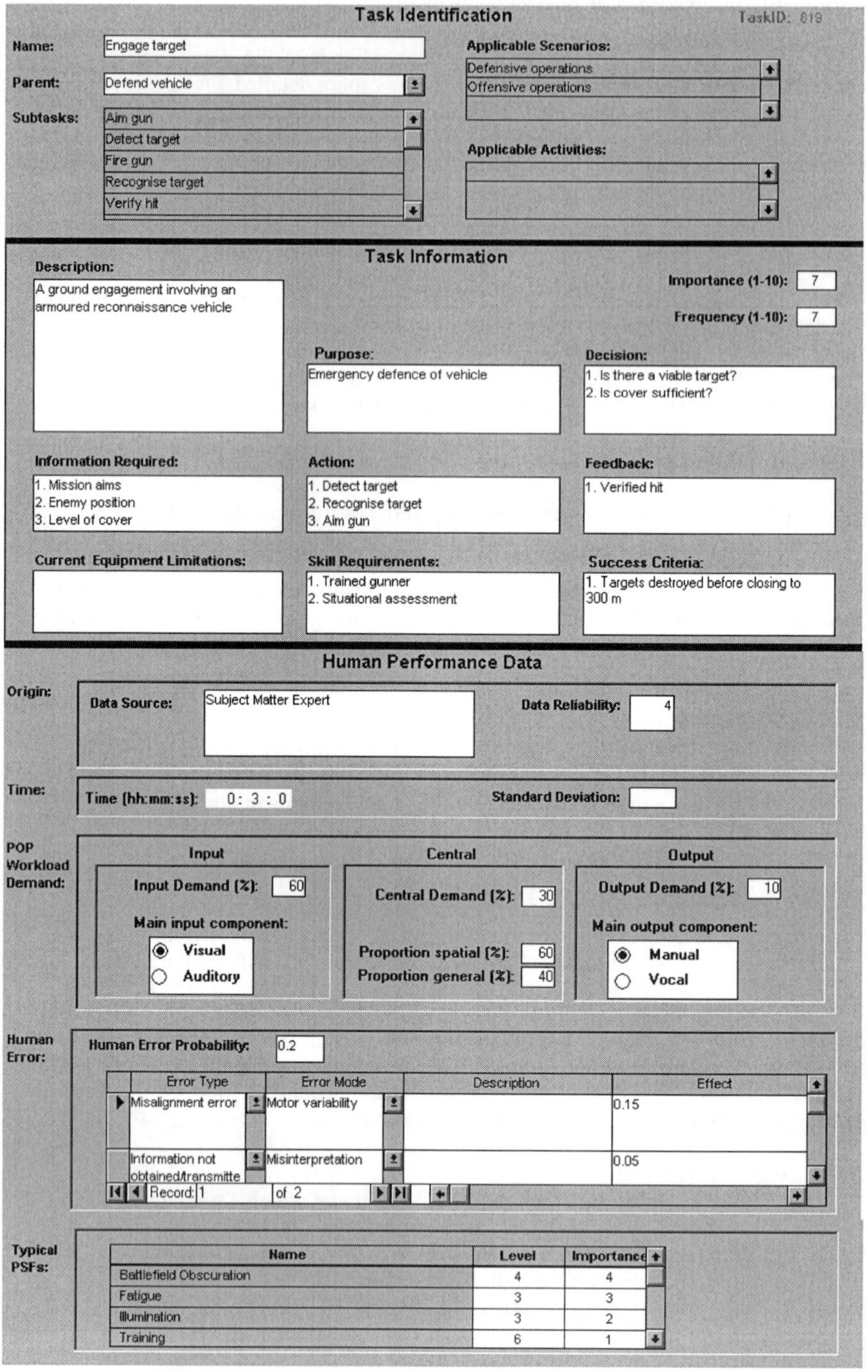

Task Identification

TaskID: 819

Name: Engage target

Applicable Scenarios:
Defensive operations
Offensive operations

Parent: Defend vehicle

Subtasks:
Aim gun
Detect target
Fire gun
Recognise target
Verify hit

Applicable Activities:

Task Information

Description:
A ground engagement involving an armoured reconnaissance vehicle

Importance (1-10): 7

Frequency (1-10): 7

Purpose:
Emergency defence of vehicle

Decision:
1. Is there a viable target?
2. Is cover sufficient?

Information Required:
1. Mission aims
2. Enemy position
3. Level of cover

Action:
1. Detect target
2. Recognise target
3. Aim gun

Feedback:
1. Verified hit

Current Equipment Limitations:

Skill Requirements:
1. Trained gunner
2. Situational assessment

Success Criteria:
1. Targets destroyed before closing to 300 m

Human Performance Data

Origin:

Data Source: Subject Matter Expert

Data Reliability: 4

Time:

Time [hh:mm:ss]: 0 : 3 : 0

Standard Deviation:

POP Workload Demand:

Input	Central	Output
Input Demand (%): 60	**Central Demand (%):** 30	**Output Demand (%):** 10
Main input component:		**Main output component:**
⦿ Visual	**Proportion spatial (%):** 60	⦿ Manual
○ Auditory	**Proportion general (%):** 40	○ Vocal

Human Error:

Human Error Probability: 0.2

Error Type	Error Mode	Description	Effect
▶ Misalignment error	Motor variability		0.15
Information not obtained/transmitte	Misinterpretation		0.05

Record: 1 of 2

Typical PSFs:

Name	Level	Importance
Battlefield Obscuration	4	4
Fatigue	3	3
Illumination	3	2
Training	6	1

Figure 1. An example of a HFTD task form.

Description of HFTD

This section briefly describes HFTD. The development work has been driven directly by the issues governing the use, sharing, dissemination, and re-use of TA data.

As discussed earlier, HFTD has a task based hierarchical structure. The TA data that can be collected includes descriptive information (*e.g.,* purpose, actions, feedback, information required, decisions made, skill required), human performance data (*e.g.,* time to perform, error rates, error types, workload ratings), and information about the conditions assumed (*e.g.,* user experience, noise, fatigue). An example task form is provided in Figure 1. Workload demand and error type are collected according to the POP (Farmer, Belyavin, Jordan, Bunting, Tattershall and Jones (1995)) and SHERPA (Embrey (1986)) classification schemes respectively. Any of these information sets can be entered into HFTD depending on the purpose and scope of the analysis.

Tasks stored in HFTD can be used as 'building blocks' to create a wide variety of operational scenarios. Scenarios define the system's demands on the operator and therefore form the basis of subsequent analyses. Tasks in a scenario can be high or low level, depending on the appropriate level of detail for the particular analysis. Like tasks, scenarios can be stored, retrieved and re-used in HFTD.

Over the last twelve months, HFTD has been used successfully on five major military projects. This work has covered the design of reconnaissance vehicles, fire control systems, infantry tasks and sonar systems. This has populated the database with 1653 tasks and 71 scenarios. Using HFTD on these projects has led to its refinement by providing opportunities to assess the data required from TA for design and operational modelling.

On-going work

Ongoing work on HFTD is occurring on two main fronts. The first is to model the effects of PSFs on performance metrics so that performance data can be extrapolated to conditions other than those that prevailed at data collection. The second is to extend HFTD into the field of human performance modelling by developing a link with the Micro Saint - Human Operator Simulator tool (MS-HOS).

Extrapolating from TA data

HFTD is currently being enhanced to provide two significant functions. The first function is to be able to store a large amount of information about the effects of a variety of PSFs on performance. (Figure 2 provides an illustration of this information.) This data will be available to the user in a library for use in analyses. The second function is to facilitate 'what if' analyses associated with changing PSF levels.

In order to achieve these two functions, a taxonomy is being developed to classify tasks in terms of their demands on internal resource (*e.g.,* visual perception, spatial processing, continuous psychomotor output). Models are also being developed to model the effects of PSF levels and their interactions on the availability of each of these resources. This should make it possible to extrapolate performance metrics for a task (*e.g.,* time and error) to all possible PSF conditions, once one set of performance data and PSFs for the task have been stored in HFTD. This use of a taxonomy avoids the need to model the effects of PSFs on the performance metrics of each task individually. In addition to this, structures are being implemented in HFTD to store multiple sets of performance data for tasks as this information becomes available. These structures will allow the extrapolation techniques to be validated and refined.

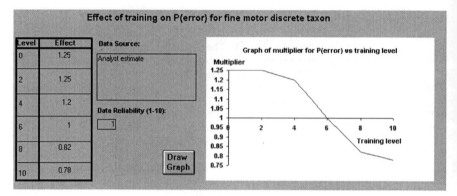

Figure 2. An example of a PSF graph.

HFTD - MS-HOS

Work is currently underway to develop an interface between HFTD and MS-HOS as part of the Integrated Performance Modelling Environment initiative (IPME) taken by the Centre for Human Sciences (DRA). MS-HOS is an extensively used Human Factors tool which models human performance in order to aid predictions of overall operational effectiveness. A MS-HOS model defines alternative sequences of task execution by connecting up tasks to form a network and using deterministic or stochastic processes to govern the route taken through the network. In addition, a network model can call on a library of micromodels to predict human performance metrics such as time, error and workload (MA&D (1993)). A MS-HOS model is therefore a logical extension of the HFTD scenario and could therefore be populated with information from the database. A HFTD - MS-HOS link consequently has potential to reduce greatly the cost of collecting task and human performance data to populate task networks and their associated micromodels.

Conclusions

Our experience of developing HFTD and using it on several projects has shown us the significant benefits associated with having a common framework for TA data. This framework has improved the application of TA data to design and has facilitated its re-use within and between programmes. The use of a database has greatly improved the cost effectiveness of performing TA as well as our ability to extract only that data relevant to a particular analysis. Work currently in progress regarding storage of data on the effect of PSFs on performance, as well as the link to MS-HOS will improve efforts to determine the effect of human performance in complex systems.

References

Embrey D.E. 1986, SHERPA: A systematic human error reduction and prediction approach. Paper presented at the International Topical Meeting on Advances in Human Factors in Nuclear Power Systems, Knoxville, Tennessee.

Farmer E.W., Belyavin A.J., Jordan C.S., Bunting A.J., Tattershall A.J. and Jones D.M. 1995, Predictive Workload Assessment: Final Report, DRA/AS/MMI/CR95100/1, DRA

Kirwan B., and Ainsworth L.K. 1992, *A Guide to Task Analysis*, 1st edn (Taylor and Francis, London) 53-60.

Micro Analysis and Design Simulation Software, Inc. 1993, *Human Performance User's Manual*.

McCracken J.H. and Aldrich T.B. 1984, Analyses of selected LHX mission functions: Implications for operator workload and system automation goals, TNA AS1479-24-84, Fort Rucker, AL: Anacapa Sciences, Inc.

WORKLOAD UNDER COMPLEX TASK CONDITIONS

Andy Leggatt and Jan Noyes

Department of Psychology, University of Bristol
8 Woodland Road, Bristol BS8 1TN

The measurement of mental workload has an increasingly important role to play in system design, and unlike its contemporary 'physical workload', it is a relatively new area of study, and poses a far more difficult problem in terms of definition and measurement. Whereas the assessment of determining the physical load on the human had its origins in the Industrial Revolution, it is only in recent years that serious attempts have been made to assess mental workload in terms of the load placed on the cognitive capabilities of the human operator. Before the establishment of cognitive psychology it was unfeasible to examine the effect of workload on cognitive capabilities; both concepts being unobservable and supposedly inaccessible to study. This paper will discuss some of the methodological problems associated with the measurement of subjective cognitive workload techniques. To illustrate the issues, a study will be cited from which conclusions will be drawn concerning the possible source of subjective workload estimations.

Defining mental workload

"For the past few years, one particular question has occupied most of my waking hours. I have been plagued by this question. Indeed, no sooner do I emerge...to attend a conference...than I know for certain that within the next hour I will again be asked the eternal question 'What is mental workload'. Experience has taught me...a blunt forthright answer, "I don't know!" (Kantowitz, 1987, p. 81).

These words of conviction were not written by a inexperienced undergraduate of ergonomics or industrial psychology but an experienced psychology academic who has studied the concept for many years. This type of exclamation is not due lack of research resources, or good minds attending to the question but due to the complexity and difficulty of the question. Why then is defining workload such a complex issue?

Mental workload encompasses every aspect of work. It can refer to actual task performance, divergence between actual and expected performance, the cost to the operator of undertaking a specified task, as well as the beguilingly simple concept of task difficulty (Gopher and Donchin, 1986). Workload is related to the *difficulty* an operator experiences when undertaking a task. However, workload cannot be easily measured. Humans are not good 'meters' which can be easily read by administering workload questionnaires, and operators do not provide constant information. Further, individual users can learn to do tasks more efficiently (Gopher and Braune, 1984), or try alternative strategies, or alter their performance criteria (Wickens, 1984), or try even harder (Yeh and Wickens ,1988). Operators are not uniform, some will have more ability/resources than others (Wickens, 1991), some will deal better with increased stress (Gaillard, 1994), and people will vary depending upon time of day and day of the month (Humphreys and Revelle, 1984; Vidulich 1988; Hancock, 1988). These findings confuse and complicate the essence of workload; however, we would argue it is the interaction between the demands placed upon an individue by the task and the individuals' ability to cope with these demands. Workload can only be invoked when the interaction of both task and the individual are considered. Gopher and Donchin (1986) correctly argued that workload is thus a hypothetical construct because it encompasses terms which are not wholly reducible to empirical terms and refers to items which are not directly observable (MachCorquadale and Meehl, 1948).

In practical terms, this makes workload measurement all the more difficult. However the discipline of psychology has long examined concepts and models, such as short term memory, which are not readily observable. It should be possible to examine the concept of workload by testing various facets of the concept empirically to determine limitations and those variables which influence its measurement. Knowing which factors, other than task difficulty, affect workload perception is important for measuring workload accurately and aiding the understanding of the concept more fully, eventually leading to further understand of humans at work. To be able to discuss which factors influence workload perception it is necessary to briefly describe workload measurement techniques.

Measuring mental workload

There are three main approaches to workload measurement:

- Objective task parameters (input data rate, number of input modes, etc.);
- Response parameters (physiological, e.g. GSR [galvanic skin response] or behavioural measures, e.g. RT [reaction time], errors, etc.);
- Subjective ratings.

Objective task parameters deal with task input constituents. Wickens (1984), for example, described a method of predicting workload by considering 21 task factors, such as familiarity of stimuli, number of concurrent task and resource competition, etc. These rating are summed to provide a task difficulty rating, which may be likened to assessing objective task parameters.

Response parameters can be further sub divided into primary task measures (measure of actual performance), secondary task performance (filler task performance) and psychophysiological measures (pupil dilation, heart rate variability, electrodermal activity, endogenous eyeblinks). There are several detailed reviews which evaluate each method on a variety of criteria such as; diagnositicity, sensitivity, reliability, and validity (see, Hancock, 1988; O'Donnell and Eggemeier, 1986; Wilson and O'Donnell 1988; Kramer, 1991).

Subjective workload ratings are commonly used to assess workload in a variety of situations (aeronautics, vehicles, ships, process control, etc.). There are numerous subjective workload measures: SWAT (Reid and Nygren, 1988), NASA-TLX (Hart and Staveland, 1988), Modified Cooper-Harper Rating Scale (Wierville and Casali, 1983), Bedford Scale (Roscoe, 1987), and psychophysical methods (e.g. Gopher and Braune, 1984). They have become widely accepted as they are easy to administer, quick to interpret and have high face validity. There are several excellent reviews of these methods (see, Kantowitz, 1987; O'Donnell and Eggemeier, 1986; Eggemeier and Wilson, 1991). They have become so popular that some researchers have gone as far as to say, "If the person tells you that he is loaded and effortful, then he is loaded and effortful whatever the behavioural and performance measures may show" (Moray, Johanssen, Pew, Rasmussen, Sanders and Wickens, 1979, p. 105). Such statements are symptomatic of subjective measures prevalence and researchers' belief that cognitive workload may be directly equated with subjective workload measures.

It is primarily because subjective workload measures have become so prominent that the authors felt it important to investigate those factors which may influence subjective workload judgements and ascertain how these relate to task performance. However, before discussion of these issues, it is necessary briefly to summarise workload theory.

Subjective workload theory

Workload as an area is renowned for its lack of theory driven research. Too much research is based upon developing or comparing workload measuring techniques with little thought given to what is being measured. Nygren (1991) provided an excellent critique of the area. Workload research based upon Wickens (1984) and ultimately Kahnman's (1973) resource theory of human information processing is the exception. There have been criticisms of this theory (Navon, 1984; Kantowitz, 1987) but it still provides a valid framework in which to consider workload.

Some interesting questions arise from how task attributes and goals may be reflected in subjective workload measures. For example, when making workload judgements, what information are operators assessing? Do individuals have access to their level of resources in use or do they recall goal success (e.g. how many occasions they hit the target) or do they recall, the number of occasions they make mistakes (e.g. they see a target but miss it), or do operators use a metric of both hits and misses? If users do make judgements of monitored success, what are the implications for workload measures of undetected errors (e.g. when operators miss a target because they fail to see it)? Furthermore, if they do use these kinds of self monitoring performance measures, how accurate is their memory for these events? How accurate is their task monitoring performance and are there systematic memory biases? How good is the relationship between workload and performance? If there is a divergence between workload and performance, why does it occur? Finally, how are tasks which are unavailable to conscious examination (e.g. motor tracking or speech production) reflected in workload judgements? Wickens (1984) suggested, for example, that tasks which require attention from the central processor (i.e. conscious operations) will largely determine subjective workload judgements.

To address some of these issues a series of research studies which involved controlled complex tasks were undertaken to examine the effect of performance monitoring, expected and actual performance, personality, mood, age, and sex influence subjective workload perception. One of these studies will be reported here.

The Study

An experimental task was constructed on a personal computer to test physiological and behavioural responses to alarms. This task required individuals to respond to auditory and visual stimuli (analogous to a complex system warning), track a moving object around a screen using a mouse and watch numeric digits presented on the moving object and respond

with a mouse click when the prespecified sequence of digits were displayed. Twenty-one pai
volunteers aged between 19 to 31 years participated in the study.

The main independent variable was the number of digits in the sequence; individuals
had to respond when there were three or four consecutive different odd numbers. This was
manipulated on a between-subjects basis. The task lasted approximately 15 minutes in whi
there were seven digit targets presented and seven auditory alarms. Individuals were trainec
separately on each of the task items before they were combined in a short practice session.

The dependent variables were as follows: 1. NASA-TLX and Bedford workload
measures; 2. digit target performance (hits, misses, false alarms, correct rejections); 3.
Estimates of the task performance on digit spotting (e.g. hits, misses, false alarms).

Results

Evaluation of workload between memory load conditions - mental demand was
significantly greater in the four digit condition (Mann Whitney ranked U test, U = 31, P <
0.0453). Unweighted TLX scores and Bedford scales failed to detect a significant differenc
between conditions.

Digit target task performance - there was a significant difference between task
performance between the working memory condition, three or four digits (Mann-Whitney
ranked U test, U = 22, P < 0.045). Individuals in the four digit target condition failed to
detect as many targets as individuals in the three digit condition.

Relationship between task performance and workload - there was a strong relationshir
between the digit target false alarm rate (responding to a digit target when there was none) a
perceived 'cognitive' workload (composed of the TLX sub elements of mental demand,
temporal demand and effort - excluding performance, physical demand, and frustration sub
elements) (t = 0.53, two tailed, P<0.013). About half of the subjective workload variance ca
be explained by the false alarm rate. This suggests that individuals were monitoring the
number of occasions they made incorrect responses.

There is a poor relationship between the number of hits (occasions when individuals
correctly responded to a digit targets) and 'cognitive' workload (composed of the TLX sub
elements of mental demand, temporal demand and effort) (t = 0.24, two tailed, not significant
Subjective workload is not explained by the number of correct digit target responses. This
suggests that workload in this case is not determined by the total amount of effort or resourct
invested to make correct responses, but is based on those occasions when individuals make
erroneous responses and are aware they have done so.

Relationship between self rated performance, workload and actual performance - if
individuals are monitoring task failures to provide workload judgements then it would
reasonable for self-rated error rates to correlate very strongly with workload judgements. Ha
of the individuals provided an indication of how many digit targets they thought they saw, the
number of false alarms they made and the number of targets they saw but to which they did
not react. Although these data were collected for only eight individuals they may provide
some insight into the way individuals go about judging workload.

There was a strong relationship between the number of false alarms and the self rated
number of false alarms (0.51, n = 8, not significant). This suggests that individuals may be
reasonably good at monitoring their false alarms rate and this information is available to then
to make workload judgements. However, individuals provided poor estimates of the total
number of targets (-0.17, n = 8, not significant). This suggests that if workload judgements
are based on the implicit memory for the number of targets presented they would be expected
to diverge from expected task difficulty (i.e. more target items presented would not necessaril

result in greater workload if the individual failed to remember them but s/he would have to work harder to complete the task).

These results may have some interesting implications for workload research. Investigating the relationship between the number of targets and the implicit memory for those targets may provide a greater insight into workload instruments. At present these data suggest workload is largely determined by error monitoring and not by successful task completion.

Conclusions

The findings of this study suggest that people are aware of their task failures, making incorrect responses, and they base their workload judgements on these aspects of performance. If individuals were monitoring the amount of resources they required for the task then it might be expected that the hit or success rate would determine workload perception. These data tentatively suggest workload judgements are derived from monitoring task failures; this would provide an interesting topic for future investigation.

References

Eggemeier, F. T and Wilson, G. F. 1991, Performance-based and subjective assessment of workload in multi-task environments. In D L Damos (ed.), *Multiple task peformance*, (Taylor and Francis, New York).

Gaillard, A.W. 1993, Comparing the concepts of mental workload and stress, Ergonomics, **36**(9), 991-1005.

Gopher, D. and Donchin, E. 1986, Workload - An examination of the concept. In K. Boff, L. Kaufman and J. Thomas (eds.), *Handbook of perception and human performance,* (Wiley, New York).

Gopher, D. and Branue, R. 1984, On the psychophysics of workload: why bother with subjective measures? Human Factors, **26**(5), 519-532.

Hancock, P. A. 1988, The effect of gender and time of day upon the subjective estimate of mental workload during performance of a simple task. In P. Hancock and N. Meshkati (eds.) *Human mental workload,* (North Holland, New York).

Hart, S.G. and Staveland, L. E. 1988, Development of NASA-TLX (Task Load Index): Results of empirical and theoretical research. In P. Hancock and N. Meshkati (eds.) *Human mental workload,* (North Holland, New York).

Humphreys and Revelle 1984, Personality, motivation, and performance: a theory of the relationship between individual differences and information processing, Psychological Review, **91**(2), 153-183.

Kahenman, D. 1973, Attention and Effort. (Prentice-Hall, Englewood Cliffs, NJ).

Kantowitz, B. H. 1987, Mental workload. In *Human factors psychology* (North Holland, New York).

Kramer, A. F. 1991, Physiological metrics of mental workload: a review of recent progress. In D.L. Damos (ed.), *Multiple task performance,* (Taylor and Francis, New York).

MachCorquadale, K. and Meehl, P. E. 1948, On a distinction between hypothetical constructs and intervening variables, Psychological Review, **55**, 95-107.

Moray, N., Johansen, J., Pew, R., Rasmussen, J., Sanders, A.F. and Wickens, C.D. 1979, Report of the experimental psychology group. In N. Moray (ed.), *Mental workload: its theory and measurement*, (Plenum, New York).

Navon, D. 1984, Resources - a theoretical soupstone? Psychological Review, **91**, 216-234.

Nygren, T. E. 1991, Psychometire properties of subjective workload measurement technique implications for their use in the assessment of perceived mental workload, Human Factors, **33**(1), 17-33.

O'Donnell, C. R. and Eggemeier, F.T. 1986, Workload assessment methodology. In K. Boff L. Kaufman and J. Thomas (eds.), *Handbook of perception and human performance,* (Wile New York).

Reid, G. and Nygren, T. 1988), The subjective workload assessment technique: a scaling procedure for measuring mental workload. In P. Hancock and N. Meshkati (eds.) *Human mental workload,* (North Holland, New York).

Roscoe, A. H. 1987, The practical assessment of pilot workload. AGARDograph no. 282, Neuilly Sur Seine, France.

Vidulich, M.A. 1988, The cognitive psychology of subjective mental workload. In P. Hanco and N. Meshkati (eds.) *Human mental workload,* (North Holland, New York).

Wickens, C. D. 1984, Engineering psychology and human performance, (Harper Collins).

Wickens, C. D. 1991, Processing resources and attention. In D.L. Damos (ed.), *Multiple tas performance,* (Taylor and Francis, New York).

Wierville, W. W. and Casali, J. G. 1983, A validated rating scale for gobal mental workloa measurement applications, In *Proceedings of the Human Factors Society 27th Annual Meeting* (Santa Monica, CA) 129-133,.

Wilson, G. and O'Donnell, R. 1988, Measurement of operator workload with the neuropsychological workload test battery. In P. Hancock and N. Meshkati (eds.) *Human mental workload,* (North Holland, New York).

Yeh, Y. and Wickens, C. 1988, Dissociation of performance and subjective measures of workload, Human Factors, **30**(1), 111-120.

Allocation of Functions

FUTURE NAVAL SYSTEMS: THE IMPORTANCE OF TASK ALLOCATION

C.A. Cook, C. Corbridge, L.H. Martin and C.S. Purdy

Centre for Human Sciences,
Defence Research Agency,
Portsdown,
Portsmouth, PO6 4AA, UK

cacook@dra.hmg.gb

DRA has been conducting research into methods of task allocation, investigating the suitability of existing methods for the Naval domain. An initial review of current task allocation methods identified seven problems with their use. A subsequent, more extensive, review of methods was carried out (Older, Waterson and Clegg, in press) and development of a new method initiated. Many methods were developed to support "dumb" automation, with the assumption that the allocation of tasks to either human or machine remains static. However, with "intelligent" knowledge-based technology there is potential for tasks to be allocated dynamically according to situational factors. This paper considers the implications of this technology for future Naval systems.

Introduction

DRA has been conducting research into allocation of function in future military systems. The aim of this work is to optimise the distribution of tasks between humans and equipment, thus maximising operational effectiveness of the total system. The Ministry of Defence is under increasing pressure to reduce the through-life costs of future platforms, with manpower being one of the largest of these costs. There is therefore much interest in identifying ways in which it is possible to reduce the number of personnel required for future ships. In terms of the individual role, the human operator performs many different tasks, which may involve several different pieces of technology. For future systems, technological constraints can exert a significant influence on any design solutions proposed since often there is a requirement to use a particular type of new technology. However at this level, the allocation of function between human and machine will act as a manpower driver for the whole ship.

The way tasks are allocated is therefore an important issue when considering the number of personnel required for future systems. A simplistic approach to this problem might be to simply automate whatever tasks that can be automated. However, this would not be a satisfactory solution for two reasons. Firstly, the allocation of function at an individual level is only one of a number of issues which impact upon the final ship complement. Figure 1 illustrates the relationship between allocation of function to human or machine at the individual level; human to human at the team level; job design for the overall ship's complement; and high level issues which are important at the fleet manning level. None of these issues can be examined in isolation, since there is a complex inter-relation between the various levels. For example, at the job design level, individuals will have a number of different roles depending upon the particular equipment fitted, the watch state, and the ship's mission. Changes at one level will affect manning at other levels, so that whilst automation of tasks at the individual level mean fewer operators are required, other roles may be left unfilled and jobs would have to be redesigned to redistribute these vacant roles.

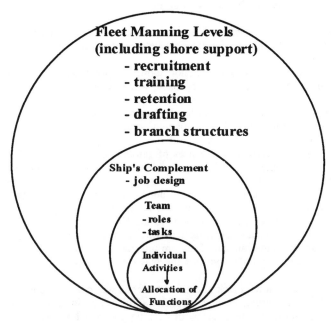

Figure 1. Relation between Function Allocation; Task Allocation in Teams; Complementing; and Fleet Manning Levels

Secondly, over-automation can lead to adverse consequences on human performance, and therefore ultimately lead to a degradation of overall system performance. Such adverse consequences may include: loss of situational awareness reduced job satisfaction and operator deskilling. An aim of this research into task allocation is therefore to bring together the previously disparate areas of research within human factors and organisational psychology, so that the two levels of individual allocation of function and job design can be considered in parallel.

Limitations of Current Task Allocation Methods

The process of task allocation can be considered to involve the assessment of alternative allocations against a number of criteria, such as cost, reliability, and workload. An initial review of available methods identified seven major problems with their use:

- The validity of comparing humans and machines. This includes methods based on Fitts List where tasks are allocated to human or machine based upon specific properties of the task, and the capabilities of human or machine. However, for many tasks neither, or perhaps both, the human and machine can perform the task, and instead their capabilities should be viewed as complementary (Jordan, 1963).
- The difficulty of identifying and quantifying criteria for determining the allocation. This may be relatively straightforward for machines, but is much less so for humans.
- Trade-offs between criteria. Frequently these may be unspecified, yet there may be complex interactions, for example between task criticality and operator workload.
- The focus of many methods on short term, individual workload. This approach is often termed the "leftover" approach, where as many tasks as possible are automated, and the operator given whatever tasks are left over, up to a maximum acceptable workload level. However, this can often lead to the operator's overall job lacking a coherent structure, requiring she/he to perform a variety of disparate functions.
- The potential adverse consequences of automation on human performance, such as poor job satisfaction, deskilling, and loss of situational awareness.
- The assumption that task allocation is a process with a fixed outcome. Many methods are based upon the use of older technology, yet with today's complex systems there is much greater potential for the use of dynamic task allocation.
- The applicability of current methods to the Naval context. Most methods were devised for use in stable environments such as manufacturing, where change is often slow and pre-planned. They are unsuitable for the rapid, uncertain, time critical environment that typifies the Naval environment.

A detailed review of task allocation methods was commissioned, and some nineteen different approaches were identified (Older et al, in press), based upon a review of the literature, and a review of existing practices both within the MOD and industry generally. These different approaches fall into four major categories, which are shown in Table 1. In practice, however, it was found that a combination of techniques is frequently applied.

In addition, the requirements for a task allocation method suitable for use with complex Naval systems were specified. These requirements covered three main areas, namely the *properties* of the method (for example, it should be usable, systematic, iterative and require minimum training); the *scope* of the method (what allocations should be covered, and the environments to which it should be applicable); and the *allocation criteria* (they should be well specified, explicit trade offs should be identified, and consideration given to the resulting job quality) for the actual allocation of tasks. It was found that none of the existing methods satisfied all the requirements. Whilst most were systematic, iterative, and covered human-machine allocation, few actually specified trade-offs between different criteria; enabled a quantitative evaluation of alternative designs; or considered human-human allocations. The need for a new method of task allocation to meet the specified requirements was therefore highlighted, and work on this method is ongoing and reported in Older, Clegg and Waterson (1996). The method will aim to address all the requirements, and also to include recommendations on dynamic task allocation.

Table 1. Current Methods of Task Allocation

Category	Details
MABA-MABA based	Men are better at ... Machines are better at. These methods tend to be based upon the original Fitts list approach (Fitts, 1957), but also includes more recent methods with greater emphasis on human abilities. The majority of methods fall into this category.
Interaction process based	Emphasis on diagrammatically mapping out the transfer of information between human and machine e.g. Job Process Charts (Tainsh, 1985).
Systems approach	Iterative, with emphasis on integration with the design cycle, and multiple design generation (e.g. Meister, 1987)
Incorporating dynamic task allocation	Approaches which have considered dynamic task allocation (e.g. Rieger & Greenstein, 1982)

Dynamic Task Allocation for Future Systems

Most methods of task allocation are based on "dumb" automation, and the assumption that during system design, once tasks are allocated to human or machine then they remain that way. However, the use of more sophisticated technology such as knowledge-based systems may enable a more dynamic or 'adaptive' allocation, dependent perhaps on certain situational factors, or by the operator's choice, whilst the system is in operation. Dynamic task allocation (DTA) seems to be especially applicable when the task can be performed equally well by the human or the machine. Task allocation can then vary according to the currently available resources (man and machine), thus being totally situation dependent. Dynamic task allocation may therefore offer the potential for optimising the allocation of tasks between human and machine by utilising the available resources in the most effective way. The technique therefore offers a potential for *optimum* manning of systems, whilst minimising the negative effects of over-automation such as deskilling and loss of situational awareness.

There are two main modes of control for the allocation of tasks (Rieger and Greenstein, 1982). If the control of allocating tasks belongs to the human operator, then this is termed *explicit* dynamic task allocation. If the control lies within the automated system, then this is known as *implicit* dynamic task allocation. One problem with explicit DTA is that the control of allocating tasks is in itself an additional task which may add to an already high workload. By the time the operator decides to allocate a task to the system, it may in fact be too late to do so. However, implicit DTA requires some mechanism to identify the available resources (human and machine) at any given time, and thus to determine the most appropriate allocation. Control mechanisms which have been used in previous research include various measures of workload, such as events-based (Debernaud, Vanderhaegen and Millot, 1992) or physiological (Prinzel, Scerbo, Freeman and Mikulka, 1995), together with queueing theory (Rouse, 1977), and predictive models of operator action or system efficiency (Greenstein and Revesman, 1986).

The feasibility of using this technique for complex systems such as those found in the Naval domain is currently under investigation. Typically, Naval command systems require the human operator to carry out a range of tasks under varying conditions of time criticality, dependent on the particular external threat. The operator may experience long periods of inactivity, or a steady workload, followed by bursts of intense activity. However, there is a requirement in these systems for manning levels to be high enough to be able to cope with the highest levels of task demands. It seems sensible instead therefore to propose that in these situations it would be more appropriate to allocate tasks to the machine instead of to the human operator, thus maintaining the operator's optimum level of workload as well as his/her situation awareness. In this way, the system could be manned by fewer operators, who would not be underloaded during normal situations yet could cope with excessive task demands through the dynamic allocation of those tasks to the system.

Planned Empirical Work

A resource allocation task within a Naval command system has been identified as an application area in which the suitability of this approach may be demonstrated. An experimental study is currently being planned to investigate this. The programme involves determining a baseline level at which the operator is overloaded; identifying those tasks which can be allocated dynamically; specifying the control of the implicit allocation; and defining appropriate measures of total system performance. An additional research question is to examine the relationship between task allocation and situation awareness.

In support of this work a more generic experimental programme is being carried out to investigate the most appropriate control mechanisms for determining the allocation of tasks to human or machine. Measures of workload are being evaluated, together with the potential utility of measures of situation awareness to control the allocation either as an alternative, or in addition to workload. This latter programme is based around the Multi-Attribute Task Battery (Comstock and Arnegard, 1992), using those tasks deemed most relevant to a Naval command and control environment.

Conclusions

It is believed that the approach presented in this paper offers a more efficient route to optimum manning. The need for a method of task allocation which is suitable for complex systems such as those found in the Naval domain has been identified, together with the potential use of dynamic task allocation which could lead to more optimum manning of systems. The development of this new method is ongoing, and findings from the research into dynamic task allocation in Naval systems will be incorporated into the method. It is hoped that the consideration of job design as well as task allocation will address the manning issues which impinge upon the overall ship complement at all levels. In this way, future targets for lean-manned ships might be achieved, without being at the expense of the skills and job satisfaction of the human operator.

Any views expressed in this paper are those of the author and do not necessarily represent those of the DRA.

© British Crown Copyright 1996/DRA
Published with the permission of the Controller of Her Britannic Majesty's Stationery Office

References

Comstock J.R. and Arnegard R.J. (1992) The Multi-Attribute Task Battery for Human Operator Workload and Strategic Behaviour Research. *NASA Technical Memorandum 104174*, Langley Research Centre, Hampton, Virginia 23665-5225.

Debernaud S., Vanderhaegen F. and Millot P. (1992) An experimental investigation of dynamic allocation of tasks between air traffic controller and AI systems. In H.G. Stassen (Ed.) *Analysis, Design and Evaluation of Man-Machine Systems*, Pergamon Press, Oxford, pp 95-100.

Fitts P.M. (1951) *Human engineering for an effective air navigation and traffic control system*, Washington, DC: National Research Council.

Greenstein J.S. and Revesman M.E. (1986) Two Simulation Studies Investigating Means of Human-Computer Communication for Dynamic Task Allocation. *IEEE Transactions on Systems, Man, and Cybernetics*, Vol 16 (5) pp 726-730.

Jordan N. (1963) Allocation of functions between men and machines in automated systems. *Journal of Applied Psychology*, 47, 161-165.

Meister D. (1987) Systems design, development and testing. In *Handbook of Human Factors*. G. Salvendy (Ed) New York: John Wiley and Sons, pp 18-42.

Older M.T, Waterson P.E. and Clegg C.W. (in press) A Critical Assessment of Task Allocation and their Applicability. Paper submitted to *Ergonomics*.

Older M.T, Clegg C.W. and Waterson P.E. (1996) A new method of task allocation: preliminary findings. *Proceedings of The Ergonomics Society 1996 Annual Conference*.

Prinzell L.J., Scerbo M.W., Freeman F.G. and Mikulka P.J. (1995) A Bio-Cybernetic System for Adaptive Automation. *Proceedings of the Human Factors and Ergonomics Society 39th Annual Meeting, San Diego*.

Rieger C.A. and Greenstein J.S. (1982) The allocation of tasks between the human and computer in automated systems. In *Proceedings of the IEEE 1982 International Conference on Cybernetics and Society*, 204-208.

Rouse W.B. (1977) Human-computer interaction in multi-task situations. *IEEE Transactions on Systems, Man, and Cybernetics* 1977, Vol 7 (5) pp 384-392.

Tainsh M.A. (1985) Job process charts and man-computer interaction within Naval Command Systems. *Ergonomics*, 28, 555-565.

BALANCING TASK ALLOCATION IN TEAMS

John Holt

CORDA, BAeSEMA,
Apex Tower, 7 High Street,
New Malden,
Surrey, KT3 4LH

When allocating tasks within a team it is important to achieve a balanced workload across the different operator roles, which is very difficult to achieve with available human factors methods. This paper illustrates how it can be done using Multi Criteria Analysis (MCA). As it is based on expert judgement, the use of MCA enables workload to be assessed without access to the operator's working environment (real or simulated), providing the great benefit of initial workload assessment early in the system development cycle, prior to more traditional methods.

Introduction

For many operational systems, the design of the operator's environment is not just concerned with individuals working on their own, but has to consider a team of people working together (e.g. Air Traffic Control, military Command and Control (C2), industrial process control). Thus, the system designer has to ensure not only that individual operators are not overloaded, but that the relative loading between different team members is not unbalanced (Stammers and Hallam, 1985).

The aim of the paper is to show how Multi Criteria (MCA) methods can be used to allocate tasks to operator roles within teams., and to describe the strengths and weaknesses of the MCA methods that were used in this study. Although MCA methods are relatively unknown in the human factors community, they have great potential utility in both individual and team workload assessment because they can be used (with appropriately guided expert judgement) without having to provide a realistic representation of the operator's task,. This is of great benefit in the early stages of system development, in role and team design, when such facilities are not likely to be available, and should reduce the requirement for expensive ad hoc team development with simulators (or prototype systems) and trained operators later in system development.

The context of this paper is a research study that was conducted into the design of Naval Command Team organisations, particularly in operating systems with significantly new technology, and was conducted over several years for the Ministry of Defence (MOD) Defence Research Agency (DRA). The ship's Command Team has to process data from a number of different sources (e.g. different types of radar, Electronic Support Measures, voice messages (from different types of radio transmissions)) in order to produce an integrated picture to enable tactical decisions to be made.

The standard human factors approaches for evaluating operator load, and that were considered in this study for assessing the balance in team load, are as follows:

- traditional human factors operator work load assessment

- the job inventory (using this method, experienced operators rate lists of tasks on aspects such as difficulty and duration, on five or seven point rating scales).

There were found to be a number of practical difficulties with using these methods, as follows.

- Workload and job inventories cannot readily be converted into aggregate measures , indicating the proportion of high versus low workload tasks that an individual operator has to perform.

- Five and seven point rating scales (used by many workload assessment methods as well as the job inventory) do not provide much differentiation between tasks

- Traditional human factors methods do not provide a systematic means of assessing how task workload can vary across different situations and scenarios.

MCA methods are well described in Watson and Buede (1987), and are typically used for assessing the preferred option from an available set. MCA requires the use of considered expert judgement to provide assessments. In this way it differs from traditional workload assessment where an operator performing a task is required to give an instant assessment .

This paper describes two MCA methods:

- Saaty's Analytic Hierarchy Process (AHP) (Saaty, 1981)
- Watson and Buede's Swing Weight Method (Watson and Buede, 1987)

AHP is a well accredited method for assessing operator workload (Vidulich, 1989). The Swing Weight Method can take into account variation in task workload across different scenarios.

For both the Saaty and the Swing Weight Method in this study, a single expert (a retired Senior Naval Rating) was used to elicit the required judgements.

Description of the MCA methods and their results

Saaty's AHP uses pairwise comparisons between tasks as the basis of its assessments. It has been subject to in-depth experimental validation as a techniqu for measuring individual workload (Vidulich, 1989). Summarising findings from number of studies, covering nine different types of tasks, with varying complexity it was concluded that this type of approach '.....can provide sensitive, reliable workload measures'.

In making comparisons between pairs of tasks using AHP for workload assessment, individual experts are asked to say if there is any difference in the loa imposed by either task. If there is, there are asked to describe the diffference usi the scale in Table 1. Thus if Task A requires greater effort than Task B and the difference is 'Strongly important' Task A gets a score 5 which means that Task A has 5 times the weighting of task B. The corresponding entry of the matrix for Task B with respect to A, is the reciprocal rating of 1/5. Saaty advocates using on the Definitions for elicitation, arguing that, from his experience, the Definitions always produce the same Scale of Difference. . A simple matrix manipulation ca be used produce task workload assessments from the pairwise comparisons (Holt, 1994).

The scale of difference in Table 1 is used in AHP to assess the differences between tasks.

Table 1: Saaty Definitions and Scale of Difference

Definition	Scale of Difference
No difference	1 times
Weakly important	3 times
Strongly important	5 times
Very strongly important	7 times
Absolutely important	9 times

In addition to assessing the weights, Saaty (1980) describes a method for assessing the consistency of the expert's assessments. If it is below the required level then either the data can be rejected, or the results can be fed back to the expe so that he can revise his assessments. Saaty also describes a method for averaging the results from a number of experts to produce a single set of consensus ratings. Vidulich (1989) describes how he has packaged the AHP approach into a compute based tool, The Subjective WORkload Dominance (SWORD) Technique, though th calculations can readily be performed by a simple spreadsheet model, as was done in this study.

In performing pairwise comparisons, to limit the number of comparisons (in this study there were about 70 tasks to be assessed), it is advisable to assess the tasks in groupings of nine or less. If related tasks are grouped together then the relative weights of the groups can be assessed using AHP (or by using a task from each group,, see Holt, 1994).

To make a proper comparison of aggregate operator workload in a given period, it is necessary to assess individual task work load, duration and expected frequency. In our analysis, the expert was asked to assess task demand (Effort x Time) in a single assessment and then later to assess task frequency (i.e. workload on its own was not directly assessed). It is more difficult to assess two variables together, and in future studies it would be better to assess these separately. Average task frequency and duration can be assessed more accurately by empirical methods (e.g. by systematic observation of the task being performed) with expert ratings being used just to assess workload. Having used AHP in the same way to assess task frequency, task demand was multiplied by frequency to give an aggregate task loading for each of the operators, as shown in Table 2.

Table 2: Summary results of Saaty analysis

Operator	Task Demand (Effort x Time)	Task Frequency	Task Demand x Frequency
1	0.63	0.49	0.31
2	0.14	0.05	0.01
3	0.12	0.15	0.02
4	0.1	0.3	0.03
Total	1.0	1.0	

Discussion of Saaty Method

Using AHP with a single expert in a short period, very detailed results were produced for about 70 tasks for 4 operators, from which the balance of tasks between operators could readily be evaluated. The method did have some minor shortcomings, in that the 1, 3, 5, 7, 9 scale appeared to the analyst to be somewhat arbitrary,. Also, how the numbers are turned into weights is not easy to explain to the experts. AHP does not assess how task workload varies across different situations, or scenarios. Because it can address these shortcomings, the Swing Weight Method (Watson and Buede, 1987) was tested as an alternative way of assessing workload for a small sub-set of tasks.

Swing Weight Method and results

In using the Swing Weight Method (SWM), the expert is asked to evaluate which task has the highest workload and then to use that task to measure the workload of the other tasks. To demonstrate simply what was done in this study, the analysis for three of the tasks (performed by the senior operator in the team) is described. These are as follows:

1. Monitor the tactical situation
2. Operate the radar controls
3. Deal with navigational system failure

Thus for the three tasks, the expert was asked to identify the three most and
least favourable situations in which each has to be performed, as shown in Table :

Table 3: Best and worst cases for tasks

	Best Case	Worst Case
1. Monitor the tactical situation	Clear picture	Misidentification has occurr
2. Operate the radar controls	Minimum picture loss	Complete picture loss
3. Deal with nav system failure	No disruption	Total disruption of the pictur

To find the task with the highest workload, the expert was asked to imagine
that all tasks were in their worst case, and that just one could be changed to its bes
case. The expert was asked to say which one he would change. The chosen task
was then assumed to have the highest weighting. The highest weighted task is the
used to assess the relative weightings of the other tasks, as follows.

Monitor the tactical situation was chosen as having the greatest weighting ar
its best case was given a weighting of 0 and its worst case 1, as indicated in Figure
1, below. To assist the weighting process, significant situations within which the
highest weighted task must be performed were elicited from the expert on the scal
shown in Figure 1, depending on their relative workload. For example, Total bla
out is regarded as having three quarters of the load of misidentification, and is thu
given a weighting of 0.75. Half the picture present has a 0.5 weighting. Radar
counter measures which have been identified (and are therefore no threat) have
only a weighting of 0.1.

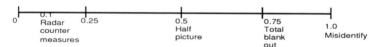

Figure 1: The range of situations for the highest weighted task

The scale can be used to identify the tasks. Dealing with Navigation System
failures (worst case) was identified as being between Total blank out and
Misidentification, though much nearer misidentification, as shown in the diagram
below. After some discussion, it was given a score of 0.96. Operate radar controls
(worst case) was seen as being between Total blank out and Half picture, though
slightly nearer Total blank and hence received a weighting of 0.66. Calculatio
of the weights from the above information, is performed by solving a set of
simultaneous equations , as described in Holt (1994).

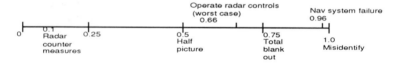

Figure 2: Task weightings

Discussion of SWM

SWM was shown to provide an alternative, judgement-based method to AHP that could be used for workload assessment. By taking the best and worst cases, and by the analyses illustrated in Figures 1 and 2, SWM was shown to account for workload variations across the whole range of scenarios in which a task could be performed, Also, the analysis demonstrated in Figure 1, using expert judgement to evaluate the effects on workload of performing a task under different conditions, was developed for the highest weighted task. The analysis could also be done for other important tasks, to provide a means of eliciting the key critical conditions in which a task is likely to be performed.

Conclusions

This paper has shown how measures of individual and team workload can be elicited, using both AHP and SWM. The AHP method has also been shown to provide a high degree of differentiation in the assessment of different tasks. It has been shown how SWM can be used to take account of the whole range of scenarios in which tasks must be performed.

AHP has already been validated as a tool for individual workload assessment. There is a requirement for a similar analysis to be performed for SWM. As individual workload tools, it would be useful to assess how consistent the methods are with each other and with traditional methods. There is also a requirement for a full assessment of the validity of the methods in team assessment.

If these methods can be fully validated, they will have considerable potential as workload tools, providing a high degree of flexibility and range of use, and addressing situations where other methods cannot be applied, in the assessment of both individual and team workload.

Acknowledgement

This work has been carried out with the support of Ministry of Defence (MOD) Defence Research Agency (DRA) Operational Studies (Naval) (OSN) Department, now the Centre for Human Sciences.

Any views expressed in this work are those of the author and not DRA.

References

Holt, J (1994). Balancing Task Allocation in Teams. Unpublished BAeSEMA Report.

Saaty, LT. (1981) *The Analytic Hierarchy Process*. McGraw-Hill. New York.

Stammers, RB and Hallam, J (1985) Task allocation and the balancing of task demands in the multi-man-machine system - some case studies. *Applied Ergonomics* 16, 4.

Vidulich, MA (1989) The use of judgement matrices in subjective workload assessment:. *Proceedings of the Human Factors Society 33rd Annual Meeting.*

Watson, SR and Buede, DM (1987) Decision Synthesis. *The Principles and Practice of Decision Analysis*. Cambridge University Press. Cambridge.

REAL WORLD CONSTRAINTS ON THE ALLOCATION OF FUNCTIONS

Michael Goom

British Aerospace Defence Limited
Dynamics Division
FPC 500, P.O. Box 5,
Filton, Bristol.
BS12 7QW

This paper examines the real world constraints that govern the allocation of functions in modern defence system. The allocation of tasks between humans and machine components is a crucial part of system design. However the requirement to allocate tasks or functions to humans is often driven by legal or political reasons rather than considerations of best performance. A method of accommodating these external constraints is proposed based around job design and workload prediction.

Introduction

This paper examines the real world constraints that govern the allocation of functions in modern defence system. It is a result of the on-going implementation of a Human Factors Integration (MANPRINT) programme within the BAe. Dynamics Division. It uses the military environment for examples but the philosophy of 'design for the user' is becoming necessary in the majority of systems. The paper is based on some of the lessons learned during the development of a methodology for the Dynamics Division of British Aerospace Defence Limited to integrate the user into the system development process.

The paper focuses firstly on the background to the current usability problems that have emerged and compromised the performance of a number of military systems. Secondly, it gives a description of a process of deciding how to allocate tasks and functions between the hardware and software (devices) and the users who operate, maintain, support and train as part of the system.

The conclusion is drawn that there are a number of methods and tools emerging that will support the efficient allocation of tasks, but these will only be really useful if they are efficiently linked into the complete system development process.

Background

In recent years, technological and performance advances in equipment design have resulted in ever increasing complexity. In the past, inadequacies in recognising the importance of the human contribution in equipment design have been partially compensated for through increased training and use of highly skilled operators and maintainers.

Demographic and political changes increasingly mean reduced numbers, and a consequent reduction in the high skill levels, of personnel being recruited into the armed forces. This factor, when combined with the trend towards more complex equipment, is leading to many more examples of systems failing to achieve their performance targets in operational use.

The United States DoD realised these problems in the early 1980s and introduced a new philosophy called MANPRINT (MANpower and PeRsonnel INTegration) into the procurement process. The UK MoD formed a tri-service cell in 1990 to ensure that due consideration is given to the capabilities and limitations of the human elements of complex systems – an area that has been largely neglected in the past. The term MANPRINT has now largely been replaced by Human Factors Integration (HFI).

The HFI philosophy should be applied from the system concept stage and must be carried on throughout the design process and beyond. It encourages the designers to consider the interaction between equipment design and the people who will use, operate and maintain the equipment and support it in service. This requires a system level approach especially to the allocation of tasks both between humans and machines and between teams of humans.

Allocation Functions Between Humans And Devices

"Why am I proposing to use human beings as part of the system I am developing?" The answer to this question can be invaluable in providing an explicit understanding of the functions the user will be required to undertake.

It could be argued that most tasks that can be clearly specified can usually be better done by machine and we certainly need to automate tasks that humans do badly such as vigilance and repetitive tasks. There are many development engineers who believe in automating the humans out of the system as they perceive the human operator as unreliable and inefficient. However they end up having to leave the operator to do the tasks that they cannot think how to automate.

One of the reasons often given for including humans in defence systems is for decision making purposes. This usually means making decisions with insufficient or uncertain information, as given all the information, the machine would probably reach a correct solution more rapidly. Hitchings [Ref. 1] building on the 'recognition primed decision' work of Klein, suggests that in most time constrained strategic decision making tasks 'satisficing' takes place. This consists of matching the current problem with one that has been encountered before and activating solutions that appeared to be effective on previous occasions. The user checks that the responses are in line with his predictions. If the responses are at variance with expectations a further matching takes place. This approach to decision making relies on the user having an understanding of how the system behaves. Indiscriminate automation can mask this essential overview and is one of the prime reasons for the current HFI approach.

One of the principal lessons that has emerged from the application of HFI has been the need to identify the jobs of the users. This must include ALL the component tasks that the users will have to undertake, not merely those of the sub-system under investigation, but also on other systems that they will be required to operate. In addition those tasks that originate from their day-to-day military duties must be considered. In many cases the allocation of system tasks to man or machine is governed by the task (and work) loading imposed by activities outside the immediate system.

Constraints on the Allocation Process

The allocation of tasks and functions between man and machine does not take place in a vacuum. The process of allocation has to recognise that certain tasks may be allocated as a result of constraints that range from technology to politics.

Technological Constraints

Technology constraints may include the need to incorporate a particular piece of equipment because the customer has made considerable investments in the item and insists that it is incorporated. The technology may be required to cope with a small proportion of cases, but because it has to be provided anyway it may have to suffice for all.

Political and Legal Constraints

With changes in legislation and associated commercial responsibility and accountability there are now many more tasks that it is not possible to allocate to the human component. The increased knowledge of toxic substances, sensitivity to public opinion and the fear of litigation is causing manufacturers to err very much on the side of caution. Many tasks that could be done 'better' by humans must now be assigned to machines.

Compatibility Constraints

Systems are now so complex and costly that the re-use of existing designs and the increased use of off-the-shelf systems is the order of the day. It is unusual to start with a blank piece of paper, and so the flexibility available in allocation of functions is immediately limited. In addition the influences of the 'outside system' tasks that the user must perform will modify the scope for allocation.

Resource Constraints

The drivers for the allocation process on many modern systems are often the skill levels of personnel available to the customer and the training time he can afford. Training is probably the most significant aspect that has been poorly represented in any of the traditional allocation exercises.

In many of the through life cost calculations, manpower and training costs can be many times the development and procurement costs. In these instances the availability of previously trained personnel and training courses is beginning to influence the allocation process on the prime equipment design.

Concurrent Engineering

There is a move within industry towards the concept of Concurrent Engineering. In the past great efforts have been made to get the requirements correct and adequately documented in such a way that the team responsible for the next phase in the system development life cycle could work from it alone. Concurrent Engineering is a recognition that for the complex systems that constitute modern systems this approach is no longer possible. The gestation period of some systems can be up to fifteen years and, with the likely technological changes, this can cause a need for modification and consequently re-allocation.

The Allocation Process

In 1951 Paul Fitts produced a table with two columns, the first contained those tasks that men could do better than machines and the second contained those where the capabilities of machines exceeded those of men. For many years those who thought about a systematic division of effort between man and machine in system development have used variations on Fitts' list. There are severe problems in trying to apply Fitts' type lists to modern systems. They have largely resulted from the vastly increased capabilities of machines. When functional allocation came into existence as a concept in the early fifties, the allocation of functions between man and machine was fixed, barring major redesign, very early in the design cycle. Systems were relatively simple and the process of allocation was fairly obvious. Many of the early lists now look like statements of the blindingly obvious.

The development of the HFI programme has identified that a different approach to the allocation problem could be beneficial. It consists of identifying those tasks for which a human is clearly 'best' or required for legal, (or other reasons), and building a coherent job structure around those tasks. In addition to providing a sensible job content, the determination of an optimum workload is probably the clearest single driver for this allocation process.

Humans and machines perform tasks in a complementary manner to fulfil functions or sub-mission goals. Whilst these higher level goals may have an obvious structure, the individual tasks that are necessary to achieve them may not in themselves have that logical structure when taken in isolation. Laughery & Laughery [Ref. 2] make the point that 'A function can be viewed as a logical unit of behaviour of a human or machine component that is necessary to accomplish the mission of the system.' When dealing with tasks the logical units may not be present. Whilst machines can operate on a task by task basis, humans faced with a random selection of tasks that have little logical connection tend not to perform very well. The automation represented by a simple allocation of tasks to the machine can remove many of the signposts from the user's mental model of the process. This in turn leads to the user's inability to provide the resource of last resort which is often his reason for existing in the automated system.

A Method of Integrating the Human Contribution

Once it is recognised that the user is an integral part of the system being developed or integrated, it is necessary to devise a formal strategy for designing the human component into the system as part of the development process. HFI goals and constraints have to be achieved by very careful allocation of tasks and functions between the user and the machine, and a continuous assessment as to whether the human workload is reasonable.

There are six basic steps needed to ensure that the human component can contribute in an optimum manner to the functioning of the overall system. These are :-

1.   Identify those tasks that <u>must</u> be undertaken by the human component.
2.   Identify who the users will be.
3.   Gain an understanding of how the user perceives the workings of the system.
4.   Assign to the users those tasks that will aid their understanding of the workings of the system.
5.   Model the user's workload to identify likely overload or underload points. This must include the degradation in task performance that will result from environmental constraints, e.g. noise, fatigue.
6.   Perform rapid prototyping experiments on the high workload points to determine if they really do exist.
*!   Iterate from 3, or 2 or even 1 if the required total system performance cannot reasonably be guaranteed with the proposed solution.

Clearly the activities that are needed for each of these steps vary from project to project as well as from company to company. Therefore only an overview of the underpinning of these steps can be given in this paper.

Identify the Human Tasks

How should we allocate tasks to people or machines? In the past a variation of the Fitts' list has been used as to decide whether to automate or let the human do the task or function. Often this activity is referred to as Allocation of Function (AoF). Many text

books give the impression that this is a distinct step in the system development process
and once the functions (or tasks) have been allocated the system can be designed.

Developments in hardware and software now mean that many of the activities
that would have been allocated to humans in 1951 are now better undertaken by
machines. It is difficult to identify many tasks where a human can outperform a
machine in terms of speed and accuracy IF the task can be fully specified and the
required information is available. The human is often used in circumstances where there
is insufficient information and a value judgement (guess) must be made. In these cases
it is essential that the user's perception of how the system operates is understood by the
system developers if the user is to be in a position to make the best decision.

Often the tasks that the user must undertake are based on factors not concerned
with performance. Many tasks must be performed by a human being for legal or
political reasons. The organisational structure of the personnel who form the users may
dictate that certain tasks are performed by certain operators. It is not feasible to court
marshal a machine yet.

Identify the Users

Who will have to use the system? This includes the operators, maintainers,
trainers, support staff, administrators, etc. All too often in the past we have been
commissioned to build systems on the assumption that people can be recruited and
trained to operate them. Unless we know who will be required to use the system it is
difficult to see how the total system can be guaranteed to perform to specified standards.

Understand how the Users Perceive 'The System'

There is the need to gain an understanding of how the user perceives the
workings of the system and its relationship to the environment. That is, to consider the
mental model the users have of the system and how they believe it mediates their world
model. All too often the way that the developer views the underlying structure of the
system is not the same as that of the user. This mismatch can lead to errors of
judgement on the part of users when they are required to reach decisions on the basis of
insufficient information. If the user is to perform this type of task, it is essential that the
development of the system considers how the users view the system and reinforce this
mental model. It is only in this manner that we can expect humans to take on tasks that
are beyond the capabilities of machines with a better than random chance of success.

Build Sensible Jobs Round the Essential Tasks

To reinforce the user's understanding it is necessary to construct groups of tasks
that form coherent 'jobs'. It may be desirable to allocate to the user some tasks that
would be better done by machine, but without which some of the tasks that the user
must perform would not make logical sense.

Determine if the Collection of Tasks can be Successfully Undertaken

When the tasks that will form the users' jobs have been identified, a workload
prediction should take place. Can an operator be expected to complete the tasks in the
correct sequence and within the time allowed? The addition of extra off-the-shelf
systems to an existing optimised system can have enormous effects on overall
performance and error rates. In most cases it is now possible to use task simulation
models and workload assessment techniques to highlight tasks, or groups of tasks that
will form bottlenecks or other points of high activity. It is these points that need to be
examined if incipient system errors and failures are to be avoided. This type of
modelling has to include the degradation that can result from environmental factors and
the decrements that occur as time on duty increases. It must be stressed that this type of
modelling can only be used to discover problem areas. It cannot provide guarantees that

the system will definitely work given the variability that is associated with human beings, but it can provide estimates and confidence limits providing the users have been adequately described ina reasonably complete and accurate manner.

Assess the Difficult Bits

Despite all the research into human performance it is often only possible to determine which tasks will not be completed satisfactorily. It would be a very bold systems engineer and an extremely foolish ergonomist who would state categorically that a particular combination of complex tasks could always be completed in a certain time period with a specified error rate. Therefore rapid prototyping of the points of high workload using true end users is highly recommended.

Need for Iteration

Designers should expect iteration and must be prepared to make changes throughout the design process, using tests and analyses, until a close match between user and equipment is reached. This iterative process can be aided by using software tools, mock-ups and computer models in order to achieve the HFI objectives with minimal expense.

Conclusions

Successful integration of any complex weapon system must include the full integration of the human components. This has been one of the lessons that have been learnt and which prompted the creation of the HFI programme. The users are as much a part of the system as a radar or power supply and their interface and performance characteristics need possibly more attention than their hardware and software counterparts if an optimal total system design is to be realised. The allocation of tasks and functions between users and devices should be performed in such a way that the tasks that are allocated to the users are logically grouped into 'jobs' that reinforce the human's understanding of how the total system responds.

It is often the case that during the integration of systems compromises have to be made that result in the users performing tasks which in ideal conditions would be performed by the machine. Methods do exist that enable the prediction of the performance of the total system, based on knowledge of the potential users. These methods have been developed and tested as part of the HFI programme and their use should be encouraged if the problems inherent in the optimal integration of complex systems are to be minimised.

References

[1]   Hitchings, D.K. (1992). *Putting Systems to Work.*(John Wiley. Chichester)

[2]   Laughery, K.R & Laughery K.R. (1987). *Analytical Techniques for Functional Analysis.* In G Salvendy, Ed., *Handbook of Human Factors.* (John Wiley & Sons Inc. New York)

KOMPASS: COMPLEMENTARY ALLOCATION OF PRODUCTION TASKS IN SOCIOTECHNICAL SYSTEMS

Gudela Grote, Steffen Weik and Toni Wäfler

Work and Organizational Psychology Unit
Swiss Federal Institute of Technology,
ETH-Zentrum, CH - 8092 Zürich

Decisions conerning function allocation are crucial for the design of reliable and safe production systems. Deficiencies of frequently used strategies for such decisions are discussed and an approach based on the idea of complementarity between human and technical system is proposed. Criteria for complementary function allocation as well as results of job analyses using these criteria are presented. Finally, a design process based on the suggested principles is outlined.

Introduction

In order to support automation strategies that further the reliability and safety of production processes, a crucial question concerns the allocation of functions between human operator and technical systems. A still frequently adopted strategy for deciding on the allocation of functions is the "leftover principle" which allocates whatever functions cannot be automated to the operator. These functions do not necessarily form a meaningful job, they might even be impossible to carry out, as Bainbridge (1982) has argued convincingly. Her criticism also applies to a second approach to function allocation, the "comparison principle" (Bailey, 1989). Based on lists comparing characteristics of humans and machines, functions are allocated to the one who supposedly can perform that function better. With increasing technical possibilities, these lists change as more and more functions can be performed better by machines, leaving monitoring and dealing with deviations and disturbances to the operator. Jordan (1963) pointed out thirty years ago, that humans and machines are fundamentally different and therefore cannot be compared and assigned functions accordingly. Instead, they are to be seen as complementing each other in performing a task together. Function allocation should allow for the support and development of human strengths and compensation of his or her weaknesses by the technical system in order to secure safety and efficiency of production as well as humane working conditions.

The shift in emphasis from the leftover and comparison principles to complementary design is most pronounced in high risk systems where it is most obvious that humans have to be 'kept in the loop' in order to be able to fulfill their final responsibility for production and safety. What is needed is the prospective design of both operator jobs and technical systems based on allocation decisions which take into account technical, organizational, and people considerations in an integral manner (cf. e.g. Ulich, 1994). However, as a number of authors have pointed out, to date these decisions are often made "unconsciously rather than by deliberation" (Price, 1985, p. 33).

Looking at existing methods and instruments developed to support a more systematic approach to the complementary allocation of functions, one is confronted with a multitude of criteria for complementarity, such as flexible and dynamic allocation, transparency, operator control, low coupling, and operative flexibility, just to name some frequently mentioned criteria. There have also been some attempts to operationalize these criteria, but to date neither a fixed set of criteria, comparable for instance to software usability criteria, nor widely accepted methods for their measurement exist (cf. also Older, Waterson, Clegg & Payne, 1995).

The KOMPASS approach to function allocation

KOMPASS is a research project concerned with developing guidelines for the complementary design of production systems. The project is part of a larger cooperative research effort in the field of advanced manufacturing systems, with a number of engineering disciplines involved in the design of highly integrated production technology.

The guidelines are intended for multidisciplinary design teams in industrial settings, supporting them in following an integral approach to systems design, with a special focus on decisions regarding function allocation. A participatory approach to the design process is envisioned, that is, future system operators should be included in the design decisions as early as possible. However, participation and the use of expert design criteria should be balanced (e.g. Grote, 1994). The instrument leads a design team through the following four design steps:

(1) Identification of furthering and hindering preconditions for complementary design
(2) Sociotechnical analysis of the existing work system
(3) Development of options for the allocation of functions between humans and technical systems as well as between humans
(4) Evaluation of options based on work-oriented criteria for function allocation, resulting jobs, and sociotechnical design.

Given the existing theoretical and empirical research that could be built on in the development of the guidelines, two main research objectives were (a) to define and operationalize a set of criteria for complementary function allocation and to adapt existing instruments for the analysis of work tasks and work systems with respect to highly automated production systems and for their use in prospective design, and (b) to develop a heuristic for the participative development and prospective evaluation of design options.

The KOMPASS criteria for complementary design

Existing criteria for the complementary allocation of functions as well as for job and organizational design were compiled and a set chosen for initial exemplary analyses of work tasks in automated production. Based on the results of these analyses, the following criteria were selected for further analyses including tests of reliability and validity (for more details see Grote, Weik, Wäfler & Zölch, 1995):
• The level of the sociotechnical system: Completeness of task of organizational unit; independence of the organizational unit; task interdependence, polyvalence of the operators; autonomy of work groups; boundary regulation by supervisors.
• The level of the individual work task: Completeness of individual tasks; planning and decision making requirements; task variety; cooperation/communication requirements; autonomy; opportunities for learning and personal development; absence of hindrances.
• Third level of analysis and design - The human-machine system: The criteria for this level are presented in more detail in Table 1. Whereas the criteria on the other

levels of analysis were mostly adapted from existing instruments (e.g. Dunckel et al., 1993; Oesterreich & Volpert, 1986; cf. Ulich, 1994), these criteria and their operationalizations were developed within the project, partly drawing on prior work by Clegg (1993), Clegg et al. (1989), Corbett (1985), and Kraiss (1989).

Table 1. KOMPASS criteria for the level of human-machine system.

Dynamic coupling
- Extent of physical coupling between operator and technical system and influence of operator on degree of coupling concerning the when, where and how of the production process
- Extent of cognitive coupling based on level of required cognitive effort and influence of operator on the amount of effort

Process transparency
- Possibilities for developing a mental model of the steps and timing of the production process
- Availability of analogous and digital feedback from the production process
- Availability of feedback conerning process quality

Decision authority
- Distribution of decision competence over controlling the production process between human operator and technical system

Flexibility
- Possibility for flexible/dynamic function allocation
- Adaptivity/adaptability (decision on changes in function allocation by technical system and/or human operator)

Technical linkage
- Dependence of the observed technical system on other technical systems

---

A model of how the criteria on the level of human-machine system are assumed to be interlinked is presented in Figure 1.The model is based on control theory as it has been proposed in psychology (cf. e.g. Thompson, 1981) as well as in systems design (e.g. Brehmer, 1993). Three different types of control - or rather prerequisites for control - are distinguished: comprehension, predictability, and influence. The criterion transparency is operationalized in terms of the possibilities for building and using a mental model of the steps and timing of the production process, which is assumed to be the basis for comprehension and predictability of the process. Degree of influence is not measured directly, but through coupling - i.e. choices provided or restricted by the technical system concerning timing, place, and method of carrying out certain steps in the production process -, flexibility of function allocation, and decision authority, ranging from complete operator authority without technical support through various steps of support to complete automatic authority. It is to be noted that the question asked in the analyses is not whether a given operator has a sufficient mental model of the process and actively influences the process, but rather whether the chosen function allocation provides opportunities for both.

The relationship between comprehension/predictability and influence is assumed to be a complex one. While being independent of each other in terms of the *possibilities provided* by a given function allocation, comprehension and predictability are regarded as preconditions for *making use* of provided opportunities to influence the process effectively and according to safety considerations. At the same time, process interventions by the operator will also further the building of a mental model. This relationship was looked at in qualitative analyses of the studied human-machine systems, which suggest that interactions between coupling and process transparency in particular need to be investigated further. Usually, research stresses the necessity of low coupling in order to provide autonomy for the operator (e.g. Corbett, 1987). With high levels of automation, low coupling may result in the loss of process transparency, however. Therefore, it is argued that not low coupling, but dynamic coupling - in the sense of Weick's (1976) loose coupling as simultaneous autonomy and dependence - is to be aimed at. This can be achieved by giving operators the possibility to influence the amount of coupling, e.g. by means of programming processes in different ways or flexibly allocating functions to the technical system or themselves.

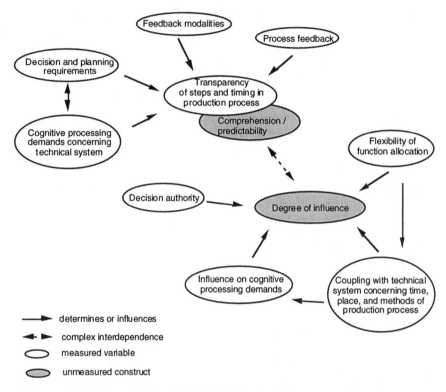

Figure 1. Model underlying KOMPASS-criteria for human-machine system.

The complete set of criteria was tested for reliability and validity by means of 29 analyses of pairs of identical jobs mostly in the machine industries. Usually two operators working with the identical technical systems on different shifts were chosen and questioned and observed by two different investigators. The operators were also asked to fill out a questionnaire containing measures for job perception and personal well-being (SALSA by Rimann & Udirs, in press) as well as a diary of all production disturbances and their management during one week. Some results of these analyses will be discussed subsequently.

The interrater reliability for the newly developed criteria on the level of human-machine system varied for the items of most criteria between .4 and .6 (Spearman's rank correlation). Lower reliabilities were found for coupling - except the extent of required cognitive effort -, decision authority, and flexibility. The latter result was mainly a consequence of lack of variance, however, as very few examples of flexible function allocation were found. When interpreting these finding, the chosen form of reliability test has to be considered: analyses of supposedly identical tasks performed by different people based on observations and interviews by different people. While some of the low reliability coefficients can be attributed to inadequate operationalizations, currently being revised, low correlations were also partly due to actual differences in the tasks, e.g. operating machines during different shifts involving different types of production orders.

Looking at the pattern of intercorrelations among the criteria, the model underlying the KOMPASS-criteria on the level of human-machine system (see Figure 1) appears to be supported, A standard regression analysis with the four variables decision and planning requirements, cognitive processing demands, feedback modalities and process feedback as independent variables explained 43% of the variance of the variable process transparency. The equation that included all the other variables as independent variables as well explained only additional 3% of the variance. No test of the assumed relationships concerning degree of influence was possible as this variable was not measured directly and the data base was insufficient for statistical methods testing relationships between latent and manifest variables.

Correlations between criteria on the different levels of analysis, i.e. human-machine system, individual work tasks, and sociotechnical system, as well as correlations with the measure of job perception and subjective well-being support the validity of the criteria. Positive correlations were found, for instance, between completeness of primary task of the work system and individual decision and planning requirements, between individual autonomy and dynamic coupling as well as between perceived qualitative overload and lack of autonomy and of dynamic coupling.

The KOMPASS design heuristic

Concerning the development of options for the allocation of functions in a planned system, a heuristic has been outlined and tried out in some pilot cases (Grote, 1994; Wäfler, Weik & Grote, 1995), but will have to be tested further in ongoing automation projects. The basic idea of the heuristic is to support the design team in following an integrated approach to system design by first developing a shared understanding of the system objectives and the contributions of technical, organizational and human aspects to successful system performance. Based on this understanding, options for function allocation are determined and evaluated by means of the KOMPASS criteria.

In order to stress the link between function allocation and job design, each function is analyzed with respect to four aspects: decision requirements, type of activity (planning, executing, controlling), importance for transparency of the production process, and automation potential. The characterization of functions in this way allows for a rough allocation into three 'slots': (a) allocation to human alone (based mainly on the automation potential), (b) allocation to technical system alone (functions that contain no decision requirements, are not important for process transparency, and technically can be automated), and (c) allocation to human and technical system (functions that contain decision requirements and planning and controlling elements, and/or are important for process tranparency). The functions in the third slot then have to be analyzed in more detail and allocation options developed for them. The different design options are then evaluated - in as much as that is possible in prospective design - on the basis of all KOMPASS criteria. Participation of operators having worked in the 'old' production system and of future system operators is crucial during the development and evaluation of design options.

KOMPASS attempts to improve the quality of design decisions in terms of production efficiency and safety by providing empirically validated criteria for integrated work system design as well as a moderation aid for the design process. Two specific characteristics of the KOMPASS guidelines further this aim - structuring the description of productions tasks independently of their allocation (for details see Grote, Weik, Wäfler & Zölch, 1995), and providing the multifunctional design teams with criteria addressing concrete technical requirements from a work psychological perspective. However, the criteria are not to be understood in the sense of a 'quick fix' leading to one straight forward design solution, but as part of a heuristic, which enables the design team to - as early as possible in the design process - reflect and if necessary correct their decisions about distributing functions between humans and technical systems from an integrative work psychological point of view.

References

Bailey, R.W. 1989, *Human Performance Engineering*, 2nd edn, (Prentice-Hall International. London).

Bainbridge, L. 1982, Ironies of automation. In G. Johannsen & J.E. Rijnsdorp (eds.), *Analysis, Design and Evaluation of Man-Machine Systems*, (Pergamon Press, Oxford) 129-135.

Brehmer, B. 1993, Cognitive aspects of safety. In B. Wilpert & T. Qvale (eds.), *Reliability and Safety in Hazardous Work Systems*, (Lawrence Erlbaum, Hove) 23-42.

Clegg, C. 1993, *Tool for the allocation of system tasks between humans and computers*, Unpublished manuscript, MRC/ESRC Social and Applied Psychology Unit, University of Sheffield.

Clegg, C., Ravden, S., Corbett, M. & Johnson, G. 1989, Allocating functions in computer integrated manufacturing: a review and a new method, *Behaviour & Information Technology*, **8**, 175-190.

Corbett, J.M. 1985, Prospective work design of a human-centred CNC lathe, *Behaviour & Information Technology*, **4**, 201-214.

Corbett, J.M. 1987, A psychological study of advanced manufacturing technology: the concept of coupling, *Behaviour & Information Technology*, **6**, 441-453.

Dunckel, H., Volpert, W., Zölch, M., Kreutner, U., Pleiss, C. & Hennes, K. 1993, *Kontrastive Aufgabenanalyse im Büro. Der KABA-Leitfaden* (Verlag der Fachvereine, Zürich; Teubner, Stuttgart).

Grote, G. 1994, A participatory approach to the complementary design of highly automated work systems. In G. Bradley & H.W. Hendrick (eds.), *Human Factors in Organizational Design and Management - IV*, (Elsevier, Amsterdam) 115-120.

Grote, G., Weik, S., Wäfler, T. & Zölch, M. 1995, Criteria for the complementary allocation of functions in automated work systems and their use in simultaneous engineering projects, *International Journal of Industrial Ergonomics*, **16**, 367-382.

Jordan, N. 1963, Allocation of functions between man and machines in automated systems, *Journal of Applied Psychology* , **47**, 161-165.

Kraiss, K.-F. 1989, Autoritäts- und Aufgabenteilung Mensch-Rechner in Leitwarten. In Gottlieb Daimler- und Karl Benz-Stiftung (ed.), *2. Internationales Kolloquium Leitwarten*, (Verlag TÜV Rheinland, Köln) 55-67.

Oesterreich, R. & Volpert, W. 1986, Task analysis for work design on the basis of action regulation theory, *Ergonomics and Industrial Democracy*, **7**, 503-527.

Older, M.T., Waterson, P.E., Clegg, C.W. & Payne, R.L. 1995, *Task allocation: A critical assessment of task allocation methods and their applicability*, Internal Report, Institute of Work Psychology, University of Sheffield.

Price, H.E. 1985, The allocation of functions in systems, Human Factors, **27**, 33-45.

Rimann, M & Udris, I. in press, Subjektive Arbeitsanalyse: Der Fragebogen SALSA. In O. Strohm & E. Ulich (eds.), *Ganzheitliche Betriebsanalyse unter Berücksichtigung von Mensch, Technik und Organisation. Vorgehen und Methoden einer Mehr-Ebenen-Analyse*, (Verlag der Fachvereine, Zürich; Teubner, Stuttgart).

Thompson, S.C. 1981, Will it hurt less if I can control it? A complex answer to a simple question, *Psychological Bulletin*, **90**, 89-101.

Ulich, E. 1994, *Arbeitspsychologie*, 3rd edn, (Verlag der Fachvereine, Zürich; Poeschel, Stuttgart).

Wäfler, T., Weik, S. & Grote, G. 1995, *Complementary design of an automated metal sheet bending cell*. Paper presented at the 7th European Congress on Work and Organizational Psychology, Györ (Hungary), April 1995.

Weick, K.E. 1976, Educational organizations as loosely coupled systems, *Administrative Science Quarterly*, **21**, 1-19.

TOWARDS A NEW METHOD OF TASK ALLOCATION: PRELIMINARY FINDINGS

Melanie T. Older; Chris W. Clegg; Patrick E. Waterson

Institute of Work Psychology,

Sheffield University,

Sheffield, S10 2TN, UK.

It has become increasingly evident that existing task allocation methods are no longer applicable to the design of complex systems. A recent review and systematic comparison of methods (Older, Waterson and Clegg, 1995) identified fifteen requirements of a method for allocating tasks. This paper reports: the revision of these requirements, through discussion with potential users; the findings from a preliminary assessment of a task allocation method presently being developed to meet these requirements; and the identification of additional requirements resulting from the method's use. The distinguishing features of the method will be its attempt to combine the literatures from the currently disparate areas of task allocation and job design; and the way in which it addresses both dynamic allocation and the allocation of tasks amongst humans.

Introduction

Task allocation traditionally refers to the distribution of tasks between humans and machines within the work organisation. With developing technology, it is becoming increasingly difficult to allocate tasks solely on the basis of limited criteria such as cost, relative performance and workload. In an earlier study (Older, Waterson and Clegg, 1995), we reviewed a range of methods, from both the literature and those used in practice, and critically assessed their applicability. We also consulted a variety of potential users of a new method, to determine their needs in the task allocation domain. From this work we produced a statement of requirements for a task allocation method applicable to future complex systems. Although some methods have been developed in an attempt to cope with advancing technology, few (if any) of these are actually adopted in practice. Our intention, therefore, is to develop a more usable method of task allocation, applicable to future systems. We have attempted to achieve this by involving potential users of the new method in its development, thereby enabling us to tailor the method to their needs.

We have used the Naval domain as an example of an environment for which a task allocation method is required. The Naval domain is particularly appropriate due to its complex nature, and may be described as: dynamic, high risk, uncertain, time-constrained, and of variable workload.

Requirements of a task allocation method

The initial statement of requirements produced in our earlier study (Older et al., 1995) was revised by interviewing a range of practitioners involved in various stages of the Naval system development process. Five human factors practitioners from different organisations were contacted by mail and telephone; and two Naval personnel, and six human factors practitioners within the Ministry of Defence, were interviewed. We presented the initial statement of requirements and requested feedback concerning the issues identified. In particular, we were concerned with: issues regarded as especially important; relevant issues that had been omitted; and issues that were not considered relevant to Naval system development.

We also consulted five Naval officers to determine any aspects of the Naval environment which might be relevant to the allocation process, such as factors which might affect dynamic allocation or the allocation of tasks between human roles. Finally, we visited a group of researchers at the Swiss Federal Institute of Technology (ETH), in Zurich, who have been developing a task allocation method as part of the KOMPASS (Complementary Analysis and Design of Production Tasks in Sociotechnical Systems) project. From this visit, we gained a better recognition of the issues important to the development of a task allocation method.

As a result of this work, we produced a revised statement of requirements for a task allocation method, as shown in Table 1.

Development of a new task allocation method

Using the requirements identified, we developed a new prototype task allocation method, applicable to the Naval domain. The structure of the method is based on the recommendations resulting from our review of the task allocation methods available in the literature, and in practice (Older et al., 1995). It consists of: a decision-aid, in the form of an iterative process (organised in a series of stages); categories of criteria, on which to base the allocation decision; and a recording device, for noting the decisions made and their rationale. It addresses both the static and dynamic allocation of tasks: between humans; between different human roles; and between humans and machines.

Table 1. Revised table of requirements for a task allocation method.

A method for allocating tasks should:
cover allocations to the human, to the machine, and to a combination of human and machine
cover allocations between humans, and examine different human roles
incorporate the concept of dynamic allocations dependent on real-time contingencies
examine the content and quality of the human's job
specify decision criteria
consider the trade-offs between the decision criteria
encourage participative use by various 'stakeholders', including the potential end-users of the system
enable the users of the method to make informed choices
enable quantitative evaluations to be made of the alternative choices
examine the whole system, as well as individual tasks and roles
be usable early in the design process
allow iterative use, throughout the design process
have a structured and systematic format
be easy to learn and usable, and require minimal training and support
be cost-effective and efficient to use
be consistent with existing methods and techniques in use (e.g. it should fit into the system development cycle, and should form the input for established system development processes)
be applicable to: environments with similar characteristics to Naval environments (i.e. high risk, composite task, variable workload, uncertain, dynamic, time-constrained); and, different systems within the same environment
be adaptable to different situations and tailorable for unique application
be capable of use in new and existing systems
cover the rationale for its use

The prototype method

The method is intended for use within a group, with the aim of promoting discussion. The group's members should come from a variety of disciplines, from within the development of the system and from the end-use of the system. Examples include: system designers, managers and purchasers; human factors experts; and direct users of the resultant system. The method guides the group through the allocation process by outlining the issues that should be identified and discussed at each stage.

The first three stages form the input to the decision-making process of the method. Here the group is required to: identify the objectives of the system, and form an overall view of how it is expected to operate; specify the system requirements that will enable these objectives to be met; and identify the tasks that are to be performed within the system.

The decision-making process itself also consists of three stages. Categories of criteria on which to base the allocation decision are specified at each stage. For every task to be allocated, the group is required to identify the criteria which may indicate whether the task is better allocated to human or machine, and if human, which human. The group members generate questions based on the criteria specified (and any other criteria which they believe are relevant). Examples are: Technological issues - for example, technical feasibility, 'is it technically feasible to automate this task?'; Internal organisational issues - for example, political issues, 'are there any internal political reasons as to whether this task should be allocated to the machine or to a particular individual?'. The recording device is used to record: the relevant criteria; any allocation decisions indicated by the questions generated; the reasons for such decisions; and other details considered to be important, such as the trade-offs between criteria.

After each task has been allocated, individual allocations are re-examined to ensure that the system objectives and requirements identified earlier have been met. Once all the tasks have been allocated, a more global re-examination is performed to ensure that the overall roles for the machine and human operators are feasible. The output of the method is in the form of a statement of the proposed allocations, and the rationale behind them; and a description of the roles for the machine and for each human within the system.

The method may be used when developing new systems, or may be applied when redesigning existing systems. The modular nature of the method enables parts of it to be applied to specific issues, for example when considering the reallocation of tasks between human roles, without the need to work through the whole method.

Preliminary evaluation

The prototype method was first evaluated during a one day workshop held at Portsmouth, in the UK. The participants of the workshop included: the five human factors practitioners, contacted regarding the statement of requirements; two former Naval officers; two practitioners from the Centre of Human Sciences (Defence Research Agency); and ourselves (three researchers from the Institute of Work Psychology). The workshop involved piloting the method using a set of design problems, identified for the purpose of the workshop by one of the former Naval officers, based on a hierarchy of Naval tasks. The workshop participants were guided through the use of the method, as if they formed the task-allocation design team. The team therefore consisted of: five human factors experts in Naval system development; two experts in Naval system operations; two experts in the task allocation domain; and three experts in the use of the method, respectively.

Preliminary findings

Preliminary evaluation of the prototype method identified a number of issues related to its use.

Benefits:

- the method enables better clarification of the requirements of the system;

- it reduces the technology-drive of system development;

- it provides a structure for contested issues;

- it provides a means for recording the rationale behind the decisions made;

- it has the potential to bridge the gap between system users and designers during the system development process;

- it can be tailored to fit into the existing Naval system development process, due to its lack of domain specificity.

Limitations:

- the terms used by the method need to be better defined to avoid confusion;

- the allocation of the constituent parts of tasks needs to be better addressed;

- the method should differentiate serial tasks from those performed in parallel;

- the order of the categories of decision criteria needs to be more flexible so that they can be rearranged as appropriate, and can aid the identification of the key criteria, relevant to the situation;

- further development is required concerning: the trade-offs between the decision criteria and between the alternative allocations; dynamic allocation; and the allocation between human roles;

- operator role configurations need to be identified for the available allocations, for example when the same task is allocated between humans, or between human and machine;

- the recording device needs to be adapted to capture more of the debate during the decision-making process;

- a facility for identifying the system's key tasks needs to be developed.

Issues for future consideration:

- further examination of the Naval system development process is required to determine the stages at which the method can be applied, and to determine other decision criteria relevant to the allocation process;

- the time- and cost-effectiveness of the method's use must be addressed and optimised;

- more details are required of the way in which the method is to be used, and how the resulting documentation is to be maintained.

Recommendations

- Further development of the method is required to meet the needs referred to above.

- The amended method will then need to be evaluated, for example comparing use versus non-use.

- Further development and evaluation will follow, with particular attention paid to the method's individual components.

Acknowledgements

The research described in this paper was funded by the Defence Research Agency, Portsdown, UK. The views expressed in the paper are those of the authors and should not necessarily be attributed to the Ministry of Defence.

References

Older, M.T., Waterson, P.E. and Clegg, C.W. 1995, A critical assessment of task allocation methods and their applicability *(Submitted for publication)*.

Drivers and Driving

WHITHER RTI, WITHER ROAD TRAFFIC SIGNS?

Alastair G Gale

Applied Vision Research Unit
University of Derby
Derby DE3 5GX
UK

Road traffic signs abound on our road systems. These range from the simple route deviation chevron markings to complex variable message signs. In progressing along a chosen route, drivers regularly update their knowledge of current route position by attending, and responding appropriately, to information in these signs as compared with some cognitive representation of the predetermined route. Somewhat similarly signs indicative of road conditions must be processed and appropriate driving action taken. With the implementation of in-vehicle navigational aids the function of these signs changes. It is proposed that future road systems may not need signs and the implications of this are considered.

Introduction

Currently all roads have a system of road traffic signs which function as primary navigational aids (route indicators) as well as advisory and mandatory devices to which the driver can decide whether to respond or not. Such signs play a key role in traffic safety as they inform drivers in advance of road conditions. Drivers voluntarily can decide whether to respond appropriately or not to particular sign information.

Whilst it may not be too important if the driver elects not to take a particular turnout but to progress to another instead, thus changing their conceptual route model, failing to respond appropriately to a mandatory sign results in the violation of existing traffic safety codes. For example it is blatantly foolish to ignore a 'stop' sign whereas many drivers will pass a speed restriction sign when entering a reduced speed zone and elect not to lower their speed appropriately.

A route can be conceptualised as a serial decision tree with a number of nodes at each of which the driver must make a route decision. Currently such decisions may be made on the basis of environmental cues (Heft, 1979; Schraagen, 1993) such as prominent landmarks such as: public houses, garages, hills, churches etc., and route signs.

Signs are designed according to current standards (e.g. traffic symbols conforming to B.S. 873) which prevail at the time. For instance, some 20 years ago the Department of the Environment (1975) recommended clear visibility distance for signs was 180 m which, as Gregg and Ingall (1986) point out, corresponds to some 6s for the driver to perceive it, assuming a vehicle travelling at the maximum design speed. Elsewhere a minimum driver preview time for signs of 3s has been described (CIE,1988).

Despite good design, signs can still be obscured, for instance by disability glare and sun phantom effects (caused by sunlight reflection), vegetation or other road users (eg HGVs) which affects such effective preview time. Such design and the placement criteria of signage emphasise the importance of visual factors such as sign conspicuity.

Drivers and Road Traffic Signs

Approaches to model driver recognition of signs typically emphasise the importance of the visual capabilities of the driver. For instance Zwahlen and Schnell (in press) stress the importance of the visual search behaviour of the driver: as a driver approaches a sign on a clear road various stages of sign information acquisition can be categorised. Initially the sign is invisible then it becomes visible as a small object growing rapidly in size. Subsequently sign colour, shape and symbol/text are all gradually recognised as the visual angle subtended by the sign at the driver's eye increases. Typically then a driver must fixate a sign so as to interpret it efficiently. Although as Avant et al. (1986) earlier demonstrated, some information present in signs can be unconsciously processed, which may not be dependant upon the driver actually attending to the sign.

Road traffic signs are important traffic safety devices but only if drivers respond appropriately to them (Thieman and Avant, 1993). To do so the driver has first to detect, recognise the sign and also appreciate it's information appropriately. Thus the driver has to visually search for signs as a necessary part of the overall driving task.

RTI and Signs

Current research in road transport informatics (RTI) are leading to the increasing implementation of various driver aids. This raises the question of how such devices affect driver behaviour vis a vis route navigation and existing road traffic signs. A driver progressing safely along a pre-determined route with different navigational information devices can be conceptualised as shown in Table 1.

With no informatics aid the driver must search constantly for sign information and compare this to some internal cognitive representation of where he/she is in relation to the destination.

Using either a map or a textual route listing then the driver must repeatedly seek route information from the environment and compare it to the paper source. Even with minimal route information there are hazards in constantly attending to such in-vehicle paper information and comparing it to external environmental signs whilst driving. With a co-driver such tasks can be delegated and essentially the driver need not necessarily search for route information per se.

A simple in-vehicle informatics device simply replaces the paper based approaches with a small computer display informing the driver of how to respond to upcoming route information. With route navigation, when the decision point is reached the driver manually updates the device which then gives brief information of how to reach the next decision point along the route. Conceptually such devices could be intelligent and recognise their route position via electronic data from the signs (or other beacons) themselves. This relieves the driver of the need to manually update. Actual recognition of the route sign itself and responding to its information may then become a secondary task.

Table 1. Informatics aids and driver visual search.

Information Source	Traffic Sign Importance	Driver sign search?
None	High	Yes
Text descriptors	High	Yes
Map	High	Yes
Co-driver	High	No
Simple system	High	Yes
Intelligent system	Medium	No
GPS	Low	No

Informatics systems which utilise satellite information (GPS) to give fairly precise position in essence have no need for route signs. Such systems would ostensibly know the vehicle position at all times. For instance whether a driver had turned off at the correct, informatics-prompted intersection or had missed it and was still proceeding along the original route and thus required updated information on how to attain the destination efficiently would be handled by the on-board system.

Future Signage

As increasingly sophisticated RTI devices are widely installed in the transportation base are road signs (of any kind) actually required?

If each road user, motorist, motor cyclist, cyclist etc. progressed on every journey with an aid then all upcoming route and road information could be provided adequately by the mobile system. The advantage of such an approach is that it could be responsive interactively to traffic and weather conditions as well as individual factors pertinent to the particular selected route of each driver. It would remove the need for drivers to reliably detect and respond to traffic sign information, thus removing one potential source of driver error. Signs could function as secondary devices to reinforce the RTI system as opposed to primary information systems.

If RTI systems are widely shown to provide errorless information or have a tolerably low error rate, acceptable to a wide range of drivers of varying ability, then signs themselves may become redundant. A future road system may then be a bare place with no route or road condition information along its length. Such information could be provided by the in-vehicle informatics system itself. There are potential cost savings in not having signs, in terms of sign construction and maintenance. Additionally some advantages accrue for drivers in reducing the need to constantly scan the environment for possible signs ahead. Whether such a future would be desirable in road safety terms or would simply further decrease the attention requirements of driving, already a task which tends not to require full attentional capacity of a driver, so decreasing road safety is a matter for conjecture.

References

Avant, L.L., Thieman, A.A., Brewer, K.A. and Woodman, W.F., 1986, On the earliest perceptual operations of detecting and recognising traffic signs, In A.G. Gale, M.H. Freeman, C.M. Haslegrave, P.Smith and S.P. Taylor (Eds.) *Vision in Vehicles*, (North Holland, Amsterdam).

CIE, 1988, *Visual aspects of road markings*, Joint technical report CIE/PIARC No. 73.

Department of the Environment Circular, 1975. *Roads 7/75, Size, design and mounting of traffic signs* (DoE/HMSO, London).

Gregg, D.J. and Ingall, C.T., 1986, Computer aided simulation of the concealment of motorway direction signs by large vehicles. In A.G. Gale, M.H. Freeman, C.M. Haslegrave, P.Smith and S.P. Taylor (Eds.) *Vision in Vehicles*, (North Holland, Amsterdam).

Heft, H.,1979, *The role of environmental features in route-learning: two exploratory studies of way-finding.* Environmental Psychology and Nonverbal Behaviour,3,172-185.

Schraagen, J.M.C., 1993, Information presentation in in-car navigation systems. In A.M. Parkes and S Franzen (Eds.) *Driving Future Vehicles,*(Taylor and Francis, London).

Thieman, A.A. and Avant, L.L., 1993, Traffic sign meaning: designer intent vs. user perception. In A.G. Gale, I.D. Brown, C.M. Haslegrave, H.W. Kruysse, and S.P. Taylor (Eds.) *Vision in Vehicles IV*, (North Holland, Amsterdam).

Zwahlen, H.T. and Schnell, T, *Driver eye scanning behaviour when reading symbolic warning signs.*

NEW CAR DISPLAYS AND DRIVER INTERACTION ASSESSMENT: A CRITIQUE

Alastair G Gale

Applied Vision Research Unit
University of Derby
Derby DE3 5GX
UK

Road transport informatics enable pertinent vehicle and road information to be presented to the driver both effectively and efficiently. To ensure that these technologies do this in an appropriate manner which does not overload the driver, either cognitively or perceptually, such devices have to be appropriately assessed. Typically part of such assessment examines the visual attentional capabilities of the driver, by measuring visual search behaviour, when driving with these systems as compared to equivalent situations without them. The rationale for this is reviewed, different eye movement recording techniques discussed and this approach assessed.

Introduction

Road transport informatics (RTI) encompass various information systems which are designed to interact with the driver and the driving task by presenting to the driver relevant information and so aid efficient progress along a particular route. Such route guidance systems of necessity must support the existing driving task itself (Sviden, 1993). It is important that these systems are designed appropriately so that they convey pertinent information to the driver in an effective and safe manner. RTI systems, ideally, would vary their information in response to the needs of the driver so that both normative characteristics as well as individual characteristics are catered for (Onken, 1993). For instance a novice driver may require a different amount of information and support (Groeger, 1993) as compared to a highly experienced one. Other factors are also important, such as the driver's age.

Driver interaction with these systems can be by several means. Driver information input can be by keyboard/keypad, touch screen or voice. Driver feedback is typically visual, auditory or possibly tactile. For instance a simple RTI head up display (HUD) may contain a light whose colour changes depending upon conditions (e.g. current speed is safe) and will indicate that the driver should stop when information indicates danger ahead. Additionally such head up indicators can display in which direction to turn next (Sviden, 1993). Other approaches feature a dashboard mounted route navigation system such as a changing map display (Wierwille et al., 1988) or can incorporate monitoring of the driver's alertness (Richardson, 1995).

RTI devices must not be distracting whilst at the same time they need to be positioned appropriately so that the driver can glean the necessary information as quickly as possible. There is a fundamental difference in the amount of time drivers have for accessing such information when they are in slow moving or stationary traffic and wish to find a way out of it or when they are travelling at speed on an unfamiliar autobahn. Thus route navigation instructions need to be kept simple (Schraagen, 1993).

RTI displays can be dashboard mounted or HUD. HUDs offer the benefit of drivers not having to refocus their eyes as the information is presented optically distant. Additionally there is no need to re-direct the line of sight (Swift & Freeman, 1986). The major advantage of HUDs over dashboard displays comes with there being no accommodation or relaxation time involved in changing from looking at the environment or display. Such times increase with age, only one of many individual driver parameters which must be considered (c.f. Taylor, 1995), and Swift and freeman (1986) recommend an IHLID display (Instrument Head Level Infinity Display). In Hughes and Cole's (1988) study of drivers in a simulation situation some 25% of fixations were located centrally around the focus of expansion of the road with 80% of the remaining fixations occurring within 6^0 of this point, possibly supporting the relevance of a HUD approach. Elsewhere Hella (1987) has shown that increasing the eccentricity of the placement of dashboard displays affects the amount of time to access them and so reduces the time available for looking at the road ahead.

Assessing driver interaction

In introducing such devices to aid the driver this very act increases the amount of visual time sharing with the existing driving task. Careful evaluation of such introduction into the vehicle is necessary (Snyder and Monty, 1986). Such methods can include measuring: the task performance; cognitive workload (Harms, 1986) for instance by using the NASA TLX (Jordan and Johnson, 1993); driving performance (e.g. lane keeping skill or speed control whilst accessing the device) and recording eye movements as attentional measures on approach which is concentrated upon here.

Visual Attention

The reason for monitoring eye movements as a measure of attention is that saccadic eye movements are the most massive instrument of selective visual attention. Road and instrumentation information is not passively acquired, but is actively searched for. It is generally regarded that where a person is measured to be looking is actually the location to which they are attending. Unfortunately this is not always the case which is an important limitation to the approach. For instance when a driver suddenly finds that the vehicle in front, which they have been safely following for some time, has in fact now stopped.

Such events occur as normally driving may not require full attention (Brown, 1965) and consequently the driver typically has some spare visual capacity. For instance, Rockwell (1972) found that drivers on low density roads often looked at objects of little relevance to the driving task. Rockwell (1988) later reported data from a six year period concentrating upon mean number of glances and average glance duration as measures evidencing that glance duration are a fairly consistent measure of a driver's in-car performance. Poor display or control design results in more glances but not longer glances. Glance frequency is age and sex dependent, task dependent and related to driver experience. Wierwille et al. (1988) evaluated an in-car navigation display system and found that single glance times were longer than those associated with most other tasks indicating that drivers found that the particular display assessed was complex.

The amount of time a driver can assign to monitoring any informatics device is related to; the traffic situation, individual factors, and display location (Hella, 1987). Their use is not always of overall benefit to the driver, for instance Zwalen et al. (1988) in investigating the introduction of CRT touch panels into cars found that their use produced significant vehicle lateral deviations. A driver may need to execute a whole series of separate glances at the system (alternating with glances outside the vehicle) in order to fully acquire the necessary information.

Eye Movements In Driving

During our everyday waking hours we exhibit a range of different eye movements depending upon the circumstances. Saccades are the most obvious example but in addition there are; pursuit, smooth, vergence, nystagmus and miniature eye movements.

In general the driver can be considered as executing a series of largely voluntary saccadic eye movements, both within, as well as out of the vehicle. These saccades alternate with fixations on particular objects in the scene. Such movements serve to bring areas of interest to fall upon the foveal, high resolution, part of the retina for detailed analysis. An individual fixation can be of varying time with a median of the order of some 250ms. It is often empirically convenient to clump together groups of close fixations into a 'dwell' and speak of 'dwell time'. Somewhat simplistically the driver is often characterised as extracting detailed visual information from the locus of the fixation whilst at the same time processing more general information from peripheral retinal input about where to fixate next. Saccades are therefore made by the driver both as a result of some cognitive search plan for particular information and in response to local and more global visual information input (c.f. Gale, 1993).

As well as saccadic movements, in looking within and outside the vehicle the driver's eyes will need to converge or diverge so as to bring instrumentation or the distant road into sharp focus. Drivers will also move their heads about to extend their field of view and may also move towards or away from the instrument panel many times during a typical journey. Such movements will generate compensatory eye movements. As static objects in the environment pass the moving vehicle then these may give rise to some nystagmus type of movements.

Monitoring Driver's Eye Movements

When monitoring the driver's eye movements it is usually not the movements per se that are of interest but parameters such as:-
- The locus of the point of regard (where the driver is fixating). The area around the point of fixation to which the driver can usefully attend is termed the useful field of vision or functional visual field. Thus a driver may not need to precisely fixate on a display in order to interpret it as long as it fell within this field. The size of this functional field may decrease as task load increases, thus more complex displays may require more precise fixation.
- Fixation or dwell times. As associated with looking at particular parts of the scene or dashboard devices.
- Mean glance dwell time. As the need to acquire information in a given time period increases then the mean gaze duration decreases.
- Sequence of saccadic movements.
- The number of glances.

To record such data the major requirements in driving are to utilise a recording technique which will:-
- allow normal head movements of the driver
- have a recording accuracy of about 1^0 or better
- not be too obtrusive
- be safe for a driver to wear (if head mounted) if used on a road as opposed to in a simulator
- be comfortable for a driver to use

Methods often applied in driving or driving simulators are based on the corneal reflection or oculometer approach (e.g. NAC or ASL). Alternatively a dashboard mounted camera is used to monitor approximate relative eye position, although this can entail considerable data analysis time and may be fairly inaccurate. Less commonly EOG has also been employed.

One problem with any systems is that they have to cope with drivers fixating on dashboard instruments and also on very distant objects, so parallax errors of movement can occur. Without appropriate calibration fairly erroneous data can result. The head mounted oculometer has the advantage of using a parallax-free approach to help overcome this. Another key problem for any eye movement method is that of potential vibration from the vehicle itself which may affect the recording process. Our own approach is to use both a head mounted and a remote oculometer as appropriate for the particular situation

Before conducting a study of drivers' visual search behaviour it is important to determine the resultant measures required and thus to ensure that the selected approach is adequate.

Conclusions

Assessing drivers' visual attention allocation by eye movement recording has a long history. The approach has been used for examining the amount of attention paid to a variety of in-car devices. The method is thorough and is useful for assessing RTI systems. Its limitations, however, must always be borne in mind and the data interpreted appropriately. The transportation researcher must fully understand the particular system which they are using, appropriate eye movement methods be employed, and the resultant data must be interpreted with caution. Otherwise the market need to introduce such new technologies will over-ride the scientific objective assessment of such devices.

References

Brown, I.D. Effect of a car radio on driving in traffic, *Ergonomics*, **8**, pp 475-479.
Gale, A. G. 1993, Human response to visual stimuli. In W. R Hendee. and P.N.T. Wells (eds.), *The Perception of Visual Information*, (Springer-Verlag, New York).
Groeger, J. 1993, Degrees of freedom and the limits of learning: support needs of inexperienced drivers. In A. M. Parkes and S. Franzen (eds.), *Driving Future Vehicles*, (Taylor and Francis, London).
Harms, L. 1986, Drivers' attentional responses to environmental variations: a dual task real life traffic study. In A. G. Gale, M. H. Freeman, C. M. Haslegrave, P. Smith and S.P. Taylor (eds.), *Vision in Vehicles*, (North Holland, Amsterdam).
Hella, F. 1987, Is the analysis of eye movement recording a sufficient criterion for evaluating automobile instrument panel design? In J.K. O'Regan and A. Levy-Schoen (eds.), *Eye Movements - from Physiology to Cognition*, (North Holland, Amsterdam).
Hughes, P.K., and Cole, B.L. 1988, The effect of attentional demand on eye movement behaviour when driving. In A.G. Gale, M.H. Freeman, C.M. Haslegrave, P. Smith & S.P. Taylor (eds.), *Vision in Vehicles II*, (North Holland, Amsterdam).
Jordan, P.W. & Johnson, G.I. 1993, Exploring mental workload via TLX: the case of operating a car stereo whilst driving. In A.G. Gale, I.D. Brown, C.M. Haslegrave, P. Smith & S.P. Taylor (eds.), *Vision in Vehicles III*, (North Holland, Amsterdam).
Onken, R. 1993, What should the vehicle know about the driver. In A.M. Parkes & S. Franzen (eds.), *Driving Future Vehicles*, (Taylor and Francis, London).
Richardson, J. H. 1995, The development of a driver alertness monitoring system. In L. Hartley (ed.), *Fatigue and Driving*, (Taylor and Francis, London).
Rockwell, T.H. 1972, Eye movement analysis of visual information acquisition in driving an overview', *Proceedings of the 6th Conference of the Australian Road Research Board*, 6.
Rockwell, T.H. 1988, Spare visual capacity in driving - revisited. In A.G. Gale, M.H. Freeman, C.M. Haslegrave, P. Smith & S.P. Taylor (eds.), *Vision in Vehicles II*, (North Holland, Amsterdam).
Schraagen, J.M.C. 1993, Information presentation in in-car navigation systems. In A.M. Parkes and S. Franzen (eds.), *Driving Future Vehicles*, (Taylor and Francis, London).

Snyder, H.L. and Monty, R.W. 1986, A methodology for road evaluation of automobile displays. In A.G. Gale, M.H. Freeman, C.M. Haslegrave, P. Smith and S.P. Taylor (eds.), *Vision in Vehicles*, (North Holland, Amsterdam).

Sviden, O. 1993, MMI scenarios for the future road service informatics. In A.M. Parkes and S. Franzen (eds.), *Driving Future Vehicles*, (Taylor and Francis, London).

Swift, D.W. and M.H. Freeman, 1986, The application of head-up displays to cars. In A.G. Gale, M.H. Freeman, C.M. Haslegrave, P. Smith and S.P. Taylor (eds.), *Vision in Vehicles*, (North Holland, Amsterdam).

Taylor, J. F. 1995, Medical aspects of fitness to drive, Medical Commission on Accident Prevention, London.

Wierwille, W. W., Antin, J. F., Dingus, T. A. and Hulse, M. C. 1988, Visual attentional demand of an in-car navigation display system. In A.G. Gale, M.H. Freeman, C.M. Haslegrave, P. Smith, and S.P. Taylor (eds.), *Vision in Vehicles II*, (North Holland, Amsterdam).

Young, L .R., and Sheena, D. 1975, Survey of eye movement recording methods, Behaviour Research Methods and Instrumentation, 7, 397-429.

Zwalen, H.T., Adams, C.C. and DeBald D.P. 1988, Safety aspects of CRT touch panel controls in automobiles. In A.G. Gale, M.H. Freeman,   C.M. Haslegrave, P. Smith and S.P. Taylor (eds.), *Vision in Vehicles II*, (North Holland, Amsterdam).

NORMAL TRAFFIC FLOW USAGE OF PURPOSE BUILT OVERTAKING LANES: A TECHNIQUE FOR ASSESSING NEED FOR HIGHWAY FOUR-LANING

Tay Wilson

Psychology Department
Laurentian University
Ramsey Lake Road
Sudbury, Ontario, Canada
P3E 2C6
tel (705) 675-1151
fax (705) 675-4823

In difficult economic circumstances, it is undesirable that decisions regarding major highway upgrading depend, in great measure, upon organized political lobbying based upon weakly grounded complaints of a some would be fast drivers and emotional response to a few well-publicized accidents. Herein, in connection with effecting a more even-handed assessment of need for four-laning a highway equipped with purpose built passing lanes, is described and applied a method of assessing aspects of the actual driving experience during normal conditions. Three conclusions resulted. First, drivers tend to wait for and use purpose built passing lanes. Second, overtaking experience in either clear-out after collision, or "normal" conditions does not indicate a strong need for four laning highway 69 north of Parry Sound. Third, although evidence here does not support the notion of increased relative frequency of dangerous overtakings when traffic is delayed by road closure compared with "normal" conditions, there is evidence that overall risk taking by traffic, assessed by estimated traffic speed, is relatively increased in the former situation.

In deciding whether to carry out major upgrading of a highway, even, in difficult economic circumstances, when social programme money must be allocated particularly wisely, there is great danger that decisions will be based in too great a measure upon organized political lobbying based upon complaints of some drivers who do not accurately interpret their experiences and to emotional response to a few well-publicized accidents.

One particularly dangerous aspect of highway driving involves overtaking. Although overtaking generates a considerable number of serious collisions, it is difficult for ordinary drivers to assess the actual overall frequencies of various types of safe and unsafe overtaking manoeuvres on highways as compared with those personally experienced, in part due to the rarity of events, and so to adjust their own driving practices if indicated. In consequence, some drivers who, unknown to themselves, overtake at the relatively infrequent and risky end of the driving behaviour distribution (some young drivers fall into this category) may interpret their experiences as indication of unacceptable danger in a particular road for the ordinary driver and press for scarce monies to be spent on major upgrading. In some cases, accidents resulting, in part, from such behaviour further fuel such pressure. In order to assist even-handed decision making, in such cases, there is need for more empirical knowledge about what drivers are actually doing and experiencing in overtaking. Some studies of actual overtaking in Briton have been carried out e.g., Postans and Wilson (1983), Wilson, Postans, and Garrod (1983), and Wilson and Best (1982). In Canada, Wilson and Neff (1995) report on overtaking experiences during traffic clearout after a traffic mishap has closed a highway.

At present, there is considerable pressure, based, in part, upon putative experience of overtaking danger and experience, to four lane the major Toronto-Sudbury trunk highway 69 north from Parry Sound. To assist decision making, a clear need was seen for assessment of actual overtaking experience during normal conditions on this stretch of road. Such observation further provided an opportunity to explore the experiential usage of, normal condition, purpose built overtaking lanes as a less costly expedient to four laning relatively low traffic highways.

Method

In order to provide the "missing" data in the form of a quantitative behavioural assessment of the overtaking experience of individuals who are driving "normally" an in-car observation technique was developed. The experimenter entered the traffic stream on Highway 69, Ontario, Canada, at the northern edge of Parry Sound, drove at the posted speed limit (90 km/h) and recorded details of all instances of overtaking and being overtaken for the 139 kilometre section ending at the outskirts of Sudbury. This section of highway was punctuated by the addition of 12 specially designed passing lanes each of which widened the north-bound direction by one full lane of the road. The approximate length of these passing lanes was as follows: eight at 2.5 km, two at 3 km, one at 1.5 km, and one at 0.5 km. Account was taken of whether overtaking occurred within or outside of these designed passing areas. Finally overtakings judged to be unusual, potentially dangerous or illegal were noted. Observation of overtakings began at 3:37 p.m. on November 26 in clear weather conditions.

Results

The experimenter took 92 minutes and averaged 90.6 km/h for the 139 km trip from the roll-over site to the outskirts of the Sudbury. During that time, he was overtaken thirteen times. A summary of the 13 observed overtakings by presence or absence of a purpose built passing lane by distance from Parry Sound appear in Table 1. In column one is recorded the distance in kilometres travelled from the Parry Sound outskirts. In column two, the road section is identified as either a designated passing lane (yes) or not (no). In column three appears the length in kms of each designated passing lane site. In column four appears the number of vehicles overtaking the experimenter at that juncture. In column five appear coded comments regarding attributes of the juncture or unusual events. In particular there were two places in which speed was reduced to 75 and 80 km/h respectively and one location in which an oncoming car crossed a double yellow line (into the experimenters lane) to overtake a vehicle. "SUDBURY" means the southern outskirts of the city of Sudbury at Richard Lake where overtaking observations were terminated. The experimenter overtook one vehicle, a truck near the 75 km/h speed area.

There was no significant difference in the distribution of total overtakings or general location of overtakings (passing lane versus non passing lane) over sections of the highway. Binomial probability comparison of the absolute frequency of passing lane overtakings to non passing lane overtakings (9/4) yielded a borderline significant probability of $p = 0.09$. However, when the number overtakings in passing lanes was pro-rated by the ratio of the total distance not occupied by passing lanes and that occupied by passing lanes (107.5/29.5 km), keeping the total number of overtakings constant, there was seen to be a significantly greater number of overtakings in the passing lane (12:1) ($X^2 = 9.3$, df = 1, $p < 0.01$). One overtaking occurred involving a vehicle illegally crossing a double yellow line; however it involved an oncoming vehicle. The experimenter had only to slow down slightly to allow the vehicle to complete its manoeuvre. Traffic flow was counted at 300 vehicles per hour, just before the experimenter entered the traffic stream.

Discussion

First, consider average traffic speed. It is folk wisdom that, although the speed limit was 90 km/h, speeding tickets would not be issued below 100 km/h and thus, most traffic drives near 100 km/h. In the approximately 1.5 hour trip to Sudbury, about 15 overtakings (13 were observed in this study) of a vehicle travelling 90/91km/h would be expected to be generated by 300 relatively evenly spaced vehicles per hour travelling at an average trip speed of 93/94 km/h. By this reckoning the actual average trip speed by vehicles in normal clear conditions (93/94 km/h) is rather slower than folk wisdom and some fast drivers might believe. This information could provide the basis of fact based driver improvement programs (see Wilson, 1991). It might be pointed out that, in non-"normal" conditions - namely in the clear out phase after a overturned truck blocked the very highway under study here, a similar analysis (Wilson & Neff, 1995) led to estimations of average traffic speed of

100 km/h. The difference between these two numbers provides some sort of estimate of delay induced driving stress on overall traffic speed on relatively uncongested roads, a figure that does not often come to hand.

Second, consider overtaking practises. Relative to their proportional length, purpose built passing lanes are well used on this highway. Overtakings on purpose built passing lanes were about double those elsewhere although passing lanes occupied only about 1/4 of the total highway mileage. Moreover, that there was little significant traffic delay on this highway is indicated; first, by the occurrence of passing lanes, on average, every ten kilometres which are signed two kilometres in advance; and second, by the finding that on only one passing lane was the experimenter overtaken by more than two vehicles. Moreover, that particular passing lane occurred shortly after an 80 km/h speed zone (not everyone drives the speed limit through this zone). Finally, in 139 km of driving at the speed limit there was experienced only one illegal double yellow line overtaking. This number is also much lower than local folk wisdom would have one believe.

Finally, compare overtaking during time delay stress induced conditions caused by traffic mishap induced road closure (Wilson & Neff, 1995) with overtaking during normal conditions on this road. Consider first overtakings rated as illegal/dangerous. In the clear-out study 3/34 overtakings were so rated and they occurred in the first half of the trip (nearest to traffic mishap) whereas, in the present study, none (0/13) were observed in northbound traffic, although one was observed involving oncoming traffic in the second half of the trip. Since this difference was not significant, it cannot be concluded, from the available data, that road closure induced delays relatively increase the frequency of risky overtakings. Consider now, in somewhat more detail, overall relative risk as indicated by traffic driving speed in the two studies. In the present "normal" conditions study, it was calculated above that following traffic averaged 93/93 km/h. The same calculation in the clear-out study yielded an average of 100 km/h indicating the adoption of a higher adoption of risk by the traffic as a whole in this situation (e.g., about a 15% increase in the cohort total energy at the higher speed - $100^2/93^2$). In the clear-out study, the experimenter noted that about 2-3 minutes passed between police releases of same direction cohorts on the single lane road area, moreover about twenty cars were behind the experimenter when his cohort of about fifty vehicles was released at the collision site. In the event, 26% of total overtakings occurred in the first 10 km and 44% in the first 20 km of the 139 km trip thus supporting the notion of a relatively high following traffic speed. Moreover, a second peak of frequent overtakings 24% of total occurred 87 km from the collision site. Noting that some traffic, at least, in the experimenter's cohort would be driving at the speed limit forcing 100 km/h traffic to slow for part of the trip until suitable passing lanes appeared, the location of the observed second peak of overtaking is perfectly consistent with a cohort of vehicles wishing to travel 100 km/h but, because of minor congestion being forced to average 95 for the first forty km until the law abiding drivers of the cohort ahead were overtaken and then averaging 100 km/h until reaching the 87 km passing lane in which the experimenter was then located.

Conclusions

First, drivers tend to wait for purpose built passing lanes. Second, overtaking experience in neither clear-out after collision, nor "normal" conditions does not indicate a strong need for four laning highway 69 north of Parry Sound. Third, although evidence here does not support the notion of increased relative frequency of dangerous overtakings when traffic is delayed by road closure compared with "normal" conditions, there is evidence that overall risk taking by traffic, assessed by estimated traffic speed, is relatively increased in the former situation.

References

Postans, R. L. and Wilson, W. T. 1983. Close Following on the Motorway. *Ergonomics*, 26(4), 317-327.

Wilson, Tay, 1991. Locale Driving Assessment - A Neglected Base of Driver Improvement Interventions. In *Contemporary Ergonomics*, (Praeger, London), 388-393.

Wilson, Tay, and Best, W., 1982. Driving Strategies in Overtaking. *Accident analysis and Prevention*, 14(3), 179-185.

Wilson, Tay and Neff, Charlotte. 1995. Vehicle Overtaking in the Clear Out Phase After Overturned Lorry has Closed a Highway. In *Contemporary Ergonomics*, (Praeger, London), 299-303.

Wilson, Tay and Postans, Rod and Garrod, Glen. Lorry and Coach Overtaking on the Motorway. *Traffic Engineering and Control*, June/July 1983, 311-314.

Table 1. Overtakings within and outside designated passing lanes (pass lane) by length of passing lane (leng) and distance from Parry Sound outskirts on Highway 69, Ontario, Canada

kms	Pass lane (yes/no)	Length (km)	Number Overtaking	Comment
1	Y	2.5	2	
5	N		1	
10	Y	2.5	0	
19	Y	2.5	0	
24	N		0	75 km/h speed
30	Y	2.5	0	
42	Y	2.5	2	
53	N		1	
56	Y	2.5	1	
68	Y	3	0	
76	N		1	
87	Y	2.5	1	
96	Y	2.5	0	
107	Y	3.5	0	
120	N	1.5	0	80 km/h speed
122	Y	1.5	3	
127	N			1 oncoming car overtakes on double yellow line
135	Y	0.5	0	
138	Y			4-Lane starts
139				SUDBURY
Total	12	29.5	13	

NEW MEASURES OF DRIVING PERFORMANCE

John R. Bloomfield
Stewart A. Carroll

Iowa Driving Simulator,
Center for Computer-Aided Design,
208 Engineering Research Facility
University of Iowa
Iowa City, 52242–1000, U.S.A.

New lane-keeping and velocity-maintenance driving performance measures, derived using ideas derived from regression analysis, are presented. A driver's lane-keeping performance can be described using a linear equation that is the line of best fit for a series of points along the track of a vehicle. The equation describes the position of the vehicle relative to the center of the lane, and provides an indication—in both straight and curved road segments—of whether the vehicle is traveling parallel to the lane, or is veering left or right. This approach also provides a measure of steering instability that can be used, along with the number of steering oscillations about the direction of travel, to indicate steering ability. Analogous velocity-maintenance measures can be derived using a second linear equation.

Driving Performance Measures

In 1961, Crawford wrote that "it has proved extremely difficult to define what is meant by driving performance and to develop adequate techniques of measuring it." More than 30 years later, Nilsson (1993) suggested that this statement still held true.

It is difficult to obtain detailed measures of the normal driving performance of a driver while he/she is driving on roads and expressways—direct observation can be used, but is a method that lacks precision. More detailed information can be obtained by using an instrumented car—for example, Lechner and Perrin (1993) used an instrumented vehicle on an 85-km route that included sections of freeway, minor roads with long straight sections, winding roads, and a town with dense urban traffic. They analyzed driving performance in terms of: velocity (percentage of time in each 20-km/h velocity range); longitudinal acceleration and braking rates; lateral acceleration and braking rates; and steering wheel movements.

If an instrumented track or a simulator is used, it is possible to obtain more precise driving performance data than was obtained by than Lechner and Perrin—although there may be some loss in validity. Investigators using simulators (e.g.,

McLean and Wierwillie, 1975; and Wierwillie and Guttman, 1978), have typically looked at longitudinal performance—i.e., velocity maintenance—in terms of mean velocity (and variability around the mean), and acceleration and deceleration rates; and at lateral performance—i.e., lane keeping—in terms of the deviation of the line of travel from the center of the lane and the number of steering-wheel reversals.

The deviation of the line of travel from the center of the lane has been the primary lane-keeping measure, while the number of steering-wheel reversals has provided a measure of steering stability. Similarly, the mean velocity and its standard deviation have been used as measures of velocity maintenance. In this paper, using ideas derived from regression analysis, we develop alternative measures for both lane keeping and velocity maintenance. In both cases, we believe the new measures provide a more accurate and more detailed description of driving performance.

Driving Performance Measures for Straight Road Segments

Lane-Keeping Behavior
This method effectively separates the previously used measure of deviation from the center of the lane into three distinct measures—two of which are lane-keeping measures (the position of the vehicle in a lane, and steering drift across the lane); while the third is a measure of steering stability. In addition, we suggest replacing steering wheel reversals with the number of crossings of the direction of travel).

A linear equation that is the line of best fit for a series of points on the track of a vehicle can be used to describe the position of the vehicle relative to the center of the lane. This equation indicates whether the vehicle is veering to the left or to the right of the lane, or is traveling parallel to it. In addition, the variability of the actual track of the vehicle around this line of fit can be used, along with the number of crossings of the direction of travel (or line of best fit), to indicate the driver's steering stability.

The method of least squares can be used to obtain a line of best fit that gives the relative position of a vehicle in a lane throughout a segment of road. First, we consider the case, illustrated in figure 1, where a driver is traveling along a straight road segment.

At any point in time, it is possible to determine the position of the center of the vehicle on a line that is perpendicular to the lane. In recent experiments conducted on the Iowa Driving Simulator (Bloomfield, Carroll, Papelis, and Bartelme, in press; Bloomfield, Christensen, and Carroll, in press), data were collected at a rate of 30 Hz—so that, as a vehicle traveled along a straight road segment, the track of the vehicle could be used to determine the position of the center of the vehicle relative to a series of perpendicular lines drawn at 1/30 s intervals.

We assume that the series of positions can be described by a linear equation: not an unreasonable assumption—since the vehicle was travelling along what was, comparatively, an extremely long longitudinal segment with a relatively narrow (3.66-m wide) lateral cross section. The equation is as follows:

$$p = a_p + b_p x \qquad (1)$$

where:

p is the point (representing the center of the driver's vehicle) at which the line of best fit crosses the perpendicular across the lane after the vehicle has traveled distance x—when p equals zero the line of best fit starts at the center of the lane.

x is the distance travelled by the vehicle in the segment of straight road.

a_p is the point where the line of best fit crosses the perpendicular at the start of the segment.

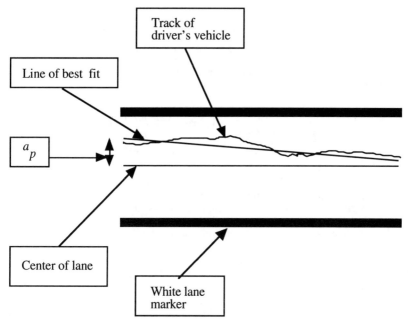

Figure 1. Schematic showing a cross section of a lane, with the track of the driver's vehicle along the lane and the line of best fit.
[Note: the cross section is greatly exaggerated compared to the distance along the lane.]

b_p is the gradient of the line—if b_p is zero, the vehicle is traveling parallel to the center line; if b_p is positive, the vehicle is moving right to left in the lane; and if b_p is negative, as in figure 1, the vehicle is moving left to right.

Because equation 1 is a linear regression equation, the method of least squares can be used to calculate the line of best fit—the method of least squares minimizes the error in predicting p from x, and allows us to calculate a_p and b_p using the following equations:

$$b_p = \frac{\sum xp - \frac{(\sum x)(\sum p)}{n}}{\sum x^2 - \frac{(\sum x)^2}{n}} \qquad (2)$$

where n is the number of data points obtained while the vehicle travels distance x, and

$$a_p = \frac{1}{n}\left(\sum p - b_p \sum x\right) \qquad (3)$$

In addition, the variability in b_p—i.e., the residual standard deviation—can be used as an estimate of I_p, the steering instability. I_p, the variability in steering occurring when the driver attempts to maintain a straight course along the line of best fit, is given by the following equation:

$$I_p = \sqrt{\left[\sum p^2 - \frac{(\sum p)^2}{n} - \frac{\left\{\sum xp - \frac{(\sum x)(\sum p)}{n}\right\}^2}{\sum x^2 - \frac{(\sum x)^2}{n}}\right] \div (n-2)} \qquad (4)$$

Equations 1 and 2 define the position of a vehicle in a straight road segment; equation 3 gives information on steering drift across the lane; and that equation 4—along with the number of steering oscillations (i.e., crossings of the direction of travel)—provides a measure of the smoothness or stability of the ride.

Velocity Maintenance

A set of equations similar to those used to describe lane-keeping performance can be used to describe the driver's ability to maintain the velocity of the vehicle. In this case, illustrated in figure 2, we obtain two velocity-maintenance measures—one a measure of the velocity at any instant, the other a measure of whether the velocity is drifting up or down—that would replace the mean velocity and its standard deviation. Also, we derive a measure of the stability of velocity maintenance, that can be used with the number steering oscillations. In these new equations, p, a_p, b_p, and I_p, in equations 1, 2, 3, and 4 are replaced by v, a_v, b_v, and I_v, respectively. Equations 5, 6, and 7 provide measures of the driver's ability to maintain velocity; equation 8 provides a measure of stability in maintaining velocity. These equations are as follows:

$$v = a_v + b_v x \qquad (5)$$

$$b_v = \frac{\sum xv - \frac{(\sum x)(\sum v)}{n}}{\sum x^2 - \frac{(\sum x)^2}{n}} \qquad (6)$$

$$a_v = \frac{1}{n}\left(\sum v - b_v \sum x\right) \qquad (7)$$

$$I_v = \sqrt{\left[\sum v^2 - \frac{(\sum v)^2}{n} - \frac{\left\{\sum xv - \frac{(\sum x)(\sum v)}{n}\right\}^2}{\sum x^2 - \frac{(\sum x)^2}{n}}\right] \div (n-2)} \qquad (8)$$

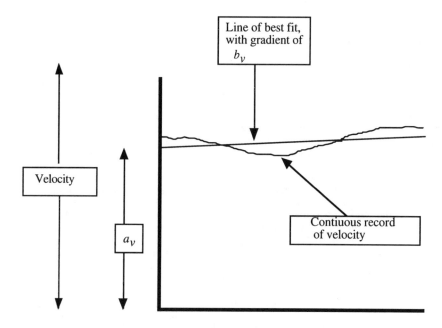

Figure 2. Schematic showing continuous record of velocity and the line of best fit.

where:

v is the velocity, indicated by the line of best fit, after the vehicle has travelled distance x.

x is the distance travelled by the vehicle in the segment of straight road.

a_v is the point where the line of best fit meets the velocity-axis at the start of the segment.

b_v is the gradient of the line—if b_v is zero, the vehicle is traveling at constant velocity; if it is positive, the velocity of the vehicle is gradually increasing, as in figure 2; and if it is negative, velocity is gradually decreasing.

I_v is the instability in velocity maintenance—it is an estimate of the extent of the velocity fluctuations that occur when the driver is attempting to maintain a chosen velocity.

Driving Performance Measures for Curved Road Segments

Bloomfield, Christensen, and Carroll (in press) used the linear equations, described above, to analyze lane keeping and velocity maintenance data. They used these measures to compare the driving performance of drivers who drove on a straight expressway segment both before and after they experienced traveling in an automated highway.

In some circumstances, a linear equation can be used to describe the track of vehicle as it travels around a curve—whether a linear equation can or cannot be used depends on the way in which lane position is recorded. If lane position is determined relative to the cross-section of the lane, then a linear equation can be used to describe a curved path.

The reasoning is as follows. If the position of a vehicle is determined relative to the cross-section of the lane, then, when the road curves, at each moment the lane position is expressed relative to a line that is perpendicular to the tangent of the curve. In an experiment conducted on the Iowa Driving Simulator, data were collected at a rate of 30 Hz, as the vehicle traveled around the curve—thus, every 1/30 s, a new tangent, and new perpendicular cross-sectional lines had to be considered. On these cross-sectional lines, a series of points was recorded, and these points represented the position of the vehicle in the lane every 1/30 s.

Now, if the cross-sectional lines are considered together—and the wedged-shaped slivers of the curve between them are ignored—the original curve can be treated mathematically as a straight line. It is then possible to determine the linear regression of the series of points representing the position of the driver's vehicle in the lane. This allows us to use the set of linear equations, presented above and derived for straight line segments, to analyze lane keeping and velocity maintenance performance when a vehicle is traveling around a curve.

If the track of a vehicle as it travels around a curve is approximately parallel to the center of the lane, then the steering drift (or gradient of the linear regression equation for lane position), will be approximately zero. In addition, if the vehicle closely maintains this parallel track, then the steering instability, will be relatively small. In contrast, the steering drift value may be positive or negative, indicating either that the vehicle veered out away from the curve—resulting in an overshoot relative to the center of the lane—or that it veered into the curve—indicating an undershoot. Also, if an overshoot or undershoot were to occur during the first portion of the curve and if, in the last portion of the curve, a steering correction were to be made, we would expect to find that the steering instability, would be relatively large over the length of the curve and that there would be few steering oscillations.

In an experiment involving travel around a curve, Bloomfield, Carroll, Papelis, and Bartelme (in press) have used these lane-keeping measures to investigate the ability of the driver to fill in for an automated system when his/her vehicle traveled through a portion of the expressway in which the capabilities of the automated system were reduced.

References

Bloomfield, J.R., Carroll, S.A., Papelis, Y.E., and Bartelme, M.J. (in press), The Ability of the Driver to Deal with Reduced Capability in an Automated Highway System, Revised Working Paper (Contract No. DTFH61-92-C-00100) to be published by the Federal Highway Administration, Turner-Fairbank Highway Research Center, McLean, VA. WDC: DOT FHWA.

Bloomfield, J.R., Christensen, J.M., and Carroll, S.A. (in press) The Effect on Normal Driving Behavior of Traveling Under Automated Control, Revised Working Paper (Contract No. DTFH61-92-C-00100) to be published by the Federal Highway Administration, Turner-Fairbank Highway Research Center, McLean, VA. WDC: DOT FHWA.

Crawford, A. 1961, Fatigue and driving, Ergonomics, 4, 143–54.

Lechner, D. and Perrin, C. 1993, The actual use of the dynamic performance of vehicles, Proceedings of the Institute of Mechanical Engineers, 207, 249–256.

McLean, R.C. and Wierwillie, W.W. 1975, The influence of motion and audio cues on driver performance in an automobile simulator, Human Factors, 17, 488–501.

Nilsson, L. 1993, Contributions and limitations of simulator studies to driver behavior research, In: Parkes, A.M. and Franzen, S. (Eds.), Driving Future Vehicles, Taylor and Francis, London, U.K.

Wierwillie, W.W. and Guttman, J.C. 1978, Comparison of primary and secondary task measures as a function of simulated vehicle dynamics and driving conditions, Human Factors, 20, 233–244.

THE ERGONOMIC DESIGN OF A HAND POWER/BRAKE CONTROLLER FOR A TRAIN CAB

Simon Layton and Colin Scott

Human Engineering Limited
Shore House
68 Westbury Hill
Westbury-On-Trym
BRISTOL
BS9 3AA

ABB Rail Vehicles Ltd.
Advanced Technology Department
Litchurch Lane
Derby
DE24 8AD

The aim of this programme of work was to design an ergonomic hand control that could be used safely and effectively by a train driver to apply power and braking to a train.

The motivation for this work was a series of cases of musculoskeletal stress that were reported when train drivers used an old style control for prolonged periods. Problems occurred primarily in the wrist and shoulder.

Background research, many design iterations and user trials resulted in a new design, which in the near future will be fitted to an advanced train cab.

The new design takes into consideration anthropometric, anatomical and kinetic factors, to bring the wrist into the 'neutral' position during operation, to distributes forces on the hand over a larger area, and to provide wrist support, whilst not compromising the ability to push or pull the control into position.

Introduction

Hand Controller Design

McCormick and Sanders (1993) give an excellent account of the principles of good hand controller design. In this section we briefly review and summarise their discussion.

When considering the design of hand controllers the anatomy of the human hand is taken into account. Lack of consideration of the complex structure of the hand leads to poor design and consequently an increased risk of accidents or injuries. As well as the single-incident traumatic type of accident and injury, such as cutting a finger, there are also less immediately obvious consequences of poor tool or hand controller design. These are known as 'cumulative traumas' and include carpal tunnel syndrome, tenosynovitis, vibration-induced white finger and epicondylitis (tennis elbow). These cumulative-trauma disorders can lead to reduced work output, poor quality work, and increased absenteeism.

Design Principles

One of the main principles in the design of a hand controller is to maintain a straight wrist. Bending of the wrist (see Figure 1), especially in palmar flexion and/or ulnar deviation, causes the tendons passing through the carpal tunnel to bend and bunch up. This can, with continued exposure, result in tenosynovitis which is inflammation of the tendons and their sheaths. Examples of motions which may lead to tenosynovitis are clothes wringing, using a screwdriver and operating rotating controls such as a motorbike throttle.

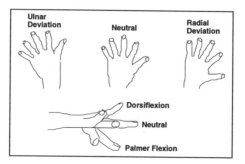

Figure 1. Movement of the wrist (Source McCormick and Sanders, 1993)

A further condition which may occur as a consequence of the extended use of controllers requiring bent wrists is carpal tunnel syndrome. The symptoms of this disorder are numbness, loss of grip and eventual muscle wastage.

It is also recommended that finger grooves should not be used on hand controller. It is not possible to provide grooves to accommodate all hand sizes and incorrect groove positioning can result in the ridges digging into the user's fingers. Care should be taken avoid repetitive finger action. A good solution to the problem of frequent use of individual fingers is a finger-strip which allows the fingers to share the load whilst freeing the thumb for grip and guidance.

What is a PowerBrake Controller

A PowerBrake Controller (PBC) is a hand operated lever within a train cab that driver pushes and pulls in order to initiate power and braking. It is operated with the le hand. There are 7 notches; Emergency Brake, Brake 1, Brake 2, Brake 3, Coast/Off, Shunt and Power. There are two buttons on the PBC; a vigilance reset button and a spee hold button. We have designed the vigilance button to be a long bar that can be operate by the fingers and must be kept depressed whilst the train is moving, or else the brakes automatically applied. The speed hold button is operated by the thumb and is pressed once to hold the speed and pressed again to release the speed hold facility.

Existing Problem

When using the existing style of control (see Figure 2) some train drivers experienced aches and pains in the wrist and shoulder. This was attributed to the fact t the old PBC forced the drivers wrist into radial deviation, and forced the elbow to be he away from the body (see Figure 3). This meant that static load was placed on the shoulder and the wrist was held out of the 'neutral' position.

Figure 2. Existing PBC in a train

Figure 3. Model of existing PBC   **Figure 4**. Model of one of the new PBC's

Rationale

Initially we had to establish a rationale for the new design. Pheasant 1988 says that when the hand is in its 'resting position', in a relaxed posture but ready for action, the fingers are slightly spread and gently flexed. Essentially, our design attempts to allow the drivers to adopt a relaxed posture in the shoulder and arm, and to encourage them to have a straight wrist (see Figure 4). Bringing the wrist closer to the neutral posture, will reduce tension on the tendons of the wrist. Bringing the elbow closer to the body and allowing the arm to be relaxed will reduce stress on the shoulder.

In a previous pilot study we established that having a horizontal PBC or a vertical PBC brings the wrist out of the neutral position. Using drivers and a concept design which embodied ergonomic principles, we established that people prefer a PBC that slopes to the side and upwards to the front, i.e. brings the wrist more into the 'neutral position'.

Three designs were then created taking into consideration the anatomy, physiology and biomechanics of the arm, wrist and hand. A lot of iterations of the designs were produced. The basic shapes evolved over a period of months and were modelled in plasticine. We performed numerous quick user trials asking people to relax their arms and hands by their sides, and then to place their hands on a desk in a relaxed posture. The PBC designs vary from a shallow open grip, through a rounder taller grip, to a thinner tighter grip (see Figure 5). The PBC's have been designed to provide wrist support, allow the wrist to fall into the 'neutral' position, distribute the forces on the hand over a large area, whilst maintaining the push/pullability of the control. The three designs were then evaluated in a trial.

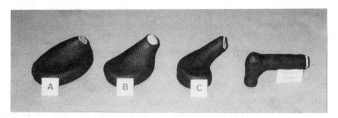

Figure 5. Photo showing the 3 new designs (on left) and the old design (on right)

Trial

There were 4 main objectives of the PBC Trial:-

• To establish the preferred angle of travel
• To establish the preferred angle of rotation about its centre
• To establish the acceptability of button location
• To establish the preferred shape

The trials were performed at the ABB site in Derby. The subjects were ABB employees. The reason for not using real drivers was that this was a physical fitting t rather than a functional trial. The range of statures was from 5th percentile female t 97th percentile male. A physical mock-up of a cab was used in the trials. This was physically identical to the geometry a real cab.

(i) Preferred angle of travel

Ideally the wrist should be kept straight, so that it is in the 'neutral' position. Force should be applied through the forearm, wrist and hand, whilst maintaining a stra wrist. This force should ideally be applied in the direction that the forearm and hand pointing (see Figure 6).

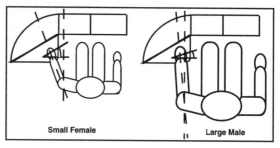

Figure 6. Diagram showing angle of approach

This angle is dependant upon stature. However, the movement should not devia too much from straight forward and backward, firstly because of historical reasons, and secondly because the PBC needs to keep its identity as having essentially a forward and backward movement. It should not be considered as a sideways moving control, which may lead to an incompatible control-effect relationship.

Subjects were asked to place their hand on the PBC, and asked if they prefer to push the PBC straight forward or at a slight angle of 10 degrees off a line perpendicula their body. 10 degrees was selected as a possible acceptable angle taking into consideration the size of both the mounting panel and the control mechanism. Anoth factor that lead to 10 degrees being selected as a possible pushing angle were the anthropometric calculations to determine the angle of approach to the PBC.

(ii) Preferred angle of rotation about its centre

The angle of approach to the PBC will affect the preferred angle of rotation (see Figure 6). It was hypothesised that a shorter person will want the PBC to be mounted a greater angle than that of a tall person.

Photos were taken of all 3 new PBC designs from directly above, while in their preferred position, as if pushing at 10 degrees angle (see Figure 7). We gave the subject some guidance about things they should consider, i.e. straight wrist, no tension on elbow arm or shoulder, elbow able to swing freely, good pushing angle. We incorporated a reliability test whereby we took two lots of photos for each person for one of the PBC (different for each person).

A grid was placed under the PBC and subjects were asked to place the PBC in thei preferred position, whilst rotating it about a central axis. The PBC's were placed on th position of the 'Power' notch as this is the position that the PBC will be in when the driver may keep his/her hand on it for long periods. By studying the photos of the PBC's in the preferred position in relation to the grid we could work out the mid-range position.

Figure 7. Photo of a subject during the trial

(iii) Acceptability of button location

Subjects were asked whether the position of the vigilance reset and the speed hold button on each of the new PBC designs was acceptable. If a position was unacceptable they were asked to state their reasons.

(iv) Preferred shape

For a number of different aspects of the design, subjects were asked whether each of the PBC's was acceptable, and to rank the PBC's from best to worst. Again the subject was asked for reasons why they found certain PBC's unacceptable and why they ranked them as they did. The aspects of the design that were considered were the push / pullability, comfort, aesthetics, and hand / wrist support. Subjects were finally asked for their overall preference.

Results

Subjects were consistent in their placement of PBC's, which in our reliability test varied typically by only 3 degrees. There was one subject who varied by 15 degrees. Therefore we decided to regard the data from this person as unreliable and withdraw it from our analysis.

(i) Preferred angle of travel

All subjects said that having the PBC at a slight angle of about 10 degrees was preferable to having it in a straight forward movement.

(ii) Preferred angle of rotation

The mid range angles of rotation X (see Figure 8) for the 3 new designs of PBC, were 53 degrees for PBC A, 48 degrees for PBC B, and 27 degrees for PBC C. The results also showed that the preferred angle of rotation for the taller users was generally less than the preferred rotation angle for the shorter users. This was attributed to the shallower angle of approach achieved by the taller users (see Figure 6).

This difference in preferred angle of rotation could be accommodated for by having an adjustable PBC to allow rotation. If an adjustable PBC was to be incorporated a safe secure fixing mechanism would need to be used, to ensure that the PBC will not rotate during use.

The large difference in the angles of rotation between PBC's was attributed to the fact that PBC C was more like a stick controller, and therefore required a shallower mounting angle for effective pushability.

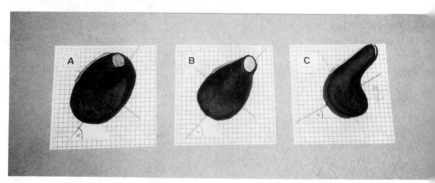

Figure 8. Photo of the 3 PBC's in the mid range angle of rotation

(iii) Acceptability of button location

All the subjects rated the button locations as acceptable for all the new PBC's.

(iv) Preferred shape

The acceptability results showed that all the subjects rated all the PBC's as acceptable for all the design aspects. However the trend that appears from the ratings show that on both push / pullability and overall preference more subjects rated B as the best and fewer subjects rated B as the worst. The other design aspects of comfort, aesthetics and hand / wrist support did not show any clear trend. The push / pullability trend is important as this directly affects the ability of the control to perform it designated function.

Conclusions

The conclusions from the trial are that PowerBrake Controller B is the recommended ergonomic solution, and it should be at an angle of rotation of 48 degrees, with an angle of travel of 10 degrees.

The study shows that the application of sound ergonomic principles can result in improvements to health, safety and efficiency.

The drivers who have seen the new PBC concepts have all been very impressed, and so it is also envisaged that there may be a significant psychosocial benefit from its introduction.

References

Pheasant, S. 1988, *Bodyspace: Anthropometry, Ergonomics and Design* (Taylor & Francis, London)

Sanders, M. S. and McCormick, E. J. 1993, *Human Factors in Engineering and Design*, 7th Edition (McGraw-Hill International)

ASSESSING THE DETERMINANTS OF DRIVER RISK: UTILITY VS. ENGINEERING

Di Haigney, Charlotte Kennett and Ray.G. Taylor

Psychology Group
Aston University
Aston Triangle
Birmingham B4 7ET

The main premise of Risk Homeostasis Theory (RHT) - that compensatory behaviour will mainly be realised through the utility mechanism - is tested. Utility and engineered safety were operationalised as experimental conditions in a pilot simulated driving task on the Aston Driving Simulators (ADS), through an adaptation of the time/distance methodology developed by Hoyes (1992) and via a damage vulnerability/invulnerability mechanism respectively. Analysis of driver performance data revealed that statistically significant differences were only recorded as a main effect of engineered safety. This result brings into question the adequacy of the 'utility driven' compensation models of driver behaviour and risk assessment against a 'passive' engineering formula. The implications of the analysis are discussed and suggestions for future work are made.

Introduction

A utility-based theory of behavioural compensation in drivers has been developed in the last decade which has been the subject of much controversy (Trankle and Gelau, 1992). The model is 'Risk Homeostasis Theory' (RHT) which was formerly referred to as the 'Theory of Risk Compensation'.

RHT was initially proposed by Wilde to account for the rate of accident loss observed per time unit of exposure, per capita and per kilometre driven (Wilde, 1989) and as a framework from which interrelations between these measures can be explained. RHT has also been posited as a potential 'general theory of behaviour' (Wilde 1985, 1989), although this paper is concerned with the theory as it relates to driver behaviour.

In RHT, the population temporal accident rate (the summed cross-products of the frequency and severity of accidents per time unit of road user exposure) is regarded as arising not from a general attempt to minimise perceived risk levels as is postulated in some theories (e.g. Naatanen and Summala, 1976), but rather to match the subjectively perceived risk against a 'target' or 'desired' degree of risk.

The 'target' level of risk is determined by the interaction of four utilities associated with those behavioural options which are regarded as either riskier or safer alternatives to the road-users current activity - and so each utility varies relative to the situation the user currently deems him/herself to be in. According to this model, road-users engaged in a driving task will, through assessing the four utilities simultaneously, experience a 'net' subjective level of risk which is deemed as being the 'acceptable' or the target level of risk. This is frequently compared with the 'perceived' level of risk at a 'pre-attentive' level of cognition (Wilde, 1982a, p.210).

Should a difference between these two variables become great enough for the discrepancy to be perceived in the road-users central consciousness, the road-user will then behave in a manner aimed at reducing this difference to a 'subjective zero', i.e. they will engage in certain 'compensatory behaviours', until the difference between perceived risk and target risk lies just below the 'just-noticeable-difference' or 'JND' level. Such compensation is not hypothesised in RHT as occurring through specific pathways (Wilde, 1989, p.277), but may be realised spontaneously through any behavioural option available to the individual, depending upon the 'net' utility associated with each alternative course of action.

Wilde regards the target level of risk as the fundamental determinant of accident rates, as all compensatory behaviour serves to subjectively match perceived risk against it. (Haigney, 1995). As a result, Wilde maintains that 'conventional engineering' road safety solutions are not effective in the long term as they do not directly address utility issues and therefore not target risk.

Hoyes (1992), argued that certain methodologies, such as the Aston Driving Simulators (ADS) provide the opportunity for 'collapsed experience',which can effectively account for the 'long term' negation of compensation in engineering interventions in the 'short term' of an experimental run. The ADS are therefore regarded as a suitable means for testing for compensation via relatively short term 'engineered' environmental interventions.

This study attempts to induce differences in both utility and 'the degree of engineered safety', so that the relative weighting of either in the compensatory process may be isolated and identified.

The operationalisation of utility was achieved through an adaptation of the methodology described in Hoyes (1992), in which participants undertaking some experimental run on the ADS were informed that they either had to 'drive' for a specified period of time or for a specific 'distance' in the simulated driving environment (all subjects in actual fact drove for the same amount of time). Hoyes (1992), argued that subjects in the 'distance' condition, would attach a greater positive utility to risky behaviours which would allow them to complete the task more quickly than otherwise, and that subjects in the 'time' condition would have no reason to attach utility to such behaviours. As a result, Hoyes (1992), asserted that any nonsignificance in ADS performance across time/distance utility conditions could indicate that utility is not necessarily the single factor governing accident rates - as claimed in RHT.

Unfortunately, Hoyes (1992) was not able to develop this means of operationalising utility further and as a result, the methodology as it stands is open to considerable criticism through the major assumption (of participant motivation and consequent action) held within it (Wilde, personal communication, 1992).

The adaptation of this methodology described in this study seeks to address this criticism through the use of penalty and reward systems as described below. Furthermore, the efficacy of these systems in inducing differences in utility across the 'time' and'distance' conditions was evaluated through post-run group discussions with subjects.

Engineering was operationalised through the provision, or non provision of information on the engineered 'robustness' of the simulated vehicle if it were to be involved in a collision reinforced through direct 'feedback' (through sound, visual stimuli and resulting performance of the simulated vehicle), after a collision was experienced on the ADS.

In this study 'compensation' has been operationally defined as those significant differences within the ADS variables logged between the experimental conditions of utility ('time' and 'distance') and engineering ('feedback' and 'non feedback').

Method

The ADS consists of a car seat, a steering wheel, accelerator and brake pedals placed so as to mimic operating conditions within an automatic car. The 'windscreen view' is represented by a computer graphic output of a road image, displayed on a terminal placed in front of the participant. When the simulation program is initiated, the 'view' also incorporate images of 'other traffic' on a single carriageway road, travelling on both carriageways. The 'other traffic' was capable of 'intelligent action', such as overtaking the user.

Participant responses, through the the use of pedals and steering wheel, are not only used to update the screen output (so that moving the steering wheel caused the 'car' to move towards one edge of the 'road', for example), but also stored as 10 driving performance variables (mean speed, mean acceleration, mean braking, number of [successful] overtakes, steering variance, collisions with others on the righthand side of the carriageway, collisions with others on the lefthand side of the carriageway, right headway, time spent on the verge on the righthand side of the carriageway, time spent on the verge on the lefthand side of the carriageway).

After a 'practice session', participants were randomised between the two utility conditions on the ADS. Half the participants 'drove' the simulator for ten minutes under a 'time' utility condition and half under the 'distance' utility condition, with both sets of participants experiencing both engineering conditions of 'feedback' and 'non feedback' conditions for five minutes each in a random order.

All participants were presented with standard instructions on the simulator screen throughout the run which informed them of the utility conditions which they would be experiencing, any penalties/rewards associated with the run overall and the vulnerability/invulnerability of the simulated vehicle as appropriate.

All subjects were informed that a monetary penalty would be levied for each collision experienced on the ADS. Subjects in the 'distance' utility condition were also informed that this penalty could be reduced by a substantial amount if they completed the driving task quickly, 'freeing up' simulator time.

When participants had completed both feedback conditions on the ADS, they were required to complete a short questionnaire and took part in a group discussion on the effectiveness of the utility operationalisation methodology.

Results

Repeated measures ANOVAs were performed for all the dependent variables provided by the ADS for both feedback conditions, between subjects ANOVAs for ADS dependent variables resulting from the utility conditions, two-way Chi squares aswell as Pearson correlations between ADS variables and questionnaire responses were calculated. Group discussions were subjected to qualitative content analysis.

Data are discussed by 'Risk' and by 'Utility'. The term 'Risk' refers to the mean values of the ADS dependent variable for the'feedback' and 'non feedback' conditions and the term 'Utility' refers to the mean ADS variable values for the 'time' and 'distance' conditions.

Mean Speed

Mean speed was found to be significant across Risk ($F[1,30]=6.749$; $p=.0144$) although it was nonsignificant across Utility ($F[1, 30]=.205$; NS) and no interactive effect was determined for UtilityxRisk ($F[1, 30]=1.53$; NS).

Mean Acceleration

A main effect of Risk was established for mean acceleration ($F[1, 30]=14.126$; $p=.0007$), though no significance was recorded for Utility ($F[1,$

30]=2.358; NS), or for UtilityxRisk (F[1, 30]=1.918; NS).

Mean Braking
Both the main effects of Risk (F[1, 30]=1.02; NS) and Utility (F[1, 30]=.012; NS) did not reach significance. The interactive effect of UtilityxRisk (F[1, 30]=.369; NS) was also nonsignificant.

Number of Overtakes
The 'overtakes' variable logs the number of times a subject has successfully managed to pass the vehicle ahead of them in the lefthand carriageway (i.e. through driving into the righthand carriageway and returning completely into the lefthand carriageway).

Risk was marginally significant as a main effect for mean braking (F[1, 30]=4.226; p=.0476), whilst both Utility (F[1, 30]=2.252; NS) and UtilityxRisk (F[1, 30]=.587; NS) were nonsignificant.

Steering
As the 'overtakes' variable logged only those overtaking manoeuvres which were successful, aborted attempts at overtaking may be deduced through reference to the 'steering' variable, which recorded the variance in simulated vehicle tracking, relative to the centre-line of the road.

No significant effects for Risk (F[1, 30]=.22; NS), or for Utility (F[1, 30]=.398; NS) were gained. A nonsignificant result for UtilityxRisk (F[1, 30]=.392; NS) was also recorded.

Collisions with others on the righthand side of the carriageway
The mean number of collisions with others on the righthand side of the carriageway was found to be significant with a main effect of Risk (F[1, 30]=5.351; p=.0278). Utility (F[1, 30]=.0057; NS) was nonsignificant as a main effect and UtilityxRisk (F[1, 30]=1.224; NS) was also nonsignificant.

Collisions with others on the lefthand side of the carriageway
The mean number of collisions with others on the lefthand side of the carriageway was nonsignificant for Risk , Utility and UtilityxRisk at: (F[1, 29]=1.044; NS), (F[1, 29]=.016; NS) and (F[1, 29]=.51; NS) respectively.

Right headway
'Mean right headway' is the value in 'simulated metres' between the subjects vehicle when in righthand side of the carriageway (such as during an overtaking manoeuvre) and of any oncoming vehicle in the righthand carriageway. It therefore allows the degree of risk taking during either successful and/or aborted overtaking manoeuvres to be established.

Utility (F[1, 28]=.242; NS), Risk (F[1, 28]=.003; NS) and UtilityxRisk (F[1, 28]=.017; NS) all failed to produce any significant results.

Time spent on the verge on the righthand side of the carriageway
During the driving task, it was possible for subjects to 'run their cars off the road' and onto the 'grass verges' on either side, generally as a result of oversteering during some collision avoidance measure.

Neither Risk (F[1, 30]=.99; NS) or Utility (F[1, 30]=1.18; NS) gave significant results for a main effect. UtilityxRisk (F[1, 30]=.017; NS) was also nonsignificant for this variable.

Time spent on the verge on the lefthand side of the carriageway
Risk was found to be a main effect (F[1, 30]=7.58; p=.0099), although Utility (F[1, 30]=2.808, NS) and UtilityxRisk (F[1, 30]=.627; NS).

Questionnaire analysis and Group Discussions

Of all the questionnaire items, only two produced any significant relationship with the ADS variables, namely question B3 ('How do you rate your overall driving performance against the average driver?') and question D5 ('It is a waste of a fast/sporty car if it is not driven fast'), both of which correlated positively with mean steering at (r=.445 and r=.423) respectively.

As noted above, the steering variable allows risk taking through successful aswell as aborted overtakes to be examined. As the questionnaire items tested for self rating of task skill and risk acceptance, it would appear that these items are useful predictors of ADS performance and risk taking propensity. The lack of any relationship with other questionnaire items is a cause for concern however, and may indicate a need for some considerable revision of the questionnaire items and format.

As subjects were run in groups of four, group discussions were held directly after each group had completed their run on the ADS and centred on evaluating the operationalisation of utility adopted in the study. As the utility conditions were balanced (i.e. two participants per run would have experienced either the time or distance condition), either condition was represented equally in each group.

The majority of subjects stated that they felt the reward/punishment systems had noticeably (in their view) affected the way in which they drove the ADS, with those under the 'time/feedback' condition, tending to emphasise the care with which they had taken to avoid collisions. Those experiencing the 'distance/non feedback' conditions reported more risk taking than other subjects.

Discussion

Whilst a number of ADS variables recorded a main effect by Risk (i.e. mean speed, mean acceleration, overtakes, collisions on the righthand side of the carriageway and time spent on the righthand verge), none of the ADS variables recorded a main effect of Utility, or an interactive effect of UtilityxRisk.

Although this result could have arisen through the utility operationalisation technique failing to create a great enough distinction in utility between the 'time' and 'distance' conditions, the group discussions would appear to confirm that the utility methodology did allow for subjects to distinguish between the conditions successfully in terms of the values attached to driving styles.

These differences are not reflected in those ADS variables which can be taken as indicating either safety acceptance (Wagenaar and Reason, 1990), or risk acceptance (Matthews, Dorn and Glendon, 1991) however, even though indications of possibly compensatory behaviours have been determined via a main effect of Risk in the ADS variables.

As the ADS have been evaluated as sensitive to changes in driving style (Dorn, 1992), non significance across Utility conditions *could* suggest that a change in a utility exceeding the JND was not realised through task relevant behaviour directly (Adams, 1988), or it may be that the subjective 'size' of the difference perceived between utility conditions was sufficiently great to warrant recognition by the subjects, but not great enough to result in direct behavioural change.

In conclusion, it would seem that the infamously 'woolly' (Haigney, 1995), concept of utility is more elusive, fluid and dynamic than has been appreciated to date. It would appear that there is scope for a 'grey area' in the utility evaluation mechanism in RHT, where although the net evaluation of the four utilities is shifting beyond the JND, subjects may not be responding with compensatory behaviour, even though they recognise a 'need' for it. On the other hand, some compensation appears to occur with little reference to the direct manipulation of utilities but is rather more 'reactive' than 'pro-active', although RHT would tend to suggest the latter. Further work is required in the area of utility in risk - especially in terms of its measurement and operationalisation, as until more progress has been made, the empirical status of utility driven theories of behaviour cannot be established.

References

Adams, J.G. 1988, Risk homeostasis and the purpose of safety regulation, Ergonomics **31(4)**, 407 - 428.

Dorn, L. 1992, *Individual and group differences in driving behaviour*, Unpublished Doctorate Thesis, Applied Psychology Division, Aston University.

Haigney, D. E. 1995, Compensation - implications for road safety, InRoads **17 (1)**, 21-33.

Hoyes, T. W. 1992, *Risk Homeostasis Theory in simulated environments*, Unpublished Doctorate Thesis, Applied Psychology Division, Aston University.

Matthews , G., Dorn, L. and Glendon A.I. 1991, Personality correlates of driver stress, Personal and Individual Differences **12**, 535 - 549.

Naatanen, R. and Summala , H. 1975, A simple method for simulating danger-related aspects of behaviour in hazardous activities, Accident Analysis and Prevention **7**, 63 - 70.

Trankle, U. and Gelau, C. 1992, Maximisation of participative expected utility or risk control? Experimental tests of risk homeostasis theory, Ergonomics **35** (**1**), 7 - 23.

Wagenaar, A.C. and Reason, J.T. 1990, Types and tokens in road accident causation, Ergonomics **33** (**10 - 11**), 1365 - 1375.

Wilde, G. J.S. 1982a, The theory of Risk Homeostasis: Implications for safety and health, Risk Analysis **2(4)**, 209-225

Wilde, G.J.S. 1985, Assumptions necessary and unnecessary to risk homeostasis, Ergonomics **28**, 1531 - 1538.

Wilde, G.J.S. 1989, Accident countermeasures and behavioural compensation: The position of Risk Homeostasis Theory, Journal of Occupational Accidents **10(4)**, 267-292.

Displays

A COMPARISON OF NEW PURSUIT DISPLAYS AND A
CONVENTIONAL COMPENSATORY FLIGHT DIRECTOR DISP

R.J. Nibbelke*, J.J. van den Bosch* and
J.J.L.H. Verspay‡

G. Kolstein

National Aerospace Laboratory (NLR)
*Man Machine Integration dept.,
‡Flight Mechanics dept.
PO box 90502, 1006 BM Amsterdam, NL
Tel: + 31 20 5113763, fax: + 31 20 5113210
e-mail: nibbelke@nlr.nl

Fokker Aircraft B.V.
ETSD/FM
P.O. Box 7600
1117 ZJ Schiphol, the Netherla
Tel: +31 20 6052375, fax: +31 20 6

Keywords: aviation, pursuit display, compensatory display, flight director

The objective of this study is to investigate two new Flight Director displays with respect to their acceptability and possible advantages. The new displays were of the **pursuit** type and aimed to increase situational awareness, by providing additional information, without a significant increase in workload. These new displays were compared to the conventional **compensatory** flight director display. For each display five airline pilots conducted part task experiments and a complex approach. Two different levels of control automation were compared, to determine the applicability of the displays in conventional manual control systems and possible future control systems with automatic inner-loop aircraft stabilisation. The results show that despite the relative unfamiliarity with the new displays, pilots can achieve an equivalent level of performance with acceptable workload. A pursuit display with a vertical and lateral displaced guidance symbol (in earth axis) appears most promising, because it showed an increase in pitch awareness at an equal level of workload.

Introduction

The objective of the overall Flight Management Concept Verification project (Kolstein, 1995 Fokker Aircraft and NLR, is to verify flight control concepts in combination with display cor tions in a future Air Traffic Management environment. This verification is to provide Fokker Aircraft with the minimum requirements for the flight control and displays in cockpits for fut regional aircraft. As part of the project this study on new flight director displays was carried

In civil aircraft a flight director system (FD) generates steering commands to the pilot to assis in tracking the desired flight path. In most 'conventional' civil aircraft, these steering comman displayed by 'cross bars' relative to the aircraft symbol (boresight) on the middle of the Prima Flight Display, resulting in compensatory tracking performance (see fig 1). A characteristic o compensatory behaviour is a lack of information regarding the source of tracking errors whicl compromise situational awareness (Wickens, 1992). An alternative way of presenting steering commands is to display the Flight Path Vector (where the aircraft is going, FPV) which has te controlled by the pilot in the direction of the desired Flight Path Vector (the so-called Flight P Director, FPD). This results in a pursuit display. By comparing to the conventional compensa display, this study aims to determine the acceptability of two different FPD-pursuit displays i of steering performance, awareness and workload (Verspay et al, 1995).

The following hypotheses were tested:

1   Pilots with only limited practice on the new displays, will be able to fly accurately with FPV and FPD.
2   Subjective ratings of situational awareness will be increased by the use of pursuit-displays
3   The overall workload will not be increased by the use of FPV and FPD.

In addition to the three displays tested, two different levels of control automation were employed. This was in order to determine the applicability of the displays in both conventional manual control systems and control systems with automatic inner-loop aircraft stabilisation (so-called enhanced manual flight control).

Display description

Display 1: Conventional Cross bars Flight Director

In the conventional cross bars flight director (figure 1) a deviation from the desired vertical path is presented by a horizontal bar moving in the direction that a pitch steering input is required (e.g. 'bar moves up' means 'pull up', the so-called "follow the needle"-principle). Any deviation from the desired roll angle, as calculated by the control system, is presented by a vertical bar moving side ways ('a move to the right' means 'turn to the right by banking the aircraft'). The deviations are all referenced to the centre of the screen (boresight square).

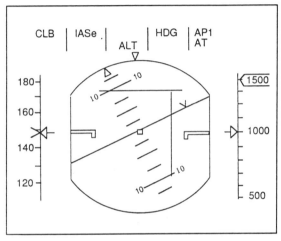

Figure 1. Conventional Flight Director display (cross bars).

Alternative display designs

Instead of the attitude of the aircraft, the new pursuit displays are based on the flight path . The Flight Path Director presents steering commands relative to the present FPV to arrive at the desired path. Both FPV and FPD are projected relative to the artificial horizon. Therefore the actual angle of climb or descend (FPV) and the desired climb or descend (FPD) can be read directly from the display. It is this additional information that is expected to increase awareness in the vertical plane.

The deviation in pitch and roll can be presented in several ways. Differences in frame of reference (earth- or aircraft referenced), rotating or non-rotating symbols, symbol size and colour were all considered in an iterative design process. Test pilots were asked to fly several complex approaches using different combinations of displays to reduce the number of display alternatives for the experiment. The two pursuit Flight Path Director displays, which were selected for the experiment, are described below.

Display 2 Cross Flight Path Director (non-rotating symbol)

For both pursuit displays a standard FPV symbol was chosen; a circle with wings and a tail. A cross was chosen as the FPD symbol because people tend to associate a cross with a target. Therefore it is unlikely that the FPV and FPD symbols will be confused. The cross FPD indicates a deviation from the desired path by a movement perpendicular to the horizon line in the direction in which a pitch steering input is required (e.g. 'cross moves up' means 'pull up'). Any deviation from the roll calculated by the FD is presented by a lateral movement of the cross sideways in a direction para to the horizon line ('move to the right' means 'turn to the right by banking the aircraft'). The deviations are all presented with respect to the Flight Path Vector symbol. Despite the fact that the cross moves in the pitch and roll direction relative to the earth's axis, during the prototyping stage pilots preferred a cross symbol which did not rotate with the earth's axis.

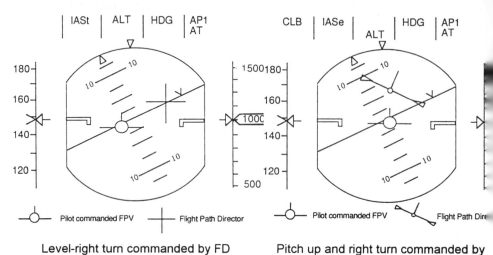

Level-right turn commanded by FD   Pitch up and right turn commanded by

Figure 2. Flight Path Vector display with a Flight Path Director (display 2: laterally moving cross)

Figure 3. Flight Path Vector display with a Flight Path Director (display 3: rotating gho aircraft)

Display 3 Ghost Aircraft Flight Path Director (rotating symbol)
A rotating ghost aircraft showing the desired flight path was expected to be the most intuitive display, because flight track deviation is presented as required roll angle instead of an indication lateral displacement. The hypothesis is that the awareness in roll will be increased as a result. A symbol similar to the one used in the Airbus 320 was chosen. To improve the detection of small re errors a 'tail' was added to the original A320 symbol. A deviation from the desired vertical path is presented by a movement of the FPD in the aircraft frame of reference. The FPD symbol moves in direction that a pitch steering input is required (e.g. 'FPD moves up' means 'pull up'). A deviation from the roll angle commanded by the control system is presented by a rotation of the ghost aircra symbol ('clockwise rotation' means 'turn to the right by banking the aircraft'). The pilot's steering task is to follow the ghost aircraft, based on the 'follow the leader principle' during formation flighing. Note that the deviations are all presented with respect to the Flight Path Vector.

Method

Subjects
Five airline pilots participated in this experiment (average age: 29, average number of flying hours 3450). All pilots had extensive experience in using the conventional flight director display, wherea their experience with flight path vectors was limited to the practice provided for this experiment.

Scenario

Pilots carried out four different types of tasks:
a. heading changes (hdg) while keeping a constant altitude,
b. altitude changes (alt) while keeping a constant heading,
c. speed changes (spd) while keeping a constant heading and altitude and
d. decelerating curved MLS approach (Erkelens, 1991).

The experiments were carried out in the NLR moving-base simulator under moderate turbulence conditions.

Subjective ratings

Pilots were asked to rate their **pitch/vertical** and **roll/lateral awareness** compared to a baseline condition (cross bars display and manual control) on a scale from -10 (very much worse) to +10 (very much better). Pilots were also asked to rate their **workload** on a one-dimensional subjective workload rating scale, called RSME (Zijlstra, 1993).

Objective measures

Steering performance was evaluated using measures of deviations from the desired path. A Root Mean Square of the deviations over the duration of the task was calculated. A within-subjects repeated measures design was used with display type and control system as independent variables. The three displays and two control systems resulted in 6 conditions. Statistical analysis was carried out in a pair-wise manner using the non-parametric Wilcoxon ranking test (at P=0.05).

Results

Hypothesis 1

For heading and speed change tasks, accuracy can be defined as the deviation in altitude (altitude should remain constant). Results are shown in figure 4. Differences between displays are not significant(P<0.05) in eleven of the twelve pair-wise display comparisons. The altitude error for enhanced manual is in the same order as achieved with the conventional control system.

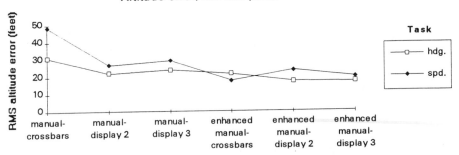

Figure 4. Altitude error for heading and speed change tasks

For altitude and speed change tasks accuracy is defined as the deviation in heading (heading should remain constant), which is shown in figure 5. These results show that the heading error is in the same order as achieved with the conventional control system (manual) and conventional flight director display (cross bars).

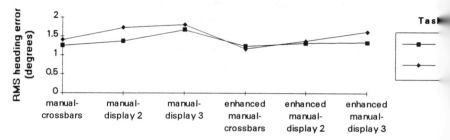

Figure 5. Heading error for altitude and speed change tasks

The results show a slightly increased heading error for the new displays. This difference, how̶
was small and is only statistically significant in three of the twelve pair-wise display compari̶
P<0.05.

Hypothesis 2

The results in figure 6 show an increase in vertical/pitch awareness for display 2. The enhanc̶
manual control system also increased the overall level of awareness, which may be attributed ̶
reduction of workload (figure 7) allowing improved scanning of the instruments. Display 3 di̶
improve the vertical/pitch awareness nor did it show the expected improvement in awareness ̶
to the roll of the aircraft.

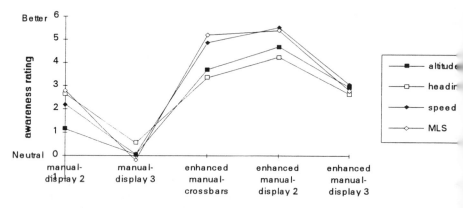

Figure 6. Pitch awareness rating compared to conventional display and manual control
Note: manual cross bars condition is not shown as this served as the baseline for compari̶

Hypothesis 3

The workload was significantly reduced by the enhanced manual control system. The higher lev̶
automation has resulted in the expected reduction of workload. For the new displays no signific̶
differences could be found although figure 7 shows an increase of workload especially for displ̶

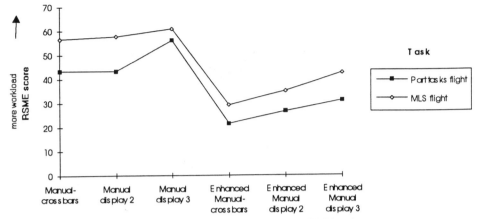

Figure 7. Workload rating using the RSME scale.

Conclusions
With the new pursuit type displays an equivalent level of performance was achieved as with the conventional cross bars display, which pilots were used to. Therefore, despite the limited experience of the pilots with pursuit displays, it can be concluded that performance does not need to be problematic when introducing the new pursuit flight director displays in the existing pilot population. The results of this study show that one of the new pursuit flight director displays (display 2) can improve the awareness in pitch while keeping a constant workload and acceptable performance. A rotating ghost aircraft did not deliver the expected roll/lateral awareness improvement nor did it improve pitch/vertical awareness. It also increased workload slightly. Display 2 appears to be a promising alternative to current compensatory flight director displays.
The enhanced manual control system, in itself, can reduce workload and increase awareness at a level of performance equal to conventional manual control.

Acknowledgements
This study was carried out in the Flight Management Concept Verification Program (Kolstein, 1995) sponsored by the Netherlands Agency for Aerospace Programs (NIVR). The authors would like to thank the subject pilots of Fokker and KLM who took part in this investigation and all people of the NLR Flight Simulator department who helped us to carry out the experiments.

References
Erkelens L.J.J. and J.H. van Dronkelaar, 1991, *Flight Simulator Evaluation of Advanced MLS Procedures*, NLR TP 91446 L, Amsterdam.
Kolstein, G, 1995 *Description and objectives of the Flight Management simulator experiment (FMt-plan)*, Fokker report R-AV94.912.
Verspay JJLH , RJ de Muynck, RJ Nibbelke, JJ vd Bosch, CAH van Gelder. 1995 *VTP Flight management concept verification, pilot experiment on control and display issues* NLR Contract Report, CR 95587 L.
Wickens C.D. 1992, *Engineering Psychology and Human Performance* (Harper Collins, NY).
Zijlstra F.R.H. 1993, *Efficiency in Workload Behaviour, A Design Approach for Modern Tools*, Doctors Thesis Technical University Delft.

THE EFFECT OF AGEING ON DETECTION OF WARNING LIGHTS DURING IMPENDING COLLISION

Ian R L Davies and David Rose

Department of Psychology,
Surrey University
Guildford, Surrey GU2 5XH

We report three experiments that investigated the effects of location uncertainty and stimulus movement on the detection of a target under conditions that simulated impending collision. Each experiment compared performance of young adults (20-25 years) with middle aged adults (45-60 years). Older subjects consistently required higher luminance levels to detect the target, but in addition the 'costs' of spatial and temporal uncertainty were much greater for the old than the young.

Introduction

Under conditions of poor visibility, driving or flying safety may depend upon the threshold for detection of warning lights on vehicles ahead. Although much is known about visual sensitivity, there has been little research on target detection in situations that simulate impending collision. In these circumstances target luminance increases with the square of time elapsed; target location may be uncertain; and if the target is not at the centre of fixation, its retinal location will drift. The threshold for detecting static targets increases with age, (Owsley et al., 1983) and the size of 'the useful field of view' (UFOV) reduces with age (Ball et al., 1988). Thus it is likely that these two factors will combine to make target detection under approach to collision particularly difficult as we age.

We describe three experiments that investigate the effects of spatial and temporal uncertainty on detecting simulated warning lights during impending collision in young (early twenties) and middle aged (early fifties) groups.

Experiment 1

Method

The stimuli were displayed on a computer monitor, viewed from 1500 mm, with a fixed head, such that their line of sight was centred on the middle of the monitor. The target stimulus was a single pixel in a 958 by 840 display mode subtending a visual angle of 0.56'. The target's luminance increased from zero to a maximum of 160 cd/m² over 30.72 seconds in proportion to the square of time against a background luminance of 0.09 cd/m². There were seven possible target locations: the centre of the screen, and three locations on each side of the centre at distances of 1.7, 3.4 and 5.0 degrees horizontally.

The subject initiated each trial: the target started to increase in luminance after a random delay of between 1 and 3 secs, and a timer started synchronously with the target onset. The subjects's detection response stopped the timer and extinguished the target. The time from the stimulus onset to the subject's response was recorded. Each subject completed four sessions on successive days. In total they performed 196 detection responses to each type of display. In addition, they performed 50 simple RT trials each day; these simple RTs were used to estimate the luminance levels of the targets when they were detected.

Results [All differences in all result sections are significant at a level of at least (<.01)]

Mean baseline RT, and mean RT for each of the seven target locations were calculated for each subject. The mean baseline RT was subtracted from the mean RT for each location, and the residual RT converted into an estimate of the luminance at the moment that the subject perceived the stimulus.

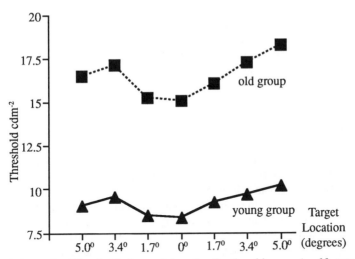

Figure 1. Detection Thresholds for each location for the old group (n=10, mean age=53 years) and young group (n=10, mean age=23 years)

Figure 1 shows the detection luminances for each location and age group. It can be seen that targets in the central location (0°) are detected at the lowest luminance levels, and that there is a small but significant increase as the target location moves towards the periphery (5°). The older group require some 5 cd/m² more luminance to detect the target, but in addition, the difference between the young and the old groups increases as the target lights move towards the periphery, particularly for targets in the right visual field.

Discussion of Results

The older group not only has an elevated detection threshold for the easiest (the central) target location compared to the younger group, but the size of this difference increases with the eccentricity of the target location. In other words, the cost of the target occurring away from the centre of fixation is greater for the old group than for the young group.

Experiment 2

Method

The characteristics of the target were the same as for experiment 1, except that the period of its increasing to maximum luminance was 10.24 seconds rather than 30.72 seconds. There were five kinds of display: (1) The baseline condition. The target occurred in the centre of the screen. (2) The horizontal array. There were three possible locations for the target which occurred with equal probabilities. The locations were: the centre of the screen, as in the baseline condition, and 2^0 26' horizontally to left and right sides of the centre. (3) The triangular array. There were also three possible locations of the target in this condition and they occurred with equal probability. They lay at the apices of an equilateral triangle of side length 4^0 52', such that each location was 2^0 36' from the centre of the triangle. The triangle was oriented such that two of the locations were in the same positions as the outermost locations for the horizontal array, and the third location lay vertically below the centre of the screen. (4) The biased horizontal array. The same horizontal array as described above in 2 was used, but now the target occurred in one location 80% of the time and the other two locations 10% of the time each. (5) The biased triangular array. This was the same array as in 3 above, but with one location used 80% of the time, and the other two locations occurring 10% of the time each.

Results

The pattern of effects for the two array shapes was essentially the same, and they will be described together. There were three main trends of interest and these are illustrated in Figures 2 and 3. First, the old group required greater luminance to detect the target. They required about an extra 6 cd/m^2 to detect the target in the baseline condition, and an extra 10 cd/m^2 when there is uncertainty as to where the target will occur (the right hand part of figures 2 and 3). In contrast, the additional luminance required by the young group when there was spatial uncertainty was only about 1 cd/m^2. Second, there was an effect of target location; both groups detected the centre target more easily than the off-centre targets, but the effect of target location was more marked for the old group than for the young group. Third, there was an effect of target probability: in the biased condition, both groups detected the high probability target more easily than the low probability target. However, the difference between the baseline condition and the high probability condition was less than 1 cd/m^2 for the young group, whereas for the old group it was about 3 cd/m^2.

Discussion of Results

The essential pattern emerging form the results is that the old group are multiply 'handicapped'. They are slower and require higher luminances to detect even the target in

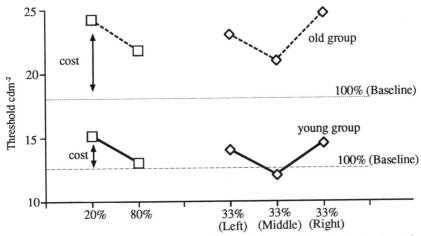

Figure 2. Costs (difference from baseline) of spatial uncertainty, for the horizontal array, for the old group (n=7, mean age=54 years) and the young group (n=7, mean age=24 years)

the known location, the baseline condition. But, more interestingly, there are additional costs of spatial uncertainty and of variations in target probability, over and above that due to differences in the simple detection task. The costs of spatial uncertainty are much greater for the old group than for the young group. The young group detect the target in the central location when the target could occur in one of three locations as easily as they detect the target when there is no spatial uncertainty. In contrast, there is a cost of about 3 cd/m² for the old group of spatial uncertainty. Further, the costs of detecting targets away from the centre of the visual field is greater for the old than for the young. The cost for the old is about double the cost for the young. There is a similar pattern in the data from the biased conditions. Even reducing the probability of where the target will occur from 1.0 to 0.8, imposes a significant cost on the old, whereas the comparable cost for the young is negligible. In addition, the cost of reducing target probability further to 0.2 is greater for the old than the young. Even the difference between targets on the left of the display compared to targets on the right of the display is greater for the old than for the young.

Experiment 3

The first two experiments have found that in addition to the general contrast detection threshold increasing with age, the size of the age effect was amplified in conditions of spatial uncertainty. The logic of experiment 3 was essentially the same as for the first two experiments; it sought to investigate whether such amplification occurs when the source of the uncertainty has a temporal dimension as well as a spatial one. Subjects were required to detect lights moving across the screen, with intensities increasing with the square of time. Such stimuli are equivalent to the stimuli that would be produced on the retinae of a driver approaching but not fixating the lights of another vehicle. Three kinds of target displays were used. A stationary target in a known location; a target moving along a single known trajectory; and a target moving along one of three possible trajectories. These three conditions enable estimates of the effects of spatial uncertainty due to motion, and spatial uncertainty due to motion plus trajectory uncertainty to be assessed. If there is a decline in the size of the UFOV with age, then as well as the general elevation in detection threshold with age, we would expect to see amplified differences between the young and old groups as a result of the sources of spatial uncertainty.

Method

There were three basic types of display: (1) Baseline. The target was stationary and could occur in one of three locations, which were the same as those in the horizontal array in experiment 2; however, within a block of trials the target always occurred at the same location and this was known by the subject. (2) Single moving trajectory. The target moved along one of three trajectories so chosen that on average the target would increase to threshold luminance when it was at one of the three locations used in the baseline condition. It moved at 32'/sec. Within a block the target always moved along a single trajectory which was known to the subject. (3) Triple moving trajectory. The targets moved along the same trajectories as for the previous condition, but all three trajectories occurred within a block and were selected with equal probabilities on each trial.

Results

The mean detection luminance levels are shown in Figure 3. There were significant effects of all three main variables. First, there was an effect of age; the older

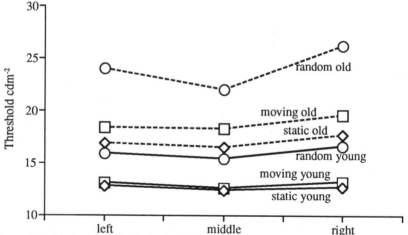

Figure 3. Detection Thresholds for each location, for static and moving targets, for old group (n=10, mean age=49) and young group (n=10, mean age=23)

group required about 5 cd/m² extra luminance to detect the targets. Second, there was an effect of spatial and temporal uncertainty; detection thresholds were higher in the triple-trajectory condition than in the single trajectory condition, which in turn were higher than in the stationary condition. Third, there was an effect of target location; the detection thresholds were lower for the central location than for either of the two peripheral ones.

More importantly, there were large interactions between the three main variables. First, the effect of target location varied across the tasks: there was relatively little effect of target location for the static condition and for the known trajectory condition, whereas, in the triple-trajectory condition, the central trajectory had the lowest detection threshold. Second, the differences between the three tasks were larger for the older age group than for the younger age group. For the younger group, the static and the single trajectory condition were virtually equivalent, whereas, for the older group, the detection thresholds were higher for the moving condition than for the static condition, for all three locations. Further, the differences between the triple-trajectory condition and the single trajectory condition were greater for the old group than for young group. Third, the effect of target location was greater for the older group than for the young group. This was most noticeable for the triple trajectory condition in which the detection threshold for the right hand target location was about 4 cd/m² higher than for the central location whereas for the younger group, the difference was only about 1 cd/m².

Discussion of Results

The basic structure of the results was similar to those for experiments 1 and 2. Overall, the old group have higher detection thresholds, but they are also more vulnerable to the various uncertainties of target location than the young group. There is effectively no cost to the young group of dealing with a moving target compared to a static target, whereas there is a small but significant cost to the old group of about 2 cd/m². Similarly, the cost of coping with the spatial and temporal uncertainty resulting from having three possible target trajectories is greater for the old group than for the young group. Lastly the variation due to location of the target is greater for the old group compared to the young group. In particular, although the right hand target location is the hardest to detect for both groups, as in the previous two experiments, this is most marked for the old group in the triple trajectory condition.

Discussion

Ageing has been shown to impair the detection of warning lights in multiple ways: sensitivity to light per se is reduced, the effects of uncertainty about where the light may appear are greater, and movement of the light makes performance worse too. Moreover, these effects sum non-linearly, making combinations of them worse than expected. The effects of these detection difficulties on stopping distances will depend upon the rate at which the weather conditions attenuate light with distance. In our experiments, the extra delays amounted to some 2 sec (experiment 1) or 1 sec (experiments 2 & 3). Older drivers may be able to develop compensatory skills in visual search that reduce the effects of the uncertainties; appropriate training programmes might therefore be designed to aid this process.

References

Ball, K.K., Beard, B.L., Roenker, D.L., Miller, R.L. & Griggs, D.S. (1988). Age and visual search: expanding the useful field of view, Journal of the Optical Society of America A **5**, 2210-2219.

Owsley, C., Sekuler, R. & Siemsen, D. (1983). Contrast sensitivity throughout adulthood, Vision Research **23**, 689-699.

THE DISPLAY OF CHINESE CHARACTERS ON A SMALL LCD

Bruce Thomas

Philips Corporate Design
5600 MD Eindhoven
The Netherlands
e-mail: c862820@nlccmail.snads.philips.nl

This paper gives a summary of some findings from an examination of the available literature on the presentation of Chinese script. This information is supplemented with the findings from interviews with native Chinese speakers. The principal findings suggest that either a vertical or horizontal format may be used, text can be scrolled but should not be split into arbitrary blocks, indicators should show when part of a text is hidden, and that a dot matrix of 15 x 15 is appropriate for the display of Chinese characters.

Introduction

Many current electronics products present information to the user by means of a small screen, typically having one or two lines, each providing space for about 12 characters. Multi-national concerns, such as Philips, develop many such products for global markets. These products are then adapted for local markets, for example by means of customising the language. In the development of products for the Chinese market, ergonomists working at Philips have noted that few recommendations were available concerning the presentation of Chinese text on small LCDs.

Approach

A brief review of the scientific literature concerning the display of Chinese characters was undertaken. This review was supplemented with interviews with four native Chinese speakers resident in the Netherlands and with correspondence with two further native Chinese speakers resident in Hong Kong. The following questions were addressed:
- How are short texts in Chinese script generally read? This should cover orientation of the text and preferred directions of reading (top to bottom, left to right, right to left).
- What rules, if any, govern the shortening of Chinese text? Should any constraints be placed on the free format of text (eg. length of text, possible limit of number of characters, splitting character combinations)?

- What are the requirements for punctuation (eg. question marks) and for spacing between characters?
- Is there a need to combine text with ASCII characters?
- Is dialect likely to be a problem?
- How large a character set is required?
- What resolution of characters is required? How small a dot matrix would be sufficient?
- Would Chinese users understand the use of symbols or icons, and to what extent can it be assumed that Chinese users are computer literate?
- How should the display be formatted with respect to aids to the user such as icons or on-screen text?

Findings

This section concerns the findings obtained from the scientific literature and interviews with native speakers. Most of the reported scientific work concerning the display of Chinese characters is based on the use of visual display terminals, an area of work which Gao et al. (1990) state is only just beginning.

Text display
Traditionally, Chinese text is read from top to bottom and from right to left, but Sukaviriya & Moran (1990) point out that the horizontal left to right Western style has become accepted. Hwang et al. (1988) also state that most Chinese now accept a horizontal format, but they point out that the horizontal format causes more visual fatigue with scrolling.
A suggestion was made by one native speaker that a constraint could be imposed to give text in a predefined format (eg. name, time, action, place). This would overcome some of the problems associated with not seeing complete text.

Shortening text
On the basis of statements made by native speakers, it is apparent that rules do exist for shortening Chinese text similar to the "telegraphese" used in English. Such text would seem to be frequently used and understood in Hong Kong.

Punctuation
In a brief overview of written Chinese material, it is apparent that the Chinese use similar punctuation marks to those used in European languages. Chinese speakers, however, have pointed out that there are some differences (eg. ∘ instead of .), as well as subtle differences in actual use. Punctuation would seem to be used only to clarify meaning, and there would appear to be no strict rules for its use. Short text may be written with no punctuation at all.

Spacing between characters
The spacing between characters would appear to be a fundamental issue in the presentation of Chinese text. Chuang (1982) studied the effects of line and character spacing on reading speed and found that line spacing had no effect, but that a spacing of 3 mm between characters produced the best results. Hwang et al. (1988), however, found an optimal character spacing of 1.2 mm, and recommend a character

spacing of three dots for characters formed with a 24 x 24 dot matrix. Chen (1989) looked at three kinds of line spacing ($^1/_4$, $^1/_2$ and 1 times character height) and five word spacings ($^1/_{16}$, $^2/_{16}$, $^3/_{16}$, $^5/_{16}$ and $^9/_{16}$ times character height) using a 16 x 16 matrix. Better performance was obtained with $^2/_{16}$, $^3/_{16}$ or $^5/_{16}$ word spacings and $^1/_2$ line spacing in terms of response time and accuracy. Chen also found that word spacing had more influence on visual performance than line spacing.

One native speaker pointed out that dividing text with spaces can make them more understandable, particularly with respect to showing which character groups belong together. This could be particularly useful for the display of partial text.

Combination with ASCII characters

From an inspection of some short Chinese text it is apparent that abbreviations in Roman characters (eg. "VDT") are used in Chinese text. Western units of measure (eg. kg., km.) and Western names also appear. Arabic numerals are used, but it is noticeable that there are subtle differences in the form of the numerals. The native speakers confirmed this finding, but noted that the recognition of Roman characters would not be so high in mainland China as in Hong Kong, Singapore and Taiwan.

Dialect

The native speakers stated strongly that dialect had no influence on written Chinese, since the written language is not associated with pronunciation. Sukaviriya & Moran (1990) point out that although a number of dialects are spoken in China, 70% of the population speak Mandarin Chinese. All literate Chinese are able to understand text written anywhere in Chinese characters. However, some characters used in mainland China are phonetic (ie. those used in the *pinyin* alphabet, see "size of character set" below) and these are subject to high error rates in their use because of local accents (Sheng, 1985).

Character font

Chen (1989) showed that visual performance reading Chinese characters on a VDT was best when the dot matrix structure was greater than 20 x 20. A higher resolution than this did not further improve performance. Tian (1987) compared four sizes of dot matrix (9 x 9, 12 x 12, 15 x 12 and 15 x 15) and found that the minimum acceptable size was 12 x 13. This produced consistent results in terms of recognition rate and subjective evaluation, but the legibility varied with the number of strokes and the presence of oblique strokes. The 15 x 15 matrix proved to be better adapted for various numbers of strokes or for oblique strokes. These results suggest that while a 12 x 12 matrix is readable, a 15 x 15 matrix would make complex characters more legible, and that ideally a 20 x 20 matrix should be employed.

Xu & Li (1989) point out that while each Chinese character has a complex and different structure, there is much redundancy in character formation. Tseng et al. (1965) carried out a study to test the possibility of reconstructing characters when some strokes were removed. They found that the average Chinese was able to reconstruct mutilated characters with up to 55% of the strokes omitted. This suggests that it might be possible to sacrifice some legibility in complex character formation if this proves to be necessary.

Size of character set

The Chinese character set is greater in size and complexity than the Japanese. The simplest character, a single stroke, means "one". The most complex character, with 64 strokes, means "verbose". Xu & Li (1989) claim that no-one really knows the exact number of Chinese characters. They state that the "*Kangxi Dictionary*" contains 47 935 characters. The "*New Chinese Dictionary*", on the other hand, contains only 8400 characters, of which only 4000 are everyday expressions. Leung et al. (1987) also seem to suggest that about 4000 characters are needed for practical use, although their paper is not entirely clear on this point.

A simplified character set is apparently in common use in Hong Kong, and this is used for telegrams. These characters are identified by means of a four digit code, which suggests that a maximum of 10 000 characters are used. It is possible that such modified characters could be adequately shown using a 12 x 12 matrix.

A simplified character set is also in use in mainland China, but this differs from that in use in Hong Kong. This character set (*pinyin*) is phonetic with characters representing syllables (Sheng, 1985). The character set has been in use for thirty years and is used in the teaching of Chinese in primary schools. Sheng argues that because of its widespread use, this alphabet would seem to be a good choice in the design of a keyboard for inputting Chinese characters. The same would probably apply for the display of a limited set of characters. In any case, it might be sensible to supplement a "standard" Chinese character set with the pinyin alphabet.

It was stated that a Chinese character set exists which can be printed with a 9-pin dot matrix printer. It was commented, however, that this was very unattractive.

Scrolling or blocking text

Hwang et al. (1988) measured the average speed of reading Chinese and found this to be 346 cpm. On this basis they studied three different scrolling speeds for Chinese text on a VDT (242, 346 and 450 cpm) and found that the scrolling speed of 450 cpm could be used to improve performance without causing additional visual fatigue. They also noted that scrolling is not superior to blocking in terms of performance, but scrolling causes less visual fatigue. Scrolling in a horizontal format was found to cause more visual fatigue than in a vertical format.

Chan & Chen (1991) found that reading performance for Chinese characters was greatest when the text was structured and presented at a rate equivalent to the reader's previously measured reading speed.

Discussions with native speakers indicated that not seeing the whole of a text could be confusing, since the Chinese do not use as many cues in word order as are used in European languages. A particular source of difficulty is that an incomplete text can appear as a complete text with a different meaning. In order to overcome this confusion, it was suggested that indicators be provided to show when parts of a text might be concealed. It was further noted that the arbitrary division of text (and not just single character combinations) can change the perceived meaning. Thus it was argued that blocking text is not an appropriate means of navigating through a text, and that scrolling is much better since character strings are not arbitrarily interrupted. Because text is not structured with such strict sequences of characters as European languages, there is likely to be a great 1need to scroll both backwards as well as forwards, and also readily to repeat the display text.

Icons and computer literacy

Single Chinese characters can be used to express some complex concepts. This has led many people to believe that Chinese characters serve the same function as icons (eg. Wood & Wood, 1987). The native speakers thought that there would be no problems associated with the use of icons, provided that they were well defined and meaningful. Sukaviriya & Moran argue that Asians generally understand icons, but the form of the icon needs to be considered. The *Sun* "mailbox" and the *Apple* "trashcan" for example do not correspond to images in the Asian culture.

The need for caution when transferring icons across cultures is inadvertently illustrated by Wood & Wood (1987) who state "...For example, the yellow diamond meta-icon informs the motorist that there is danger ahead....Even if the motorist cannot recognise the pictograph the meta-icon (shape and color) conveys information, and will probably evoke the desired response of caution." Wood & Wood argue from an American context. The sign they describe has some similarity with the European "right of way" sign. A European driver seeing the sign they describe might not, therefore, react in the way they desire. If transferring symbolic meaning creates problems between Western cultures, then these problems will inevitably be much greater when attempting to transfer meaning to an Oriental context.

Wang (1986, 1987) reports that computers are being used increasingly in mainland China industry, and Gao et al. (1990) also note the widespread use of visual display terminals. This suggests a level of computer literacy which could conceivably compare with European levels, an observation which was confirmed by the native speakers.

Additional factors

Zhang & Jin (1988) argue that luminance contrast is a major factor in assessing the readability of Chinese characters on VDT screens.

Sukaviriya & Moran (1990) point out that "each culture has its idiosyncrasies - some of which are factors necessary to consider in building an appropriate user interface for that culture". This means that not only the display of characters must be addressed, but also some cultural stereotypes need to be questioned.

Anecdotal evidence suggests that in oriental cultures the quality of a product is more important than in Western cultures. A product is perceived as a "gift" from the manufacturer, and features of the product which emphasise gift like characteristics make the product more acceptable to its user. Such features might include easy and direct access to function and an attractive presentation. Anything suggesting cheapness (even if the product is inexpensive) thus makes the product less acceptable.

Conclusions

The implications for the display of Chinese text on small LCDs are as follows:
- A vertical format is preferred, but a horizontal format is acceptable.
- Text can be scrolled but should not be blocked. A scrolling speed of about 346 cpm may be appropriate.
- Indicators should be used to show whether part of a text is hidden.
- The space between characters should be about $\frac{1}{4}$ - $\frac{1}{3}$ of the character height.

- A set of 8500 characters would cover almost all modern Chinese usage, and even 4000 characters may be sufficient.
- A 15 x 15 matrix allows adequate recognition of all characters. A 12 x 12 matrix is acceptable but has limitations.
- The *pinyin* character set should be included if the product is to be sold in mainland China.
- Roman characters and arabic numerals should be included in the character set.
- Punctuation marks should be included in the character set.
- Any icons used should be drawn up by a Chinese designer.
- Short "technical" text can be as little understood by Chinese users as similar text in English by Western users.

References

Chan, K.T. and Chen, H.C. 1991, reading sequentially-presented Chinese text: effects of display format, Ergonomics, **34**, 1083-1094.

Chen, J. 1989, Experiments on effects of dot-matrix size and its format of Chinese characters upon VDT visual performance. MA Thesis, Hangzhou University.

Chuang, C.R. 1982, The effects of Chinese text layouts on reading speed (in Chinese)
In: H.S.R. Kao and C. Cheng (eds.) *Psychological aspects of Chinese language,* (Crane, Taipei) 219-226.

Gao, C., Lu, D., She, Q., Cai, R., Yang, L. and Zhang, G. 1990, The effects of VDT data entry work on operators, Ergonomics, **33**, 917-923.

Hwang, S.-L., Wang, M.-Y. and Her, C.-C. 1988, An experimental study of Chinese information displays on VDTs, Human Factors, 30 461-471.

Leung, C.H., Cheung, Y.S. and Wong, Y.L. 1987, A knowledge-based stroke-matching method for Chinese character recognition, *IEEE Transactions on Systems, Man and Cybernetics*, SMC-17/6, 993-1003.

Sheng, J. 1985, A pinyin keyboard for inputting Chinese characters, Computer, **18**, 60-63.

Sukaviriya, P. and Moran, L. 1990, User interface for Asia, In: J. Nielsen (ed.) *Designing user interfaces for international use,* (Elsevier, Amsterdam) 189-218.

Tian, Q.H. 1987, An experimental study on legibility of the dot-matrix sizes of Chinese characters, BA Thesis, Hangzhou University.

Tseng, H., Chang, L. and Wang, C. 1965, An informational analysis of the Chinese language (in Chinese), Acta Psychologica Sinica, **4**, 281-290.

Wang, Z. 1990, Recent developments in ergonomics in China, Ergonomics, **33** 853-865.

Wood, W.T. and Wood, S.K. 1987, Icons in everyday life, In: G. Salvendy, S.L. Sauter and J.J. Hurrell (eds.) *Social, ergonomic and stress aspects of working with computers* (Elsevier, Amsterdam) 97-104.

Xu, L. and Li, W. 1989, The recent development of inputting systems for the Chinese characters and its psychological foundation, In: G. Salvendy and M.J. Smith, *Designing and using human-computer interfaces and knowledge based systems* (Elsevier, Amsterdam) 675-680.

Zhang, H.Z. and Jin, Q.C. 1988, Effects of illuminance and VDT luminance contrast on visual performance, Acta Psychologica Sinica, **20**, 243-252.

TOWARDS SPECIFYING THE DESIGN CRITERIA FOR THE VISUAL FRONT-END OF TELE-PRESENCE SYSTEMS

Jörg W. Huber (1) and Ian R.L. Davies (2)

(1) School of Life Sciences,
Roehampton Institute, West Hill, London SW15 3SN
(2) Department of Psychology,
University of Surrey, Guildford GU2 5XH

Tele-presence systems provide the operator with visual input from a
remote video camera. We report a series of experiments that aimed (i) to
establish the extent to which providing the observer with the 'natural' link
between their own movement and transitions in the visual field, improves
perception of the spatial layout of the scene, and (ii) to identify what are
the essential components of the movement-generated information. The
first two experiments indicate that depth judgements were more accurate
in the viewing conditions that included movement-generated information.
However, the way the image-transitions were produced, i.e. the nature of
the perception-action coupling, had no detectable effect. The third
experiment points to the presence of learning effects which may prevent
utilisation of movement-generated information. Future research should
study tele-presence with more complex stimuli, and consider the role of
learning and training in the use of movement-generated information.

Introduction

Tele-presence systems provide operators with visual input from a remote video
camera which may allow them to control some kind of mechanical manipulator in a
hazardous environment with relative safety. There is a variety of technologies for
presenting the information captured by the camera, but the most common (and the
simplest) is to use a standard CRT screen. It is therefore important to study the
effectiveness with which pictures on CRT screens can be used to perceive accurately
spatial layout in the remote environment, focusing on the importance of movement-
generated information.

In pictures, photographs and flat-screen displays the depth cues of motion parallax,
accommodation and convergence all suggest flatness while perspective information
suggests three-dimensionality. These two conflicting sets of cues are the basis of the
'dual reality' of pictures (Sedgwick & Nicholls 1993) and the cue-conflict reduces the
sense of depth produced by the display. Video images produced by a camera whose
movements mimic the head movements of the observer are one way in which the conflict

between motion generated information and perspective information may be removed. Such systems link the video image to the head-motion of the observer, so that the changes in the video image viewed by the observer emulate the changes that would have occurred in the visual field if the observer was viewing the scene directly.

We report here a series of experiments that aim to establish the extent to which providing the observer with the 'natural' link between their own movement and transitions in the visual field improves perception of the spatial layout of the scene. They also aim to identify what are the essential components of the movement-generated information.

Experiment 1

Observers viewed video images produced by a camera whose movements mimicked the head movements of the observer. The stimuli consisted of the video images of two rods produced by a camera and viewed on a monitor, as shown in Figure 1. The rods used on any given trial were selected from a set of rods that varied in diameter; thus the observer was not able to use the relative widths of the rod-images as a reliable indicator of relative distance. A mask on the monitor prevented the observer from seeing the ground plane and the top of the rods. The distance between the two rods along the z-coordinate varied from trial to trial. In each experiment, depth judgements of the rods were made i) in a stationary condition where the camera was fixed and ii) in a moving condition where the camera could rotate in the arc shown in Figure 1a. The distance between camera and the centre of movement was ca. 1850 mm. All observers (n=12) took part in all viewing conditions.

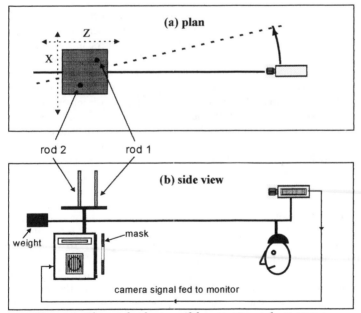

Figure 1. *Plan and side view of the experimental apparatus.*

Results

The results (Figure 2.) showed a strong advantage for the movement condition ($F(1,11)=35.87$; p<.0005). Further, this advantage increased as the distance between the two rods increased ($F(4,44)=5.57$; p=.001). The very low numbers of correct judgements in the stationary condition are due to the fact that the width of the bars suggested an incorrect depth order. In general the experiment confirmed the pick-up of motion parallax information.

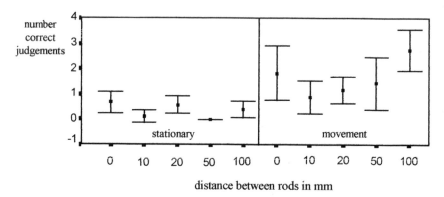

Figure 2. *Means and 95% confidence intervals for number of correct judgement.*

Experiment 2

This experiment compared the judgements of an active observer with those of i) a passive observer and ii) an observer with reversed action–perception coupling. The image information was equivalent for the first two observers, however the passive observer could not control the image transitions. The passive observer was seated in front of a slave monitor without being able to see what the active observer did. The observer with reversed action–perception coupling saw the mirror-image. Reversing the image horizontally disrupts the natural link between observer movement and image-motion, but leaves intact the transitional image information for relative distance. All observers viewed the stimuli both in the stationary and the movement condition. The presention order was counter-balanced. Distance between rods varied from .75 to 4.5 inches; six trials were carried out for each distance resulting in a total of 48 experimental trials for each observer.

Results

The mean scores for the stationary and moving condition for the passive, active and reversed-picture observers are shown in Figures 3a, b and c respectively. In all three cases performance was better with the moving image ($F(1,36)=10.87$; p<.005), but there were no significant differences between groups ($F(2,36)=1.58$; ns): the passive observers did as well as the active observers; and those with the reversed head–motion/image–motion link did as well as those with the natural link. Therefore the advantage for the moving images over the static image is not influenced by the nature of the coupling between observer, camera and picture. This finding differs from that of Smets,

Overbeeke and Stratman (1987) who found a significant advantage for the active observer over the passive observer.

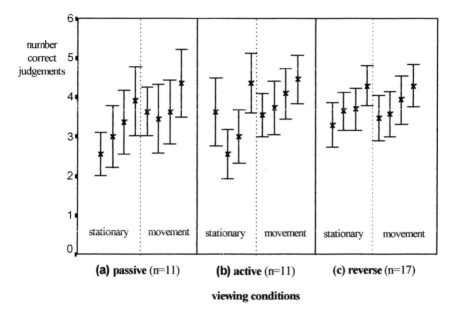

Figure 3. *Means and 95% confidence intervals for number of correct depth judgements. The theoretical maximum number for each mean is 6 correct judgements. Results are based on 3 groups of observers. Within each viewing condition distance between the rods is shortest on the left and increases towards the right (.75, 1.5, 3.0 and 4.5 inches).*

Experiment 3

The first two experiments looked at the effectiveness of tele-presence in judging static environments. The third experiment included spatial judgements in a moving world. Subjects were required to control a vehicle and place it at the same distance away as a target object. The vehicle moved along a track parallel to the line of regard of the camera. The target was at a fixed position, offset from the line of regard. The apparatus for this experiment simulated tele-presence in a more realistic way by minimizing any non-visual feedback resulting from the observers adjustment operations. The sample consisted of 13 observers who were randomly allocated to a group starting with head movement and a group starting with head stationary. Each observer carried out a total 30 experimental trials, which were equally split between conditions.

Results

Observers who started with the head movement condition were more accurate in the movement condition than in the stationary condition (F(1,11)=4.95; p < .05). Figure 4b shows the group which started with head movement; they performed significantly better in the movement condition than in the stationary condition. In contrast the group which started with the stationary passive condition (Figure 4a) did not improve in the movement condition. This suggests the presence of short term learning effects which are counter-productive for the group which started with the stationary condition. However, the group which started with the movement condition performed as expected: movement lead to better performance compared to stationary viewing.

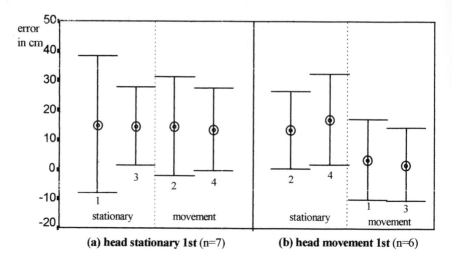

(a) head stationary 1st (n=7)   **(b) head movement 1st (n=6)**

Figure 4. *Mean signed errors as a function of presentation order. Means and 95% confidence intervals for observers starting with static condition first ('head stationary 1st') and those starting with movement condition first ('head movement 1st'). Numbers underneath confidence bars give order in which blocks of trials were presented.*

Discussion and Conclusion

In the first two experiments, judgements of relative depth were more accurate in the viewing conditions that included information provided by changing the camera's view point, i.e. movement-generated information. However, the way the image transitions were produced – the nature of the perception-action coupling – had no detectable effect on the accuracy of depth judgements: the passive observer, and the observer with reversed coupling did as well as the observer with the natural perception-action link. These results are not consistent with those of Smets et al. (1987) who found an advantage for the active observer, but support those of Beall, Loomis, Philbeck and Fikes (1995) who found that motion parallax is only a weak cue to depth in both real and virtual environments.

The final experiment points to the presence of short term learning effects which could be interpreted as the development of perceptual 'functional fixity' (Duncker 1945): this fixity prevents the observer who is initially stationary and looks at a static picture from utilising movement-generated information in later trials. It is possible that such effects could be overcome with appropriate training on tele-presence systems.

The stimuli used so far in the experiments have been very simple. For such stimuli the findings suggest that motion parallax is only a weak cue to depth. However, it is possible that perception of more complex scenes would lead to a clearer advantage for the active observer. This possibility could be understood in two ways: either (i) motion parallax is more effective in more complex scenes, perhaps in combination with occlusion and dis-occlusion of objects, or (ii) tele-presence systems provide control of the station-point of the camera. The latter means that the observer can explore the scene by shifting the station-point of the camera. This form of 'exploratory behaviour' could furnish depth information not based on motion parallax. Instead changes of the station-point could reveal depth information by turning depth into 'breadth'.

This is shown in Figure 5 where the station-point of the camera moves from position 1 to 2: 'invisible' depth along the Z-axis is turned into 'visible breadth'. This suggestion is reminiscent of Berkeley's conception of depth which Merleau-Ponty (1994/1962) succinctly summarised as: "depth is tacitly equated with *breadth seen from the side*" (p 255, italics in original). The underlying idea is that spatial extend is constructed from depth seen from the side. A possible implication of these considerations for the design of tele-presence systems is that it maybe sufficient to supply the observer with a number of views from different station-points of the camera rather than continuous movement information.

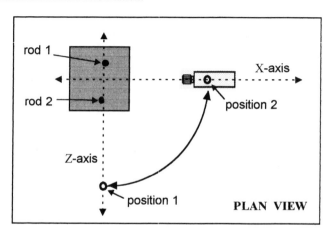

Figure 5. *"Depth is ... equated with breadth seen from the side"; for details see text.*

References
Beall, A.C., Loomis, J.M., Philbeck, J.W. & Fikes, T.G. (1995). Absolute motion parallax weakly determines visual scale in real and virtual environments. *SPIE Human Vision, Visual processing, and Digital Display* VI, **2411**, 288-297.
Duncker K. (1945). On problem solving. *Psychological Monographs*, **58**, No. 5, 1-113.
Gibson J. J. (1979). *The ecological approach to visual perception.* Boston: Houghton Mifflin.
Merleau-Ponty M. (1994/1962). *Phenomenology of perception.* London: Routledge.
Sedgwick H.A. & Nicholls A.L. (1993). Cross talk between the picture surface and the pictured scene: effects on perceived shape. Sixteenth European Conference on Visual Perception, Edinburgh, 25-29 August 1993. *Perception*, **22**, Supp., 109.
Smets G., Overbeeke C. & Stratman M. (1987). Depth on a flat screen. *Perceptual & Motor Skills*, **64**, 1023-1034.

Medical Ergonomics

MINIMISING HUMAN ERROR IN THE DETECTION OF BREAST CANCER

Helen Cowley, Alastair Gale

*Applied Vision Research Unit,
University of Derby,
Mickleover, Derby DE3 5GX*

The PERsonal perFORmance in Mammographic Screening (PERFORMS©) self-assessment scheme has been widely used by UK breast screening radiologists. Data obtained from this programme reveal several important factors concerning skill in identifying mammographic abnormalities. These include: those mammographic features which were undetected or misinterpreted by the participating radiologists and any improvement or deterioration in detection ability over the two completed rounds. Areas of individual difficulty have been identified and this information fed back to the radiologists. Training subsets concentrating on these specific features have subsequently been developed. These subsets have been shown to improve radiologists' cancer detection performance.

Introduction

Breast cancer is the most common form of cancer among women in the UK. It claims the lives of nearly 16,000 women every year and one in 12 will be affected by it at some time in their lives. The UK has one of the highest mortality rates from breast cancer in the world. Incidence and mortality both increase with age, with most cases occurring in women over 50. Breast cancer needs to be detected early in order to increase the chance of the woman concerned surviving. It is for this reason that the breast screening programme was established.

The National Health Service Breast Screening Programme (NHSBSP) was established in 1988 following a report by Sir Patrick Forrest. This report stated that deaths in women aged 50 to 64 could be reduced by one third or more if they were offered screening by mammography. Mammography involves a low dose x-ray of the breast, which can detect small changes in breast tissue long before they are seen or felt. The programme offers free breast screening every three years for women over the age of 50 living in the UK. Women under the age of 50 are not offered screening because the breast tissue is denser and this makes it more difficult to detect cancer.

The success of the NHSBSP depends largely upon the skill of the radiologists involved in detecting and appropriately interpreting the information available on the mammograms. Due to the nature of screening it is difficult for individual radiologists to assess how well they are actually performing and to gain insight into any areas of personal difficulty. This is due to the relatively low incidence of the disease, pressure to detect small cancers, ie < 1 cm, and the subtle signs associated with these small cancers. The programme has nationally agreed quality assurance parameters and some of these can be used as individual performance indicators. These data however tend to be incomplete as sensitivity is dependent on interval cancers and the number of these are unknown for several years.

The PERFORMS self-assessment scheme was devised as part of the quality assurance of the NHSBSP. It was set up in response to a request by the Breast Screening Programme and the Royal College of Radiologists. The two rounds of this scheme that have been completed have shown that PERFORMS can and does highlight areas relating to individual and national factors where human error is prominent.

PERFORMS

The aim of this scheme is to directly measure an individual radiologist's performance and compare it with that of their peers. These data give regional and national mean performance values that are based on performance at that point in time and can be used repeatedly to assess improvement or deterioration in film reading ability.

PERFORMS has been running since 1991 and two full rounds have been completed. Over 90% of UK breast screening radiologists have participated in both. It consists of a standard mammographic film reading task that contains 120 pairs of mammograms which illustrate a wide range of radiological abnormalities covering normal, benign and malignant features. The set is heavily weighted towards recall and malignant cases. The cases chosen are particularly difficult, in order to encourage false positive and negative errors. Pathology data are available for all the cases that have had biopsy. Each radiologist is invited annually to report on the set of mammograms, for which a 'radiological standard' opinion has been agreed upon. This radiological standard was compiled by a group of breast screening radiologists who had a great deal of expertise.

For each case the radiologist marks any abnormal feature on a diagram of a pair of films, rates their confidence in the presence of any abnormality on a five-point rating scale (absent to present) and classifies the film (normal/benign to malignant). A film classified as normal/benign would be returned to normal screen, any other rating would result in the case having further assessment.

Following participation in PERFORMS, results are given concerning individual as well as anonymous mean regional and national performance. The information returned shows how appropriately the cases were classified and how suitably various radiological features were detected. In addition, information concerning the number of missed malignant cases and whether the malignant features involved were undetected or misinterpreted is also included. This information can highlight areas of individual error performance for radiologists. Individual data are then compiled to form regional and national data. Having completed two rounds of PERFORMS, it is possible to see how the performance of individuals and regions has changed.

ROC Analysis

Results from PERFORMS take many forms, one of these being receiver operating characteristic (ROC) curves and the area under these curves (A_z values). ROC analysis is based on the principles of signal detection theory and is frequently used in diagnostic radiology as a measure of performance. In PERFORMS, each case is classified as recall or non-recall and the classification ROC curve is generated from the number and rating of cases in each of these two conditions. The detection ROC curve is obtained from the combined absent to present rating data for each feature. ROC curves are plotted by fitting true and false positive pairs (CR and IR respectively) by maximum likelihood estimates of its parameters.

Classification Results

In PERFORMS, individual radiologists have to classify each case as to whether they would recall it for further assessment or whether they would return it to normal screen. From these data, results are initially examined as; correct and incorrect recall and correct and incorrect return to screen decisions for all the cases. Individual classification results are also given in the form of a ROC curve.

These individual data are combined in order to give regional and national mean classification performance values. These national values show the change in performance over the two rounds. Figure 1 shows the mean national increase in correct recall performance for the second round in comparison to that for the first.

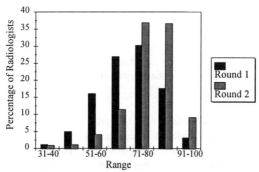

Figure 1. Percentage range of correct recall values and the percentage of radiologists scoring in each of these ranges.

This figure shows a significant increase in correct recall performance from the first to the second round (p<0.001).

Malignant Cases

Breast screening radiologists are most interested in whether they have missed any pathologically malignant cases from the PERFORMS scheme. In the first round, the national mean for the number of missed malignant cases was five, this dropped to a mean of three in the second round. Figure 2 shows the mean number of missed malignant cases for each of the UK regions over the two rounds and the national mean (NM). It can be seen that the number of malignant cases missed has significantly decreased in the second round (p<0.001). Some regions have not participated in both rounds.

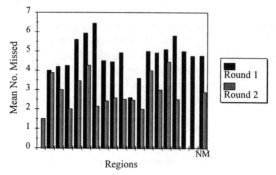

Figure 2. Mean number of malignant cases missed by each of the 18 regions over the two rounds.

A mean of five pathologically malignant cases missed in the first round corresponds to an error rate of about 21%. This error rate does however correspond to performance for a difficult film set. This observer error rate is also demonstrated in other areas of radiology. The mean error rate drops to around 13% in the second round.

Features

PERFORMS also identifies which radiographic features individual radiologists have problems detecting and interpreting. Following participation in this scheme, individual radiologists receive a ROC curve showing their combined detection data for all the features. They also receive data concerning the numbers of each feature undetected and misinterpreted. These data can be combined and this provides national results.

There are two types of false negative (misses) observer errors in PERFORMS. These are misinterpreted or undetected features. When a feature is noted, but recorded as another feature and then misclassified; or when a feature is noted, correctly recorded, but then incorrectly classified, then these features are misinterpreted. Other features are not seen by the radiologist and therefore not recorded, these features are undetected. False positive (over-reading) errors also occur. These are due to misinterpretation of the mammograms and the features present.

These errors also occur in the breast screening programme. False negative errors often result in the woman presenting with an interval or incident round cancer. Whereas false positive errors result in unnecessary stress for the woman and expense for the breast screening programme. Both types need to be kept to a minimum.

PERFORMS results concentrate on the false negative observer errors. Figures 3 and 4 show the mean number of features that were undetected and misinterpreted for all radiologists in the second round. Figure 3 shows those features that were undetected and misinterpreted for all the cases. This figure clearly shows that asymmetry, parenchymal deformity and ill defined mass are most frequently undetected. Figure 4 shows the features undetected and misinterpreted in the malignant cases alone.

Calcification is most frequently misinterpreted in the malignant cases; with asymmetry, parenchymal deformity and ill defined mass again being most frequently undetected.

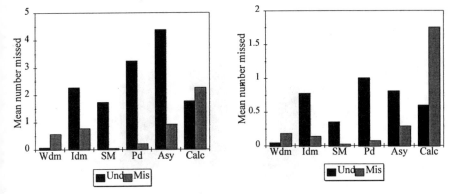

Figures 3 and 4. Features undetected and misinterpreted in the recall and malignant cases.

These figures show which features cause the most problems for radiologists. The UK interval cancer database also shows that these features are most frequently missed in the breast screening programme.

Training Subsets

These have been developed following results from the two rounds of the PERFORMS self-assessment scheme and from the UK interval cancer database. These measures of performance indicate that certain abnormal features found in the breast cause more problems for individual and national radiologists than other features.

The training subsets each concentrate on one specific feature; namely parenchymal deformity, masses, calcification and asymmetry. To date, the parenchymal deformity subset has been tested and over 40 radiologists have completed this set. Over 70% of UK screening radiologists are interested in participating in this subset. Other subsets are in various stages of completion.

The parenchymal deformity training subset comprises a training and an assessment part. In the training part, the participating radiologist examines a case, completes a reporting form for that case and then compares their diagnosis to that of a radiological standard 'answer'. The radiological standard consists of a completed reporting form. They then complete a brief questionnaire before proceeding to the next case. The radiological standard has been compiled from a group of 25 radiologists' individual diagnoses for each of the cases. In the subsequent assessment section the individual simply records his/her diagnostic decisions.

Following participation in the subset, the radiologists receive data concerning their classification and parenchymal deformity performance in the training and assessment part of the subset as well as previous performance in PERFORMS. They receive these data in the form of two by two tables as well as ROC curves.

Results

Mean results from the radiologists who have already completed this subset show that there is a significant increase in parenchymal deformity detection performance following training. Figure 5 shows the A_z values for the three sets of data.

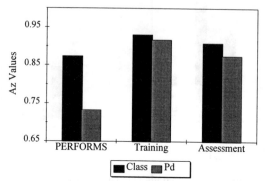

Figure 5. The mean A_z values for PERFORMS and the training and assessment parts of the parenchymal deformity training subset.

These results suggest that following training, parenchymal deformity detection performance does improve. Performance will however need to be compared in the subsequent round of PERFORMS in order to say whether training has a long term effect.

Conclusions

PERFORMS can and does measure individual and national radiologists' breast screening performance and does highlight areas of individual difficulty. It has shown that the mean performance of radiologists has improved over the two rounds, but that there is a wide variation between individual radiologists. Training subsets concentrating on the specific features that cause problems for radiologists have been shown to improve cancer detection performance.

References

Patnick, J. 1993, *NHS Breast Screening Programme Review.* NHSBSP Publications, Sheffield

Savage, C.J., Gale, A.G., Pawley, E.F. & Wilson, A.R.M. 1994, *To err is human, to compute divine?* Digital Mammography.

Gale, A.G., Savage, C.J., Pawley, E.F., Wilson, A.R.M. & Roebuck, E.J. 1994, *Breast screening: visual search and observer performance.* In: Image Perception, H L Kundel (Ed.) Proc. Spie 2166 pg 66-75

Gale, A.G., Wilson, A.R.M. & Roebuck, E.J. 1993, *Mammographic screening: radiological performance as a precursor to image processing.* In: Biomedical Image Processing and Biomedical Visualisation, R S Acharya and DB Goldof (Eds.) Proc. Spie 1905 pg 458-464

Green, D.M. & Swets, J.A. 1974, *Signal Detection Theory and Psychophysics.* (Wiley, New York)

Metz, C.E. *Basic principles of ROC analysis.* Semin Nucl Med 1978; 8:283-298.

ERGONOMICS OF KEYHOLE SURGICAL INSTRUMENTS - PATIENT FRIENDLY, BUT SURGEON UNFRIENDLY?

N A-M Crombie and R J Graves

Department of Environmental & Occupational Medicine, University of Aberdeen

Minimal Access Surgery (MAS) has gained the distinction of being patient friendly and resulting in significant cost savings. However, the rapid development of these procedures has meant that the physical ergonomics of the situation appears to have been largely overlooked. The present study assessed the risks of discomfort and injury to the surgeon from the nature of the tasks and the design of the instruments used during Laparoscopic Cholecystectomy. The task elements were analysed for biomechanical risk factors and a questionnaire survey was undertaken to obtain surgeons' opinions regarding discomfort and/or injury. A number of ergonomic risk factors in instrument design were found in relation to tasks undertaken by the surgeon. Sixty four percent of the survey respondents reported pain / discomfort associated with the use of specific instruments.

Introduction

Minimal Access Surgery (MAS) or keyhole surgery is intended to reduce trauma associated with traditional techniques by eliminating the large incision needed to access the operative field (the internal body cavity). Laparoscopy is an approach which involves accessing the operative field via several ports through the abdominal wall. Purpose designed instruments and a miniature camera are introduced through hollow sleeves called cannulae and are manipulated outside the body, their operation guided by a surgeon watching the internal view of the patient on television monitor(s).

Laparoscopic Cholecystectomy (LC), the surgical removal of the gall bladder by laparoscopic techniques, is one of the most well established MAS procedures. There are a number of reasons for the rapid progression of this procedure since it was first carried out in 1987. These include minimal cosmetic disfigurement, fewer post-operative complications and pain (Soper et al, 1992), earlier release from hospital and earlier return to full activity (Zucker et al, 1991). A further reason is pressure from the Department Of Health to cut costs. It has been estimated that the cost saving per patient resulting from reduced hospital stay is £900 (Cuschieri et al, 1991).

The introduction of LC has also brought with it many implications for the surgeon. First, surgeons must learn to overcome restrictions imposed by the two dimensional view and master co-ordination skills needed whilst using the uniquely designed instruments. Although training issues are now being addressed, very little attention appears to have been paid to the implications of the procedure in relation to the physical aspects of surgeons' tasks. There is evidence to suggest that the ergonomic design of laparoscopic instruments can cause surgeons a number of problems. For example, there have been a number of reports of paresthesia in the distribution of the lateral digital nerve of the thumb (see Kano et al, 1993). These have been attributed to inadequate handle design of certain instruments. It has also been suggested that problems experienced by surgeons may be more widespread than this, occurring at other locations than the hand. A study by Graves et al (1994) indicated that there are potential musculoskeletal risk factors associated with awkward postures adopted by laparoscopic surgeons. It was suggested that such postures may lead to discomfort or fatigue, perhaps even to permanent disorders over a period of time. Such fatigue or injury may lead to an increased risk of error, with implications for patient safety.

Changes in instrument design and the way they must be used have had a drastic effect on the physical ergonomics of surgeon's tasks. MAS techniques, however, including those involved in LC, have developed rapidly, with little opportunity for a gradual evolution of instruments, apparently leading to their ergonomics aspects being largely ignored. The aim of the present study (Crombie, 1995) was to assess the potential risk of discomfort / injury to the laparoscopic surgeon by examining ergonomic aspects of the design and use of instruments used during L C.

Method

This study relied on observational techniques which have the advantage of allowing analysis of any system without changing it. The need to ensure patient safety limited the type of measurements that could be made.

The first stage of the project was to carry out a detailed description of a typical, complication-free, LC. The procedure was observed on a number of occasions in theatre. A video recording of an example of the procedure was later analysed in detail. The task was broken down into a series of sub-tasks which reflected different stages in the procedure. Using a variation of Gilbreth's Work Method Analysis each sub-task was further broken down into a series of elements. Task elements associated with force, repeated/sustained exertions and awkward postures were recorded to provide a means of assessing biomechanical risk factors associated with the task.

A further stage involved carrying out an instrument evaluation to determine whether there were any ergonomic risk factors inherent in the design of the instruments. Instruments were selected on the basis that they were used the most intensively throughout the LC procedure, as confirmed by a consultant surgeon. A checklist was developed and used to determine ergonomic aspects of design features of the instruments. The checklist was based on recommendations from the Health & Safety Executive (1990), Putz-Anderson (1988) and Fraser (1980).

An additional stage involved obtaining opinions from surgeons regarding difficulties they had encountered with instruments and of any discomfort / injury they had experienced. A pilot questionnaire was modified on the basis of feedback from various medical personnel and sent to 63 surgeons. These were surgeons at National Health Service (NHS) hospitals throughout Scotland and England chosen at random from an NHS directory. The criterion for completion of the questionnaire was that the surgeon was currently carrying out LC.

Findings And Discussion

The aim of a cholecystectomy is to detach and remove the gall bladder. The specific technique examined in this study can be divided into six stages. The first stage involves insufflating (inflating) the peritoneal cavity with carbon dioxide to allow space for manipulation of organs and instruments within the cavity. This cushion of gas, termed Pneumoperitoneum, must be maintained throughout the procedure to ensure that the surgeons view of the cavity is never obstructed. The purpose of the second stage is to establish access to the operative field. A cannula is put in place and the tiny camera / light source is inserted and orientated through this. Three further cannulae for additional instruments are then inserted. The surgical dissection begins in stage three where the cystic duct and artery are freed from surrounding tissue, the cystic pedicle. Having grasped and orientated the gall bladder with the grasping forceps in the left hand, the dissection is carried out using the electrosurgical hook and/or the dissecting forceps in the right hand. Once completed, the duct and artery are clipped using a clip applicator in stage four. The gall bladder is dissected free from the liver bed in stage five. The final stage involves extracting the gall bladder from the abdominal cavity.

Postural analysis identified that the use of certain instruments was associated with awkward postures and a number of these had static, forceful and repetitive components. The instrument checklist indicated that many of these risk factors can be explained with reference to the ergonomic design of the instruments.

Grasping Forceps

These are used continuously throughout all stages of the procedure. This instrument is operated as conventional forceps with the thumb and middle finger inserted through ring handles. The length (395 mm) and handle design of this instrument could affect the ease with which the instrument can be manipulated, the degree of loading on the surgeon's fingers and thumb and in some circumstances lead to awkward postures. For example, the position of its pivot at a position where approximately two thirds of the length of the instrument is beyond the handles adds to forces being applied to the finger and thumbs. Evidence of risk from the design is supported by several reports in the questionnaire where pain and discomfort were identified in the thumb and middle finger, possibly related to pressure from the narrow handle edges and small rings. The latter were 22x25 mm at their largest which does not provide adequate room for larger users (Pheasant, 1992).

Another factor relates to the surgeon having to change his grip on this instrument. Normally he has the tissue grasped in its jaws with the fingers extended with the edge of the instrument pinched between them. This means that the instrument cannot be held and operated at the same time. Holding the instrument in this way could be fatiguing due to pinching with extended fingers. The forceps are held in this manner because of the way the ratchet mechanism is designed to eliminate the need continually to exert force on the handles to keep the jaws of the instrument closed. The design and location of this ratchet means that it can be inadvertently actuated thus releasing the jaws' grasp on the tissue. If the surgeon is not aware of the open jaws, they could cause damage to the patient, so the surgeon tends to maintain this static grasp as a safeguard. Twelve respondents to the survey, i.e. almost half, indicated they had difficulties holding the grasping forceps in place reinforcing the suggestion that this type of ratchet is inappropriate.

Tasks using the forceps appeared to lead to awkward postures. For instance, in one particular task element, the shoulder was abducted with the forearm being pronated and the elbow flexed at approximately 70 degrees, elevated and held away from the body. The wrist was held in a neutral position, with the fingers in a pinch grip holding the edge of the forceps. Over a period of 33 minutes the lower arm postures were adopted 35 times for 2-3 seconds at a time, while the shoulder was held almost continuously in its abducted posture. Holding the shoulder in this manner for this length of time is likely to lead to static loading on muscles. Overall, the combination of repetition and extreme postures are likely to contribute to the general fatigue of the surgeon.

The forceps must be manipulated through various angles. It was observed that the angle of the handles (115°) resulted in awkward postures. Although angled handles can reduce awkward postures when an operator works in one plane, they can exacerbate awkwardness in other planes such as illustrated in this situation.

Dissecting Forceps

At various intervals throughout stage three the surgeon uses this other type of forceps in the right hand. This device is used to separate and divide tissue. It is 421 mm in length with angled handles. Using the instrument typically involves a semi-pronated forearm with the elbow flexed and held away from the body. The wrist is in slight radial deviation and the fingers hold the forceps in a precision grip. Because the left arm orientates the gall bladder for the right hand to operate on, the posture of the right wrist and forearm are not extreme and are unlikely to cause much harm. However, the shoulder is held in a static posture, thus discomfort or pain in the shoulder / neck area may occur on the surgeons right hand side.

The use of the forceps is likely to lead to excessive pressure on the fingers caused by narrow edges of the rings and inappropriate sizing of the rings. In addition, the forceps is operated repetitively by the thumb to open and close the jaws. This action produces pressure on the back of the thumb after each snip to open and close the jaws. There were seven reports of pain in the fingers in the questions associated with the dissecting forceps.

Electrosurgical Hook

This device is operated by electricity with two modes of operation to cauterise and dissect tissue. The surgeon holds the shaped cylindrical handle mainly to direct the static functional tip and operates it by two foot pedals which are located to his right. Asymmetric whole body posture was observed during the use of this instrument. This is caused by the surgeon having to twist the lower half of his body to use the foot pedals whilst his upper body faces the direction of the instrument and screen. As it is often in prolonged use, it is likely that the surgeon will experience some general discomfort. Indeed, three respondents from the survey reported experiencing pain/discomfort in the knees whilst using this instrument; others reported non-specific pain/discomfort in the leg and lumbar region and the neck/shoulder region which may also be attributable to this device. This mode of operation is at the very least inconvenient, as it interrupts the procedure because the surgeon often has to stop what he is doing to relocate the foot pedals.

The hook is held in an internal precision grip where the stem rests between the middle and ring finger, the thumb and two fingers oppose each other at or near the end of the handle. Because the wrist is often flexed and the fingers extended there is poor contact with the surface of the handle and limited sensory feedback. It appeared that this is a consequence of too short a handle, some 26 mm below the recommended minimum of 100 mm (Putz- Anderson, 1988). A precision grip can contribute to fatigue of the hand as the weak muscles of the finger are working at a mechanical disadvantage.

Suction /Irrigation Device

This device is used often throughout the procedure to irrigate the operative site and remove excess fluids. The instrument rests along the length of the index and middle fingers with the thumb operating its push buttons. Although not likely to be particularly uncomfortable or painful, (indeed, there were no reports of pain / discomfort whilst using the device), this grip does not provide good support of the instrument. In turn this can cause the surgeon to exert more force to increase stability of the instrument. The instrument is held in this way because there is no handle and the surgeon must grasp onto the housing of the push buttons. A suitably designed handle is a basic ergonomic requirement of tool design to enable adequate grip and control.

General Risks

In this specific surgical procedure, there were approximately 30 changes of instruments. Any one change of instrument involved removing the instrument from the cannula, inserting it into the instrument holder, removing the replacement instrument from its holder and then inserting it into the cannula. This series of actions can involve awkward postures, typically elbow and wrist flexion, due to the location and length of the instruments. Moreover, inserting these long instruments into the narrow cannulae openings with diameters of 11 mm, introduces a need for precise actions.

The awkwardness of the postures of the surgeons left arm are exacerbated by the location of the camera operator. Because the operator must stand close to the cannulae access, the surgeon often has limited clearance for manoeuvring and needs to bridge his arm over the camera operator leading to a further static loading. It appears that the pains reported in the left shoulder, forearm and elbow and wrist (6 in total) reflect such awkward postures combined with static loading.

Conclusions

The analysis of the LC showed that there were a number of important task factors which could lead to fatigue and awkward postures. In addition the results of the instrument evaluation indicated that many of these risk factors can be attributed to failings in the ergonomic design of the instruments. Evidence for surgeons experiencing pain/discomfort as a result of using these instruments was confirmed by the survey results. Sixty four percent of respondents experienced pain/discomfort in one or more parts of the body while using specific instruments during this procedure.

In addition to pain/discomfort being associated with using specific instruments, there was also a number of instances of non-specific pain. These may be due to the asymmetrical posture caused by the electrosurgical hook. Other factors, such as repeated changes of instruments and the overall precision demands of the task, are also likely to contribute to general fatigue. It may be that such fatigue will increase risk of injury to both surgeon and patient.

As this study was a preliminary investigation, the survey was based on a small sample, and the analysis focused on one specific procedure, care must be taken before generalising these initial findings to other MAS procedures. The results of the survey have been used to interpret some of the problems identified in the work methods analysis. It should be noted, however, that direct causality cannot be inferred because of slight variations in design between different makes of instrument. However, it would appear that there is a need for further ergonomic research into the MAS surgeons' tasks and instruments. As the number and variety of MAS procedures are set to increase, more surgeons are likely to be exposed to the types of risk factors identified in this study, unless ergonomic design of instruments is improved.

References

Crombie, N. A-M. 1995, *Ergonomic Evaluation Of Instruments Used During Laparoscopic Cholecystectomy*, MSc Ergonomics Thesis, (University Of Aberdeen, Aberdeen)

Cuschieri, A., Dubois, F., Mouiel, J., Mouret, P., Becker, H., Buess, G., Trede, M. and Troidl, H. 1991, The european experience with laparoscopic cholecystectomy, *The American Journal Of Surgery*, **161**, 385-387.

Fraser, T. M. 1980, *Ergonomic Principles In The Design Of Hand Tools*, Occupational Safety And Health Series, (International Labour Office, Geneva)

Graves, R. J., Gardner, J. M., Anderson, M. Seaton, R., Ross, J. and Porter, R. 1994, *The Development Of An Ergonomic Methodology For The Evaluation Of Operating Table Systems, CSA final report*, (University of Aberdeen, Aberdeen)

Health & Safety Executive 1994, *Work Related Upper Limb Disorders : A guide to prevention,* (Health & Safety Executive, Sheffield)

Kano, N, Yamakawa, T. & Kasugai, H. 1993, Laparoscopic surgeon's thumb, *Archives Of Surgery*, **128**, October, 1172.

Pheasant, S. T.1992, *Bodyspace : Anthropometry, Ergonomics and Design*, (Taylor & Francis, London)

Putz-Anderson, V. 1988, *Cumulative Trauma Disorders : A Manual For Musculoskeletal Diseases Of The Upper Limbs*, (Taylor & Francis, London).

Soper, N. J., Stockmann, P. T., Dunnegan, D. L., Ashley, S. W. 1992, Laparoscopic Cholecystectomy: The new 'gold standard'? *Archives Of Surgery*, **127**, August, 917-923.

Zucker, K. A., Bailey, R. W., Gadacz, T. R., Imbembo, A. L. (1991) Laparoscopic guided cholecystectomy, *The American Journal Of Surgery*, **161**, January, 36-44.

DEPTH PERCEPTION AND STEREOSCOPIC SYSTEMS FOR MINIMALLY INVASIVE SURGERY

Anthony H. Reinhardt-Rutland[1] and Walter H. Ehrenstein[2]

Psychology Department, University of Ulster,
Newtownabbey, BT37 OQB, Northern Ireland[1]
Institut f. Arbeitsphysiologie, University of Dortmund,
Ardeystrasse 67, D-44139 Dortmund, Germany[2]

In minimally invasive surgery (MIS), the surgical team views the site of operation indirectly at a TV monitor. MIS has had problems, some of which reside in the loss of information for depth. The development of stereoscopic viewing systems is intended to reduce this loss. However, stereoscopic viewing differs in important ways from conventional, full-cue viewing, both because information for depth remains incomplete, and because of particular difficulties entailed by stereoscopic viewing. Such issues require careful research, if the patchy reputation of MIS is to be improved. It may be preferable to concentrate on non-binocular pictorial cues, in conjuction with careful selection of surgeons to perform MIS.

Introduction

Minimally invasive surgery (MIS), often known as "keyhole" surgery, has had a major impact on health care in the last few years. The general technique has obvious potential in reducing patient pain and post-operative infection. The resulting more efficient use of resources should lead to financial savings, although this must be set against increased equipment costs. The crucial feature of MIS is that incisions need be much less extensive: trocars are inserted into the patient at the site of the operation, containing remotely-operated surgical instruments and a miniature camera, via which the surgical team views the operation externally on a TV monitor. MIS has been applied most notably to cholecystectomy; other examples include hernia, stomach, prostate and gynaecological procedures (Grace and Bouchier-Hayes, 1990; Leahy, 1989).

Unfortunately, the use of MIS has had problems, with persistent evidence of increased complication rates requiring post-operative intervention; generally, this intervention entails conventional surgical techniques, thus negating the advantages of MIS (Antia, 1994; New York State Department of Health, 1992).

One general area of concern with MIS must reside in the information-processing difficulties inherent in viewing and operating indirectly. The issue of viewing conditions - the focus of the present paper - has prompted the recent development of real-time systems based on the stereograms known since Wheatstone in the 1830s (Julesz, 1971; Rogers and Collett, 1989). One such system parallels a successful research tool for investigating binocular interactions: the crucial feature is the rapid switching between the eyes of the scene viewed from slightly different positions (see Wills, 1993). Wenzl, Lehner, Vry, Pateisky, Sevelda and Hussein (1994) outline their experience with such a system applied to MIS. A lens system installed in an endoscope transmits slightly different views of the site of operation to a pair of cameras. At the TV monitor, the outputs from each camera are presented in rapidly-alternating sequence. To exploit the stereoscopic information, spectacles employing liquid-crystal technology are worn. One eye views one camera's output at one instant and the other eye views the other camera's output at the next instant. The resulting combined images are reported as stable and convincingly three-dimensional.

Nonetheless, experience with stereograms suggests that there may be important drawbacks: what may be satisfactory for research into binocular interactions may not be satisfactory in the more demanding context of surgery.

Normal depth perception

In normal viewing conditions, as might apply in conventional surgery, there are many sources of information for depth.

Binocular information includes retinal disparity. The two retinal images differ because of the separation of the eyes in the head: closer features of the scene lead to bigger retinal differences. Convergence is also notable: this entails rotation of the eyes to maintain perception of a fused image. Convergence increases as the distance ot the viewed scene decreases.

Monocular information includes monocular parallax, associated with depth-related ocular adjustments, such as adjustment of lens curvature. Information arising from the observer's motion is also available. Features at different distances from the observer move visually across the retina at different rates: the rates of visual motion are greater for nearer features than for more distant features. This phenomenon is known as motion parallax.

Pictorial information refers to two-dimensional representations of depth, such as photographs. It includes elements such as edges; these signal discontinuities in depth. Higher-order examples are occlusion - a nearer object may obscure a more distant object - and relative visual size - two objects of similar physical size will have different visual sizes, if they are at different distances from the observer. Luminance and contrast may be pertinent: bright, high-contrast objects are seen as closer than dull, low-contrast objects (Ittleson, 1952). Pictorial information can be highly subtle;

for example, observers make judgments of the orientation-in-depth of triangular surfaces on the basis of their visual widths (Reinhardt-Rutland, in press).

There is considerable redundancy in depth information. For this reason, the approximately 10% of the population lacking stereoscopic vision (Sachsenweger and Sachsenweger, 1991) seem to perform many depth-dependent skills adequately; for example, casualty-rates for permanently monocular drivers are not reported to be higher than for drivers with normal vision (McKnight, Shinar and Hilburn, 1991).

Depth perception and "conventional" MIS

Single-camera "conventional" MIS must in the main be dependent on pictorial information (Reinhardt-Rutland and Gallagher, 1995). That laparoscopic cholecystectomy is popular is partly because pictorial depth is relatively good: although the perceptual problems should not be underestimated, the "target" material is relatively well-defined in relation to surrounding tissue, so, for example, edge information should be relatively good.

The reliance on pictorial information might seem precarious, since it can mislead. For example, difference in brightness and contrast in two parts of a scene might be explained by lighting conditions, rather than difference in distance between the two parts of the scene. However, pictorial cues are important in normal vision and can override monocular and binocular cues (Gehringer and Engel, 1986; Reinhardt-Rutland, 1992, 1995; Stevens and Brookes, 1988).

Of other sources of information available in MIS, motion information may have a role (Rogers and Graham, 1982), although motion of the image in MIS must be restricted. A more important limitation is that motion information requires correlated kinaesthetic information to be fully affective: under normal perceptual conditions, visual motion arises because of the observer's self-initiated motion (Gibson, Gibson, Smith and Flock 1959). The lack of useful kinaesthetic information has other ramifications, as considered below.

Stereoscopic depth perception and MIS

Stereoscopic viewing systems should represent an advance in the performance of MIS. However, depth information is still partial; for example, motion of the viewing observer has at best a limited and indirect effect on the viewed scene. Also, stereoscopic viewing introduces conflict in depth information: monocular information and convergence still convey the flatness of the TV monitor's screen. Furthermore, research suggests that stereoscopic viewing has complex and somewhat unpredictable effects on convergence (Logvinenko and Belopolskii, 1994). These points are in accord with observations by Wenzl et al (1994) that the surgical team requires a period of familiarisation before operating, and that switching between stereoscopic viewing and normal viewing during an operation is difficult. Anyone familiar with conventional stereograms would expect this. They would also report that the eyes tend to feel fatigued after stereoscopic viewing, almost certainly because of the perceptual conflicts noted above.

Another limitation of stereoscopic viewing is that very careful manipulation is required if *absolute* depth is to be conveyed, i.e., the viewer is not only able to determine that point a is behind point b - indicating *relative* depth perception - but that point a and point b are both at precise numbers of distance units from the observer, so point a is at a precise number of distance units behind point b. Because of the opposing requirements in MIS of a sizeable field of view - requiring "zooming-out" of the site of operation - and of magnification - requiring "zooming-in" to the site of operation, binocular disparity becomes much altered. In normal viewing, the same effect arises if the observer moves towards and away from the viewed object. However, in that case the change in binocular disparity can be scaled by reference to kinaesthesis: for a given distance of the viewed object, there is a systematic relationship between change in binocular disparity and kinaesthesis. It can be concluded that the scaling necessary for absolute depth perception is not feasible in stereoscopic viewing systems for MIS.

The importance of perceiving absolute depth in the kind of procedures for which MIS should be advantageous is not known. However, if MIS is to be used for ever finer and more complex operations, the question becomes crucial and should be carefully researched.

A final point resides in recent evidence that differences in luminance may be misperceived during stereoscopic viewing (Taya, Ehrenstein and Cavonius, 1995): these authors report that some differences in luminance are diminished because of processes of neural assimilation, while other differences are exaggerated because of processes of neural contrast. Since differences in luminance act as pictorial depth cues (Ittleson, 1952) and since - as noted earlier - pictorial depth information can override binocular depth information (Gehringer and Engel, 1986; Stevens and Brookes, 1988), misperception of depth due to this factor may be predicted during stereoscopic viewing. Informal observation with conventional stereograms supports this prediction. Clearly, this is another issue requiring research.

Proposals for a research programme

The development of stereoscopic viewing systems for MIS is potentially valuable, since binocular disparity is undoubtedly important in fine depth perception. Furthermore, the technology is undergoing rapid development (Coghlan, 1995). However, we have alluded to some problems which should counsel caution before such systems become widely available. Hopefully, the sometimes patchy history of single-camera MIS will encourage such caution. In summary, stereoscopic viewing for MIS is not equivalent to normal viewing for the following reasons:
(a) information for flatness remains and is likely to cause delays in adjustment and eye fatigue;
(b) absolute depth is not conveyed;
(c) misperception of brightness and contrast may adversely affect perceived

depth.

Among further issues that require investigation are:

(d) the extent to which the lack of absolute depth perception limits the type of operation for which stereoscopic viewing systems may be appropriate;

(e) the extent to which misperception of luminance during stereoscopic viewing contributions to misperception of depth.

Last but not least, some points are directed to *selection* and *training*. We intimated earlier how stereo-deficient individuals adapt by exploiting those sources of depth information which remain available to them. Given the redundancy of depth information, this should not be difficult. This has implications for training to perform MIS. Viewing limitations imposed by TV monitors may not be insuperable, provided we known how best to facilitate the surgeon's adaptation to the reduced information. However, there is an important caveat. The stereo-deficient individual does not suffer the degree of competition between perceiving depth and flatness that must afflict the normally-sighted surgeon performing MIS. Furthermore, away from the operating theatre, that surgeon has usual levels of depth information: we do not know how much this inevitable switching between viewing conditions might compromise training.

An interesting and intuitively unexpected point concerns procedures for selecting surgeons to perform MIS. For "conventional" single-camera MIS, it may be that surgeons with poor stereoscopic vision are better-suited than those with good stereoscopic vision. This assumes, of course, that monocular acuity remains adequate. For stereo-deficient surgeons, the discrepancy between their normal day-to-day depth perception and the demands of MIS would be smaller. Before the wholesale adoption of expensive stereoscopic viewing systems, such avenues should be explored.

Acknowledgments. We thank Judith Annett, Mervyn Gifford and Mary McClean for helpful comments. The project was supported by grants from Deutscher Akademischer Austauschdienst and the British Council to AHRR.

References

Antia, N. H. 1994, Keyhole surgery, Lancet, **344**, 596-597.

Coghlan, A. 1995, Keyhole surgeons enter new dimension. New Scientest, **148:2007**, 27.

Gehringer, W. L. and Engel, E. 1986, Effect of ecological viewing conditions on the Ames' distorted room illusion, Journal of Experimental Psychology: Human Perception and Performance, **12**, 181-185.

Gibson, E. J., Gibson, J. J., Smith, O. W., and Flock, H. 1959, Motion as a determinant of perceived depth, Journal of Experimental Psychology, **58**, 40-51.

Grace, P. A. and Bouchier-Hayes, D. J. 1990, Laparoscopic cholecystectomy: the implications for surgical practice and training, Journal of the Irish Colleges of Physicians and Surgeons, **19**, 180-181

Ittleson, W. 1952, *The Ames Demonstrations in Perception* (Princeton University Press, Princeton NJ).

Julesz, B. 1971, *Foundations of Cyclopean Perception* (Chicago University

Press, Chicago).

Leahy, P. F. 1989, Technique of laparoscopic appendicectomy. British Journal of Surgeons, 76, 616.

Logvinenko, A. D., and Belopolskii, V. I. 1994, Convergence as a cue for distance, Perception, 23, 207-217.

McKnight, A. J., Shinar, D. and Hilburn, B. 1991, The visual and driving performance of monocular and binocular heavy-duty truch drivers, Accident Analysis and Prevention, 23, 225-237.

New York State Department of Health Memorandum, Laparoscopic surgery. Soris 29-20, June 12, 1992.

Reinhardt-Rutland, A. H. 1992, Primary depth cues and background pattern in the portrayal of slant, Journal of General Psychology, 119, 29-35

Reinhardt-Rutland, A. H. 1995, Perceiving the orientation in depth of real surfaces: background pattern affects motion and pictorial information, Perception, 24, 405-414.

Reinhardt-Rutland, A. H. (in press), Judging the orientation-in-depth of real triangular surfaces, Perception.

Reinhardt-Rutland, A. H. and Gallagher, A. G, 1995, Visual depth perception in minimally invasive surgery. In S. A. Robertson (ed.), Contemporary Ergonomics 1995, (Taylor & Francis, London) 530-536.

Rogers, B. and Collett, T. S. 1989, The appearance of surfaces specified by motion parallax and binocular disparity, Quarterly Journal of Experimental Psychology, 41A, 697-717.

Rogers, B., and Graham, M. 1982, Similarities between motion parallax and stereopsis in human depth perception, Vision Research, 22, 261-270.

Sachsenweger, M., and Sachsenweger, U. 1991, Stereoscopic acuity in ocular pursuit of moving objects, Documenta Opthalmologica, 78(1-2), 1-133.

Stevens, K. A., and Brookes, A. 1988, Integrating stereopsis with monocular interpretations of planar surfaces, Vision Research, 28, 371-386.

Taya, R., Ehrenstein, W. H., and Cavonius, C. R. 1995, Perception, 24

Wenzl, R., Lehner, R., Vry, U., Pateisky, N., Sevelda, P., and Husslein, P. 1994, Three-dimensional video-endoscopy: clinical use in gynaecological laparoscopy. Lancet, 344, 1621-1622.

Wills, C., 1993, The Runaway Brain (Harper-Collins, London).

COMPUTERS IN THE GP'S SURGERY

Derek Scott, Ian Purves and Robin Beaumont

**Sowerby Unit for Primary Care Informatics,
Medical School,
University of Newcastle,
Newcastle-upon-Tyne NE4 1AA**

Previous research looking at the effects of computers in the surgery suggest that this technology has a minimal invasive effect on the doctor-patient relationship. The present study therefore aims to examine what areas of the traditional doctor-patient relationship might be impaired by a more demanding---in terms of inputting and outputting (e.g. reading the screen)---software system. Such increasing demands on the doctor may negatively effect the patient's feelings of importance, their faith in the doctor, the perceived infallibility of medical advice, and hence compliance with prescription taking. Adapting a three-way triadic model (doctor-computer-patient) this paper describes some of the proposed avenues of study aimed to examine what areas of the traditional doctor-patient relationship might be impaired by more cognitively demanding software.

Introduction

The main impetus for computers in the doctor's surgery came more than ten years ago with the Micros for GPs Scheme, funded by the Department of Trade and Industry, and evaluated under the guidance of the DHSS and the medical profession (General Practice Computing, HMSO, 1985). The vast majority of practices are now equipped and making increasing use of the facility. Typically they provide facilities for patient registration, recall, screening, and repeat prescriptions. Further uses would include providing information (e.g. on prescribing), prompts for opportunistic preventive measures, and providing treatment protocols to use for particular conditions. Whilst there are obvious and undeniable benefits, any possible detriment to the doctor-patient relationship needs to be borne in mind.

General practice prescribing in England accounts for over 10 per cent of the entire NHS budget and is the largest single item of the NHS expenditure after staff costs. In the year 1992-1993, for instance, approximately 400 million prescriptions were issued in general practice in England and Wales, at a cost of some £2.6 billion (Purves and Kennedy, 1995). The value of reducing this expenditure is therefore clear.

The aspects of the study described here form only part of the much larger UK PRODIGY project (Prescribing RatiOnally with Decision-support In General practice studY). The UK and the Netherlands have one of the most computerised general practices in the world. In both countries, the systems have been supported by government sponsorship and have developed along a clinical as well as an administrative direction. Both countries are trying to dip their toes in the water and create software which aids clinical decision making. It is clear that, with the ever increasing amount of information with which doctors need to practice in conjunction with increasing patient expectations, some form of support is needed. Computers are no doubt the most effective way of presenting information which helps solve information dilemmas. The PRODIGY project involves collecting data from 150 practices in England for whom five suppliers of software/hardware are presently installing their respective versions of a computerised prescribing aid system. Further details of the wider scope of the project may be found in Purves (1995).

One of these five suppliers provides both DOS and Unix based systems, and therefore there are six practices, distributed throughout the length and breadth of England, each to be employing one of these systems, were chosen for more detailed analysis of the effects of the system on the doctor-patient consultation. Hereafter, the various (relatively simple) software systems currently in situ within the various practices shall be referred to as the "old" software, and these six more complex and sophisticated software systems awaiting installation shall be referred to as the "new" software.

The Doctor-Computer-Patient (DCP) triadic model

Whereas much has previously been written respectively describing doctor-patient relationships, physicians' use of the microcomputer, and human-computer interaction generally, such relationships have been discussed within a two-way or dyadic perspective only. It is important to consider the Doctor, Computer, Patient elements within a three-way interactive (DCP) model where each component has a undeniable effect on the relationships between the other two. The doctor-patient dyadic relationship has traditionally been viewed as one of compliance with the direction of influence within the DCP model as from Doctor *to* Patient. However, more recently, the patient has been regarded as an active consumer rather than merely a passive recipient of treatment countering the relationship into a two-way dyadic one. This is especially so in primary care where attendance at the GP's surgery is the patient's own prerogative, and where, generally speaking, those attending are keen to continue to be involved in decisions about their treatment. There has therefore been increasing concern over measuring patients' satisfactions with the services given.

An analysis of videotapes made by Fitter and Cruickshank (1982) suggests that the computer does have a substantial impact at a detailed level on communication between doctor and patient. One cognitive area which is effected is direction of attention. The actual use of the computer in retrieving or inputting information requires the doctors attention and gaze to be directed towards the terminal, thereby seemingly distancing themselves from the patient and not giving them their direct attention. Although this pattern is similar to that used to refer to written notes, it is suggested that it appears to result in more attention spent on the computer rather than with the patient. Fitter and Cruickshank conclude that: "The doctor is required to switch his/her attention between patient and computer. For the doctor, communication with each is a very different task needing different skills and requires an additional mental load to manage the alternation of attention.... The patient must not be allowed to come off worst in a competition with the computer for the doctor's attention" (p. 91).

Video analysis

The present study therefore aims to examine what areas of the traditional doctor-patient relationship might be impaired by a more demanding---in terms of inputting and outputting (e.g. reading the screen)---software system.

Research plan

a) Various questionnaires have been prepared for administration to agreeing patients before and after consultation with their GP. The same questionnaires are to be administered on both of two visits to the surgery (different patients, naturally). In addition, various methodologies of quantifying the interactions, and for finely analysing audio-video recordings of the doctor-computer-patient consultation.

b) A researcher will visit six GP's surgeries to administer questionnaires to patients and to organise remote videoing of sessions with those patients who have absolutely no reservations about this. At this stage, "old" software is in situ. Questionnaires are administered before and after the consultation, some of these questionnaires being specifically designed (e.g. equivalent versions) to tap changes in opinions between before seeing the GP (with either "old" or "new" software) and afterwards.

c) After "new" software has been installed, and some five to six months later, the researcher returns to do the same again.

d) After video material has been analysed and questionnaires scored, statistical analyses are made on the data. These will take the form of a) comparisons in changes between before and after measures in the "old" condition versus before and after changes in the new condition, and b) differences in "old" versus "new" measures conducted after the consultation.

Data will be analysed in terms of a) differences between After measures on the "old" versus "new" conditions, and b) differences in changes between Before and After on the "old" condition versus changes between Before and After measures for the "new" condition.

Questionnaire data

The following lists the questionnaires which will be administered to the patients before and/or after the consultation.

Patient Intentions Questionnaire (PIQ)
Fifteen items from Salmon and Quine's (1989) questionnaire relating to what primary health care patients intend to receive from or expect from a consultation. Administered before consultation.

Expectations Met Questionnaire (EMQ)
Fifteen items corresponding to the above 15 (Williams, Weinman, Dale, & Newman, 1995), largely rephrased into the past tense. This scale is given after the consultation as a comparison measure of whether patients feel that their wishes and expectations were met.

Perceptions of Ideal/Real Doctor
Thirteen bipolar scales (e.g. Not listening versus Listening), derived from Cruickshank (1985), given both before (as patients' perceptions of the "ideal" doctor) and after the consultation (as an appraisal of patients' perceptions of the doctor just seen).

Computer Attitude Questionnaire (CAQ)
Twenty items derived from Cruickshank (1984) and Rethans, Hoppener, Wolfs, & Diederiks (1988) measuring negative or positive attitudes towards computers. The four subscales consist of: Time/Resources (financial) Issues, Technical Issues, Confidentiality Issues, and (doctor-patient) Relationship Issues. It is administered as a split-half questionnaire before and after the consultation. Data obtained can be analysed globally as a 20-item scale, as four subscales, or as a measure of attitudinal change globally, on subscales, or on conceptually similar single items as administered before and after the consultation.

Audio-video tape analysis of the DCP Interaction

Direction of Gaze
Total time of consultation will be recorded as displayed automatically within the video frame, or as timed from the audio-tape. From video playback, it will be possible to record timed instances of, e.g., eye contact between patient and doctor, and doctor using computer.

Time Talking
This is possibly the simplest form of analysing who is directing the conversation. Overall time each participant spends talking during the meeting will be timed and expressed as an index of the entire consultation time. This measure can also be expressed as a ratio of Doctor:Patient time.

Pauses Between Speech

Pauses in conversation are normal, but within the busy context of a doctor' surgery, are likely to be more meaningful. Pauses represent a disruption within what is usually smooth, unconscious turn-taking behaviour. An fairly arbitrary figure of three seconds might be taken as constituting the threshold for pauses to be recorded. Pauses can be made: a) whilst one speaker is waiting for a response to what they have just said, or b) where they pause to deliberate over what they are next about to say or because their attention is taken up on operating the computer.

Patient Centredness

The traditional training of medical students might be described as centred around the "clinical agenda", as opposed to giving major consideration to the views of the patients themselves. In general practice, as opposed to within undergraduate training, GPs are taught to understand the patient's agenda. It should be possible to detect this change of paradigm and, particularly with regard for the use of two different software systems, to see whether the new system has any effect on the level of "patient centredness". Henbest and Stewart (1989) have devised as scoring system to assess the patient-centredness of the doctor-patient interaction. It is intended to look at whether the doctor interacting with a more computer-driven system will be less "patient-centred" than with a system demanding less interaction from the GP. As the patient-centred approach is less directive and requires more time for open debate, there may be a positive association with overall consultation times.

NUDIST Qualitative Assessment of Clinical Style

The five highest and five lowest scoring consultations, on the relatively quantitative patient-centredness inventory will be taken for more detailed Qualitative assessment. This procedure aims to combine qualitative and quantitative methods to produce a description of the interaction, "the sum of which may be greater than the two parts". It should be possible to tease out exactly what it is which characterises the patient-centred clinical style. Several computerised dialogue packages are available, with NUDIST (Replee P/L, 1992) being one of the most recent developments). NUDIST (Nonnumerical Unstructured Data Indexing Searching and Theorizing) has been specifically designed for the management, organisation and support of qualitative data analysis. As with other similar programmes, it allows users to create "nodes" (tags or categories), and to search for and reorganise nodes. This programme also allows for the addition of user notes and memos about the ideas and theories that emerge during projects. "The functions provided by NUDIST take it beyond being simply a code-and-retrieve system, allowing the user to store and develop their understanding of the data within the computer environment. The processes and functions are supported by the creation of two systems: the document system and the indexing system" (Gorely, Gordon, & Ford, 1994).

Assessment of General Practitioner's Performance

The purpose of the study by Cox and Mulholland (1993) was to identify those important characteristics of doctors' and patients' behaviour which distinguish between "good" and "bad" consultations when viewed on videotape, and to use these characteristics to develop a reliable instrument for assessing general practitioners' performance in their own consultations. It must, however, be borne in mind that the reliability and validity of this tool only holds when used by those qualified to assess the performance of those trained or in training as GPs; i.e. GPs themselves. It is not, in this form, established as suitable for use by non-medically trained people. For instance, "The doctor lacks up to date knowledge" and "Physical examination of the patient is inadequate/examination inappropriate" are items not suitable for unqualified observers. However, Cox and Mulholland do tabulate the ten statement pairs most often volunteered as distinguishing "good" from "bad" doctors. It can be seen that they relate more to social psychological factors observable in the two-way doctor-patient interaction. It is intended to attempt to validate this shortened form for use by non-medically qualified raters (e.g. psychologists) by having three raters (including one GP) assess a large number of consultations and analyse for inter-rater reliability by Cronbach's alpha.

References

Cox, J. and Mulholland, H. (1993) An instrument for assessment of videotapes of general practitioners' performance. British Medical Journal, 1043-1046.

Cruickshank, P.J. (1984) Computers in medicine: patients' attitudes. Journal of the Royal College of General Practitioners, **34**, 77-80.

Cruickshank, P.J. (1985) Patient rating of doctors using computers. Social Science and Medicine, **21**, 615-622.

Henbest, R.J. and Stewart, M.A. (1989) Patient-centredness in the consultation. 1: A method for measurement. Family Practice, **6**, 249-253.

HMSO (1985) *General Practice Computing.* London: HMSO.

Fitter, M.J. and Cruickshank, P.J. (1982) The computer in the consulting room: a psychological framework. Behaviour and Information Technology, **1**, 81-92.

Gorely, T., Gordon, S. and Ford, I. (1994) NUDIST: A qualitative data analysis system for sport psychology research. The Sport Psychologist, **8**, 319-320.

Purves. I.N. (1995) *PRODIGY: Information for Participants.* Sowerby Unit for Primary Care Informatics, University of Newcastle upon Tyne.

Purves. I.N. and Kennedy, J. (1995) *The Quality of General Practice Repeat Prescribing.* Sowerby Unit for Primary Care Informatics, University of Newcastle upon Tyne.

Replee P/L (1992) *NUDIST: Qualitative data analysis solutions for research professionals.* Aptos, CA: Aladdin Systems.

Rethans, J-J., Hoppener, P. Wolfs, G. and Diederiks, J. (1988) Do personal computers make doctors less personal? British Medical Journal, **296**, 1446-1448.

Salmon, P. & Quine, J. (1989) Patients' intentions in primary care: Measurement and preliminary investigation. Psychology and Health, **3**, 103-110.

Williams, S., Weinman, J., Dale, J. & Newman, S. (1995) Patient expectations: What do primary care patients want from the GP and how far does meeting expectations affect patient satisfaction? Family Practice, **12**, 193-201.

Manual Handling

ARE GENDER-FREE PHYSICAL SELECTION CRITERIA VALID PREDICTORS OF MAXIMUM BOX LIFTING?

MP Rayson

*Centre for Human Sciences,
DRA Farnborough,
Hampshire GU14 6TD, UK.*

The objective of this investigation was to determine if 'gender-free' physical selection tests may be used to predict box lifting capability. Three hundred and seventy nine trained soldiers, mean age 23.5 (sd 4.45) yr, height 1.76 (sd 0.063) m and body mass 73.6 (sd 10.01) kg undertook a box lifting task (BL) with progressive load increments. They also underwent measurements of body size and composition, muscular strength and endurance, and aerobic power. Mean maximum BL loads were 65.7 (sd 7.0) and 36.3 (sd 9.0) kg for the men and women, respectively. The highest simple correlate with BL scores was fat free mass (FFM). The best multiple regression model involved FFM, maximum lifting on the incremental lift machine divided by body mass and gender as predictor variables. A gender-free model was not valid. The level of agreement between measured and predicted task scores was moderate.

Introduction

It is the operational effectiveness of the work force which should provide the basis for physical selection criteria for occupations where performance is of paramount importance. Such occupations include the Armed Forces, the Fire Service, and to a lesser extent the Ambulance Service, the Police Force and heavy industry. Operational effectiveness is ultimately dependent upon the ability of an individual to perform the job. If the job demands approach or exceed the physical capacities of employees, efficient and effective performance is precluded and the risk of injury is increased (Chaffin 1974; Chaffin et al 1978). However, at present only minimal efforts are made to screen applicants to these physically demanding occupations.

Ideally, suitability of job applicants would be assessed by their performance on relevant tasks. However, the practicality of testing unskilled applicants on role-related tasks is problematic. Therefore, physical selection tests which are quick, easy to administer and relatively safe are often used in their place.

It has been shown via job analyses in the United States (Sharp et al 1980), in

Canada (Allen et al 1984), and in the UK (Rayson et al 1994) that handling materials is by far the most common physically demanding military activity. Such activities may demand a combination of strength, anaerobic power and endurance, and aerobic power and capacity, depending upon the nature of the task. Anthropometry may also be important.

The EEC Directive (1976) on equal treatment for men and women, which superseded all previous national legislation has provided an incentive to implement objective entry criteria to jobs. It stated that "there shall be no discrimination whatsoever on grounds of sex either directly or indirectly ... including selection criteria, for access to all jobs or posts, whatever the sector or branch of activity". The development and validation of role-related physical selection standards should provide a legally defensible means of physically matching worker to role.

Several previous studies have investigated the relationships between maximal box lifting from floor to 1.35 m or higher and physical tests in men and women (Sharp 1980, McDaniel et al 1983, Myers et al 1984, Teves et al 1985, Nottrodt and Celentano 1987). Some have employed constrained lifting techniques (eg Sharp 1980) which have since been shown to be gender biased (Stevenson et al 1990), whilst others have employed a variable end height of lift which is related to the individuals stature (eg Myers et al 1984).

Methods

Three hundred and seventy nine trained soldiers (304 males and 75 females) with mean (SD) age 23.5 (4.35) yr, height 1.76 (0.063) m, body mass 73.6 (10.01) kg took part in the study. Informed consent was given by all subjects. All undertook a progressive box (dimensions 48 x 20 x 19 cm) lifting task from the ground to a height of 1.45 m, up to a maximum load of 72 kg. Subjects were advised on correct lifting techniques, but essentially the lift was freestyle. On a separate occasion subjects also underwent a large number of anthropometric measurements, percentage body fat and fat free mass (FFM) estimated via skinfold measurements (Durnin and Womersley 1974) and two electrical impedance devices (Bodystat and ELGll). Isometric strength tests included 38 cm and 85 cm upright pull, 90^0 isometric arm flexion, hand grip and ankle plantar flexion. Dynamic strength/power tests included a hydro-dynamic lift (Grieve 1993) and a progressive maximum lift on an incremental lift machine (ILM). Muscular endurance tests included a box hold, biceps curls, deltoid raises, sit-ups, push-ups and pull-ups to a bar. Aerobic power was estimated via the Multistage Fitness Test.

For all statistical tests significance was set at $p < 0.05$. Product moment correlation coefficients were calculated between BL and test scores. The relationships between the tasks and the predictor tests were estimated using stepwise multiple linear regression. Where maximal BL scores could not be assured for a minority of individuals who achieved the load limit, a maximum likelihood linear regression procedure was used to take account of the modified distributional form associated with the censored observations. All of the relationships were estimated for males and females separately, and the hypothesis that the relationships were the same was tested in the standard manner by breaking the hypothesis down into successive tests of parallelism (equal slopes) and identity (co-incident intercepts given equal slopes).

Results

Mean BL load for females was 36.3 (sd 9.0) kg and for males, 65.7 (sd 7.0) kg. The female scores ranged from 14 to 52 kg with a modal load of 39 kg. The distribution was negatively skewed. The male scores ranged from 52 to 72 kg with a modal score of 72 kg. 44.5% of males achieved the load limit of 72 kg and hence were not necessarily assessed to maximum. The true shape of the male distribution could not therefore be ascertained. The female to male ratio was 0.55. The 90th percentile female (48 kg) corresponded to the 2nd percentile male.

The estimates of FFM produced the highest correlation coefficients on the uncensored data for the pooled sample (r=0.86) and for male (r=0.64) and female (0.62) sub-samples. ILM test scores also featured prominently in the pooled sample (r=0.80) and for the genders separately (r=0.57-0.59)

Stepwise regression revealed the three test variables of FFM and ILM145/WEIGHT to best predict BL in the pooled gender sample. Post hoc tests revealed no gender difference in the slope but a significant difference in the intercepts, necessitating the addition of gender as a predictor variable. The equation had a standard deviation of 7.16 kg and an R^2 value of 0.87.

BL (kg) = (1.237*DWFFM) + (7.986*ILM145/WEIGHT) - (5.921*GENDER) - 12.0.

Discussion

The means and ranges of BL scores fell within the reported range in the literature (Teves et al 1985). As for most physical performance data, but particularly marked in performance related to upper body strength, a distinct bimodal distribution was apparent between the genders with only a small overlap in values. The 90th percentile female scored the same as the 2nd percentile male. Similar results have been reported in other studies using strength-related tasks (eg Sharp et al 1980). These gender differences are underpinned by the widely reported differences in male and female strengths (Laubach 1976; Wilmore 1979; Pheasant 1983).

The true maximum lift capacity of males in this study will have been underestimated, by the enforcement of a 72 kg maximum load. Nearly half of the males achieved this load. It is also likely that an increase in height of lift from 1.32 m cited in the studies by Sharp (1980) and Teves et al (1985), to 1.45 m in the present study had a greater effect on female capability compared to male capability. This is due to the shorter stature of the average female, requiring both greater dependence on upper body strength (which has a smaller female to male ratio than mid and lower body strength), and requiring the female to lift through the weaker range of movement at shoulder height and above (Stevenson et al 1990).

A number of authors have reported female to male strength ratios on box lifting tasks to be between 0.49 and 0.61 (Ayoub et al 1982; Myers et al 1984; Teves et al 1985; Beckett and Hodgdon 1987; Stevenson et al 1990). Although this study showed female to male strength ratios to be within the quoted range (0.55) this ratio will have overestimated female to male lifting potential, by a substantial proportion of the male sample achieving the maximum load.

FFM was the single most important predictor test in the pooled gender sample and for males and females separately. FFM was a stronger predictor than any of the dynamic and static lifting tests. From amongst the studies in the literature which considered FFM, and dynamic and static lift tests, this finding is in line with only one previous study (Nottrodt and Celentano 1987). A study by Teves et al (1985) showed both FFM and ILM scores to be equally well correlated with maximum lift load. Other studies have shown ILM test scores to be superior predictors to FFM (Myers et al 1984; Ayoub et al 1982; Beckett and Hodgdon 1987). It was wrongly anticipated that adoption of recommended ILM procedural modifcations (matching the heights of lift on ILM and BL tests, allowing freestyle lifting, coaching subjects in power lifting techniques, administering lighter starting loads and smaller increments for females (Stevenson et al, 1990)) would have enhanced the predictive strength of the dynamic lift tests above that of FFM.

The superiority of FFM as the best single predictor test for BL might be explained by the fact that FFM, as a measurement requiring no skill on the part of the subject, provides a more transferable indicator of lifting capacity than the two skill-dependent simulated dynamic lifting tests that were employed. Perhaps box lifting is better predicted by generic tests such as FFM as the movement and muscle recruitment patterns of other lifting tests are too discrete. These findings strongly support Vogel's contention (1992) that minimum FFM standards have an equally important place as body fat standards in military performance.

Multiple regression techniques produced a reasonable model between BL scores and test variables. The selected model included tests of FFM and ILM145 divided by body mass. FFM and ILM test scores were also the most significant contributors to a box lifting model derived by Nottrodt and Celentano (1987). Post hoc analysis of our equation showed a "gender-free" model to be invalid which necessitated the addition of "gender" as a predictor variable. Although the model could account for 87% of the variance, the sd of 7.16 kg was substantial resulting in a 95% CI of 13.4 kg. This could lead to a high proportion of misclassification errors if the tests were to be for screening, for example.

To contribute to the debate as to the relative merits of dynamic versus static strength tests as predictors of maximum lifting capacity, in this study, the dynamic measures provided by the ILM were superior single predictors of BL than were the static lift tests in both the pooled and separate male and female samples.

In conclusion, estimates of FFM were found to be the single highest correlates of BL score in both genders. BL score could be modelled using a gender-related equation requiring FFM, ILM score divided by body mass and GENDER as predictor variables, but the level of agreement between measured and predicted task scores was moderate, at best. The level of agreement might be improved by modification of task protocols. Further work is planned in this area.

References

Allen CL, Nottrodt JW, Celentano EJ, Hart LEM, Cox KM (1984). Occupational physical selection standards for the Canadian Forces summary report. DCIEM.

Ayoub MM, Denardo JD, Smith JL, Bethea NJ, Lambert BA, Alley LR, and Duran BS. (1982). Establishing physical criteria for assigning personnel to air force jobs. Inst of Ergonomics Research. Lubbock Texas. Contract no F49620-79-C-0006.

Beckett MB and Hodgdon JA (1987). Lifting and carrying capacities relative to physical fitness measures. Report No. 87-26 Naval Health Research Center, California.

Chaffin DB (1974). Human strength capability and low-back pain. *Journal of Occupational Medicine*, vol **16**, p 248-254.

Chaffin DB, Herrin GD, Keyersling WM (1978). Pre-employment strength testing: an updated position. *J Occupational Medicine*, **20**, 403-408.

Durnin VGA and Womersley J "Body fat assessed from total body density and its estimation from skinfold thickness - measurements on 481 men and women aged 16-72 years" *Brit J. Nutr.* vol. **32**, 1974, p 77-97.

Grieve DW (1993). Measuring power outputs of bi-manual dynamic lifts using a hydrodynamometer. *Int Soc of Biomechanics XIV Congress*. Abstracts 1(A-L),510-511.

Laubach L (1976). Comparative strength of men and women; a review of the literature. *Aviation, Space and Environ. Med.* vol **47**. no 5. pg 534-542.

McDaniel J, Skandis RJ, and Madole SW (1983). Weight lift capabilities of Air Force basic trainees. Report No.TR 83-000 Air Forces Aerospace Medical Research Lab.

Myers DC, Gebhardt DL Crump CE, and Fleishman EA (1984). Validation of the military entrance physical strength capacity test. US Army Research Institute for Behavioral and Social Sciences. Tech Rep. 6 10.51.

Nottrodt JW and Celentano EJ (1987). Development of predictive selection and placement tests for personnel evaluation. *Applied Ergonomics*, vol **18**. no 4, p 279-288.

Pheasent S (1983). Sex differences in strength- some observations on their variability. *Applied Ergonomics*, vol **14**, no 3. pg 205-211.

Rayson MP, Bell DG, Holliman DE, LLewelyn M, Nevola VR, Bell RL (1994). Physical selection standards for the British Army. APRE report 94R036.

Sharp DS, Wright JE, Vogel JA, Patton JF, Daniel WL, Knapik J, Korval DM (1980), Screening for physical capacity in the US Army. An analysis of measures predictive of strength and stamina. USARIEM T8/80.

Stevenson JM, Bryant T, Greenhorn D, Smith T, Deakin J and Surgenor B (1990). The effect of lifting protocol on comparisons with isoinertial lifting performance. *Ergonomics*, vol **33**, no 12. pg 1455-1469.

Teves MA, Wright JE, Vogel JA (1985). Performance on selected candidate screening test procedures before and after army basic and advanced individual training. USARIEM Rep.T13/85.

Vogel JA, (1992). Obesity and its relation to physical fitness in the US military. *Armed Forces and Society*. Vol **18**, 4, 497-513.

Wilmore JH (1979). The application of science to sport: physiological profiles of male and female athletes. *Can. J. Appl. Spt. Sci.*, vol **4**. pg 103,1979.

Physiological Evaluation of a Shoveling Task Using a Conventional and a Two-Handled Shovel

RS Bridger* N Cabion**, J Goedecke**, S Rickard**, E Schabort**, C Westgarth-Taylor** and MI Lambert**

* Dept of Biomedical Engineering, UCT Medical School, Observatory 7925, South Africa, ph +21 406 6541, fax +21 448 3291.
** MRC/UCT Bioenergetics of Exercise Unit, UCT Medical School

There is evidence that the two-handled ("levered") shovel is advantageous compared to the conventional tool from a biomechanical point of view. The aim of this experiment was to determine whether less energy was consumed while shoveling a load of sand with this shovel compared to a conventional one. Subjects (n = 10) shoveled 1815 kg sand in the laboratory using either a conventional or a levered shovel . Heart rate and oxygen consumption were measured continuously during the trial. Although total energy expenditure was similar under both conditions (120 ± 20 and 125 ± 25 kcal; conventional vs two-handled shovel), average heart rate was 4% higher when the two-handled shovel was used ($p < 0.05$).

Introduction

Digging and shoveling are common tasks in a range of industries including construction, farming and horticulture as well as in leisure pursuits such as gardening. Both are a form of manual handling using purpose built hand tools. A large amount of research has been carried out, both on manual handling and on the design of hand tools in manufacturing industry and interest in the design of gardening equipment appears to be growing (Kumar, 1995).

Sen (1984) reported that shovel design can be improved by fitting a second handle to the neck of the shovel, thereby reducing the need to stoop. Freivalds (1986b) reported that the second handle was more of a hindrance than a help due to usability problems. Neither Frievalds nor Sen reported data to support their claims for or against the alternative design. The rationale for the addition of a second handle appears to be that it enables the user to lift the loaded blade with the hand vertically above the neck of the shovel while reducing the need to stoop. Degani et al. (1993) evaluated a design with two perpendicular shafts. Lumbar paraspinal EMG activity was significantly less when the two-handled shovel, was used. Furthermore, in a field study, subjects had a lower rating of perceived exertion when digging with the levered shovel, compared to the conventional one.

Bridger et al. (1995) compared a conventional spade with a levered spade based on Sen's design (Sen, 1984) and studied subjects' lumbar positions, velocities and accelerations in three axes as well on their foot forces and posture while they used each type of spade. Bending was reduced by 40% when the levered spade was used. Using the model of Marras et al. (1995), digging with the conventional spade was classified as a task with a high probability of inclusion in the high-risk group for low-back injury. Substitution of the conventional spade by the levered spade reduced the risk for low-back injury by almost 9%. However, the reduction in low-back injury risk bought about by the large reduction in stooping was partly offset by an increase in lumbar twisting.

Any new design of hand tool which can be shown to enable the same amount of work to be done with a lowered rate of energy expenditure would be of particular interest for application in developing countries where there is still concern about the workload of daily tasks in relation to the aerobic fitness of the population (Varghese et al., 1995) and the availability of food. Rogan and O'Neill (1993) point out that human effort provides some 70% of the energy required for crop production in developing countries. In some countries (Diaz et al., 1989, for example) labourers enter into a state of negative energy balance during seasons of the year in which there is a shortage of food. Brun et al. (1991) report that various technical interventions have enabled the same amount of work to be carried out using less energy. Clearly, in many parts of the world, a range of better designed, more efficient tools would leave more energy for other tasks or for leisure activities (see Rogan and O'Neill, 1993, and Bridger, 1995). It may also delay the onset of fatigue and reduce the risk of injury as is illustrated in Figure 1, which is derived from an idea by Professor S McGill of the University of Waterloo. According to this model, human tissues have no absolute injury threshold, rather injury becomes more likely as fatigue sets-in and previously innocuous loads can become threatening to the fatigued worker.

Method

Eleven healthy male subjects volunteered for the study. One day prior to the trial, subjects reported to the laboratory to familiarize themselves with the task. Trials were performed between 8h30 and 10h30. On the morning of the trial, at least 2 hours before their trial commenced, subjects consumed a breakfast containing 454 ± 7 kcal energy ($69 \pm 1\%$ carbohydrate) (Foodfinder, Medical Research Council, Cape Town, South Africa). No caffeine was consumed.Subjects were randomly assigned to start the experiment using either a conventional or levered shovel. Subjects' mass and stature were measured when they arrived at the laboratory. Then skinfold thicknesses and muscle girths were measured for the determination of body composition and muscle mass according to the procedures of Durnin and Womersley (1974) and Martin et al (1990) respectively.

Subjects started their first trial with the allocated shovel. They shoveled 1815 kg sand at a predetermined rate of 25 strokes per minute, paced by a metronome. This relatively high rate was selected to ensure that the subjects' work load was high

(Freivalds, 1986a and b). The sand was shoveled from one large, low container into another.During the trial, oxygen consumption, minute ventilation and respiratory exchange ratio were calculated every 10 seconds (Oxyconsigma, Mijnhardt, Bunnik, Netherlands) and heart rate (Sports Tester PE3000 heart rate monitor, Polar Electro, Finland) was recorded every 5 seconds.

The total oxygen consumption of each subject while they shoveled 1815 kg sand was calculated using the trapezoidal method. The subject's average heart rate, minute ventilation and respiratory exchange ratio were calculated for the duration of the trial. The energy expenditure was calculated for each subject using the respiratory exchange ratio and volume oxygen consumed (Weir, 1949).

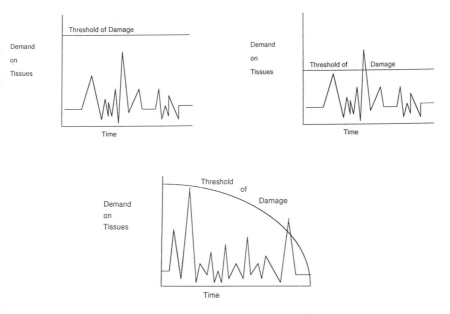

Figure 1. Relationship between injury threshold and fatigue (after S. McGill)

The number of scoops, mass of sand per scoop and total time taken to shovel the loads were also determined. After the first trial, the subjects had a rest interval of less than 1 hour (29 - 59 minutes) to allow heart rate to return to within 10 beats per minute of the starting heart rate, before starting the next trial. The protocol was repeated for the second trial, but the alternate spade was used.One subject did not complete the trial because of an aggravation of a back injury and was excluded.

Results

Table 1 presents the results of the analysis. Average heart rate was significantly higher during the trial with the levered spade compared to the conventional spade (p < 0.05). Subjects had a heavier average load per scoop with the conventional spade compared to the levered spade (p < 0.05).

Two subjects had higher heart rates when digging with the conventional tool. Seven subjects had higher average heart rates with the levered shovel and one subject's average heart rate was the same with both. Four subjects had a higher VO_2 with the levered shovel and six subjects had a higher VO_2 with the conventional one. Five subjects had a higher V_E digging with the levered shovel, four subjects had a higher V_E with the conventional shovel and one subject had the same V_E for each shovel. Three subjects expended more energy while digging with the conventional shovel. Four subjects expended the same energy and three subjects expended more energy digging with the levered shovel.

Energy expenditures were 10.51 kcal/min and 10.43 kcal/min for the conventional and levered tools respectively. In both cases, this corresponds to "very heavy work" (see Astrand and Rhodahl 1977, for example). Both the rate of oxygen consumption data and the energy expenditure data are in the upper part of the range found for various forms of hard manual labour. Thus, at a high shoveling intensity, no statistically significant differences in energy expenditure were found between the conventional and levered shovels, although heart rate was higher with the levered shovel and the amount of sand shoveled/scooped was approximately 4% less.

Table 1. Variables measured (mean, s.d.) while shoveling with conventional and levered shovels (n=10).

Variable	Conventional	Levered
Time (minutes:seconds)	11:25, 1:35	11:59, 1:57
Total oxygen consumption (lO_2)	24.5, 4.3	25.4, 4.4
Average minute ventilation ($l.min^{-1}$)	64.1, 16.1	63.5, 13.6
Average respiratory exchange ratio	0.89, 0.06	0.92, 0.08
Total energy expenditure (kcal)	120, 20	125, 22
Average heart rate (b.min^{-1})	136, 21	141, 20*
Average mass/scoop (kg.)	7.0, 0.7	6.6, 0.6*
Number of scoops	263, 30	278, 28

Values are means ± SD. *Significant difference between conventional and levered spa‹ (P<0.05)

Discussion

Although posture and lumbar motion were not measured in the present study, it was clear that subjects bent less when they used the levered spade. Despite this, no statistically significant reduction in energy expenditure was observed when the levered shovel was used. In fact, heart rate, was slightly higher which may be due to postural factors. When the body is in an erect position, the hydrostatic head, against which the heart must work, is increased compared with the stooping posture (Rowell, 1986).

There are several possible explanations for the present findings. The addition of a second handle increases the weight of the tool by 43% and of the loaded shovel by approximately 20% when using low loads such as the damp sand used in the present study. Additionally, the average mass of sand transferred was 100g lower when the levered shovel was used (approximately 4% less sand per scoop was transferred). A likely explanation is that when the conventional spade is used, the shaft is grasped firmly in both hands when it is thrust into the sand, both hands being in line with the blade and able to guide it into the material. When the levered shovel is used, only the hand on the main shaft is really involved in thrusting the blade into the soil - essentially, the user is digging with one hand only. For this reason, it would be predicted that in shoveling tasks with loose material, smaller loads would be transferred when a levered shovel is used. In other tasks, different techniques may be employed to load the blade (for example, one foot may be placed on the blade and body weight used to penetrate the soil) and this may not be a problem.

Conclusions

The physiological cost of digging with the levered shovel is the same as that incurred when digging with a conventional shovel. The previously reported biomechanical benefits of the levered shovel can be thus obtained without incurring a physiological penalty due to the weight increase caused by the addition of a second handle. Superficially, the addition of the second handle appears to greatly reduce the risk factors associated with digging and shoveling. However, the present findings, taken together with previous findings of the first author, indicate that the objective benefits of the new design are much less than they appear to be

References

Astrand, P.O. and Rodahl, K., 1977, *Textbook of Work Physiology* (McGraw-Hill Inc, New York).

Bridger, R.S., 1995, *Introduction to Ergonomics*, 1st edn (McGraw-Hill, New York), 205-223.

Bridger, R.S., Sparto, P. and Marras., W.S., 1995, The ergonomics of digging and the evaluation of a novel design of spade. In, *Contemporary Ergonomics, 1995*, SA Robertson (ed.) (Taylor and Francis, London), 391-397.

Brun, T.A., Geissler, E. and Kennedy, E., 1991, The impact of agricultural projects on food, nutrition and health, *World Reviews of Nutrition and Dietetics*, **65**, 99-123.

Degani, A., Asfour, S.S., Waly, S.M. and Koshy, J.G., 1993, A comparative study of two shovel designs, *Applied Ergonomics*, **24**, 306-312

Diaz, E., Goldberg, G.R., Taylor, M., Savage, J.M., Sellen, D., Coward, W.A. and Prentice, A., 1989, Effect of dietary supplementation on work performance in Gambian labourers, *Proceedings of the Nutrition Society,* **49**, 45.

Durnin, J.V.G.A. and Womersley, J., 1974, Body fat assessed from the total body density and its estimation from skinfold thickness: measurements on 481 men and women aged from 16 to 72 years. *British Journal of Nutrition 32: 77-97.*

Freivalds, A., 1986 b, The ergonomics of shoveling and shovel design - a review of the literature, *Ergonomics,* **29**, 3-18.

Freivalds, A., 1986, The ergonomics of shoveling and shovel design - an experimental study, *Ergonomics,* **29**, 19-30.

Grandjean, E., 1971, *Fitting the Task to the Man,* Taylor and Francis, London, 50

Kumar, S., 1995, Electromyography of spinal and abdominal muscles during garden raking with two rakes and rake handles. *Ergonomics,* **38**, 1793-1804.

Marras, W.S., Lavender, S.A., Leurgans, S.E., Fathallah, F.A., Ferguson, S.A., Allread, W.G. and Rajalu, S.L., 1995, *Biomechanical risk factors for occupationally-related low-back disorders,* Ergonomics, **38**, 377-410.

Martin, A.D., Spenst, L.F., Drinkwater, D.T. and Clarys, J.P., 1990, Anthropometric estimation of muscle mass in men. *Medicine and Science in Sports and Exercise,* **22**:729-733.

Rogan, A., and O'Neill, D., 1993, Ergonomics aspects of crop production in tropically developing countries, *Applied Ergonomics,* **24**, 371-386.

Rowell, L.B. 1986, *Human circulation regulation during physical stress.* (Oxford University Press, New York).

Sen, R.N., 1984, Application of ergonomics to industrially developing countries, *Ergonomics,* **27**, 1021-1032.

Varghese, M.A., Saha, P.N., and Atreya, N., 1995, Aerobic capacity of urban women workers in Bombay, *Ergonomics,* **38**, 1877-1883.

Weir, J.B. de V., 1949, New methods for calculating metabolic rate with special reference to protein metabolism, *Journal of Physiology (London),* **109**: 1-9.

MIDWIFERY -
MANAGING THE MANUAL HANDLING RISKS.

Sue Hignett

Ergonomist
Nottingham City Hospital
Hucknall Road, Nottingham NG5 1PB

Midwifery is a specialised area of health care provision which has not been generally recognised as being high risk for low back injuries. As part of the manual handling risk management programme in a large acute NHS Hospital Trust it became apparent that there were manual handling risks which were unique to midwifery (for example assisting with breast feeding and delivery of babies). In order to address these risks a twelve month exploratory study was set up with the aim of generating a model which could be used as the framework to plan, implement and evaluate an ergonomic programme to reduce the risk of musculoskeletal injury,

Introduction

Midwifery has not been identified as a high risk area for occupational musculoskeletal injury, in fact relatively little research seems to have been published looking at the physical demands of midwifery work. Much of the literature focuses on three issues: (1) position of delivery and assisting with breast feeding; (2) maternal choice of position and location of delivery of care; and (3) autonomy and the role of the midwife. As with many areas of health care provision there have been a considerable number of central government publications issued including guidelines for good practice and setting standards for provision of care (Department of Health, 1992, 1993).

In a previous exploratory study (Hignett and Richardson, 1995) a model (figure one) was developed which challenged the definition and application, by many researchers, of the concept of 'task' for animate manual handling in Health Care of the Elderly (HCE) nursing work. It was found that the sub-tasks of patient handling (e.g. transferring from bed-to-chair, wheelchair-to-toilet) did not correlate with the nurses' perceptions of patient handling activities. The identification of high risk manual handling operations in Midwifery presented the opportunity to test the reliability of the HCE model in a different area of health care provision. For Midwifery it was anticipated that the concept of 'task' would be more clearly perceived, for example the task of delivering babies.

Figure One Outline of model for Health Care of the Elderly

```
                      ┌──────────────┐
                      │ Organisation │
                      └──────┬───────┘
                             │
                      ┌──────────────┐
  ┌──────────┐        │    Manual    │        ┌──────────┐
  │ Patient  │────────│   Handling   │────────│  Worker  │
  └──────────┘        │  Operations  │        └──────────┘
                      └──────┬───────┘
                             │
                      ┌──────────────┐
                      │  Work place  │
                      └──────────────┘
```

Aims

The study aimed to generate a model which could be used as the framework to plan, implement and evaluate an ergonomic programme to reduce the risk of musculoskeletal injury. A secondary benefit was identified, during the study, of facilitating the objective of the Department of Health (1993) report by identifying the interaction of the midwives with respect to the organisation, training and other professional groups.

Procedure

A twelve month exploratory study was set up to explore the perceptions of midwives with respect to manual handling operations in tasks encountered in all aspects of midwifery work and locations of delivery of care. Midwives were recruited by structured convenience sampling to ensure that the sample included staff with a range of experience (senior sisters, sisters, newly qualified and students), age, and working locations (wards, labour suite, community, ante-natal clinic, low dependency unit). All staff were given an information sheet and signed a consent form prior to data collection.

Data collection

Initial group sessions were used to orientate the researcher with respect to the risks of assisting with breast feeding and delivery. Twelve individual sessions were arranged when the researcher observed midwives at work in the different locations, immediately followed by a semi-focussed interview using a work study matrix (Aberg, 1981, Hignett, 1994) as a tool to focus the interview. The matrix included themes from the HCE model, the Manual Handling Operations (MHO) Regulations (Manual Handling, 1992) and the researcher's preconceptions (from the group sessions). It was anticipated that there would be clear categories reflecting the different tasks and locations of delivery of care, so these categories were also included in the matrix. These interviews were audio-taped and transcribed to be analysed using NUD*IST (Non-numerical Unstructured Data Indexing Searching and Theorising, Richards and Richards, 1991) to assist with data management.

The data then underwent preliminary open coding to give a total of sixty-eight categories, these were rescrutinised during secondary (or pattern) coding and a model (figure two) was constructed which was presented at five sessions to forty-two midwives. The expected outcome was discussed and interpretation challenged by the researcher to extract further elucidation from the midwives on specific issues, in particular the application of the model to the wide range of tasks and locations of delivery of care encountered in midwifery work.

Findings

The model (figure two) was unanimously accepted with all the categories being relevant to a greater or lesser degree in all areas of work. There were found to be a number of distinct differences to the HCE model and examples of these are given.

Figure Two Midwifery Model

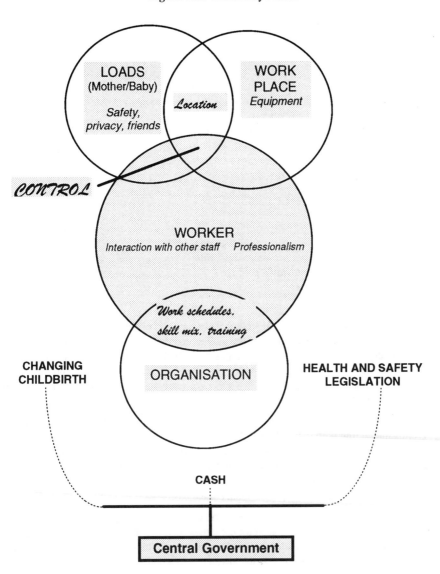

The first factor which is unique to this area of health care is that the midwives are always dealing with two loads. This changes the focus of MHO for the midwives and places constraints on the type of handling techniques and equipment which can be used. The second factor is the difference between the obstetric client group (usually fit, independent young women) and clients receiving other types of health care services as expressed in the extract below:

*M: you know, We're not treating, you know, some geriatric that you know has broken a leg or something, they're healthy young women that are probably fitter than what I am, and you know lugging them around....

The position chosen by the mother for delivery, breast feeding or ante/post natal checks determined, for the most part, the working posture of the midwife. It was found, to the researcher's surprise, that the same issues or perceived constraints were identified regardless of the location of service delivery (labour suite, home, ward, low dependency unit, clinic).

Throughout the categories of 'Clients, Worker and Workplace' there was an element which has been labelled 'control'. This attempts to reflect the constant negotiation which was taking place between midwife and the mother (and accompanying personnel) with respect to the location (hospital v. home, choice of room in the home), position (for service delivery in breast feeding, birth and ante/post natal checks) and the workplace. The midwives expressed the notion of being a visitor both in the client's home and low dependency unit. In the labour suite perceived work space ownership imposed constraints (extract below):

*M You can't set a room exactly how you want it, because you come back in and find that the visitors have moved it. When you're in there all the time, you can negotiate, you can get your room set round a good working position and a comfortable position for mum and visitors, but if you're not in there all the time, ...it's difficult.

Overwhelmingly the concept of 'mother first' was expressed in a variety of different ways, for example one midwife thought that she would probably jump into the birthing pool (regardless of her own safety) if the mother and baby were perceived to be at risk. If a client decided to deliver in a particular position which the midwife, for personal reasons, did not feel able to assist with, this then raised issues of continuity of care and personal standards as the midwife might feel that she had already developed a relationship with the mother and so was less inclined to pass her care onto another member of staff.

During the study it was observed that there were a number of other professional groups interacting with the midwives, in particular the medical staff. When a doctor was requested to assist with the birth the perceived control of the work environment and negotiating relationship with the mother shifted from the midwife to the doctor. This reinforces the need for medical staff, in particular, to consider the effect of their actions (or systems of work) on their professional colleagues.

The equipment used to deliver obstetric care ranges from beds through to breast pumps and surgical equipment used for caesarean sections. In the different locations for delivery of care there was a wide range of equipment to which the midwives had to adjust themselves, this could be from a double divan bed in a home delivery or an electric bed with removable sections in labour suite. The common issues about design, usability, suitability, compatibility and availability were all raised.

The 'organisation' category included issues which were out of the immediate control of the ward staff, for example the work scheduling, skill mix, manual handling training and staff uniforms. This was also the entry point for most of the external central government influences from the cash available (money from purchasers of health care services), to Health and Safety legislation and the provision of a service at a standard recommended for child birth care. The difficulties for an organisation in trying to balance the different agendas from each of these issues was acknowledged by many of the midwives, although not always sympathetically.

Discussion

In 1991 Imada suggested that the conservative nature of the scientific paradigm which underpinned ergonomics was not well suited for creating change, and that the way to overcome that was not necessarily more of the same, but rather a need to devise different solutions and look to new paradigms. Richardson and Hignett (1994) followed up this idea by challenging the use of the scientific model of enquiry and quantitative methodologies in the context of risk management for manual handling operations. They suggested that models based in the positivist paradigm did not include a methodology which supported real world evaluation and cited examples of manual handling operations in the health care industry.

Although there is no clear and accepted set of conventions for analysis of qualitative data corresponding to the 'scientific' management of quantitative data, there are ways in which qualitative data can, and has been dealt with systematically (Robson, 1993). The approach taken for this study was to use a framework from previous work and data management tools to assist with the quantity of text to facilitate the analysis and interpretation whilst retaining the richness, depth and chronological flow. It was found in the HCE study (Hignett and Richardson, 1995) that this approach lead to a fuller understanding of all the factors which were perceived to constrain the nursing staff when carrying out manual handling operations.

The midwifery model contains the same basic four categories as the HCE model: they both omit 'task'. However major differences were identified in nature of the service, with the midwives finding that it was very difficult to plan their work: they responded to the needs or demands of the mother and baby. This issue was not identified at all in the Care of the Elderly study where the criticism that *'the services seemed to have been designed to meet the needs of the professionals rather than the women'* (Department of Health, 1993) might be have been more appropriately aimed. The exploratory study found little evidence of this, in fact it seemed that the services had been designed to meet neither the needs of the women nor the midwives. This should be an area of great concern when there are central government edicts which may impose further constraints of midwifery practice, for example the 'Department of Health (1992) consultative document 'Health of the Nation' puts forward the goal of increasing the proportion of breast feed infants (at birth) from 64% (1985) to 75% by the year 2000. It is possible that the impact of increased in-put by midwives may be measured in terms of increased prevalence of musculoskeletal injury as well as the percentage of breast fed babies.

The ability to genericise categories, as described in the midwifery model, across a range of services and locations of delivery of care again brings into debate the relevance of assessing manual handling operations for human loads as 'tasks'. If the variability introduced by the load(s) has the potential to be so great that a risk assessment for the task, for example 'assisting with breast feeding', cannot be generic then the value of adopting this approach (basing risk assessment on identified tasks) must be questioned.

Underlying all these issues is the question which was raised in the Department of Health report (1993) - how much of current practice is based on measures known to be effective and how much on tradition? There seem to be two principles areas of conflict, (1) the safety of the mother and baby versus the safety of the midwife, and (2) the autonomy of the midwife versus the choice of the mother with respect to the location, type and personnel involved in the delivery of care. When 'control' of the workplace and type of care delivered is constantly being negotiated there must be an increased risk of musculoskeletal injury as the manual handling operations cannot always be predicted or planned.

References

Aberg, U. 1981, Techniques in redesigning routine work. In E N Corlett & J Richardson (eds) *Stress, Work design and Productivity* (John Wiley & Sons) 157-163

Department of Health. 1992. *The Health of the Nation - A summary of the strategy for health in England* (HMSO)

Department of Health. 1993. *Changing Childbirth* (HMSO)

Hignett S. 1994. Shifting the emphasis in patient handling, Occupational Health 46 4 127-130

Hignett, S and Richardson, B. 1995. Manual handling human loads in a hospital: an exploratory study to identify nurses perceptions, Applied Ergonomics 26 3 221-226

Imada A S. 1991. The rationale and tools for participatory ergonomics. In: K Noro K and A S Imada (eds) *Participative Ergonomics* (Taylor and Francis, London)

Manual Handling. 1992. *Manual Handling Operations regulations 1992 Guidance of Regulations L23* (HSE Books, London) ISBN 0 11 886335-5

Richards, L and Richards, T. J. 1991. The transformation of Qualitative method: computational paradigms and research processes. In N Fielding and R Lee (eds) *Using Computers in Qualitative Research* (Sage Publications) 38-53

Richardson, B and Hignett, S. 1994. Risk Assessment - Myth or Method?, *Proc Ergonomics and Health & Safety Conference*, (The Ergonomics Society), Bristol

Robson, C. 1993. *Real World Research* (Blackwells)

A PRELIMINARY EVALUATION OF THE CHINESE WHEELBARROW

Ted Megaw and Babu Ram Gurung

Industrial Ergonomics Group
School of Manufacturing and Mechanical Engineering
The University of Birmingham
Edgbaston
Birmingham B15 2TT, UK

This study investigated the use of the Chinese wheelbarrow as an alternative to the conventional Western one for carrying large loads over long distances. Five young male subjects pushed the wheelbarrow over a predetermined course for 10 minutes at a speed of 3.5 km h^{-1} with additional loads of either 50 or 75 kg. On half the trials, shoulder straps were worn in an attempt to increase the controllability of the task. Results demonstrated that the physical workload under all conditions could be classified as moderate. The use of the shoulder straps did not make the task of the operators any easier. The same subjects also performed dynamic and static tests separately to reveal substantial negative extra-cardiac costs and very small positive extra-oxygen costs. A number of design factors to improve the stability of the Chinese wheelbarrow are suggested for further investigation.

Introduction

In the rural areas of many industrially developing countries, successful transportation of food, water and other materials cannot be achieved by powered vehicles. The inhabitants of these countries cannot afford to purchase these means of transport and often live in areas not serviced by a suitable network of roads. A general requirement of alternative methods of transport is that they should make use of materials and components which are readily available to the local inhabitants and can be assembled with a minimum of cost and technology. In reality, the most critical element from the cost and the engineering point of view is the availability or the manufacturability of the wheel/s. In the case of flat terrain where wide tracks are available, transport makes use of animal-drawn carts and sledges, hand-drawn carts with two or three wheels, and bicycles and tricycles with panniers or trailers. In the case of both hilly and narrow tracks, materials are frequently carried individually often using shoulder poles or head mounted devices, in addition to using animals. By way of human-powered vehicles, wheelbarrows (which can

be regarded as one-wheel carts) would appear to offer a viable means of transport
provided the terrain is not too steep. However, the very high static workload associated
with the conventional 'Western' wheelbarrow means that it is not suitable for the
transportation of heavy loads over long distances.

An alternative to the conventional wheelbarrow is the Chinese wheelbarrow where
a single wheel is positioned directly under the load to be transported (see Figure 1) unlike
in the conventional one where the wheel is towards the end away from the driver. Not
only does this mean the load for the driver is less with the Chinese wheelbarrow, but by
using a comparatively large diameter wheel, it can be used for uneven terrain. The
disadvantage with the Chinese wheelbarrow is the problems encountered in controlling it
because of its high centre of gravity. It is possible to lower the centre of gravity by
allowing the wheel to pass through the carrying surface, but this restricts the type of load
that can be transported. Alternatively, a smaller diameter wheel can be used but this
increases the pushing forces.

Figure 1. A view of the Chinese wheelbarrow.

The main aim of this preliminary study was to confirm the suitability of the Chinese
wheelbarrow for carrying loads as high as 75 kg over long distances f rom the point of
view of acceptable physical workloads. A subsidiary aim was to see whether the physical
workload could be predicted by simply adding together the dynamic and static components
of the task (Sanchez, Monod and Chabaud, 1979) so that the total workload could be
calculated from existing data. Additionally, an initial attempt was made to increase the
controllability of the wheelbarrow by introducing a system of shoulder straps.

Method

Subjects

Five healthy male subjects, aged between 23 and 26 years, took part in the experiment.

Equipment

The Chinese wheelbarrow was provided by IT Transport Ltd of Oxford and is shown in Figure 1. The wheelbarrow weighed 33 kg unladen and had a wheel diameter of 55 cm. The width of the tyre tread was 6 cm with a tyre pressure of 1.27 kp cm^{-2}. The carrying surface of the wheelbarrow was 100 cm long and 60 cm wide. The wheel was centred 40 cm from the front end. The handle height could be adjusted to the standing knuckle heights of the subjects. The arrangement of the straps is shown in Figure 2. The straps only transmitted some of the load to the shoulders when the wheelbarrow became unstable. Bags of gravel distributed over the wheelbarrow surface were used as the load to be transported.

Figure 2. A view of the Chinese wheelbarrow illustrating the use of the straps.

During the course of the experiment heart rate was continuously monitored with a Polar Sport Tester heart rate monitor and the volume of expired air was sampled every minute using a Kofrany-Michaelis respiration gas meter. A Taylor Servomex oxygen analyser was used to estimate the percentage of expired oxygen. Questionnaires were administered to identify body discomfort and subjective controllability of the wheelbarrow.

The wheelbarrow trials were conducted in a large room. The course to be tracked was marked out on the floor and involved a number of 90° and 180° turns. It was not possible to control the temperature, humidity and air speed in the room. The separate dynamic and static tests took place in the ergonomics laboratory, the dynamic tests taking place on a treadmill.

Experimental design and procedure

Each subject took part in all the experimental conditions. The first part of the experiment involved subjects pushing the wheelbarrow at approximately 3.5 km h^{-1} around the predetermined course for 10 minutes under 4 experimental conditions given in a different random order for each subject. The four conditions were -

- load of 50 kg with no straps (50-only)
- load of 50 kg with straps (50-straps)
- load of 75 kg with no straps (75-only)
- load of 75 kg with straps (75-straps).

With a load of 50 kg to be transported, the total static holding weight was adjusted to 12 kg (approximately 6 kg at each handle) by altering the positions of the bags of gravel. With a 75 kg carrying load, it was 16 kg. These values were set when the subjects lifted the wheelbarrow with both their arms fully adducted. Standing resting values for heart rate and oxygen consumption were obtained before each experimental condition. Subjects were given practice runs before completing the main trials.

Separate dynamic and static trials were subsequently held. The dynamic tests involved walking on a treadmill at speeds of 0 and 3.5 km h^{-1}. Static tests involved holding weights of 0, 12 and 16 kg (0, 6 and 8 kg in each hand) with arms fully adducted. Each test lasted 10 minutes and both heart rate and volume of expired air were monitored.

Results

A summary of the results obtained from the main wheelbarrow experiment are given in Table 1. The data have been averaged over the 5 subjects. The highest average heart rate of 115.2 bpm (beats min^{-1}) was obtained under the condition of 75 kg load with the straps and the highest average energy consumption of 5.7 kcal min^{-1} was recorded under the condition of 75 kg load without the straps. When resting levels are taken into account, the results in Table 1 show that the heart rate component due to the work was highest (36.6 bpm) under the condition of 75 kg load without the straps and that the work calories was highest (4.4 kcal min^{-1}) under the condition of 75 kg load with straps. Two-way ANOVAs on all the sets of data shown in Table 1 yielded the same results. All values were significantly greater for the 75 kg load than for the 50 kg load (p.<0.05). There were no significant effects of using the straps and there were no significant interaction effects between static load and the use of the straps.

Table 1. Results from the wheelbarrow trials.

	50-only	*50-straps*	*75-only*	*75-straps*
Heart rate, bpm	106.2	106.3	110.2	115.2
Oxygen consumption, 1 min^{-1}	1.01	0.93	1.19	1.17
Energy consumption, kcal min^{-1}	4.8	4.5	5.7	5.6
Work heart rate, bpm	29.7	29.7	36.6	34.6
Work energy consumption, 1 min^{-1}	0.75	0.69	0.90	0.92
Work energy consumption, kcal min^{-1}	3.6	3.3	4.3	4.4

Body discomfort scores indicated that the largest amount of discomfort was experienced in the left and right forearms. Other regions of the upper body also yielded substantial scores of discomfort, but the scores for the lower body parts were negligible. Statistical analyses failed to reveal consistent differences between the four main experimental conditions even for the forearm scores.

To establish if there was any extra-cardiac or extra-oxygen costs associated with pushing the wheelbarrow, combined work values (i.e. where resting values have been subtracted) were obtained from the separate dynamic and static tests and these combined values were compared with the work values obtained from the main wheelbarrow experiment. In relation to heart rate, the results showed over all conditions an average negative extra-cost of 10.5 bpm. This corresponds to a saving of 33% in terms of work heart rate values. In relation to oxygen consumption, a slight positive extra-oxygen cost was found equivalent to $0.07 \, l \, min^{-1}$ which corresponds to an increase of 8% in terms of work values.

There was no indication from the subjective rating scores that the straps made the task of balancing the wheelbarrow any easier and this conclusion was confirmed by other informal comments from the subjects.

Discussion

In general the results have demonstrated that pushing the Chinese wheelbarrow with loads of either 50 or 75 kg can be classified as representing 'moderate' workloads (Christensen, 1964) and, therefore, such work should be able to be performed for comparatively long periods of time without having to resort to the introduction of frequent rest pauses. This confirms the advantage of the Chinese wheelbarrow over the traditional one. However, it should be remembered that the subjects in this study were young males who were comparatively fit and that, in industrially developing countries, work of this kind is frequently performed by female workers. Also all the subjects did report considerable discomfort in the forearms and this study does not reveal how this discomfort might have progressed with longer work periods. With the current design of the wheelbarrow, the carrying load could have been reduced further by centring the carrying load nearer the front of the carrying surface. However, as the load becomes lighter, balancing the load becomes more difficult.

A more general limitation of the study relates to its ecological validity. Pushing a wheelbarrow on a perfectly flat surface is obviously very different from the conditions met in the natural environment. An uneven terrain is likely to increase the physical workload by introducing continuous changes in the pushing forces as well as the changes in the distribution of the static load from one side of the body to the other.

It was hoped that the inherent problems of maintaining the balance of the carried load would be alleviated by the use of the shoulder straps. However, this study failed to reveal any advantage of them. While there is no doubt that balancing the load is a skill which requires a considerable time to acquire, there are still a number of design factors that need to be investigated further. These include the location of the handles, the position of the wheel, the distribution of the carrying load, the diameter and tread width of the wheel, and the shape and area of the carrying surface. Unfortunately, any improvement in controllability of the wheelbarrow is likely to be at the cost of increasing the physical workload, in relation to either the pushing forces or the static loading. The various payoffs need to be investigated in future studies.

Regarding the observed considerable negative extra-cardiac costs and very small positive extra-oxygen costs for the combined components characteristic of the wheelbarrow task, the results are in agreement with those reported by Sanchez, Monod and Chabaud (1979) in respect of combined holding and walking, particularly when carrying light loads. They suggest that this is because the increase in the trunk's postural load can be more than compensated by the accompanying decrease in the vertical oscillation of the centre of gravity. It should be pointed out that the assumption has been made in this study that the wheelbarrow task can be reduced to two components - the carrying of a static load and the forward walking activity. That is to say no account has been taken of the pushing forces. If a more theoretical understanding of the workload is required, it will be necessary to include this small but possibly important component.

References

Christensen, E.H. 1964, L'Homme au Travail, *Securitè, Hygiène et Médecine du Travail, Series No 4*, (International Labour Office, Geneva).

Sanchez, J., Monod, H. and Chabaud, F. 1979, Effects of dynamic, static and combined work on heart rate and oxygen consumption, Ergonomics, **8**, 935-943.

Acknowledgements

We are indebted to IT Transport Ltd not only for providing the Chinese wheelbarrow but also for the invaluable advice offered by Ron Dennis.

BACK PAIN IN THE NURSING PROFESSION: MEASUREMENT PROBLEMS AND CURRENT RESEARCH

Diana Leighton and Thomas Reilly

*School of Human Sciences,
Liverpool John Moores University,
Mountford Building, Byrom Street,
Liverpool, L3 3AF*

Current research has indicated that the prevalence of back pain in the nursing profession has increased by approximately 40% during the past decade. There has not, however, been a concomitant increase in the amount of sickness absence from work due to symptoms. Back pain was noted to be equally prevalent amongst a non-nursing sample. Nurses attributed their back pain to patient handling activities; specific occupational tasks perceived by nurses as stressful to perform have been identified and the compressive loads imposed on the spine following lifting activities established. It is implied that back pain is ubiquitous within society and that the perceived causes of back pain are not wholly responsible for symptoms. There is a need to differentiate between individuals who do/do not display signs of disability. Differentiation between the mode of onset of back pain (acute vs insidious) may also aid the identification of predisposing risk factors.

Introduction

Musculoskeletal disorders of the back pose a great problem throughout the industrialised nations; they constitute the most frequently reported occupational disease in the United Kingdom. Associated with the magnitude of the problem are the disability caused and the economic consequences of symptoms. In recognition of this, the quantity of research conducted to investigate possible risk factors, preventive measures and the effectiveness of treatment for back problems has escalated over the past two decades. Nurses comprise one occupational group frequently reported as having high prevalence and incidence rates of back problems (Videman et al., 1989; Caboor et al., 1995). It is the object of this paper to i) illustrate the progress that has been made in our research programme at Liverpool John Moores University to identify possible risk factors associated with back disorders amongst nursing professionals and, ii) outline some of the issues relating to the difficulties encountered when conducting research in this area.

Epidemiological investigations

Epidemiological research conducted in Liverpool has examined the incidence and prevalence of back pain and the accompanying disability caused by symptoms (Leighton and Reilly, 1995a). Questionnaire surveys were conducted amongst: i) a nursing population and, ii) non-nursing members of a sample of the general population. The data were compared to previous figures reported by Stubbs et al. (1983). This was possible as specific questions designed to elicit information enabling the calculation of figures relating to the prevalence and incidence of back pain in the Liverpool study were modelled upon those in the earlier questionnaire. It was established that there has been approximately a 40% increase in the point and annual prevalence of back pain amongst nursing personnel over a ten year period. There was not a concomitant increase in the quantity of absences from work due to back pain. The information collected from the non-nursing sample indicated that back pain was equally prevalent amongst this group as amongst nurses (Figure 1). The sickness absences caused by symptoms in terms of days off work due specifically to back pain were also comparable. The nurses attributed their back pain to the performance of patient handling and lifting activities routinely undertaken during the course of a work-shift. The perceived causes of back problems amongst the non-nurses were not established. It may be implied that whilst nurses do experience back pain, prevalence rates are similar to those of the general population. These results do not belie the fact that nurses demonstrate a relatively high prevalence of back disorders. Moreover, it emphasises the necessity to research further the causes and classifications of disorders of the back within specific and general populations.

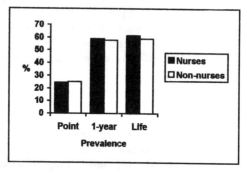

Figure 1.   The prevalence of back pain in nursing personnel and non-nursing members of the general population (Leighton and Reilly, 1995a) (Point, annual and lifetime prevalence).

There are many individuals experiencing symptoms whose back pain is termed idiopathic. It is particularly with this group of nurses in mind that research into back pain amongst nurses has been conducted in Liverpool. The epidemiological work has been extended to interview nurses with regard to the tasks/manoeuvres perceived as the most stressful to perform. The information collected was not just in relation to the lower back but identified how stressful tasks were, the anatomical location where stress was felt and identified reasons why the tasks were stressful. The results of this study provided interesting points to pursue in further research. The tasks of moving a patient up the bed, the transfer of a patient from bed to chair and static postures were perceived as the three most stressful to perform, respectively (Table 1.). However, the reported anatomical sites of stress included the upper back, neck and shoulders for the former two tasks which has implications for the prevalence of other occupationally related musculoskeletal disorders.

Table 1.   The anatomical sites where most stress was perceived to occur during the performance of the ranked tasks.

Rank	Task	Anatomical site
1	Move patient up the bed	Neck, shoulders, upper back
2	Carry patient from bed to chair	Multiple sites of upper body
3	Static postures	Lower back and buttocks

In association with the previous questionnaire surveys and with the knowledge that the non-nursing sample experienced back pain symptoms, it could be suggested that the perceived causes of pain by the sufferers may not actually be those wholly responsible for the onset of symptoms. These perceived causes should not be disregarded, but because nurses are made aware through training of the risks associated with poor lifting technique, this may evoke the automatic perception that lifting has caused pain. Although the physical consequences of lifting may place individuals at risk, it is probable that these are only part of the multifactorial aetiology of back pain. The validity of this statement and the sources of other risk factors still need to be determined.

Quantification of the loads imposed on the spine (*in-vivo*)

Investigations of the compressive loads imposed on the vertebral column have utilised the technique of measuring changes in stature (Eklund and Corlett, 1984). These studies have aimed to i) quantify the loads following isometric and dynamic lifting tasks and an occupational lifting activity and, ii) establish the diagnostic application of the measurement of stature changes in relation to the prevalence of back pain symptoms.

Dynamic lifting tasks, transferring a load from the floor to a height of 76 cm, induced greater loss of stature than either isometric lifting or an asymmetric lifting activity. The lifetime prevalence of back pain amongst nurses (back pain at some time during their life) did not reveal differences in response to spinal loading compared to asymptomatic individuals. In a separate study, the compressive loads imposed by the repetition of the task of transferring a patient from bed to chair were investigated (Leighton and Reilly, 1994). Loss of stature was not influenced by the existence of chronic back pain symptoms (symptoms at least once a month) amongst nursing personnel.

These investigations have demonstrated that the prevalence of back pain symptoms is not indicative of pathological abnormality of the intervertebral discs. It has been suggested that greater loss of stature is associated with the increased elasticity of degenerated discs (Hutson, 1993). It is acknowledged that disc degeneration occurs in the absence of pain and that not all back pain is associated with pathological abnormality/degeneration.

Muscle strength training and assessment of the trunk musculature

The studies referred to above have been associated with the identification of factors presdisposing individuals to symptoms of back pain. The influence of the trunk musculature has been the focus of additional work (Leighton and Reilly, 1995b). Improvements in trunk muscle strength and manual handling skills were observed following a 10-week period of physical training (Figure 2). The results demonstrated the beneficial effects of physical training programmes for personnel involved in occupations demanding manual handling. The future implementation of longitudinal and intervention studies incorporating physical training to improve trunk muscle strength may have implications for attempts to reduce the incidence of back pain.

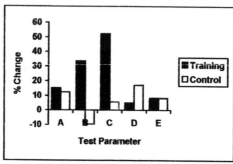

Figure 2.   Percentage change in test parameters between baseline measurements and following the 10-week physical training programme (Leighton and Reilly, 1995b). (A=Maximal isometric lifting strength, B=1-repetition maximum C=Maximal acceptable weight of lift, D=Peak torque extension, E=Peak torque flexion)

Measurement problems

Research into back pain, and in this instance occupationally related disorders, is notoriously difficult. A problem inherent within epidemiological studies of back pain is the subjective reporting of both the symptoms and mode of onset of pain. It is difficult to differentiate between pain that occurs gradually and pain which suddenly starts following the performance of a particular activity; there is clearly some overlap within these categories. If the mode of onset of pain can be distinguished, the risk factors associated with symptoms may also be identified.

Retrospective epidemiological studies do rely upon the recall of episodes of pain. Respondents may also wish to register the existence of symptoms despite the occurrence of pain outside time periods specified by the questionnaire. Buckle (1987) suggested that where surveys have generated prevalence data of more than one type (e.g. annual, point, lifetime, one-month prevalence), the validity of the data increases.

Psychological and social parameters constitute complex variables associated with individuals in whom persistent back pain symptoms exist. Considerable amounts of disability may be demonstrated by these individuals who have probably also been forced to adapt their lifestyles in accordance with the pain (Troup, 1995). It is important to differentiate between these individuals and the sample populations representative of the back pain which appears to be ubiquitous within societies. The pain behaviour and disability caused by symptoms may lead to erroneously high prevalence and disability rates being derived from survey data.

The identification of risk factors associated with the onset of back pain symptoms is just one aspect of research conducted to reduce the prevalence of back pain amongst the nursing profession. There are problems not only in classifying and assessing the natural history of the condition, but in making the results applicable to the workplace environment. Compliance by nursing staff in the use of approved lifting techniques or in the use of assistive devices (e.g. hoists) is a considerable problem within the profession. It was established from the questionnaire survey that 44% of nurses, representing a range of specialities, did not have assistive devices on their current ward. This demonstrates that either nurses are not being trained in the use of such pieces of equipment or nurses are reluctant to use them. Factors which could influence their use include i) time and, ii) the space required for the safe and efficient movement of patients on and off the hoists. The unpredictability of patients and restrictions in the work environment are also factors which are difficult to measure directly and integrate into an overall preventive framework for the reduction of back pain within the nursing profession.

Conclusions

It is apparent that that disorders of the back are not just restricted to occupations where manual handling is performed; the problem is far more widespread than that. The increase in the prevalence of symptoms over the past decade amongst nurses is of great concern. However, there has not been an equivalent increase in the disability caused by symptoms. The loss of stature induced by the compressive loads imposed by lifting activities was not differentiated by the existence of back pain symptoms and the performance of physical training has implications for reducing the incidence of back pain symptoms amongst manual handling occupations. The problems associated with the measurement of back pain amongst populations include the selection of samples in whom disability is associated with symptoms, the mode of onset of pain and work-related factors such as compliance and the work environment.

References

Buckle, P. 1987, Epidemiological aspects of back pain within the nursing profession, *International Journal of Nursing Studies*, **24**, 319-324.

Caboor, D., Zinzen, E., Van Roy, P. and Clarys, J.P. 1995, The effects of nursing tasks on spinal shrinkage. In K. Häkkinen, K.L. Keskinen, P.V. Komi and A. Mero (eds), *Proceedings of the XVth Congress of the International Society of Biomechanics*, 146-147.

Eklund, J.A.E. and Corlett, E.N. 1994, Shrinkage as a measure of the effect of load on the spine, *Spine*, **9**, 189-194.

Hutson, M.A. 1993, *Back Pain: Recognition and Management*, (Butterworth-Heinemann, London).

Leighton, D. and Reilly, T. 1994, Effects of repetitive lifting on female nurses with and without low back pain, In S. Robertson (ed.), *Contemporary Ergonomics 1994*, (Taylor & Francis, London) 106-111.

Leighton, D. and Reilly, T. 1995a, Epidemiological aspects of back pain: the incidence and prevalence of back pain in nurses compared to the general population, *Occupational Medicine*, **45**, 263-267.

Leighton, D. and Reilly, T. 1995b, Trunk muscle strength and manual handling skills: the effects of training, In S. Robertson (ed.), *Contemporary Ergonomics 1995*, (Taylor & Francis, London) 397-402.

Stubbs, D.A., Buckle, P.W., Hudson, M.P., Rivers, P.M. and Worringham, C.J., 1983, Back pain in the nursing profession I: epidemiology and pilot methodology, *Ergonomics*, **26**, 755-765.

Troup, J.D.G. 1995, Fear-avoidance: the natural history of back pain and its management, In R.M. Aspden and R.W. Porter (eds.) *Lumbar Spine Disorders: Current Concepts*, (World Scientific, Singapore) 155-163.

Videman, T., Rauhala, H., Asp, S., Lindström, K., Cedercreutz, G., Kämppi, M., Tola, S. and Troup, J.D.G. 1989, Patient-handling skill, back injuries and back pain, *Spine*, **14**, 148-156.

Work induced professional diseases of the spine - development of a portable stress measuring system

Rolf Ellegast, Jürgen Kupfer, Dietmar Reinert, Stefan Busse and Wolfgang Reis

Berufsgenossenschaftliches Institut für Arbeitssicherheit (BIA)
53754 Sankt Augustin,
Germany

The described research project has the aim objectively determining the physical strain of the spine at workplaces. The basic idea is the recording of the external physical stress, e.g. body postures and the carried loads directly at the workplace using a portable data acquisition system carried by the worker. The data acquisition should be possible for a typical working time of 8 hours. During the data acquisition the data should be filtered on-line so that data storage is limited to the minimum, i.e. body postures together with the carried loads.

Introduction

Spine diseases may have different reasons (i.e. genetic, behaviour during leisure time, occupational strain). The compensation of occupational diseases caused by lifting and carrying heavy loads, assumes the proof of the causality between external stress and internal strain (especially the segment L5/S1 of the vertebral column; Bongwald, Luttmann and Laurig 1995). In this context the determination of the temporal lapse for extreme body bending and for carried loads is of great interest (Kupfer and Christ 1996). Currently, the BIA examines within the framework of a „case study", the development of a data acquisition system, which can be used for recording body postures and the carried load for a complete working day. In this study internationally approved procedures are considered (Marras, Sudhaker and Lavender 1989 and others).

Method

Classification of body postures

In ergonomic research it is common to detect body postures at the workplace by manual observation, video systems or in the laboratory by three dimensional records. After that follows a normally time-consuming evaluation of the data. Subjective elements may counterfeit the results of the first two procedures. A method that allows the

estimation of the physical strain of the spine both by lifting or carrying heavy loads and in case of extreme trunk bending was developed in the Finnish steel manufacturer OVAKO (**OWAS**-method (Karhu, Kansi and Kuoriuka 1977) - **O**VAKO **W**orking Posture **A**nalysing **S**ystem). This method enables us to classify working postures and estimate them relative to their influence on human health. A working posture is described by basic postures of three parts of the body:
- back   straight, bent, twisted or bent sideways, bent and twisted
- arms   both arms below acromion, one arm above acromion, both arms above acromion
- legs   sitting, standing upright, standing upright on one leg, standing bent on both legs, standing bent with one leg, kneeling, walking.

In addition, the carried work load is classified in smaller than 10 kilograms, between 10 and 20 kilograms and bigger than 20 kilograms. A four digit code allows the numerical classification of the working posture (see figure 1). From 84 possible OWAS-basic postures we have chosen 12 body postures for the classification of the external stress of the spine in a first approach (without taking into consideration the arms) (Kupfer and Christ 1995 and 1996 see figure 2).

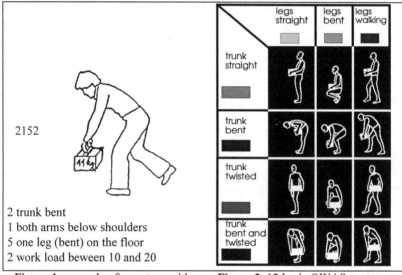

2152

2 trunk bent
1 both arms below shoulders
5 one leg (bent) on the floor
2 work load beween 10 and 20

Figure 1. example of a posture with   **Figure 2.** 12 basic OWAS postures
OWAS-code

Each of these leg and trunk postures is identified by a number according to the OWAS-code or a colour code (for a quick visual detection by on-line data acquisition). With the help of these codes all working postures can be described by means of the matrix of figure 2. If the working postures are recorded together with their temporal appearance these characteristical body postures can be linked to the time.

Experimental setup for the recording of the working postures
      With the help of angle and angular velocity - sensors that are worn on the person (see figures 3 and 4) the following body angles are measured:

- the trunk bending angle (sensor: inclinometer, measuring the angle to the vertical)
- the trunk torsion angle (sensor: gyrometers, measuring angular velocities)
- the hip angle (angle between trunk and thigh; sensor: potentiometer, measuring the angle)
- the knee angle (angle between thigh and tibia, sensor: potentiometer).

The temporal range of the angles is stored and evaluated in a controller worn on the person.

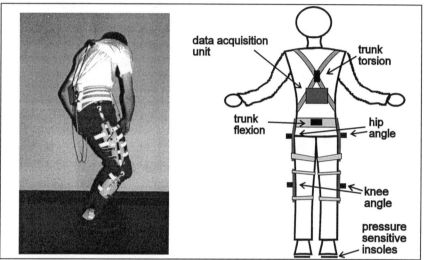

Figure 3, Figure 4. laboratory sensor mounting

Load estimation using the foot reaction force

The temporal range of the foot reaction force of the worker is recorded with a foot

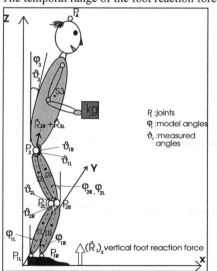

Figure 5. biomechanical chain model for load estimation

pressure sensor system (novel 1994; see figure 3). Similar to the principle of a balance this system is used to determine the body weight and the additional carried load in static positions. To calculate the carried load also during typical dynamic working procedures an on-line estimation method on the basis of a biomechanical model was used (Ellegast 1995). In this model the temporal synchronously recorded body angle data serve as input values. Figure 5 shows the data used for the model. The parts of the body are considered as fixed mass points that are connected via joints (chain model; Jäger 1987). The model is used in the first step for the prediction of the temporal foot reaction force that is calculated from the dynamical signals of the angle sensors. By comparison of that data with the measured foot reaction force an estimation of the temporal development for the carried load during static and dynamic working procedures is possible.

First results

It could be shown in principle with the above described experimental setup that the sensors (angle- and foot pressure measurements) and the model calculations are suitable for determining the load dynamically (Ellegast 1995 and Reis 1995). An example for the estimation of the body posture is shown in figure 6. In the motion process the traditional leg lift was included (Ayoub and McDaniel 1971; see figure 6, pictogram and table 1). Figure 7 shows the accompanying graph for the leg and trunk postures of the traditional leg lift. With the aim of future data reduction only the alteration of the three digits OWAS-code together with its time stamp will be recorded on-line.

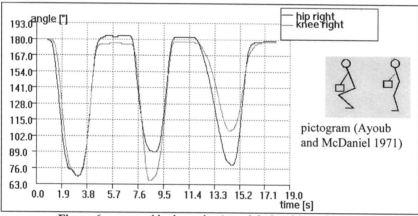

pictogram (Ayoub and McDaniel 1971)

Figure 6. measured body angles (e.g. right leg: hip and knee angle) in a typical action

Table 1. chronological order of the described movement

beginning	end	action
0 s	1 s	standing upright
1 s	4.5 s	taking a load of 10 kg from the floor (traditional leg lift)
4.5 s	7 s	standing upright while carrying load
7 s	10 s	knee bend
10 s	14 s	standing upright while carrying load
14 s	16 s	putting the load onto the floor
16 s	18 s	standing upright

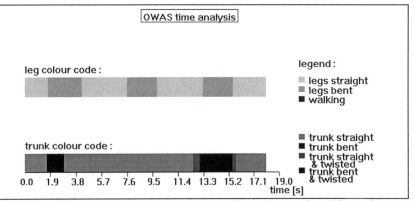

Figure 7. OWAS-postures (measured simultaneously to Figure 6)

Figure 8 shows the synchronously measured signal of the foot reaction force and the calculated foot reaction force based on the biomechanical model for the above described motion process. The good agreement between the graphs confirms our model assumptions. In the period between 4,5 s and 14 s one can clearly see the carried load. A definite algorithm for the estimation of the carried load that is based on the chain model and which will allow an on-line load estimation is in preparation.

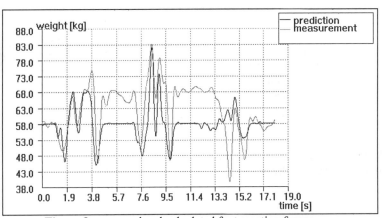

Figure 8. measured and calculated foot reaction forces

Future implications

The first results encourage us to estimate body postures and carried loads synchronously at workplaces using standard sensors. Today problems arise for the reproducible mounting of the sensors on typical working clothes and for the calculation of the carried loads if no static time periods can be identified by the system. The reason for that is the different time constant of the different sensors so that the biomechanical model is not sufficient for high dynamic body motions. Up to now the discussed results have been measured in the laboratory. A field study is in preparation.

Acknowledgement

The authors wish to thank the Institute of Applied Physics of the University of Bonn, especially Wolfgang Urban, for their support of this project.

References

Ayoub, M.M and J. W. McDaniel 1971, *The Biomechanics of Pushing and Pulling Tasks* (Texas Tech. University, Lubbock,)

Bongwald, O., A. Luttmann and W. Laurig 1995, *Leitfaden für die Beurteilung von Hebe- und Tragetätigkeiten. Gesundheitsgefährdung, gesetzliche Regelungen, Meßmethoden, Beurteilungskriterien und Beurteilungsverfahren*(study for the HVBG, Inst. für Arbeitsphysiologie of the University Dortmund, Abt. Ergonomie)

Ellegast, R. 1995, *Entwicklung eines biomechanischen Modells zur Bestimmung des gehandhabten Lastgewichtes während dynamischer Arbeitsvorgänge* (thesis at the Institute of Applied Physics of the University of Bonn)

Jäger, M. 1987, *Biomechanisches Modell des Menschen zur Analyse und Beurteilung der Belastung der Wirbelsäule bei der Handhabung von Lasten,* (VDI-Verlag Düsseldorf)

Kupfer, J. and E. Christ 1996, *Heben und Tragen/extreme Rumpfbeugehaltungen - Ergonomische Kennwerte als Grundlage der Prävention,* Die BG, **2**, (being printed)

Kupfer, J. and E. Christ 1995, *Lifting and carrying heavy loads, working in an extreme body bending position: a concept for investigating physical strain at the workplace,* (poster at conference: from research to prevention - Managing Occupational and Environmental Health Hazards, 20-23 March 1995, Helsinki, Finland)

Marras, W. S., Sudhaker, L. R. and Lavender, S. A. 1989, Three dimensional measures of trunk motion components during manual materials handling in industry. In The Human Factors Society(ed.): *Proceedings of the Human Factors Society:* 33rd Annual Meeting, Denver, Colorado 662-666

Karhu, O., Kansi, P. and Kuoriuka, I. 1977 *Correcting working postures in industry: A practical method for analysis,* Applied Ergonomics, **8**, 199-201

novel 1994, *Technical Specifications, Capacitive Sensors for Pressure Measurements.* novel GmbH, München, No. 6

Reis, W. 1995, *Entwicklung eines portablen Meßsystems zur Registrierung und Analyse von Körperhaltungen* (thesis at the Institute of Applied Physics of the University of Bonn)

PORTABLE TELEMETRY AS A METHOD OF MEASURING ENERGY COST DURING WORK IN HIGH LEVEL SQUASH

Mark K. Todd[1], Craig A. Mahoney[2], William FM. Wallace[3].

[1] Department of Physical Education and Sports Studies, Worcester College of Higher Education, Henwick Grove, Worcester WR2 6AJ
[2] School of Leisure and Sports Studies, Leeds Metropolitan University, Beckett Park, Leeds LS6 3QS
[3] School of Biomedical Sciences, The Queen's University of Belfast, 97 Lisburn Road, Belfast BT9 7BL

The Cosmed K2 (Rome) is a portable telemetric gas analyser which continually measures oxygen intake and ventilatory parameters. The apparatus was modified for better vision and fitted with a mouthpiece rather than the facemask. The Cosmed is attached to the body together with a battery pack and the total system weighs approximately 800 g. Twelve county level squash players were fitted with the K2 and a heart rate monitor (Polar), and performed work in the form of a warm-up and two common squash routines which were shown to differ significantly in energy cost from heart rate and blood lactate measurements taken during the work. The Cosmed K2 was found to be an accurate tool for measurement of energy cost during non-laboratory exercise, and confirmed trends in work intensity as indicated by heart rate and blood lactate readings associated with the three progressive levels of squash activity. It was found to be practical and easily accommodated by the subjects.

Introduction

Squash is a sport involving moderate to high intensity intermittent activity where almost 80% of the rallies last for 10 seconds or less (Mercier M, Beillot J, Gratas A, Rochcongar P, Lessard Y, Andre AM, Dassonville J., 1987), and players are active 50 to 70% of the playing time (Montpetit, 1990). The aims of the study were firstly to accurately measure the energy cost from $\dot{V}O_2$, recorded every fifteen seconds, during isolated patterns of play specific to squash and during a game situation using the K2 portable telemetric gas analyser, and secondly, to corroborate the findings using heart rate and blood lactate measurements.

The intensity of play was related to $\dot{V}O_{2max}$ and HR_{max} which were measured previously on the treadmill. Blood lactate profiles from samples obtained during the court test were also analysed to examine the anaerobic component of squash at provincial level.

Cosmed K2 system

The Cosmed K2 is a portable telemetric system that measures the volume of expired air in $l.min^{-1}$ ($\dot{V}E$) and the volume of oxygen consumed in $l.min^{-1}$ ($\dot{V}O_2$). There are no CO_2 measurements, and therefore, $\dot{V}O_2$ is calculated from the $\%O_2$ and the $\dot{V}E$ assuming a respiratory quotient (RQ) of 1.00 (Lothian F, Farrally MR, Mahoney C., 1993;

Lucia A, Fleck SJ, Gotshall RW, Kearney JT., 1993; Kawakami Y, Nozaki D, Akifumi M Tetsuo F., 1992). It has been found that the treadmill test is not a good predictor of squash performance since the nature of continuous running is not similar to movement around the court (Steininger and Wodick, 1987) therefore the use of the Cosmed K2 in monitoring squash players on court has great potential.

The K2 has proved to be an effective tool for measuring $\dot{V} O_2$ during field tests an in game situations. This is especially important for sports which are not easily mimicked b traditional laboratory ergometry. It has been used for sports such as table tennis (Faccini P Faina M, Scarpellini E, Dal-Monte A., 1989); stunt flying and parachute jumping (Dal-Monte A, Gallozzi C, Faina M, Tranquilli C, Biffi A, Menchinelli C, Fiaccarini L., 1989); wheelchair basketball, fencing and table tennis (Bernardi M, Canale I, Felici F, Marchettoni P., 1988) and football (Kawakami et al., 1992). It has been found reliable when compared to a mechanical lung (Crandell C, Taylor S, and Raven P. 1993), the Douglas bag (Lucia et al., 1993; Kawakami et al., 1992), though it has been reported to differ from the Quinton on-line system by 22% at peak values (Lothian et al., 1993).

On Court $\dot{V} O_2$ measurement in squash

The estimation of oxygen consumption on court was pioneered by Montpetit R, Leger L, Girardin Y.,(1977) and Montpetit RR, Beauchamp L, Leger L., (1987). The studies involved the collection of expired air on court using a Douglas bag for one minute at three points during a 25 minute game and the equivalent energy cost was measured from the respiratory exchange ratio. The authors reported values of 57% $\dot{V} O_{2max}$ for recreationa players, and 62.5% $\dot{V} O_{2max}$ for elite players. On court measurements of $\dot{V} O_2$ have been estimated from heart rate recordings during a competitive match against a player of similar standard where $\dot{V} O_2$ at the game heart rate was measured on a treadmill (Van Rensburg JP, Van der Linde A, Ackermann PC, Kielblock AJ, Strydom NB., 1982; Loots and Thiart, 1983). Values of 83% $\dot{V} O_{2max}$ for nationally ranked players, and 75.7% $\dot{V} O_{2max}$ for University league players have been reported during a game.

A game of squash, when played for 20 minutes or more, against an opponent of similar standard, is sufficient to raise the heart rate to between 150 and 185 b.min^{-1} (Mercier et al., 1987; Beaudin P, Zapiec C, Montgomery D., 1978; Loots and Thiart, 1983; Van Rensburg et al., 1982; Montpetit et al., 1987), and the mean heart rate for the game increases as the standard of the opponent improves.

The game of squash invades the anaerobic energy system periodically with mean game blood lactate concentrations ranging from 2.5 - 3.5 mmol.l^{-1} (Mercier et al., 1987; Garden G, Hale PJ, Horrocks PM, Crase J, Hammond V, Nattrass M., 1986; Van Rensburg et al., 1982; Beaudin et al., 1978; Noakes TD, Cowling JR Gevers W, Van Niekerk JP deV., 1982), and peak blood lactate levels have been recorded by individuals as high as 8.04 mmol.l^{-1} (Mercier et al., 1987), or above 5 mmol.l^{-1} (Garden et al., 1986).

Methods

Subjects

The subjects were division 1 players (n=12), ranked in the Ulster top 20 as determined by league and competition performance. The anthropometrical and physiological data of the players were collected in a Human Performance

Laboratory prior to the court testing (Table 1). Written informed consent and a medical questionnaire were obtained from each subject prior to any testing.

Table 1: Anthropometrical and physiological data for the Ulster Provincial squash players

Variable	Squash Players (x ± SD)
Age (yrs)	33.1 ± 6.3
Height (cm)	176.2 ± 7.1
Weight (kg)	77.5 ± 9.4
Body Fat (%)	14.5 ± 3.5
$\dot{V} O_{2max}$ (ml.kg.$^{-1}$ min.$^{-1}$)	55.2 ± 8.3

Court Test

The subjects were fitted with the Cosmed K2 on arrival at the laboratory and a pre-test lactate sample was obtained from the ear lobe prior to the start of the test. One author (MKT) and the subject performed a 'normal' squash warm-up on court for three minutes after which a blood sample was obtained. Then followed three minutes of a 'boast and drive' routine, during which the author boasted and the subject hit the ball straight down the side wall returning to the 'T' after each drive. A further blood sample was obtained. A 'boast, drop and drive' routine followed for three minutes with the subjects boasting on their forehand and three minutes on their backhand side, after each of which, blood samples were obtained. Finally, two 3 minute games 'as normal', with a blood sample collected after each. The blood samples were all collected within a 30 second period and were analysed within 30 minutes of collection. Table 2 outlines the court test protocol.

Table 2: Court Test Protocol

Stage	Time (mins)	Work load
1	0 - 3.0	Warm-up
2	3.5 - 6.5	Boast and Drive
3	7.0 - 10.0	Boast, Drop and Drive
4	10.5 - 13.5	Boast, Drop and Drive
5	14.0 - 17.0	Game
6	17.5 - 20.5	Game

The facemask was felt to be unsuitable for the purposes of this study since there was a possibility of gas leakage, vision was obscured and the subjects could not play properly. A mouthpiece was therefore developed to be used in conjunction with a nose clip. This was considered more practical and comfortable by both the subjects and the authors.

Treatment of data

The statistical significance of the court test measurements on heart rate, oxygen intake and blood lactate concentration taken at four points during the court routine (Table 3) was determined using an analysis of variance with repeated measures and a post hoc Scheffé test was used to determine the direction of significant differences.

Results

On court ergonomic demands

For purposes of analysis, the court test was divided into 4 parts. Three of these parts were perceived intensities for squash, all of which occur to varying degrees within a game. These intensities are:- i) Warm-up; ii) Boast and drive routine; iii) Boast, drop and drive routine. The 4th and final stage consisted of a game.

Table 3: Mean $\dot{V}O_2$, energy cost, heart rate, and blood lactate concentration during the court test.

Test Phase	$\dot{V}O_2$ (ml.kg^{-1}.min^{-1})	Energy cost (kJ.min^{-1})	Heart rate (b.min^{-1})	Blood lactate (mmol.l^{-1})
Warm-up	29.4 ± 3.9	42.2 ± 5.5	120 ± 19	1.8 ± 0.6
Boast & drive	41.2 ± 4.4	59.1 ± 6.4	153 ± 15	2.7 ± 0.7
Boast, drop, drive	48.1 ± 5.7	69.0 ± 8.3	168 ± 12	4.1 ± 0.9
Game	42.2 ± 4.6	60.7 ± 6.5	160 ± 15	2.9 ± 0.9

On-court oxygen demands

$\dot{V}O_2$ values for the three levels of simulated play were all significantly different from each other ($p<0.05$), while the mean $\dot{V}O_2$ for the game was similar to that of the boast & drive phase.

On-court cardiac demands

Heart rate during the warm-up was found to be significantly lower ($p<0.001$) than the boast and drive, boast, drop and drive, and game which were not significantly different from each other.

On-court blood lactate concentration

There was a significant difference between blood lactate concentration during the self determined warm-up phase and the boast and drive, boast, drop and drive and game phases ($p<0.05$). There was a significant difference also between the boast, drop and drive phase and all other phases of the test ($p<0.05$), though the differences between boast and drive and game were not significant.

As can be seen from Figure 1, the mean values obtained on the treadmill lie on the same line as the on-court values for the $\dot{V}O_2$ and heart rate contributing to the validity of the present study.

The energy cost of provincial level squash

The method used in expressing energy expenditure was MET's, where 1 MET = 1.2 kilocalories per minute (Sharkey, 1990) equivalent to equal an oxygen intake of 3.5 ml.kg.$^{-1}$ min.$^{-1}$(Nieman, 1990).

During the warm-up period, the mean energy cost to the Ulster players, as calculated from $\dot{V}O_2$, was 42.2 ± 5.5 kJ.min^{-1} (10.1 ± 1.3 kcal.min^{-1}). This increased to 59.1 ± 6.4 kJ.min^{-1} (14.1 ± 1.5 kcal.min^{-1}) during the boast and drive routine. The most taxing to the energy system was the boast, drop and drive routine which demanded $69.0 \pm$

8.3 kJ.min^{-1} (16.5 ± 1.9 kcal.min^{-1}). During the game, which incorporated all of these intensities to some extent, the mean energy cost to the players was 60.7 ± 6.5 kJ.min^{-1} (14.5 ± 1.6 kcal.min^{-1}).

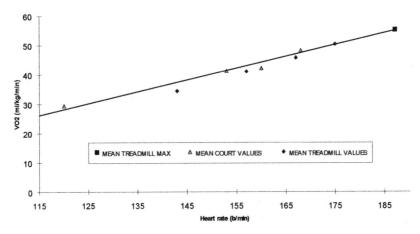

Figure 1 : Linear relationship between court measurements and treadmill measurements for the Ulster players.

Discussion

The energy cost of squash for recreational players was found to be 2828 kJ.hr.$^{-1}$ (47 kJ.min.$^{-1}$), and 3380 kJ.hr^{-1} (64.7 kJ.min.$^{-1}$) for elite players, (Montpetit et al., 1987; 1977). The energy cost of squash for a 68 kg man has been reported by Paish (1989) as 10 kcal.min^{-1} (41.9 kJ.min.$^{-1}$), while Brookes & Fahey (1985) report the cost of squash to be 14.4 kcal.min.$^{-1}$ (60.3 kJ.min.$^{-1}$).

The Ulster players reported an energy expenditure ranging from 42.2 kJ.min^{-1} to 69.0 kJ.min^{-1} The mean cost of the game was 60.7 kJ.min.$^{-1}$, which is comparable to the elite players of Montpetit et al., (1977), but greater than the values reported for recreational players (Montpetit et al., 1987).

Summary

The energy cost of squash using the Cosmed K2 analyser has been measured for the first time in the present study to provide measurements every 15 seconds during the activity. The energy cost of squash for the Ulster provincial squad ranged from 42.2 ± 5.5 kJ.min.$^{-1}$ during the warm-up phase, to 69.0 ± 8.3 kJ.min.$^{-1}$ during the intense boast, drop and drive phase with the energy cost of the game averaging 60.7 ± 6.5 kJ.min.$^{-1}$. Other studies have estimated the energy cost of squash to range from 42 kJ.min.$^{-1}$ to 65 kJ.min.$^{-1}$ during a game situation (Brookes and Fahey, 1985; Montpetit et al, 1977, 1987; Paish, 1989).

References

Beaudin P, Zapiec C, Montgomery D. Heart rate response and lactic acid concentration in squash players. *The research quarterly* 1978; **49**(4): 406-412

Bernardi M, Canale I, Felici F, Marchettoni P. *Field evaluation of the energy cost of different wheelchair sports.* International Journal of Sports Cardiology; 1988; **5**(2): 58-61

Brooks GA, Fahey TD. *Exercise Physiology Human Bioenergetics and its Applications.* 1985; MacMillan Co. NY

Crandell C, Taylor S, Raven P. Evaluation of the Cosmed K2 portable telemetric oxygen uptake analyser *Medicine and Science in Sports and Exercise* 1986; **26**(1): 108-111

Dal-Monte A, Gallozzi C, Faina M, Tranquilli C, Biffi A, Menchinelli C, Fiaccarini L. Fisiologia degli sport avaitori. *SDS Rivista di cultura sportiva*; 1989; **8**(16): 39-44

Faccini P, Faina M, Scarpellini E, Dal-Monte A. Energy cost in table tennis. *Scuola-dello-sport*; 1989; **8**(17): 38-42

Garden G, Hale PJ, Horrocks PM, Crase J, Hammond V, Nattrass M. Metabolic and hormonal responses during squash. *European Journal of Applied Physiology* 1986; **55**: 445-449

Kawakami Y, Nozaki D, Akifumi M, Tetsuo F. Reliability of measurement of oxygen uptake by a portable telemetric system. *European Journal of Applied Physiology* 1992; **65**: 409-414

Loots SL, Thiart BF. Energy demands of league and social squash players. *S.A. Journal of Research Sport* 1983; **6**(1): 13-19

Lothian F, Farrally MR, Mahoney C. Validity and reliability of the Cosmed K2 to measure oxygen uptake. *Canadian Journal of Applied Physiology* 1993; **18**(2): 197-206

Lucia A, Fleck SJ, Gotshall RW, Kearney JT. Validity and reliability of the Cosmed K2 instrument. *International Journal of Sports Medicine* 1993; **14**(7): 380-386

Mercier M, Beillot J, Gratas A, Rochcongar P, Lessard Y, Andre AM, Dassonville J. Adaption to work load in squash players: laboratory tests and recordings. *Journal of Sports Medicine* 1987; **27**: 98-104

Montpetit RR. Applied physiology of squash. *Sports Medicine* 1990; **10**(1): 31-41

Montpetit RR, Beauchamp L, Leger L. Energy requirements of squash and racquetball. *Physiology of Sports Medicine* 1987; **15**(8): 106-112

Montpetit R, Leger L, Girardin Y. Racquetball, le squash et la condition physique. *Racquetball-Canada* 1977; **3**(2): 6-11

Nieman DC, *Fitness and sports medicine:an introduction.* 1990; Bull publishing. USA.

Noakes TD, Cowling JR Gevers W, Van Niekerk JP deV. *The metabolic response to squash including the influence of pre-exercise carbohydrate ingestion* SA Medical Journal; 1982; **62**: 721-723

Paish W. *Diet in Sport.* 1989; AC Black London

Peel C, Utsey C. Oxygen consumption using the K2 telemetry system and a metabolic cart. *Medicine Science Sports Exerc* 1993; **25**(3): 396-400

Sharkey BJ. *Physiology of fitness.* 1990; Human Kinetics USA

Steininger K, Wodick RE. Sports-specific fitness testing in squash. *British Journal of Sports Medicine* 1987; **21**(2): 23-26

Van Rensburg JP, Van der Linde A, Ackermann PC, Kielblock AJ, Strydom NB. Physiological profile of squash players. *S.A. Journal of Research Sport* 1982; **5**(2): 25-26

Usability

A HUMAN FACTORS STUDY OF TELEPHONE DEVELOPMENTS AND CONVERGENCE

Martin Maguire

HUSAT Research Institute,
The Elms, Elms Grove,
Loughborough, Leics LE11 1RG
m.c.maguire@lut.ac.uk

This paper reports on work undertaken to explore the interface styles being adopted acr‍ telephony products in the fixed, cordless and mobile environments. It reviews the impa‍ of technological developments on the usability of different telephone types. Considerati‍ is also given to the potential for convergence of user interfaces across these types to increase their ease of use. It is recognised that in order to improve the usability of telephones, attention needs to be focused on 'families' of products that belong together serving similar users, for similar purposes, in similar contexts of use. However the mc‍ important requirement in order to improve telephone usability is conformance to good general human factors principles.

Introduction - basic telephone usage patterns

Since the introduction of the first telephone service in the UK, technological developments have had different impacts upon the market and the end user. The introduction‍ of the early telephones established a basic pattern of understanding and use. As numbers and‍ letters were dialled (the area being represented by letter codes), auditory feedback was received from the switching system that a connection was being made to the intended receive‍ of the call. Of course there were additional actions that sometimes needed to be performed that also became part of established practice. For public call boxes the sequence was to dial a number, put in the money and press 'button A' (or button B to abandon the call). For shared telephones (party lines), if the line was free, it was also necessary to press a button to make a connection. With this simple level of operation established, subscribers were able to use telephones fairly easily. The basic action of lifting the receiver, listening for a dial tone, dialling the number and waiting for either a ringing tone or engaged tone became the expected sequence of actions. In particular the handset was regarded as the instrument for initiating and completing a call, or abandoning it if a dialling error was made

Telephone developments and usability

Fixed phones
As fixed telephones advanced and provided push buttons to support functions other than simply entering a number, and possibly a small display, this offered more scope than a simple physical 'dial'. Also as the range of telephone numbers that people used on a regular basis increased, and the numbers themselves lengthened, the use of a personal telephone directory became necessary. An important innovation then was to be able store numbers in an electronic form within the telephone itself. However problems which restrict people from using this function include: being unsure which button sequence to press to store and recall a number, storing the name in alphabetic form by using a numeric keypad, and difficulty in viewing and editing the electronically stored telephone number list. Many of the services now

becoming available to the general public via the fixed network are similar to the advanced features available on business PABXs. It has been known for some time however that these features are little used by many business customers (Clarke et al, 1984). Patterns of use often show that some apparently useful features are little used. In a study performed by Clarke et al, a series of site visits to commercial organisations were performed. It was found that only partial use was made of the telephone systems provided. The facilities found to have a high percentage of use were:- call transfer 95%, call pick-up 86%, enquiry call 66%, system abbreviated dialling 65%, and follow me diversion 51%. Other facilities such as save number dialled, group pick up, conference call and various types of call parking were used by less than half the sample of users, a result either of lack of utility or the ease with which they could be accessed. Table 1 presents the results of a small survey of 26 telephone users (Maguire, 1996) in an organisation. The table lists the range of functions available together with the number of people who used each one, the number who do not, and the number who use the function but need the manual to hand while doing so. As before, some functions which might seem valuable and improve telephone efficiency (e.g. save last number, short code dial) appear low down the list.

Table 1. Usage survey of PABX facilities

Telephone function	Use	Don't use	Use but need to check
Call Transfer	24	2	8
Call Pickup	20	6	3
Night service	18	8	3
Enquiry call	17	9	5
Divert calls	16	10	8
Conference call	9	17	8
Camp-on	7	19	3
Save last number	7	19	4
Short code dial	5	21	4
Call park	1	25	0

Cordless telephones

Cordless phones have found a niche market for people who have disabilities and are unable to move freely around the house, since they can keep the phone near at hand rather than move to a fixed point each time the telephone rings. Other people find then convenient for use in certain locations, away from the base station, such as in the bedroom or garden. The sequence of operations for a cordless phone shares some features of both fixed and mobile phones. To make a call, the user typically extends the aerial and presses a 'talk/standby' button to make the phone ready (as with a mobile). They then enter the number which dials and then rings automatically (as with a fixed phone). To answer a call, the user again presses 'talk/standby' as with a mobile phone

Mobile telephones

The development of mobile telephones has brought about new methods of operating telephones which may initially confuse users but which may also have a positive influence on the development of all telephone types. The provision of a display, now a standard feature, means that the user gets good feedback as they enter a number. The display can also prompt the user when storing or retrieving numbers from its memory. While mobile phones are based upon cellular networks providing traditional telephone facilities, another development was the launch of 'Telepoint' services in the UK allowing users to make calls using a portable handset by standing within range of a Telepoint base station. Despite positive UK market research predictions, the service was not well accepted as it was costly, users had to find a Telepoint to make a call, and could not receive calls. However Telepoint-like systems such as Northern Telecom's 'Companion', with a range of 1 million square metres, have been very successfully applied to large organisations such as hospitals and department store.

Current trends

The late 1980's and early 1990's saw considerable growth of cellular phones, largely for business use by people. However the domestic mobile market has taken off rapidly in th UK and it was projected that 750,000 mobile telephones were purchased leading up to the Christmas 1995 period. There are now more domestic user in the UK than business users a shown in Figure 1, below (Mail on Sunday, 1995). However it has been reported that 20% or more of new subscribers drop or change the telephone service they initially take-up. This may be related to factors of cost, quality of service, coverage or usability. While financial an technical factors may tend to override usability considerations, if a product is intuitive and easy to use, the subscriber will certainly be encouraged to retain it. Improving usability may also promote a higher level of confidence in telephone products, generally helping to increase total market size.

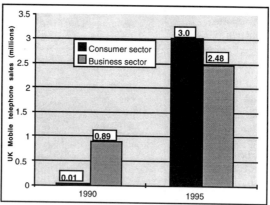

Figure 1. UK Mobile phone subscribers - comparison of business and domestic users

Convergence and divergence in telephone user interfaces

Various studies have been performed at HUSAT on the operation of different telephones models and categories (e.g. fixed, cordless and mobile). Some similarities appear to exist between the various categories of telephone unit which may enhance usability in general. The following sections discuss a range of areas where convergence may be a possible benefit or where further research is needed.

Auto-dialling versus 'enter and send'

Normally on fixed phones, the number is transmitted as it is dialled, requiring the user to dial it correctly or to halt the process completely by replacing the handset. However with mobile phones, where a display is standard, the user can check and, if necessary, correct the number before sending it. This offers benefits in terms of error tolerance that are likely to become more important as telephone numbers become ever longer. As some fixed and cordless phones now include displays, the question arises as to whether it would be desirable for all phones (with displays) to move towards the mobile method. To distinguish between the use of the handset and the operation of setting up a telephone call, would (i) support new ideas for handsfree usage and (ii) assist the process of convergence with computer linked telephone systems. Set against this change would be the fact that most people are familiar with the process of lifting the handset and simply dialling the number and to change this would not easily be accepted. Further research may be needed to examine this issue further and to see in which usage contexts this change might be appropriate.

Actions to receive a call

With fixed phones the action of receiving a call is also well established i.e. simply lifting the receiver and holding it the ear. For mobile phones the user either presses a specific button ('Send' or handset raised *☎*) or alternatively any button on the keypad. However for cordless phones two different possible actions are required. If the handset is on the base unit, the user simply raises it. If the handset is positioned elsewhere, they need to press a TALK key to take the call (normally labelled TALK/STANDBY to toggle the connection on and off). This raises the issue of whether there should be a move towards convergence whereby a button is pressed to take a call, either instead of lifting the handset (i.e. in handsfree mode) or together with lifting the handset (LUSI, 1995). Again this step could not to be taken lightly given people's normal expectations of simply lifting the handset to take a call. However the button press concept opens up the possibility for the user to take some other manual action. Combined with caller identification, it is conceivable that the user might take an action such as responding with one of a series of specific pre-programmed message e.g. 'please phone back in 5 minutes'. Again convergence needs to be considered in terms of particular user requirements in different usage contexts.

Icons, labels and codes

The ETSI/ITU standard (ETSI, 1993) proposes a number of service and videophone symbols but informal feedback indicates that they not necessarily completely clear or self explanatory. One problem is that the ideas being communicated are fairly abstract and so are difficult to represent graphically:

| Call pick-up | Repeat last call | Call waiting |

Figure 2: Examples of ETSI/ITU icons for telephone services

The key to further recommendations is perhaps to decide what functions can easily be represented as icons and which require textual labels. The use of symbols such as a raised handset *☎* on mobile phones seems appropriate when receiving a call, but may be counter intuitive when used to make a call, since the user expects to lift the handset before entering a number not afterwards. Regarding textual labels, a variety of names are used to refer to storage locations for telephone numbers including MEMORY, STORE, and DIRECTORY. Other terms such as DIAL may have little meaning to users not familiar with the old style circular dial and have only experienced push buttons on phones. Similarly, in a previous study, a button labelled RING, used to set the ring tone, was thought to ring a number. It is also of interest to note that some labels are taken more literally than intended and for one cordless phone considered, some users (perhaps not unreasonably) thought it was necessary to hold down the HOLD button. Further work could thus be done to offer advice on use of these terms in appropriate contexts, based on further feedback from users as to what they currently understand by different labels. For users who have more advanced telephone facilities, access to them via */# codes has been found difficult to remember, so an alternative presentation of telephone services could be recommended such as soft keys with a display and possibly also with some dedicated keys. However for basic telephones that only provide

*/# keys, command sequences should be re-designed to make them intuitive, customisable by a user organisation, and possibly standardised.

Sequences of actions and prompting

Different phone types often require different action sequences to be performed to store numbers in memory. The picture is further complicated if the user can also enter a name along with the number to help retrieve it. This is now available on mobile phones in general and on some fixed telephones Further research is required to determine what is a natural sequence of actions to perform to store numbers. Closely tied in with this is the development of basic mental models or structure around which user facilities such as directories, call handling, voice mailboxes, 3-way and conference calling can be based. Having developed such structures (which should also be described in the product's documentation) the sequences of actions required to perform telephone tasks becomes much more intuitive. Finally, with the growing use of displays, suitable prompts to guide the user can greatly improve the level of usability of a device. However such messages need to be clear and understandable to the user. Similarly soft function key labels (e.g. BACK, FORWARD, MODIFY, DELETE, CLEAR for a directory) need to be clear and to offer a clear path to performing an operation. A study of the most effective methods of providing user guidance and feedback on a limited sized display could lead to guidelines to assist designers in this area. It is apparent that while mobile phones made good use of visual feedback to indicate signal strength, cordless devices tend to provide less information even when a display is present. There is clearly scope for offering this type of feedback on all phones using wireless technologies to make their usage easier.

Relating convergence to user requirements and hardware constraints

This section has offered some suggestions as to where convergence may be considered between telephone types. Nevertheless convergence should be based on a knowledge of the requirements of typical user groups, their task areas, and hardware constraints. For example, a business user may require 100 memory locations on their telephone whereas a domestic user may not. As stated in the ETSI/ITU (1993) standard:

"Throw away any, or indeed all, of the ... design guidelines, in the interest of developing a user interface to your product or system, which has a PROVEN HIGHER LEVEL OF USABILITY for the full spectrum of people it is intended for, and across the essential range of the intended user tasks"

Discussion - Creating usable telephone products in the future

This paper has discussed the diversity of interface operation between core functions. However, these services are only a fraction of those planned for the future which will include the transmission of video, data and voice transmitted to specialist telephone terminals or general purpose PCs as well as smarter ways of handling calls (e.g. BT's Call Minder service). How, then, can telephone companies manage usability in the future?

Developing a Consistent Standardised Interface

The development of a design approach that is both Human Factors best practice and could be consistently applied across all products is perhaps a 'holy grail'. To have a serious impact, prescriptive standards would have to be applied across an ever widening array of services and modes of interaction. They are the province of a wide array of standards bodies who would approach it from different perspectives and no doubt argue for different standards Even were a coherent set of standards to be agreed, companies may not follow them, faced with a competitive market where 'ease of use' becomes a significant market differentiator. There is also a conflict between consistency and fitness for purpose. Where attempts are made to apply standards to novel features or new modes of interaction, e.g. animation instead of text, video instead of graphs, etc., the result is often a convoluted form of interaction which may sustain consistency but loses fitness for purpose. The search for universal consistency may not actually be needed and in fact users are very accepting when similar functions, e.g. entering a number into a telephone or a personal computer, are performed in different ways in different contexts. It may be that the contexts of 'fixed' and 'mobile' are sufficiently different for users to cope with different modes of interaction. This topic, as well as the ease of use of the functions themselves, needs empirical investigation. One approach is the development of standards for low level user interface elements for consumer products as was carried out on the Esprit 6994 FACE project (Maguire et al, 1994).

Adopting a User-Centred Development Process
The alternative to prescriptive standards is to adopt a process in development which maintains a user focus throughout the development process. The aim is to ensure that the user requirements for the specific product are understood and are met. Many computer suppliers who a few years ago were relying on strongly enforced housestyles to achieve usability are now adopting this kind of process as is being incorporated in the multi-part international standard on usability ISO 9241. It is a recognition that user requirements are very varied and that to meet them successfully will need an interactive design and evaluation process, which maintains the user focus. It is a mechanism which allows design teams to balance the conflicting demands of consistency and ease of use. On this basis, different foci for convergence may be identified. These might include situations where (1) the user is likely to use the products in the same context, (2) where the hardware or functions are similar or (3) where there is a strong cultural stereotype or metaphor which dictates the user's expectations.

Conclusion

Our previous work at HUSAT suggests that divergence in interface styles in telephones is occurring at all levels e.g. labelling, sequencing of operation etc., and that as products grow more sophisticated, even greater divergence can be expected. It may be concluded that complete convergence around a single set of standards for telephone interfaces is neither desirable or achievable. This does not negate the need for convergence around good human factors principles since, without them, there is every sign users will struggle with the array of functions the telephone can supply and market resistance will occur.

Acknowledgement: Thanks are due to Professor K. Eason, M. Ashby, G. Allison and S. Hannigan of HUSAT, and T. Hewson (Northern Telecom) for their inputs to this paper.

References
Clarke, A.M., Hannigan, S., Poulson, D.F., and Titley, N., 1984, Human factors and the End user of Advanced Telephone Systems, *Contemporary Ergonomics, Proceedings of the Ergonomics Society Conference*, E. D. Megaw (ed.), 2-5 April, Exeter, 253-256.

ETSI/ITU, 1993, ETSI Handbook: Human Factors guidelines for ISDN terminal equipment design, European Telecommunications Standards Institute Secretariat: F-06921 Sophia Antipolis Cedex, France, Issue 2, May.

LUSI, 1995, RACE project R2092, Likeable & Usable Service Interfaces, Annual Review Report, HUSAT Research Institute.

Mail on Sunday, 1995, Mobile users call changes, Financial Mail, June 18, p12.

Maguire, M.C., Butters, L.M. and McKnight, C., 1995, Usability Issues for Buyers and Users of Home Electronic Products, *Proceedings of the 7th European Conference on Cognitive Ergonomics*, Oppermann, R., Bagnara, S. and Benyon, D. (eds), GMD, ECCE 7 Human-Computer Interaction, Bonn, Germany, 5-8 September, 1994, 117-133.

Maguire, M.C., A small survey of telephone usage, HUSAT Memo 874, January 1996.

USABILITY EVALUATION IN THE FAR EAST: CROSS CULTURAL DIFFERENCES

Linda Herman

HF & HCI Consultant, Technology Group
Information Technology Institute,
National Computer Board,
11 Science Park Road, Singapore Science Park II
Singapore 117685.
Email: Linda@iti.gov.sg
100255.3231@compuserve.com

This paper reports a preliminary study of the cultural effects on the reliability of objective and subjective usability evaluation of software user interfaces. Subjects used in the study were drawn from actual users involved in a real system development project. The results obtained indicated that cultural effects not only exist, but exert a strong influence on the outcome of a user interface evaluation. Implications of the findings for practitioners are discussed in the paper. In particular, to account for cultural effects the need to modify 'western' usability evaluation methods for application in the Far East, is highlighted. In this respect, some initial recommendations are proposed in the paper.

Introduction

Although objective measures of usability traditionally generate more reliable data than subjective measures, they are expensive on resources. Consequently, commercial practice frequently prefers subjective assessment (such as questionnaires, interviews, etc.) rather than objective evaluation. This tendency towards subjective usability assessment is also observed in the Far East.

Such a tendency involves several implicit assumptions:

(1) That subjective evaluation constitutes an equally reliable measure of usability as objective evaluation. As such, it may be assumed to be a suitable substitute for objective evaluation. Although there is some indication that a positive correlation may exist between the types of evaluation (e.g. Herman, 1991), the research is far from conclusive. Consequently, a common wisdom among practitioners is that complete substitution of objective usability evaluation is ill-advised. Unfortunately, commercial system development practices tend to succumb to the temptation of substitution as a result of resource, time and cost constraints.

(2) That the design information generated by subjective and objective evaluation is equally rich. It can not be over-emphasised that the objective of user-interface evaluation is not only to rate a software on its usability, but also to uncover sufficient information to support design modification. An evaluation should therefore provide both diagnostic information (e.g. incidence and nature of problems) and prescriptive information (that supports design reasoning to solve user problems). Research comparing the strengths of subjective and objective evaluation in this respect has not been extensive. Among practitioners and researchers, however, there seem to be an informal agreement on objective evaluation being better for deriving diagnostic and prescriptive design information. This conclusion follows probably from the more systematic and controlled test conditions entailed by an objective evaluation.

(3) That no other factors exert disproportionate influences that differentiate further the reliability of subjective and objective evaluation, e.g. cultural effects associated with subjects involved in the evaluation.

The study reported in this paper focuses primarily on the last concern (i.e. item (3) above). Specifically, cultural factors may affect significantly the correlation between subjective and objective evaluation (i.e. concern (1) above), and the relative richness of information generated (i.e. concern (2) above). Without adequate understanding of such factors, appropriate decisions can not be made on the type of usability evaluation to undertake.

In the next section of the paper, the case study scenario is discussed. This is followed by an account of the case-study observations. These observations form the basis for addressing cultural implications for the application of subjective evaluation. Some advice for practitioners involved in usability evaluation in exotic places is summarized in the conclusion.

Case Study Scenario

The software concerned was developed by a group of software engineers initially with little or no human factors input. At later stages of software development, human factors practitioners were recruited. However, their design recommendations following an expert walkthrough were largely not implemented, as a result of frozen design inter-dependencies and resistance of designers to late changes. Due to proprietary sensitivity, further information can not be given on the product and organisation.

Usability testing was conducted on a real system developed for a specific group of professional users. Subjects used in the study were drawn from actual users. The objective of usability testing was to evaluate the ease of use and learnability of new functions and concepts associated with the software. Although the subjects had attended formal training courses in computing and had been exposed a number of times to the new software, they may still be considered occasional users.

A usability test session comprised the following activities:

(1) an introduction and demonstration of the software
(2) a short pre-test training session to familiarise subjects to test instructions and typical task scenarios
(3) questionnaire completion
(4) informal interview

Subjects were tested individually and later in pairs with an observer constantly in attendance. The tests were also video-recorded for analysis.

Objective evaluation data was derived by scoring user performance according to typical measures such as task completion, extent of help referral, errors, understanding of application semantics and concepts, effective use of appropriate functions and efficiency of observed interaction sequences (e.g. use of short cuts). Subjective evaluation data was extracted from questionnaire and interview responses.

It was anticipated that the results of subjective evaluation would be consistent with those of objective evaluation.

Case Study Observations

The behaviour of subjects tested individually may be characterised by two groups. One group sought substantial help from the observer in clarifying functions and tasks, and requested "hand-holding" throughout the testing session. The other group was rather introverted and did not seek help until a disastrous situation developed. In this group, two subjects eventually refused to complete the tests and one subject actually broke down and cried. Four subjects refused to complete the questionnaire. The observed overall response from the subjects was therefore rather negative.

Subjects tested in pairs may also be characterised by two groups. One group comprised subjects who were novices, while the other group comprised a novice-expert pair. The novice pair communicated frequently with one another, explored many functions, and were very vocal in commenting on the user interface design. Their comments provided very useful information to support the identification of necessary modifications of the software. The novice pair exhibited collaborative learning. They were generally comfortable with the software and were more successful in completing the tasks.

The expert-novice pair communicated poorly with each another. In particular, the expert grew impatient and restless whenever the novice performed the test. As a result, the expert-novice pair developed different perceptions: the expert, having observed the difficulties of the novice, perceived the product to be rather complex. In contrast, the novice, having observed the expert's ease of performance, perceived the product to be easy to use, and attributed personal difficulties in performance to personal inadequacy as opposed to design problems with the software.

This account completes a description of the qualitative observations made during the usability evaluation. An account of the evaluation results follows. Objective evaluation results of subject performance may be ranked in order from good to bad as follows:

> novice-pair
> expert in the expert-novice pair

individual subjects

novice in the expert-novice pair

Generally, the overall performance of the subjects with the software could not be considered satisfactory.

Subjective evaluation responses obtained from the subjects were generally reserved. The subjects were too polite and few ventured to make negative comments about the software. Despite unsatisfactory objective performance, subject comments were either uncommitted or pre-disposed towards a neutral or positive response. This inconsistency was highlighted most clearly by a chance incident following the usability test. The incident involved the subject who broke down and cried during the test. Having failed to complete the required tasks, she refused to complete the questionnaire. Although her performance was clearly poor, she was surprisingly very positive about the software when she was interviewed and given press attention! This outcome may be interpreted as a variant of the Hawthorne effect.

Conclusion

To summarise, the results of objective and subjective evaluation correlated poorly in this study. Specifically, the results of subjective evaluation tended towards the positive despite clear indications of poor user performance. This anomaly may be attributed to cultural effects, namely the subjects in the study were less vocal, were exceedingly polite and were disinclined to express negative comments in front of observers. This contrasted starkly with subjects in the West. Perhaps it is considered culturally unacceptable in the Far East to criticise too openly or directly, as this may cause the designers hurt or loss of face (see Craig, 1993).

In view of the strong cultural effects, it is recommended that particular care be taken when conducting subjective evaluation in the Far East. To avoid undue cultural influences on the outcome of a user interface evaluation, it is advised that:

(1) Subjective evaluation be augmented by objective evaluation. It should not be applied in isolation or used to substitute objective evaluation. As such, subjective evaluation techniques such as questionnaires served in-person, should be applied and interpreted carefully as their reliability is highly susceptible to cultural effects.

(2) The presence of an observer during the test may significantly affect the results of the evaluation. Alternative test arrangements should be considered, e.g. unobtrusive observations.

(3) Testing subjects individually should be avoided if possible as little information may be derived.

(4) To avoid reticence in commenting on software, verbal protocol may be encouraged as opposed to the elicitation of written feedback (e.g. via a questionnaire). In addition, verbal protocol techniques seem to work best if the tests are conducted using subject pairs. The pair of subjects should preferably be familiar with each other. Furthermore, the competence level of the individuals comprising the pair should be matched carefully. In particular, expert-novice pairs should be avoided. By ensuring these test conditions, the subjects are more likely to complete the tests and be forthcoming in verbalising their reactions to the software. Richer information may thus be extracted from an audio/video recording of their responses and discussions with each other during the test.

Due to the complex issues that may be involved, the preceding observations of cultural implications for usability evaluation should be considered preliminary. More extensive studies should be conducted to substantiate the findings.

Acknowledgements
The above study was conducted during usability evaluation in Asia. The author would like to thank all those who assisted in the study, in particular, the users, technicians, and software developers. Views expressed in the paper are entirely those of the author.

References
Craig, J M, 1993, Culture Shock! Singapore, *Times Books International.*
Herman, L, 1991, A Technique for Correlating Objective and Subjective Measures for Usability Assessment. In E J Lovesey (Ed.), Contemporary Ergonomics 1991, *Taylor & Francis.*

DEFINING THE USER MODEL TO ENHANCE PERSONAL READER SYSTEMS FOR THE BLIND

P J Simpson

Department of Psychology,
University of Surrey,
Guildford,
Surrey GU2 5XH

Computer based personal reader systems give access to text based information by transforming it into synthesised speech. However synthesised speech lacks auditory features present in normal speech. Personal reader systems provide only a limited set of search and review tools to aid comprehension. These tools depend on graphic 'visual signposts' which are indirectly related to the auditory features in a speech stream which mark linguistic and conceptual structure. By analyzing the task and information presented to the user, we can enhance the interface tools to provide better support for the processes involved in the comprehension of text presented in the form of synthetic speech.

Introduction

A number of systems are available which provide a synthesised speech representation of text encoded as ASCII characters. When combined with an optical character reader, the resulting systems offer visually impaired and blind individuals access to printed material which can be flexible and responsive to their individual day to day needs. These 'personal reader' systems can avoid delays associated with the production of audio tapes of books and articles, and the difficulties of arranging human readers.

Over a period of months I monitored two individuals' use of one personal reader system to give them access to printed materials - books, articles and research papers - which were required reading for their degree courses. The two individuals had contrasting responses to the system. One was an enthusiastic user who valued the flexibility and independence the system offered. The other user became irritated by the limitations of the system and preferred to work with audio tapes prepared by human readers. This observation of differing reactions to a reader system prompted me to consider what features of a personal reader system and synthetic speech could presented problems for the visually impaired user.

Properties of speech derived from text

Obviously the pathway from printed text to synthesised speech includes a number of steps which can go wrong. The optical character reader system can run into difficulties with certain font designs, degraded text and multi column layouts. Adaptive recognition procedures and verification routines working at the word level can overcome some of these problems, Evans (1995). Assuming however that there are no visual recognition and layout problems, the user of a personal reader system is presented with a synthesised speech representation of lines of printed text. The translation from text to a 'spoken' version is a complex process. This is especially true when the language concerned contains many exceptions with respect to spelling and pronunciation, Edwards (1991). In the best systems a large pronunciation dictionary and good software design can minimise this problem.

The result of these recognition and encoding operations is a synthetic speech representation of the text. The speech produced will differ with respect to a number of features found in normal speech. Since it will be derived from written text, it is likely to be more grammatically complete but also involve more complex syntactic forms than are found in normal speech. It will lack prosodic features used in normal speech. Prosody entails the use of stress, rhythm , timing and intonation to mark structural (syntactic) features in the speech stream and convey meaning and intention. In written text some prosody encoded features are expressed by punctuation while other features are expressed through sentence content. When expressed as words in synthesised speech lacking intonation and stress, punctuation lacks the impact of the prosodic based features found in normal speech, Crystal (1987).

The lack of normal prosodic cues makes detection of the underlying linguistic structure more difficult. Wingfield and Butterworth (1984) have argued that when speech lacks normal timing, pauses and amplitude variation or stress, the speech is perceived as it were a list of unrelated words rather than a structured speech stream. Wingfield and Butterworth found that the lack of prosodic features markedly changed subjects' strategy with respect to the size of perceptual unit subjects attended to when ask to remember verbal material.

Reference was made earlier to the greater grammatical complexity of printed text relative to speech. This claim must be qualified in the light of the fact that the writer of text can vary the grammatical complexity of textual material to fit the intended readers' reading skills. However in the present context in which the personal reader system was used to encode degree level reading material, the complexity of sentence structure poses a problem for the user. The use of relative clauses, material in parenthesis, nested sentences, and so on in the original text will pose problems of comprehension when presented as synthesised speech which lacks prosodic features to indicate punctuation and structure. Moreover, problems of lexical(word reference) , structural (syntactic) and referential (pronoun) ambiguity can present further difficulties in achieving the understanding of synthesised speech encoded text.

Strategies in reading and listening

Studies of eye movements during reading suggest that when faced with a complex sentence or ambiguity, the reader will stop the normal left to right saccadic eye movements and engage in a regression to fixate earlier material, Rayner, Carlson and Frazier (1983). During these operations it is claimed that the pattern of fixation is governed by linguistic factors reflecting a need for further processing of the sentence content. Similarly the amount of time spent fixating an individual content word (nouns and verbs) is dependent on their familiarity.

In the case of synthetic speech encoded text, the listener can adopt a passive listening strategy equivalent to a succession of left to right eye movements along a line. With this strategy the listener could adjust the speaking rate of the speech synthesizer to allow sufficient time for understanding and comprehension.

However text into synthesised speech systems also allow the use of an active strategy to access speech encoded text. The user can choose to listen to the current, next or prior character, word, sentence, line, or paragraph. The cursor controlling the speech output can also moved to the beginning and end of the line and top or bottom of the screen. Options exist for choosing between saying or spelling a word, or phonetically spelling a character to aid understanding.

Contrasting structures for text and speech

Designers of personal reader systems have foreseen problems leading to failure of recognition and understanding at the word level, and at the level of the sentence and paragraph. Thus they have provided a series of 'tools' to help the user review words, sentences and paragraph in order to achieve comprehension.

The facilities for moving around the screen at the level of character, sentence, line and paragraph are based on the recognition of features of the text which are encoded 'naturally' in the ASCII representation i.e. spaces, full stops, commas, colon characters, etc as well as soft returns, and hard returns.

The features available to the user to 'signpost' a directed analysis of the synthesised speech relate to the graphic structure of the original text encoded in ASCII form. But the 'features' of character, word, sentence, and paragraph are indirectly related to the synthesised speech version of the original text which the user is actually trying to comprehend.

This point is clearly illustrated by the feature 'line'. It is possible to ask the personal reader system to 'say' the next, current or prior line, or 'move cursor' to the end or the beginning of the line. But this feature has doubtful significance for the user who wishes to understand a speech representation of the text. Sentences do not obligingly live on single lines !

Matching the perspective of the blind user

It seems likely that the personal reader 'tools' provided for reviewing selected parts of text expressed as synthesised speech is shaped by a graphics 'text on a VDU screen' perspective. But the user of a personal reader system could be working with a different model. This model will reflect the linguistic and conceptual structure represented in the speech stream rather than graphic features and layout of the original text.

The ability to request that the system 'says' individual sentences, or move through a synthesised speech representation to link individual sentences, will of course serve the 'linguistic' needs of the user concerned to comprehend the synthesised speech. But could this facility for review and analysis be improved ?

Would it be possible to provide the facility to request that the personal reader identify and 'say' individual clauses or phrases within a sentence ? The problem would involve finding an ASCII character based feature which would define the boundaries of the clause within a sentence. Commas are a possible feature but in many instances there are no reliable character based feature. However, before worrying more about which ,if any, ASCII character string could be used, it is important to review the principle involved here.

It has been suggested that when text is encoded into synthesised speech, the resulting representation will pose problems of comprehension by the user. These problems will arise from the restricted nature of the speech synthesis and therefore representation, the complexity of sentence structures, and ambiguity. Facilities should be made available to the user to allow review and analysis of the content of the synthetic speech content which are analogous to regression and review in reading. It is noted that facilities to support this operation are provided in existing systems. They rely on features encoded in the ASCII text representation which identify a character, word, sentence or paragraph. However no provision is made to access individual phrases or clauses although these are distinct 'units' which have to be related to achieve a coherent interpretation.

It is further suggested that with current systems which allow review, the user is expected to relate to the synthesised speech in terms of its original textual representation and layout rather than its synthesised speech form. Thus the 'tools' provided for analysis and review are based on an inappropriate model of how the LISTENING user relates to information presented as a synthesised speech representation.

In practice, the structure of any text will vary with the individual author's writing style and choice of sentence construction, as well as conceptual topic and content. Thus the listener's choice between selecting a previous word, sentence or sentences, or previous paragraph when facing difficulties of comprehension may reflect a exploratory heuristic decision rather than a directed search based on a clear grasp of how previous and present sentences relate together.

Enhancing personal reader systems

It is often argued that problems of sentence complexity and ambiguity, which emerge from linguistic analysis, assume less importance in the context of language understanding guided by prior knowledge and expectation in a 'top down' model of language perception, Best (1995). It should be noted however that the top down process is only likely to be efficient when the reader/ listener has access to appropriate topic relevant knowledge in his/her memory. When the individual is reading (listening) to learn, the relevant knowledge and expectations may be lacking. In this situation a clear and unambiguous representation of the material is crucial.

When considering the case of reading, it has been recently argued that we should consider this activity as one involving constrained reasoning intended to achieve a coherent and consistent interpretation, Stanovich and Cunningham (1991) . The same case can be made for the user of a personal reader working with synthesised speech derived from text. We should acknowledge the need to lessen the interpretive problem posed by a degraded speech representation.

To this end, tactics used in the design of teaching texts could be employed. Care could be taken to avoid complex explanatory material. A summary of the text could be provided to establish the scope of the topic. Information could be given about the narrative structure of the text. For example it should be possible to insert a phrase indicating a new paragraph encoded in a voice pitch different from that used for the main text. An extension of this idea could take the form of a summary of the layout of each page - if pages are being encoded - describing the number of paragraphs included on each page and the clear identification of titles when included. This kind of system implies a limited form of linguistic processing to extend the information available to the listening 'reader'. At the level of an individual sentence, it might be possible to have the option to request a summary of its structure including its constituent clauses.

These suggestions appear reasonably straightforward. They arise from the idea that a visually impaired user requires an extended set of review 'tools' and guidance to aid comprehension when attempting to make sense of a synthetic speech version of text. These exploratory and review tools provided should support the strategies required to achieve comprehension on the basis of incomplete auditory and linguistic information. These facilities might be extended to include commentary on grammatical as well as the 'narrative' structure of text.

Computer based systems able to classify the linguistic structure and content, and then generate a commentary for the blind user will not be available within the foreseeable future. This is a task for human expertise. However deciding what information and commentary to provide to aid comprehension of degraded speech brings into focus the question of how far an integrated and implicit perceptual activity can be 'instructed'. Users of personal reading systems will have evolved strategies to aid the comprehension of everyday 'normal' speech. An understanding of how these work could provide a more appropriate framework to guide the development of personal reader interface facilities.

References

Best, J.B. 1995. *Cognitive Psychology,* 4th edn (West Publishing Co, Minneapolis/St Paul)

Crystal, D. 1987. *The Cambridge Encyclopedia of Language* (Cambridge University Press, Cambridge)

Edwards, A.D.N. 1991. *Speech Synthesis Technology for Disabled People.* (Paul Chapman Publishing Ltd)

Evans, J. 1995. *Personal Computer Magazine* August 1995

Rayner, K., Carlson, M. and Frazier, L. 1983. The interaction of syntax and semantics during sentence processing: Eye movements in the analysis of semantically biased sentences, *Journal of Verbal Learning and Verbal Behaviour,* **22,** 358-374.

Stanovich, K.E., and Cunningham, A.E. 1991. Reading as Constrained Reasoning. In R.J. Sternberg and P.A. French (eds), *Complex Problem Solving:Principles and Mechanisms,* (Lawrence Erlbaum Associates, Hove and London) 3-60.

Wingfield, A, and Butterworth , B. 1984. Running memory for Sentences and Parts of Sentences: Syntactic Parsing as a Control Function in Working Memory. In H. Bouma and D.G. Bouwhuis (eds), *Attention and Performance X,* (Lawrence Erlbaum Associates, London) 351-363.

THE POTENTIAL USE AND MEASUREMENT OF ALTERNATIVE WORK STATIONS IN UK SCHOOLS

J. A. Taylour* and Dr Joanne Crawford**

*Back In Action, PO Box 1457, Bourne End, Bucks, SL8 5YU

**Dept. of Work Design and Ergonomics, School of Manufacturing and Mechanical Eng., University of Birmingham, Edgbaston, Birmingham, B15 2TT

Direct observations were made to record sitting behaviour and activities undertaken in the classroom using three different styles of chair and desk. Questionnaires were also used to gauge levels of satisfaction and comfort. Field trials suggest that on those children tested, alternative designs to those currently found in schools, significantly improve sitting behaviour and are better received in terms of general satisfaction. The different activities were also found to effect sitting behaviour both negatively and positively in all conditions. Trials were extended in a laboratory environment to include the influence of training on sitting behaviour. On those children tested, training appeared to have a positive effect. The techniques and analysis were also useful in exposing product weaknesses and strengths and certain recommendations are outlined.

Introduction

Concepts behind conventional school seating with a horizontal work surface, are increasingly being challenged in favour of more ergonomically designed furniture. Comparative studies suggest that for working at a desk, the higher forward tilting seat with slanting desk encourages an upright posture and reduced forward bending of the back, Mandal (1994), Aagaard-Hansen and Storr-Paulsen (1995), Marschall, Harrington and Steele (1995)

There is also increasing evidence to suggest that children's back health is decreasing as a result of prolonged exposure to existing furniture, dimensional mismatch and a lack of postural training. Fish in 1984, as cited by Wilson (1994), studied 500 American teenagers and showed that 56% of teenage males and 30% of females suffered from x-ray evidence of the spinal degenerative condition known as Scheuemann's disease. The cause was thought to be due to long periods of sitting when children under go growth spurts and is associated with back problems in later life.

Auckland College of Education now have in place a full ergonomics teaching programme for those training to become teachers in keyboarding subjects having noted the high rate of injuries associated with VDU work (Demeter 1993).

The use of anthropometric data and physiological techniques alone are clearly not enough to determine the most appropriate furniture for children in schools. Research commissioned by the Department of Education and Science in the UK (1976) demonstrates the value of established techniques in analysing complex activities. The results of the activity sampling, for example, showed that the duration of sitting increased dramatically with age from approximately 5% for five year olds to over 70% for sixteen year olds.

Storr-Paulsen and Aagaard-Henson (1994) involved teachers in a Danish school to record various working positions of the school children and found only half the sitting time was spent leaning backward and making use of the backrest. The study also recognised the demands and influence of the school time table and organisation of the teaching on school ergonomics.

Linton (1994) studied the effects of ergonomically designed school furniture by using a combination of questionnaires and behavioural analysis. Subjective feedback from the experiment group revealed significantly higher ratings of comfort and satisfaction with the new furniture. They also felt significantly more knowledgeable about correct sitting posture and symptoms were significantly less compared with the control group.

The aim of the study was to examine and refine the more relevant measurement techniques and pilot them in the field to help recognise some of the unique characteristics facing UK schools e.g. movement of children, variety of tasks and group work, as well as identifying realistic alternative design solutions to those currently found in the classroom. Trials were extended in the laboratory to measure the effects of training on the sitting behaviour of participants.

Method

Field study

Three male and three female 11 year old pupils were screened for height for compatability with size mark 4 furniture of the dimensional standards and were observed for a week in a Birmingham Primary school using three different types of furniture. Furniture type A as supplied to the school consisted of a polypropylene chair with separate seat and backrest and steel frame and a conventional four legged double desk with a horizontal laminate top. Furniture type B was a conversion of the above with a raised wedged shape seat to alter the slope and a forward sloping work surface mounted on the desk. Furniture type C was a commercial Danish school work station representing the ergonomic 'ideal' with adjustments possible for the angled work surface and a raised chair with forward sloping seat at the front.

Two of each style of furniture were arranged in pairs to enable participants to work together or share books when required. Participants experienced all furniture types for at least two lessons before assessment commenced. Observations took place during normal academic class conditions where the pupils were based mainly at their chairs and desks. Throughout testing periods the posture of participants was recorded from a lateral view every five minutes by way of categorising their position into one of a number of pre set positive or negative classifications, e.g. slumped, neck flexed, back erect forward, back erect back. The activity being undertaken was also noted at the same time, i.e. writing, reading, listening, drawing, group work or other.

Over 100 observations were made per participant using the three types of furniture and the frequencies and scores for each sitting behaviour were added together for statistical analysis. At the end of each test day participants were asked to fill in questionnaires regarding the furniture they had been using for that day with reference to ease of use and comfort and any specific discomfort or pain encountered.

Results from the sitting behaviour scores were averaged and ranked both in terms of the level of positive and negative sitting per furniture type as well the individual type of sitting per activity. Non parametric statistics were used to establish significance differences in the scores. The same statistical analysis was applied to the scores formulated from responses to the post trial questions.

Laboratory study

Five randomly selected boys of secondary school age took part in the experiment, aged 11 to 15. All were screened for height for compatibility with size mark 5 furniture as recommended in the dimensional standards for their age range. One standard ISO workstation was used (type A), three modified versions (types B1, B2 and B3) and the commercial Danish chair and desk to BS EN 1729 (type C).

Analysis of results from part 1 suggested that the two most frequent types of activity (listening and writing) would provoke a variety of postures and tasks required to assess positive and negative interaction with the furniture. This took the form of watching television which simulated listening to the teacher and filling in puzzle books which simulated reading and writing.

The times allocated to sitting and breaks reflected typical duration in secondary schools and were repeated on three separate days over a one week time span. After each session participants filled in post trial questionnaires and then swapped positions so that all five furniture types were experienced in the same day.

Modified versions of the sitting behaviour task analysis and questionnaires were used to evaluate sitting behaviour and participant satisfaction with the different types of furniture. The second day commenced with a 15 minute training session to explain the differences between good and poor posture and some of the health justifications for adopting good posture. At the beginning of the third session participants were tested on what they could remember from the training. Sitting behaviour was then observed and recorded as per the previous days trials. Non parametric statistics were used as above to help detect any significant differences as a result of training and to the scores formulated from post trial questions for each furniture type.

Results

Field study

The individual participants demonstrated a significant positive trend in sitting behaviour using furniture types A, B and C (Page's L trend test L=82 C=3 N=6 p<0.01). 28.5% of the sitting behaviour observed with furniture type A was considered positive, 48% for furniture B and 50% for furniture C. A similar trend was displayed with responses to the post trial questionnaires (L=81 C=3 N=6 p <0.01).

A dominance of negative sitting was recorded from writing and drawing tasks while listening and reading induced more positive sitting. These four activities made up 78% of the total time observed and furniture A displayed the least frequent amount of positive sitting in all tasks undertaken while type B appeared to dominate with regard to positive sitting while listening.

A hunched sitting position was the most frequently observed sitting behaviour type at nearly 30% followed by a flexed neck. Slouching and chair rocking occurred a lot less frequently than anticipated. Writing tasks revealed a high proportion of neck flexion and hunched sitting compared with listening which showed a dominance in backrest supported erect sitting.

With regard to results from the post trial questionnaire all participants responded favourably to furniture type C in particular. The ability to change the angle of the top and a footrest for the chair and desk were especially liked although in actual observation the top was rarely adjusted if at all.

Laboratory studies

The first trial period displayed the least positive collective sitting when no instruction was given at 28%. Trial two which commenced with a short training programme displayed the most amount of positive sitting at 46% as might be expected while positive results on the third day tailed off to 40% when no further prompting was given.

The individual participants demonstrated significantly different sitting behaviours when using all the furniture types (Friedman test N=5 C=5 Xr2=8.68 p<0.05) but there was not a significant trend between types A through to C as expected (Page's L test N=5 C=5 L=239.5 p>0.05). Responses to the post trial questionnaires did not reveal significant differences or trends either (N=5 C=5 Xr2=2.72 p>0.05, L=231 p>0.05).

With regard to positive sitting observed between conditions during the two activities the trend from furniture type A through to C while writing was significant (Page's L trend C=5 N=5 L=253 p<0.01) but not for listening (L=215.5).

Discussion and future work

The modified and ergonomic workstations when used by the selected children in the field resulted in improved sitting behaviour and were better received in terms of general satisfaction and comfort. The sitting behaviour analysis method seemed to give a reasonable indication of the types of sitting induced by different furniture styles and between activities undertaken by the user.

Activity sampling seemed to provide a good insight into the movement and tasks undertaken by children in schools. Negative and positive performance induced by the furniture during different activities can help highlight areas for product design improvement and specific training awareness.

Training in laboratory conditions appeared to have a positive effect on sitting behaviour for the selected children although the poor level of recall should be noted for future training development programmes. Results from the field were difficult to replicate in the laboratory and it is suggested that care should be taken when assessing furniture in isolation of the school environment.

Results indicate that the Danish 'ergonomic' furniture (work station C) may not be practical in some secondary schools due to lack of space, restricted budgets and the style of working (i.e. having to share text books, no dedicated classrooms for the pupils etc.). A certain amount of value engineering is required to bring the cost down and improve the ease of adjustment to enable hourly changes in height if it is to be considered for typical UK state secondary schools.

With regard to the modified workstations (type B), all enhancements were achieved within the dimensional limits of the current BS and ISO standards and were better received than the existing furniture. Features such as a less negatively raked seat pan, forward sloping desks, flexible back rest and raised heights, etc. are not reliant on the implementation of the BS EN draft. Prototypes tested were not developed solutions, however, and more extensive trials would have to take place with refined products to understand the overall impact of the demands typical in the school environment .e.g. stackability, durability, mobility, ease of maintenance etc.

The current British and International dimensional standards have influenced the nature of furniture found in most UK schools. There are now greater incentives to pilot ergonomic alternatives partly due to the proposed inclusion of such furniture in the new draft European Standard BS EN 1729 (draft). It should be possible to develop and combine the above techniques to assist in larger scale analysis across the whole school age range. This could then be used as a predictive tool to help understand the effects of furniture on different age groups as well as the effects of certain subjects and activities. Attention could also be given to appropriate preventative training for both child and teacher particularly in subjects involving a lot of writing, drawing or computer work.

Acknowledgements

We would like to thank the Furniture Industrial Research Association for the opportunity to explore this topic. Thanks are also extended to B.D.S. Interiors and Hostess Furniture for loan of their furniture and to the children and schools concerned for their participation and support during the study.

References

Aagaard-Hansen, J. and Storr-Paulsen, A. 1995, A comparative study of three different kinds of school furniture, *Ergonomics*, **38**, (5) 1025-1035

British Standards Institute, B.S. 5873 (1980) Educational Furniture Part 1, Functional dimensions, identification and finish of chairs and tables for educational institutions. Building Bulletin

Comite European de Normalisation (CEN), BS EN 1729 (DRAFT) Doc 94/507742. Chairs and tables for educational institutions - functional dimensions, safety requirements and testing.

Demeter, B. 1993, Teaching ergonomics in schools *5th conference of New Zealand Ergonomics Society.*

Department of Education and Science (1976), School Furniture: Standing and Sitting Posture, *Building Bulletin 52.* Her Majesty's Stationary Office, 49 High Holburn, London WC1V 4HB.

International Organisation for Standardisation, ISO 5970 (1979) Chairs and tables for educational institutions.

Linton, S. J. et al. 1994, The effects of ergonomically designed school furniture on pupils' attitude, symptoms and behaviour, *Applied Ergonomics* 235(5) 299-304.

Mandal, A. C. 1994, The prevention of back pain in school children, In Lueder, R. & Noro, K.(ed.), *Hard facts about soft machines,*

Marschall, M., Harrington, A. C. and Steele, J. R. 1995, Effects of work station design and sitting posture in young children, *Ergonomics* **38**, (9), 1932-1940

Storr Paulsen, A. and Aagaard-Hensen, J. 1994, The working positions of school children, *Applied Ergonomics*, **25** (1) 63-64

Wilson, A. 1994, *Are you sitting comfortably? A self-help guide for sufferers of back pain, neck strain, headaches, R.S.I. and other associated problems,* Optima, Little, Brown and Co.(UK)) 41

RESEARCH INTO THE HUMAN FACTORS DESIRABILITY OF A TRULY PAPERLESS CONTROL ROOM AT BNFL

S.Briers*, Reed, J.^ & Stammers, R.*

* Department of Human Psychology, University of Aston
^ Human Factors Group, British Nuclear Fuels plc.

This investigation examined the feasibility of the paperless control room in the context of the emerging technological objective to computerise all current paper based control room tasks. The investigation focused on the operators' perspectives and performance and therefore considered the desirability and practicality of different paper and electronic systems in the context of control room operations. The study involved a literature review, interviews within the operational environment and controlled experiments using mock-up facilities.

Introduction

This programme of work was initiated by BNFL Advanced Engineering Research and Development Programme to explore whether the paperless objective was ideal in light of the tendency for paper to be introduced into control rooms which were originally designed to be paperless. This generic investigation examined the feasibility of the paperless control room. It involved plants and operators from British Nuclear Fuels (BNFL), including the new and highly automated Thermal Oxide Reprocessing Plant, (THORP) at Sellafield, and the New Oxide Fuels Complex, (NOFC) and the more dated Integrated Dry Route Process Plant (IDR Plant) at BNFL Springfields.

The operation of processes within the nuclear industry largely involves the automatic execution of pre-programmed sequences of actions controlled remotely from a control room. The operators are provided with process information from which they are required to monitor, schedule and control the entire process, intervening to maintain safety and efficiency where necessary.

An entire control system can involve any combination of computerised and paper alternatives. Paper systems include: standard reports; paper manuals and notice boards. Possible alternative systems include: VDU control systems; input devices, *eg* keyboards, trackerballs, specific function keypads; electronic documentation, *eg* operating procedures; communication networks and speech recognition systems.

Most of the literature available in this area focuses on paper presentations compared with electronic documentation and VDUs usage. Studies have focused on memory tasks and demands, searching, navigation, manipulation of information and reading performance and strategies. A split has been revealed which shows paper and alternative systems can advantage tasks in different ways depending on their exact nature. It is not always clear from research papers where the advantages actually lie.

Different requirements are made of human cognitive skills in the control room with the utilisation of different systems. Increased technology has been linked to increased use of problem solving abilities and a more concurrent style of problem solving (Brownsey and Zajicek, 1994). However paper holds several advantages, concerned with the execution of control room tasks: its portability and manipulability; its lessor constraints on task pace and physical position; its more personal nature; and its constancy of global context. Alternative systems have a number of different advantages, *eg*: distribution and retrieval of information is quicker and easier; ease of updating, controlling and securing documents; reduction in volume of documentation; templates and demarcation techniques; such documentation has also been found to be more efficient and logical, with a consequent reduction in workload. Therefore, it can be seen that there is no clear cut decision basis as to whether to provide a paper or electronic system in order to facilitate the best operator performance.

Operator Interviews

The main objective of the interview based investigation was to determine operators' preferences and requirements for paper and electronic systems in the control room and to reveal any subtle advantages either system may provide during real control room operations. The investigation identified: all the paper and electronic systems used in the control room, including their functionality, reasons for introduction and the expression of preferences. A representative sample of 15 operators and 3 Shift Team Managers from different shifts and plant areas were interviewed in THORP and NOFC control rooms.

Results

The interviews revealed that there is still a great deal of paper used throughout control room operations. These tend to be supplementary, to aid the use of an electronic system or address an additional need in operations. These can evolve to become permanent systems. Paper is also used when information needs to be displayed to personnel over a period of time, different locations or when a portable, flexible, handy and unofficial system, *eg* scrap paper, is needed. Several reasons were suggested by the operators for these paper applications, including: the delay and cost of developing electronic systems and reliance on others for development. Finally, paper is necessary in some circumstances to meet Site License requirements.

It was evident from the interviews that the majority of the electronic systems were introduced at the design stage of the plant as their requirement could be predicted in advance and they are viable regarding cost, skill availability and compatibility with existing systems. They are rarely used to supplement existing systems.

Operators noted that electronic systems can be a very powerful tool in the control room, capable of increasing the speed and accuracy of many tasks. This is demonstrated in the success of some electronic control systems. Electronic databases are also useful as information is easy to update quickly in an organised, structured and controlled manner. Information can be held securely in a small area and distributed easily with feedback regarding its receipt. The retrieval of the electronically stored information can be facilitated by available search facilities, hypertext and a good interface. A hard copy remains intact whereas paper copies may be misplaced or become tattered. However, paper systems do have advantages in that information can be displayed, in any location, regardless its size and to any number of people regardless of training and experience.

Whilst there was an initial slowness and reluctance reported by operators in accepting electronic systems, with increased experience their usage and acceptance has increased. Those which have been accepted and used to their full potential, tend to have effectively reduced the time or requirement for control room operations, whilst increasing, (or at least having no detrimental effects on) the accuracy and reliability of operator performance.

Experimental Investigation of Operator Performance I

This investigation was performed to assess operators' comparative performance, when using paper reports and an electronic system to present plant data on the generic control room task of interpreting plant data. This was used to identify which system was the most practical and desirable for performing the task from the operator's and performance perspective. The comparison addressed a number of issues: speed, accuracy and nature of performance: subjective mental workload and operator preference. The experiment adopted a within subjects design and counterbalancing was employed to vary the order in which the experiment was completed. The plant data differed in each case but the events were similar in number, nature and complexity. The independent variable was the display format (paper or electronic), and the dependent variables included performance measures (*eg* speed, navigation), accuracy, task responses, workload and operator preferences.

Five operators performed both the paper and the electronic version of the task scenario. Despite the small number of subjects these were adequate for the tests used and produced significant results. The operators were experienced with the simulated process and the electronic system used. The electronic VDU system was used to dynamically simulate the plant status over a 10 hour period. The following displays were available: a process mimic page showing all the relevant parameters: parameter trends of the last 10 hours; alarm messages; source, details and priority. The paper version comprised tables showing times and parameters values.

The NASA TLX, Task Load Index (Version 1.0, Paper and Pencil Package; Hart and Staveland, 1988), was used to measure the operators' subjective mental workload. Each participant was presented with the information to be used during the experiment, including: the plant data (paper reports or VDU) and relevant prompt sheets. The participants read the information and queries were answered.

A series of questions were asked of each operator and their spoken answers recorded by the experimenter. The experimenter noted navigation techniques used. The participant was then given the NASA-TLX questionnaire. There was a short break before resuming the experiment using the other source of information (i.e. paper or electronic). The NASA-TLX was then undertaken a second time for the different system and a de-brief interview delivered.

Results

Both the paper and electronic systems elicited broadly similar responses in the detection, diagnosis and response to an event. The overall score for paper slightly exceeded that for the electronic system, originating in the detection of events. However, the Wilcoxon Sign test used to compare these scores revealed that the electronic system was significantly faster, ($z = 0.894$, two tailed probability $z > 0.371$). The information presentation led to different approaches being taken to the task. Paper reports elicited a more ordered approach with parameters being examined sequentially, few interactions were checked but few parameters were missed. The electronic system approach involved process related prioritizing, some interactions being checked but some parameters were missed or checked twice. The Wilcoxon Sign Test for paired samples compared the overall mental workload on the paper and electronic systems for each participant, finding no significant differences. Paper systems were rated as having lower mental workload by only two operators.

Interviews revealed four of the five preferred the electronic system to paper in their overall preference, with 61% of all preferences expressed favouring the electronic system. This suggests a favourable outlook for the increasing introduction of electronic systems to the control room, although it was highlighted that it takes time to adjust to new computer systems. Paper systems were reported as being untidy and often difficult to read, taking up a lot of space and time, although preferred in fear of computer system failure.

It appears that operator performance and subjective mental workload may not be adversely affected by the incorporation of electronic systems. Some advantages for electronic systems are evident in that the time taken to complete the task of interpreting plant data was significantly reduced and the task approach appeared more process oriented. This suggests that systems used for the interpretation of plant data may benefit from electronic systems. Operator preferences favoured electronic systems suggesting their usability and desirability in the interpretation of plant data. However, accuracy was advantaged by paper systems both in terms of preferences and performance measures.

Experimental Investigation of Operator Performance II

This experiment investigated operators performance in a generic control room task - form filling (ie. a shift log), using paper or electronic reporting systems. This was used to investigte three more generic factors: whether paper or electronic forms are the most practical, and desirable for performing such control room tasks and how the nature of the task differs.

The experiment adopted a within subjects design whereby each participant used the paper or electronic log to record plant data from a control screen. Counterbalancing was employed by varying the order in which the experiment was completed. The independent variable was the shift log form, (paper or electronic) and the dependent variables included: performance measures, (speed, accuracy and navigation techniques); operators' strategies and operators subjective opinions.

Six operators carried out the experiment. All were experienced with the simulated process, the control system package and the task required. Each subject was presented with the information to use: the plant data system; and the paper shift log and pen or electronic shift log with keyboard and mouse. In the case of the electronic shift log, the subjects were then instructed on how to use the system and allowed a short practice period. Any questions were answered and a video recorder was started. During the investigation, observations were noted regarding the order in which the shift log was completed. On completion of the task, time was recorded and the completed shift log stored before repeating the experiment with the other version of the shift log (paper or electronic) and finally, the de-brief interview was conducted.

Activity sampling was used to identify the operators' different tasks and how the nature of the task changed. Categories of activities involved in completing a shift log were identified and checked in a pilot study to include: searching in the shift log; searching in the control system screens; moving through the log; moving through the control system screens; filling in the shift log; cross-referencing between log and control screen and other activities. Every five seconds during the main task activities were recorded, identifying the frequency and sequencing of tasks.

Errors which may occur in completing a shift log were identified to include: incorrect parameter; copying errors; missing a parameter; mis-ordering of readings.

Results

The total errors made when using the electronic systems exceeds that of the paper system only in the category of missing a parameter. A Wilcoxon Sign test revealed no significant differences in the accuracy of performance between paper and electronic systems, (z=0.5, two-tailed probability z > 0.61). The nature of the errors suggest the operators had a greater need for checking and less complacency on the electronic system.

Time taken on the electronic systems was consistently longer and a Wilcoxon Sign test demonstrated this difference to be significant (z=2.04, two tailed probability z > 0.61). Activity sampling revealed the task strategy was sequential with the exception of 2 operators who, at first adopted a strategy which was more suited to usage of the VDU control system. This implies that a strategy was developed and re-applied with no adaptations. However, this may change with increased experience of electronic shift logs.

Inter-operator correlations were weak on the electronic log (r = 0.106 - 0.604), but strong on the paper log (r = 0.888 - 0.964). Little relationship was shown between the task strategy on the paper and electronic log (r = 0.252 - 0.738). This difference is mainly attributable to increased time spent searching and filling in the electronic log, which is expected to decrease with experience.

It is evident from the interviews that all the operators had similar prior experience with computers, including extensive experience and training in VDU based environments. No age related trends in task performance or strategy were found. No preference was evident for 39% of the task characteristics investigated in the interview, a promising start for the electronic shift log as there is substantial evidence that a preference may develop for electronic systems once they become more experienced with it. Features identified as outstandingly advantageous for paper systems were related to navigation, input and trust in the system, suggesting their basis is in past experience. These features are achievable and can potentially be improved by technology, a good interface, training and obtaining experience. The ability to correct mistakes is favoured by electronic systems and several operators reported more care and checking on the electronic shift log and the liklihood of easily becoming accustomed to an electronic shift log. However, operators claimed paper systems were always needed as a back up in case the computer system fails.

The results demonstrate that the paper shift log holds several advantages in comparison to the electronic shift log including speed, consistent task strategies and subjective preferences and is still the most practical and desirable option. However, there is substantial evidence of the potential for electronic shift logs and associated operator performance with increased experience of such electronic forms. This would certainly involve an increase in the speed of completing electronic forms, the development of a more consistent and efficient task strategy, and in turn gain operators confidence with such systems. This is supported by operators comments in the interview, which described a positive outlook for electronic systems in the control room provided adequate training and time to adjust and adapt control strategies was provided.

This leads to the conclusion that the task of completing forms in the control room has the potential to be a paperless task provided the discussed factors are dealt with.

Conclusions

Is the paperless control room feasible? Yes, but the key question is: is it desirable? The acceptance of electronic systems is slow and this has been identified as being caused by lack of awareness by the operator of the capabilities and features of such systems. Their acceptance and desirability with operators tends to be slow when their established and highly usable paper systems are replaced by an alternative system which requires extra training and learning. However, the eventual acceptance of the electronic system is inevitable and can be facilitated by such methods as training and promoting the system.

Electronic systems have the potential to change the nature of control room tasks and in turn change the role of the operator. It is evident from the studies that where this has been the case, user acceptance of paperless systems has been quicker. A major concern of operators is that a computer system will fail. Acceptance and therefore success of systems will be hindered if functionality problems are evident.

Considering the practicality of paperless systems in the control room, no performance decrement has been identified which cannot realistically be overcome by increased experience and some training on paperless systems. The paperless systems deliver many practical advantages in the context of control room objectives. These include: information can be updated quickly in an organised, structured and controlled manner; quick, easy and secure storage and distribution of information; the ability to readily identify information due to potential system architectures and display techniques; and there is the potential for computers to perform calculations and produce reports reducing cognitive load and error potential.

Many practical advantages, not associated with past experience, which paper systems provide to the control room environment can be replicated on paperless systems with the incorporation of state-of-the-art technology. An electronic notebook could be incorporated onto a VDU system and potentially enhance the concept of a paper notebook by allowing neat and structured updating of information and also allow other personnel to leave messages. Alternatively, where paper cannot feasibly be replaced, electronic systems could incorporate a paper system, *eg* automatically print a report required to be kept on paper for regulatory reasons or to be taken to an alternative location.

In summary this leads to the conclusion that the paperless control room is a feasible objective. However, in order to achieve this, the electronic systems installed should accommodate all the advantages of paper systems using increased technology. In addition, the design of the paperless system should not be aimed at mirroring the paper systems but designed to support the operator in meeting and performing control room objectives. The implementation of these electronic systems should be carefully managed so that operators receive all the training which is required and are allowed a period to adapt to their new system and develop new strategies. The advantages of electronic systems should be highlighted to the operators and pitfalls minimised or compensated in the design. The development of a paperless control room, should be an iterative process during which the progress of the electronic systems is evaluated and any potential improvements are tested and incorporated where possible to benefit the operator. If paper does reappear in the control room, investigations should be made as to how this paper system supplements the electronic system and this should be addressed by technology.

References

Baber, C., & Tate, N., 1994, Mini Project - Electronic Document Management System on THORP, BNFL.

Brownsey, K. & Zajicek, M., 1994, *Contemporary Ergonomics*, 1995.

Hart & Staveland, 1988, Development of NASA-TLX: Results of empirical and theoretical research. In P. Hancock and N. Meshkati (eds), *Human Mental Workload*. Amsterdam: North - Holland.

CASE STUDY: USABILITY EVALUATION
OF THE SYSTEMS AVAILABILITY MODEL

Janette Edmonds

*Human Factors Group, British Aerospace Defence Limited,
Dynamics Division, FPC 500, PO Box 5, Filton, Bristol, BS12 7QW*

The Systems Availability Model is a suite of computer programmes
specifically designed to predict the Availability, Reliability and
Maintainability of single systems or combinations of systems. The model
started life over 22 years ago using punched (Hollerith) cards and the result
of numerous modifications has resulted in a system which is difficult to use.
This study was performed in support of a proposal to the Ministry of
Defence to win the contract for modifying the programme during its transfer
to a new operating platform. The study identifies future users of the system,
the task requirements and assesses the usability of the current model, using
structured interviews and user interface evaluations. Three options for
solution were identified and presented to the customer.

Introduction

The Systems Availability Model (SAM) is a complex computer based modelling
tool, used to predict the Availability, Reliability and Maintainability (AR&M) of single
systems or combinations of systems. In particular, it was designed to take account of the
interaction between systems, and between systems and the various phases of a mission,
for example, during a ship's mission.

SAM has been continually developed and extended over the past 22 years. It
started life on an IBM mainframe computer and made use of punched (Hollerith) cards
for data entry. The SAM programme progressed through a number of hardware
upgrades before being ported to a Digital Equipment Corporations VAX system. In the
history of the model, little attention was paid to the user interface design, leading to
several usability problems.

The model is currently being transferred to run on the Open System UNIX
workstation, using the XWindows (Motif) user interface, OSF 1991. This latest transfer
brings with it the opportunity to rectify some of the inconsistencies and the highly
repetitive task of data entry. It also highlights the responsibility to avoid further
complication of the interface. As such, this study was commissioned with the aim of
improving the usability of the SAM user interface.

SAM was designed to be used at different stages in the design process to model and to compare design options and indeed, for verifying data gained from trials and in-service records. SAM is, therefore a computer tool designed to support the decision making process of system procurers, designers, modellers and maintainers.

The user interface of SAM is mainly concerned with the input and editing of data, the submission of computational runs and the requisition of outputs. Inputs and editing of data are related to the use of three different charts:-

Availability Dependency Chart (ADC); The ADC uses logic statements to describe the dependency of the system functions on constituent equipment and on support functions provided by other systems. It also provides for the entry of equipment Reliability and Maintainability (R&M) data.

Model Equipment List (MEL); The MEL is only used where the baseline R&M data is at a lower component level than that being modelled on the ADC. The SAM programmes aggregate data entered on the MEL to provide data at the correct level for entry to the ADC.

Mission Profile Chart (MPC); The MPC uses similar logic statements to that used on the ADC, but in this case they are used to define the dependency of mission phases on the system functions defined on the ADC's. Data relating to the length of each mission phase and the effects of success or failure of each phase are also entered here.

These charts, especially have become complex and difficult to use due to several piecemeal additions. In particular, data entry tasks are unnecessarily repetitive, information on the screen is extremely compact and a huge reliance is placed on abbreviations which are not self explanatory and sometimes rather obscure. There is no on-line error checking facility (this must be performed as a separate task) and the user loses the overall perspective of the charts due to the need to move between several pages of information. As a result, the current user base extends mainly to just a handful of experts.

The usability study was divided into three parts:-

Firstly, the future end users of SAM were identified and a task analysis was performed to highlight the user requirements of the system.

Secondly, the current user interface was evaluated by three members of the Human Factors group and structured interviews were performed with current users. The objective of these two activities was to identify usability problems which current users had experienced.

Finally, the findings were reviewed and options for modification were generated.

For the purpose of this study, 'usability' is defined as a multi-dimensional property of a user interface, comprised of five attributes; Learnability, Efficiency, Memorability, Satisfaction and Error avoidance, Nielson 1993.

User Requirements

Target Audience Description
The aim of target audience description was to identify the future users of SAM. Based on the identification of current users, these would include safety and reliability engineers, system procurers, designers, modellers, support staff, software installers, software maintainers and system administrators. However, as the usability of the system improves, further target audiences could be envisaged, to include for example, technical staff concerned with the maintenance or construction of systems or sub-systems. In this case, users may not have as detailed a knowledge of AR&M engineering principles or similar computer familiarity.

For the purposes of this study, it is assumed that users have some familiarity with the windows interface format. It is also assumed that users will fall into one of three categories; the novice user, including new and very infrequent users of the system, the casual user, who uses the system at irregular intervals and the expert user, for those making frequent and extensive use of the system. It is recognised that the system must be able to accommodate all categories of end user.

Task Analysis
A task analysis was undertaken to understand and describe the user requirements of the system. It was intended to provide the basis for evaluating the system and to aid software designers in considering the user interface design in a systematic manner. For brevity, the task analysis is not reported in any further detail in this paper.

Usability Assessment

Usability Evaluation
A usability checklist was devised based on the usability heuristics specified by Nielson, 1993. The key aims for devising the checklist were to maintain some broad consensus between the assessors and to ensure adequate coverage of usability issues.

The usability evaluation highlighted a large number of usability problems. These were grouped and reported under nine headings:-

- Information presentation
- Demand on the user's memory
- Consistency
- Feedback to the user
- Navigation problems
- Shortcut facilities
- Error handling and feedback
- Help and documentation
- Additional problems

Structured Interviews
The structured interviews were based around eight key areas and elicited information concerning the background of the user, including past experience of SAM, general questions about SAM, inputting data, structuring dependency charts, analysing system performance using AR&M predictions, simulating mission patterns, obtaining and using the output of SAM and any additional comments which users wished to include.

The reasons for structuring the interview were twofold. Firstly, it enabled a means of triggering the thoughts and memory of interviewees. Secondly, it helped to maintain some consistency in the format of the questioning.

The structured interviews also enabled a large number of usability problems to be highlighted. These were reported in conjunction with the information gathered from the usability evaluation.

Study Recommendations

It was recognised that some usability problems would naturally disappear through the move of SAM to the new operating platform.

For those problems remaining, three options were generated in response to the information gathered about the end users, the task analysis and the usability problems highlighted by the usability evaluation and structured interviews.

A number of software constraints were recognised which increased the cost of implementing a solution which would solve every usability problem identified. For this reason, different levels of solution were proposed to offer options depending on different budget allocations. This therefore allowed a reasonable platform for negotiation with the customer.

The levels of modifications were derived through consultation with the software engineers responsible for implementing the system on the new platform. The agreed options were a 'minimal change' solution, a 'usability optimised for current users' solution and a 'usability optimised for a larger user base' solution.

The Minimal Change Solution
The 'minimal change' solution retains the chart format as depicted in the previous version of SAM, but places it within the graphical user interface format of Motif. In conjunction with this a help system is incorporated. (The charts used in the previous version of SAM used tabular formats of information of up to 132 columns width).

This option highlights the removal of usability problems by the move to a graphical user interface format. For example, easier traceability from the use of overlapping windows, the decreased memory demand for recalling commands and the improved navigation of the system using a menu system and scroll bars for the charts. It also highlights the ease of editing the complex charts by having facilities such as 'cut and paste', not present in the previous version of SAM.

This option removes other usability problems, such as the poor layout of menus and forms, poor error handling and poor feedback to the user.

The 'Usability Optimised for Current Users' Solution
This option builds on the improvements highlighted for the 'minimal change' solution. It provides additional facilities such as the use of split screens, on-line error checking in the charts, context sensitive error handling and a help system incorporating easily accessible context dependent information.

The actual format of the chart editor, again remains the same, but the user has the option to use simple 'forms' to input and edit information on the tabular charts. This improves data entry and editing by making the interface easier to understand and follow, especially as all abbreviations are replaced by their actual function name, for example 'Failure Defect Probability', as opposed to 'FX1'. The tabular format also allows more effective grouping and layout of functions.

The 'Usability optimised for a larger user base' Solution

This option involves a re-write of SAM within the windows environment, but incorporates the usability solutions already outlined by the previous two options.

The radical difference with this option is that users are able to model systems by interacting with iconic representations of equipment and their respective links to other equipment. Information would be input via forms which guide the user through the task of modelling. The other key addition is that of a library of intelligent icons which can be used to build new systems without the need to build equipment and sub-systems from scratch.

Conclusions

The 'minimal change' solution offers the cheapest and quickest option to implement and makes a number of usability improvements. It is noted, however, that the usability of the system is not optimised and would be unlikely to be usable by people without a knowledge of AR&M modelling.

The 'Usability optimised for current users' solution does not constitute significant costs over and above those quoted for the 'minimal change' solution (costs not included due to company confidentiality requirements). However, the usability is significantly improved. It was recommended that this solution could be used to extend the use of SAM to a user group other than just AR&M modelling specialists.

The 'Usability optimised for a larger user base' is significantly more expensive and time consuming to implement. The usability however, is maximised and the possibility of increasing the user base further is greatly enhanced.

Follow-up

Following the proposal outlined in response to the usability study, the customer awarded the contract to British Aerospace Defence Limited, Dynamics Division. The 'Usability optimised for current users' solution was chosen.

The software design of the system has been underway for the past year. Throughout this time, advice has been sought of the author regarding specific interface designs and also in conjunction with performing design reviews. The results are presently being viewed by the customer and it is anticipated that full user trials will be taking place in the near future.

Acknowledgements

The author would like to acknowledge the support and co-operation of fellow colleagues. Comments and suggestions were very gratefully received.

References

Open Software Foundation 1991, *OSF/Motif Style Guide,* Revision 1 (PTR Prentice-Hall, Inc.)

Nielsen, J. 1993 *Usability Engineering* (Academic Press, Inc.)

This paper is a summary of a project undertaken at British Aerospace Defence Limited, Dynamics Division (BAeDef, DD). The views expressed are those of the author and do not necessarily represent those of BAeDef, DD, MoD or Her Majesty's Government.

USABILITY TESTING OF AN INTERFACE FOR TELE-ACTIVITIES

M.J. Rooden, H.Kanis, W.M. Oppedijk van Veen   **A.C. Bos**

Faculty of Industrial Design Engineering   *BOS Industrial Design Consultants.*
Delft University of Technology.   *'s Gravendijkwal 71,*
Jaffalaan 9, 2628 BX Delft,   *3021 EE Rotterdam,*
the Netherlands   *the Netherlands*

This paper serves to discuss some methodological issues in executing a usability test of an interface for tele-services. As we wanted to observe as many interaction problems as possible, we selected subjects on age as an indication for experience. Older people were expected to have less experience with similar products and consequently were expected to show more problems in interacting with the interface. We describe the way we translated the problem-data into valuable information to assist designers improve the design of the interface.

Introduction

This paper covers some methodological aspects of the usability testing of an interface for tele-activities. The interface, which is still in the process of being designed, is intended to embody various applications. It can be used in combination with a television, so that viewers can react on what they see on the screen. They can, for instance, pick their favourite contestant in a quiz or ask for information about a product or service shown in commercials. Viewers are enabled to do this by using the tele-interface, which is a plastic box with buttons and a display (because the development is still in progress, we can not show a picture of the product). The users' choices are transmitted to a central computer via the telephone system. Apart from this so-called response television, users can access a broad range of tele-services, e.g. telebanking. After connecting a service, voice response guides the user to perform the intended tasks.

The interface is used in combination with a chipcard and a personal pincode to make it serviceable for different users. Each application comes with a specially designed template, i.e. a sheet of paper with holes in it for the buttons. By putting a template on the product buttons are 'functionalized' by adjacent text on the template. Only those buttons which have a function in the particular application the template is designed for are visible and operable. The templates also contain instructions for use.

The scheme below (figure 1) shows the designed procedure the users should follow for a successful interaction.

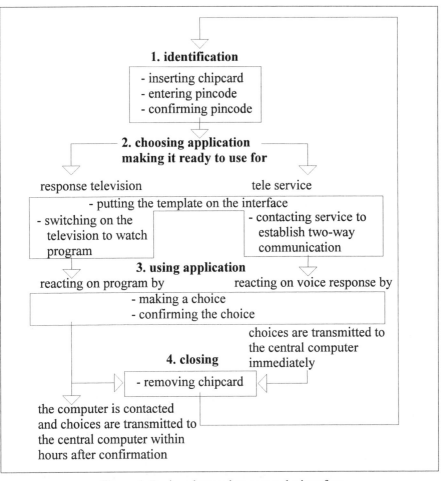

Figure 1. Designed procedure to use the interface

One difference between response television and voice-responce applications is the moment of contacting the computer and transmitting the choices, as can be seen at the bottom of the scheme. With response television the computer is contacted after step 3 is completed. Contact with the computer may be delayed to avoid overcrowded lines or to use cheaper rates at night. When using voice-response services, the computer is contacted in step 2. Choices are transmitted immediately and feedforward and feedback is presented by means of voice response.

This product is totally new for the Dutch market, and the developers have realised that it is difficult to anticipate users' behaviour without involving future users in the design process. As part of an overall plan to assess functionality (does the product perform the functions it should), and usability (are users able to successfully interact with the product) we carried out a usability test with a prototype. The results of this usability test were to be used in order to improve the design. In the phase of the design process, the plastic box could still be adapted to some extent. The chip, auditory and visual feedback (LCD-display) and the templates, could be changed substantially without problems.

The aim of the research was to find out if the interface was used as anticipated in the design. If not, we wanted to know what problems potential future users experienced and if there was a link between the experience of the participants and the problems they encountered. Ways of use anticipated by the designers are indicated by cues like feedforward in display and through voice response, indications where to insert the chipcard, and how to position the templates. The question was whether users notice, understand and follow these indications.

Method

Menbers of a consumer household panel (Tan, 1992) (n=19) were invited to the laboratory where they performed tasks on three different applications, one being a voice-response application, because of differences in the designed procedure for use. First they received a verbal introduction to the product and its applications. The participants were then asked to perform the three tasks. No manual was supplied, and the participants were only helped by the observer/test leader when they had no clue at all about what to do. After each task, uncertainties about what the subjects had done or thought were clarified. For two applications video-fragments were prepared to serve as stimuli. For the third application voice response was designed. After the tasks an interview was held to trace the participants' past experience.

As we wanted to observe all possible interaction problems and to find critical aspects of the interface, in our recruitment of subjects we included people who were expected to have interaction difficulties. We expected experience with similar products positively to affect the use of the interface. Because it is difficult to assess 'experience with similar products', we used age as an indication for experience. We assumed older people to have less experience and thus more interaction difficulties. We wanted to see young and middle-aged people as well. Although we expected them to have less problems we wanted to verify this and to see what problems they still experiencd. We ended up with three categories (15-25, 35-45, 55-65). The experience was later checked in the test by allowing the participants to perform some tasks with the remote control of their own television, which we had asked them to bring along.

The only other selection criterium we used was familiarity with one out of three proposed television shows. This was done to ensure that the participants were part of the potential user group. Response television is seen as the most important application by the developers.

Results

All participants showed interaction problems. As it was a completely new product they had to work out how to use it.

First we elaborate on the assumed link between age, experience and interaction problems. One participant had no experience. With the others a distinction was made between good or poor performance on two tasks; good performance indicating ample experience, poor performance indicating limited experience. The participants with ample experience revealed less problems (n=13, median = 13 problems) than participants with limited experience (n=5, median = 22 problems). No analysis of the seriousness of these problems was made. Four of the five participants with limited experience were from the 55-65 category, which supports our assumption.

Observed difficulties in use

All sessions were video-taped. The tapes were the starting material for the analysis. The first step was to transfer the information to paper, because information is much easier and quicker to use on paper than on tape. Thus all actions of the participants were written down and relevant comments were recorded. The interviews were analysed as well.

From the summaries for each subject, a long list .of problems was drawn up. The structure was a pragmatic one, a mix of protocol-order and some product characteristics. The problems were classified under thirteen headings, to facilitate searching for certain problems during the analysis. We used headings such as chipcard, pincode, the applications, switching off and correcting errors. In all, 201 problems were identified. For each problem a record was madeof which participants experienced the problem. Because the list was in chronological order it resembled in part the use-procedure as shown in figure 1. In figure 2, some of the problems are inserted in this scheme to give an impression of how difficulties are spread over the list of use difficulties. It can be seen that problems which are very much alike, such as forgetting to confirm a pincode and forgetting to confirm a choice, are put in different places, which blurs the identification of use difficulties as to their common origin.

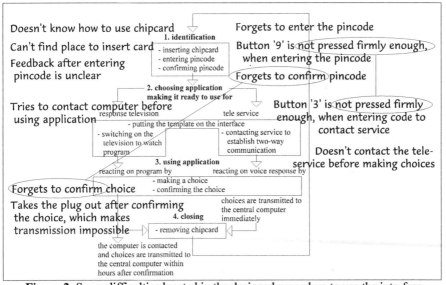

Figure 2. Some difficulties located in the designed procedure to use the interface

The list we compiled contains all information from the test, but it does not help to understand the problems or to improve the design. People unfamiliar with this type of research might be tempted to make use of this list to attempt to reach quantative conclusions. We restructured the information to reach design-relevant conclusions.

Users' difficulties in a design context

To reach to design-relevant descriptions of the difficulties, the emphasis, was switched from user to product. The long list showed the problems mostly in terms of use difficulties. In fact, we wanted to describe the problems in a design context. Problem descriptions were grafted onto featural and functional product characteristics, such as feedback or stability. This process can best be described by showing some examples.

- All problems that had to do with the chipcard were split into problems with finding where and how to insert the card, and problems with feedback. The designers are designing an interface, not a chipcard. By presenting information related to the interface, rather than the chipcard, the designers can start improving the product, by for instance, changing the place to insert the card or by giving better feedback. In this way for each problem the relation to the product has been made clear.
- A lot of problems were initially listed under the names of the three applications. By comparing and combining problems with these three applications more general problems were identified. For instance the problem emerged that it was difficult for some participants to distinguish between response television and the voice response application. This led to various errors. The application-related problems were regrouped, in order to get a close product related description.

We ended up with thirty-one problem types. The need for confidentiality prevents us from presenting them all. Some focused on specific details of the product and are less interesting to discuss here. Other problems were of a more general nature. To make these clear to the designers would also help them to focus on these aspects when designing products in the future. A few of the problem types are described below.

- The place to insert the chipcard was difficult to find.
  When the interface was lying on a table the seated user couldn't see any indication of where to insert the card.
- Feedforward and feedback were insufficient or unclear, and feedback was interpreted as feedforward and vice versa.
  As the display was very small, the amount of visual feedback and feedforward was limited and often coded, which made it difficult to understand.
  Because feedforward was sometimes presented immediately after an action was completed, it was interpreted as being feedback.
- Instructions on the templates didn't tempt the user to follow these step by step.
- The product concept of 'interactivity' was difficult to grasp.
  Some participants expected feedback on the television screen, which prevented them from completing the interaction.
  Also, the moment of transmission of the choices was wrongly interpreted by most of the panel members. Because they thought the choices were transmitted immediately, they tended to disconnect after completing the interaction.

The results of the test have been reported to the designers, who are currently implementing the findings in a redesign. A distinction was made between problems which concerned the physical parts of the product, like the plastic box and the templates (hardware) and the parts of the product which have a direct communication function like the software and graphics, because different groups of people are designing these different parts. An example of a hardware problem is the positioning of the templates. An example of a software problem is the clarity of the feedback. These problems were translated into product specifications.

Due to market development which took place parallell to the usability test, new technical functionalities are to be integrated into the existing interface. Prototypes that result from the redesign will probably be presented to members of the consumer household panel.

Discussion

The usability test led to information which the designers could use to improve the interface. We started from the user side, by analysing the use procedure and observing users. We did this to be open to all possible outcomes of the test. Then we increased the value of the information in the design context by relating the problems to featural and functional characteristics of the product. In the future it might be a good idea to start from use cues (Kanis and Vermeeren, 1996) which are known, or supposed to be designed in order to provoke certain user activities. Then it is possible to focus more specifically on these product characteristics in observing users. This may facilitate the step from observations in the field to (re)design considerations. Researchers, however, should always be open to unexpected problems showing up during a users' trial, such as users adhering to 'use cues' which were certainly not designed as such.

References

Kanis, H. and Vermeeren, A.P.O.S. 1996, Teaching user involved design in the Delft curriculum, *Proceedings of the Ergonomics Society's 1996 Annual Conference*, (Taylor & Francis, London)
Tan, A.H.L. 1992, *Gebruikershandleiding PEL-panel*, Vakgroep Bedrijfskunde van de produktontwikkeling, Technische Universiteit Delft.

HCl

THE SPECIFICATION, ASSESSMENT AND ACCEPTANCE
OF THE HUMAN FACTORS ASPECTS OF
COMPUTER BASED SYSTEMS

Dr Mike Tainsh
DERA, Portsdown, Hants, PO6 4AA

A general framework for the specification, assessment and acceptance of the human factors aspects of computer based systems is presented. It takes as its origin eight Technical Areas (TAs) derived from the human factors domains. Associated with each TA is a set of Technical Issues (TIs) and models which can be used to structure them and derive tests and criteria for assessment purposes. The TIs can be used to generate requirements statements and hence agreed characteristics which may be used for the definition of tests and testing procedures. An example is provided to show how one might derive TIs and hence requirements statements and acceptance tests to ensure that there is an effective human factors contribution to the procurement process for computer supported systems.

Introduction

This work arises from the requirement on all MoD project managers to run Human Factors Integration Programmes (HFIPs) covering the six currently agreed human factors domains (personnel and manpower, human engineering, training, environmental ergonomics, habitability and accommodation, and health and safety), and industry to supply equipment in accordance with the contractual arrangements negotiated with the project manager (Tainsh 1995).

General national and international documents such as DEFSTAN 00-25 and ISO 9241 apply to general human factors aspects of a broad range of systems and equipments but in many ways they are inadequate in the detailed specification, assessment and acceptance of military equipment and especially large scale computer based systems. It is not possible to use them as source material for either :

(a) the specification of critical issues to be used during requirements specification, assessment or acceptance, or

(b) assessment, evaluation or acceptance procedures.

Aim

RN authorities and Project Management Teams (PMTs) need to specify requirements, assess design options and programme progress, and ultimately accept the characteristics of the systems under procurement..

From a technical point of view the requirements specification, assessment and acceptance processes are closely related processes. These processes are described, in outline, for those portions of MoD(PE) concerned with Royal Navy in a set of three Sea System Controllerate Publications: SSCPs 10, 11 and 12. The first is concerned with the whole ship issues while the latter are concerned with computer based systems (command and marine).

The current results from the work on TAs and TIs

The work on the general characteristics of this class of computer based systems has lead to the following:

(a) The identification of a set of eight Human Factors TAs appropriate to computer based systems ;
(b) A set of TIs associated with, and hence defining the scope of each of the TAs;
(c) The understanding that models can be created which would structure the TIs in terms of variables, criteria and their relationships;
(d)The belief that requirements for computer based systems can be expressed in terms of TAs, TIs and models, and that these can be used to assess proposals, design options, developments and implementations. Hence they can be used for acceptance of products which is the final assessment procedure based on Agreed Characteristics derived from the statement of requirements.
(e) The belief that effective decision strategies can be defined to enable accurate, reliable and valid assessments and evaluations.

Rational strategies for assessment and evaluation.

The basis for the decision-making component of assessment or evaluation rests on the application of a strategy for processing the information on the selected variables associated with the TIs. Tainsh (1991) put forward the concept of a rational decision-strategy based on selecting between task designs.

Comparisons between design options, comparisons against reference models can be achieved by a staged process where validity improves in the following order i.e one to four (Tainsh 1995):

(1) presence of user support functionality within the equipment.
(2) operability criteria
(3) performance on critical tasks
(4) conformance with all aspects of human factors policy.

The decision making strategy requires that appropriate variables are declared and assessments made of the presence of each, the information from this process is then fed into a process determined by the assessment strategy and the overall assessment or evaluation results.

The framework

It is necessary here to consider how the assessment strategy fits with the combination of information from the TAs and TIs. A framework of relationships is proposed and shown in Diagram One.

The HFIP domains defines the total scope of the technical work, areas and issues. The TAs and TIs must lie within this boundary. The relationships between the TIs may be understood in terms of models that help in the derivation of test criteria.

The assessment strategy is driven by the need to demonstrate compliance with the requirements in the various domains as interpreted by the TAs and TIs. The requirements for the system and the ultimate Agreed Characteristics (ACs) are derived directly from the TIs. The test procedures and criteria derive from the requirements, or ACs depending on the phase of the procurement cycle. (The ACs are produced as part of the development and build phases.) The test results and subsequent assessment and evaluation all flow from the definition of the test procedures, and the implementation of the assessment strategy.

This particular item of work had its origins in the specification of requirements statements for future computer based systems. It was considered that:

(a) The six HFIP domains, and associated standards, were insufficiently specific when providing guidance to PMTs on the technical areas and issues be addressed;

(b) It was not possible to move directly from the six domains to statements of requirements and their associated assessment or acceptance procedures.

The aim of this work has been to develop a framework for specifying, assessing and accepting the human factors aspects of computer based systems and in particular to seek to confirm the usefulness of the set of Technical Areas (TAs) and specify some of the Technical Issues (TIs) within them.

Approach

The programme
DERA has ongoing programmes on the requirements specification, assessment and acceptance of the human factors aspects of computer based systems. Part of this work has been a substantial effort on the development of a framework.

The approach taken in this programme has been influenced by that used in the development of a warship which is managed in terms of Warship Design Areas (WDAs). These are agreed TAs that have been commonly understood by technical experts and operational personnel for many years. The WDAs define characteristics of the ship, its personnel and its working arrangements. Each encompasses a set of well know issues. Hence it was considered useful to investigate whether one could identify an analogous set of TAs for the human factors aspects of computer based systems.

The identification of TAs and their associated set of TIs
The concept of TAs was introduced by this author (1992) as a basis for grouping together sets of assessment tests and procedures in a systematic way. This concept has been adopted and developed in the course of work by DERA for MoD(PE) and by DERA with EDS (1994). The set of TAs identified as a result of all this work are:
(a) Scenario, mission and task context;
(b) Team/work organisation;
(c) User characteristics and manpower;
(d) Mission critical task performance and operability;
(e) Equipment spatial design and layout;
(f) Lines of maintenance;
(g) Training;
(h) Environmental conditions, health and safety.
Each TA has was hypothesised as having two major characteristics: firstly, that they help form a comprehensive technical description of the user working as part of a computer based system; and secondly, they can be used in a management context for procurement purposes. Each TA is considered as being formed from a set of TIs.

EDS (1994) proposed the concept of models. They suggested that the TIs may be related formally through models. The models are structured sets of relationships. They may be considered as being defined by variables. The variables may be quantified and associated with criteria specific to the requirements.

The human factors domains may be mapped on to the TAs which in turn encompass the TIs. Both the TAs and TIs may be seen as the technical expressions of the domains.

Diagram one: Framework of relationships between Technical Areas and Issues, and the HFIP domains, including requirements statements, Agreed Characteristics, assessment strategy and test characteristics.

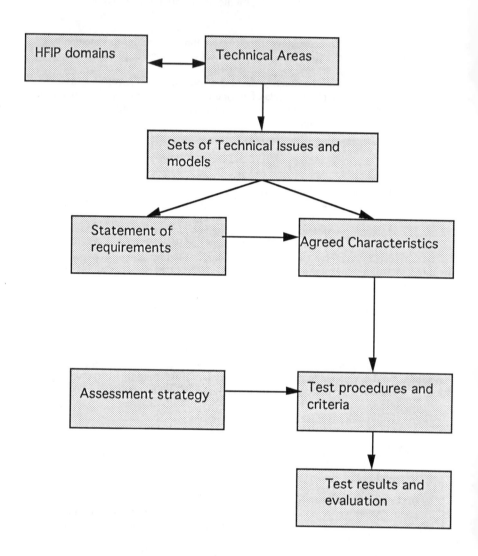

Usefulness of the TAs

Work as been carried out for MoD(PE) to populate the TAs with TIs for the human factors aspects of current computer supported systems. This was achieved through the use of structured interviews which were carried out with a wide range of operational, management and technical personnel (twenty in all).

The interview was structured around the six human factors domains. Those interviewed were asked about with their work and the operations, events and constraints that influenced them in terms of each of the six domains. Typically the interview were carried out individually or in groups of two. Two interviewers were present and they both wrote down the replies of the interviewees. Their replies were categorised using standard content analysis techniques. It was found that there was a considerable correspondence between the TAs (categories) described in Table One and those hypothesised from earlier work. The allocation of interviewees replies to categories was carried out by the two interviewers with a high degree (greater than 95%) of agreement.

However, although the categories i.e. TAs were found to conform to the general set hypothesised, the issues found here were specific to the systems involved. Hence, it was believed that the TAs and associated TIs could be used as a basis for a set of requirements specifications and thus assessment and acceptance tests: they could be used in the management of the procurement process.

Conclusions and Way Ahead

It is believed that the set of TAs could be elaborated and verified for the more classes of systems. This would lend additional support to the general framework, which is presented here but also inform specific cases. Clearly there is still a great deal to understand about assessment strategies. They feed directly to the definition of test procedures, test results and evaluations. These in turn will contribute to the evaluation outcome.

Additional work is particularly required on the nature of the models that structure the TIs within the TAs.

Acknowledgements

I gratefully acknowledge the assistance of many DERA, RN and MoD(PE) colleagues and supporters who have contributed to this programme in terms of technical and operational expertise, and financial support. Dik Gregory (Gregory Consultants), and EDS (Ed Spencer and Mike Burnett) contributed in many important ways through the development of concepts when under contract to DERA.

References

EDS 1994, *Human Factors Principles for Acceptance Tests (Final Report)*
Tainsh, M.A. 1991, On the Assessment and Evaluation of Designs for Human-Computer Interaction. Contemporary Ergonomics. pp 346 - 351.
Tainsh, M. A. 1992, *Acceptance Procedures for Human Factors Aspects of Interactive Computer Supported Systems.* DRA TM(CAD5) 92009.
Tainsh, M.A. 1995, Human factors contribution to the acceptance of computer-supported systems. Ergonomics. Vol. 38(3) pp 546-557.

Table One: A summary description of the TAs and TIs for computer supported systems

Technical Areas (TAs)	Description of Technical Issues (TIs)
Operational Scenario	This is a description of the operational tasks and the systems contribution to it, in terms that are relevant to role and task descriptions for the users.
Organisation	This covers all aspects of the user's jobs roles and their interconnectedness. Hence it covers watchkeeping routines, system integration and definition of jobs and roles.
User characteristics	This covers both the manpower requirement determined from operational scenario, task definition and equipment availability and personnel knowledge, skills and experience meeting that requirement.
Human-computer interaction and operability	The focus here is on the user working at his equipment. This includes the functional and non-functional aspects, operability, interoperability and task performance.
Equipment layout	This covers the overall space requirement and arrangement of equipments within compartments. It includes the users' working space while taking account of traffic flow.
Lines of maintenance	The roles of user and maintainer need consideration as they may be executed by a single individual. The maintenance requirements including task design training and manpower. The provision of facilities to enable record keeping and the execution of support activities must be considered.
Training	Training is a major area of interest including the division between various facilities, individual and team, and quality assurance. The resources and time available are major issues.
Environment, health and safety	These items are grouped together as they may be handled this way by a project. Clearly the environmental conditions and habitability require detailed attention. The impact of national legislation in the field of safety is a major consideration.

COGNITIVE MAPPING AS A TOOL FOR REQUIREMENTS CAPTURE

Adam Pallant [1]*, Peter Timmer [1], and Scott McRae [2].

[1]	[2]
Ergonomics and HCI Unit,	BT Development Unit,
University of London,	BT Laboratories,
26 Bedford Way,	Martlesham Heath,
London WC1H OAP	Suffolk IP5 7RE

email (Adam Pallant):
< zap@factor.netkonect.co.uk >

This paper describes the application of cognitive mapping techniques, for eliciting and representing cognitive structure, to the problem of user requirement capture for the design of an electronic database. The research is based upon the assumption that organising information so as to be compatible with the structure of users' knowledge will enhance information retrieval by supporting more effective navigation behaviour. Although several researchers have noted the potential of cognitive mapping techniques, previous research has generally failed to address serious methodological problems associated with their application to design. A methodological approach is devised which addresses these problems, and thereby enhances the utility of the techniques as tools for user requirement capture.

Introduction

Information Retrieval

In recent years, there has been an enormous increase in the quantity and variety of information stored in electronic databases. As they become increasingly large and complex, the task of retrieving information from these databases is becoming increasingly difficult. Such information retrieval problems are likely to be associated with low task quality and high user (and system) costs; including slow and error-prone task performance, and low user satisfaction.

The take-up of database-centred electronic services, such as teleshopping, is therefore likely to depend, at least in part, upon customers being able to find the desired - or **target** - information quickly and easily. Users may adopt one of a number of retrieval strategies (such as 'searching' and 'browsing'), each of which depends to varying degrees upon effective navigation around the database. Database 'navigability' is therefore likely to be an important user requirement for the design of the user interface (UI) to large databases of product information supporting electronic teleshopping.

---

* Now with **Racal-Thorn Wells**.

The Design Problem

Electronic databases are commonly organised into hierarchical menu structures. The process of partitioning large amounts of complex and varied information into discrete categories presents two overlapping problems:

• **Fuzzy Groups**: real-world objects may often be grouped in a variety of ways depending upon the context, and the groups so formed are often over-lap. For example, a 'vege-burger' may be categorised either as an instance of a burger (along with hamburgers etc.), a frozen food (along with pizzas etc.), or a vegetable product (along with carrots etc.). Indeed, several studies have demonstrated that people take significantly longer to select the correct menu category when the target item is a less familiar example of the category (e.g. Somberg and Picardi, 1983).

• **Fuzzy Labels**: labels for large and ill-defined groups of items are often obscure and inaccurate (over or under-inclusive). For example, Lee et al. (1984) reported that fuzzy category labels such as 'miscellaneous' created confusion and uncertainty among users.

Compatibility

A number of researchers have recognised the importance of achieving a 'fit' between the design of the user interface to an electronic database, and the structure of users' knowledge of the content of the database (e.g. Roske-Hoestrand and Paap, 1986). The present research is based upon the assumption that organising a database UI such that it is compatible with users' structural knowledge - or **cognitive structure** - will enhance information retrieval by supporting more effective navigation behaviour. In addition, it is assumed that rapid and error-free retrieval of product information will enhance user satisfaction and may thereby increase the uptake of database-centred electronic teleshopping services among the potential user population.

Cognitive Mapping

Techniques for eliciting and representing cognitive structure are based upon the assumption that 'similarity data' - describing the relationships between a set of stimuli - provides an index of the organisation of these concepts in human memory (e.g. Fillenbaum and Rapoport, 1971). The validity of the representations generated by these techniques - or **cognitive maps** - is, however, compromised by several methodological limitations. These limitations must be overcome if cognitive mapping techniques are to provide a useful tool for informing design.

The methodological problems associated with the application of cognitive mapping techniques, together with the steps adopted in the present study to resolve them (*in italics*), are outlined below.

• **stimulus selection** - previous research has generally failed to provide explicit a priori criteria for the selection of a comprehensive and representative stimulus set.

An analytic framework for characterising the grocery shopping domain was devised to provide an explicit rationale for stimulus selection.

• **contextual effects** - since similarity judgments are likely to be sensitive to contextual variables, different cognitive maps may be elicited in different situations.

A relatively simple, rapid and unconstrained elicitation procedure - 'free card sorting' - was employed, leaving subjects free to determine their own criteria of correspondence.

• **choice of representational model** - different statistical models for representing similarity data are based upon different assumptions, tend to reveal different aspects of cognitive structure, and may have 'Procrustean properties' which may impose inappropriate structure upon similarity data (Fillenbaum and Rapoport, 1971).

*Both **spatial** [multidimensional scaling] and **network** [hierarchical clustering] analyses were employed to provide contrasting representations of cognitive structure.*

• **interpretation** - although elicitation techniques are associated with formal analytic tools which generate visual representations of similarity data, the process of interpreting cognitive maps remains largely subjective.

*The card sorting procedure was combined with more 'qualitative' elicitation techniques, including concurrent and retrospective **verbal protocols** and direct observation, to provide a fuller picture of cognitive structure and facilitate interpretation.*

Stimulus Selection

The validity of a cognitive map depends critically upon the selection of a representative and comprehensive set of stimuli. This in turn depends upon an understanding of both cognitive structure, and of the domain of real world objects to which this knowledge corresponds.

Cognitive Structure

After Rosch (e.g. Rosch et al., 1976), grocery object categories can be characterised as having both a 'horizontal' within-category and a 'vertical' between-category structure. Within-category structure is assumed to be based on a hierarchy of levels of abstraction, with a **basic level: "that** is appropriate for using, thinking about, or naming an object in most situations in which the object occurs" (Rosch and Lloyd, 1978). Between-category structure is thought to be based upon typicality; with typical category members (or **exemplars**) being the most familiar, the most easily and accurately classified, the most likely to be elicited as category instances in free recall, and the most likely to be used as cognitive reference points in comparisons.

The stimulus set should therefore consist of basic level exemplars sampled from each of the principal super-ordinate categories of grocery object.

Rosch has pioneered the use of a **free-association** procedure for generating category exemplars. The technique is based on the assumption that the items cited most frequently as examples of a particular category are the most typical category members.

The use of a free association procedure for stimulus selection demands that the principal super-ordinate categories of grocery object be first identified. An analysis of the domain of grocery objects was therefore conducted in an attempt to elucidate the nature of grocery categories.

Domain Analysis

After Dowell and Long (1989), the grocery shopping domain can be characterised as consisting of **grocery objects** (e.g. apples and bacon), and **people's shopping needs** (e.g. fully or partly specified objects on an implicit or explicit shopping list). The foregoing analysis also addresses the supermarket worksystem; consisting of **shoppers** (users) and **supermarket devices** (e.g. displays and trolleys).

Grocery objects are defined by a number of attributes [A][1] (see table 1).

In the context of the supermarket worksystem, grocery objects also have the attribute of location [A]; which is determined by the interaction of the attributes of grocery objects with those of worksystem devices. For example, frozen (nature [A]) grocery objects are located in freezers; while fresh (nature [A]) grocery objects are located in gondolas and chillers.

Grocery shopping needs are specified in terms of the attributes of grocery objects.

---

[1] In order to enhance clarity of the foregoing discussion, the symbol "[A]" is employed to qualify every reference to **an attribute of a grocery object.**

Table 1. Principal Grocery Object Attributes [A]

[A]	example:
appearance	red tin
brand	Heinz
name	tomatoes
nature	canned
price	40p.
quantity	12 oz.
type	vegetable

Each grocery object attribute may have a range of possible values: for example, nature [A] may have the values of either **frozen, fresh, dried** or **canned** [where 'canned' includes all grocery objects which are packaged in tins, bottles, jars, and tubes]. The value of a grocery object's type [A] is expressed in terms of the 'class name' of the grocery object. Analysis of a domain instance in the supermarket worksystem revealed 23 values of grocery object type [A] (see table 2.).

Table 2. Values of the type [A] of grocery objects

bakery products	fruit	ready meals
biscuits	fruit juice	salad
cereals	herbs & spices	sauces
cooking ingredients	hot beverages	snacks
dairy produce	meat	soft drinks
delicatessen products	pickles	soup
desserts	poultry	vegetables
fish	preserves	

Different attributes correspond to different levels of abstraction in the structure of grocery object categories. A grocery object's **name** [A] (e.g. tomatoes) will generally correspond to Rosch's basic level of abstraction - especially for highly differentiated grocery categories (such as 'fruit' and 'meat'). Grocery object **type** [A] (e.g. vegetable) and **nature** [A] (e.g. canned), in contrast, will generally be associated with the more abstract, super-ordinate levels of category structure.

Procedure

Free Association Task

A group of 10 subjects were asked to generate an instance of each of 23 values of grocery object type [A] and the 4 values of grocery object nature [A]. The single most frequently cited instances of grocery object type [A], together with the 3 most frequently cited instances of grocery object nature [A], were selected - giving a total of 35 (23 + 12) grocery object exemplars.

For example, 'bread', 'rolls', 'pastries', and 'cakes' were all cited as instances of **bakery products**. Since 'bread' was cited by a total of 7 subjects, it was deemed to be the most prototypical exemplar and was therefore selected as a stimulus item.

The names of the selected grocery objects were printed on pieces of white card to be used as the experimental stimuli.

Card Sorting Task

One hundred subjects were asked to sort the 35 stimulus cards into as many groups as they wished, putting "similar ones together". Subjects were asked to describe the reasons for their similarity judgements (concurrently), and to label the groups and sub-groups so formed (retrospectively).

Data Analysis

A group similarity matrix was derived from the card sorting data and transformed into a chi-square dissimilarity coefficient matrix for input into SPSS-X 6.1. Group solutions were then derived for both standard non-metric Multi-Dimensional Scaling (MDS) and between-group linkage Hierarchical Clustering (HCA) procedures.

Results

SPSS provides several numeric formal measures of 'fit' (see table 3.) which suggested that the 3D solution provided the most appropriate solution - explaining over 96% of the variance in the similarity data (RSQ) with a stress value of only 8%. However, these fit metrics are only heuristics, and the most important consideration is clearly the interpretability of the representations.

Although subjects exhibited little difficulty in arranging the stimuli into groups, articulation of the reasons for these judgements and labeling the groups proved more problematic. Nevertheless, grocery object **type** [A] appeared to be a much more important sorting criteria than was nature [A].

Table 3. Measures of Fit for MDS Solutions (D= 1 to 3).

solution	stress	RSQ
1 D	0.346	0.695
2 D	0.148	0.906
3 D	0.078	0.965

The 2D MDS solution was interpreted in terms of 4 distinct clusters; labeled **'animal products'**, **'meals'**, **'snacks and drinks'**, and **'fruit & vegetables'**. The 3D MDS solution was interpreted in terms of the 3 continuous dimensions; labeled **'animal - plant'**, **'solid- liquid'**, and **'natural-processed'**.

The dendogram of the HCA solution was interpreted in terms of 5 principal clusters; labeled **'fruit and vegetables'**, **'animal products'**, **'liquids & sweets'**, **'staples'**, and **'meals and snacks'**. Factor analysis of the similarity data supported a 5 cluster interpretation. These upper-level clusters could, however, be further decomposed; for example, the **'fruit and vegetables'** cluster consists of distinct 'fruit' (apples and raisins) and 'vegetables' (lettuce, tomatoes, carrots and onions) subclusters.

The good statistical fits of the MDS solutions suggests that a single 'group' solution provided an adequate representation of the structure of subjects' grocery object knowledge. Indeed, only one subject reported having employed idiosyncratic criteria of correspondence in their similarity judgments (i.e."ones that I like/dislike").

Discussion

The results of the cognitive mapping procedure were used to generate a set of user requirements for the organisation of grocery product information on the UI to an electronic database supporting grocery teleshopping. For example, people expect to find [frozen] fish fingers, [fresh] cod and [canned] tuna fish together. Extrapolation from the HCA solution suggests that grocery product information can be organised into 5 distinct categories, based on grocery object type [A]. The use of prototypical category exemplars as descriptors may also enhance category labeling (e.g. 'carrot' may be appended to the 'vegetables' category label) to facilitate information retrieval.

Although the nature [A] of grocery objects is an important determinant of their organisation in the supermarket worksystem, this is not reflected in the cognitive maps derived in the present study. Clearly, the organisation of groceries in a supermarket is affected by a host of worksystem constraints and marketing considerations which may not be relevant to a teleshopping environment. The supermarket metaphor may therefore not provide an appropriate structure for the organisation of grocery product information.

Although the application of cognitive mapping tools advances the problems of organising database information, some issues remain unresolved and need further interpretation. Thus highly differentiated grocery object categories - such as 'sauces' - remain difficult to classify and label. For example, 'chocolate sauce' and 'tomato ketchup', though of common type [A], are generally used in very different contexts. Indeed, when people are actually shopping, for example, the nature [A] of grocery objects may become a much more salient attribute than their type [A].

Conclusion

Resolution of the methodological problems associated with the application of cognitive mapping techniques yields a set of tools which are potentially very valuable for informing design. Although the present study has been concerned with a specific instance of a teleshopping database, these tools are potentially applicable to a wide range of design problems in a variety of contexts. Their utility is likely to be enhanced through the use of more sophisticated elicitation and representational procedures (including 'framed-' and 'multi-' sorting; and Roske-Hoestrand and Paap's (1986) Pathfinder algorithm).

The utility of cognitive mapping techniques, however, depends not only upon the extent to which they provide an accurate representation of subjects' cognitive structure, but also upon the extent to which cognitive compatibility itself facilitates information retrieval. Future research must attempt to assess the validity of cognitive maps through comparative evaluation of these taxonomies against those generated by expert opinion.

References

Dowell, J. and Long, J. (1989). "Towards a conception for an engineering discipline of human factors." Ergonomics 32, 1513-1535.
Fillenbaum, S. and Rapoport, A. (1971). "Structures in the subjective Lexicon." New York: Academic Press.
Lee, E., Whallen, T., McEwen, S. and Latremouille, S. (1984). "Optimising the design of menu pages for information retrieval." Ergonomics 27, 1051-1069.
Norman, D. A. (1988). "The Psychology of everyday things." Basic Books, Harper Collins: USA.
Rosch, E. and Lloyd, B. B. (eds.). (1978). "Cognition and categorization." Hillsdale, NJ: Erlbaum.
Rosch, E., Simpson, C. and Miller, R. S. (1976). "Structural bases of typicality effects." Journal of Experimental Psychology: Human Perception and Performance 2, 491-502.
Roske-Hoestrand, R. J. and Paap, K. R. (1986). "Cognitive networks as a guide to menu organisation: An application in the automated cockpit". Ergonomics 29, 11, 1301-1311.
Somberg, B. L. and Picardi, M. C. (1983). "Locus of information familiarity effect in the search of computer menus." in Proceedings of the Human Factors Society 27 th Annual Meeting, pp. 826-830. Santa Monica, CA.: Human Factors Society.

GROUP DIFFERENCES IN ABILITY TO USE VERBAL ROUTE GUIDANCE AND NAVIGATION INSTRUCTIONS

Paul Jackson

Centre for Transport Studies,
Department of Civil Engineering,
Imperial College,
London, SW7 2BU

Safety concerns have prompted designers of car-based route guidance and navigation systems to make more use of the auditory sense as a receiver of verbal instructions in addition to those displayed on a computer screen. This paper reports on a series of experiments which consider ability to process route guidance information received aurally. In particular, whether the same information should be provided to all system users, or whether different levels would be more suitable. Contrary to previous research we find that there are differences between demographic groups in their ability to use the information presented aurally to them via route guidance systems. Results suggest that different types and levels of information should be made available to enable users to select the combination most suitable for them.

Introduction

Car-based route guidance and navigation systems are already available as standard in some top of the range executive cars. Research has considered the advantages of using such systems: to the individual driver; to the road network; and to the transport system as a whole. As yet however, little research has considered how this additional information will be processed. Is information acquired from a computer based system comparable to that acquired from direct experience? Is this information mentally represented in the same way as that acquired from personal experience?

The potential for distraction and the necessity for the driver to look away from the road ahead to a screen based system has prompted system designers to make more use of verbal instructions. This paper looks at the provision of additional route guidance through the use of verbal messages and considers the effects of receiving this information upon our ability to acquire mental representations of journeys. In addition, the paper compares the ability of different demographic groups to process and use information that they hear whilst attending to a primary task: does it help or does it hinder?

Method

The study uses video footage of three interconnecting routes of moderate length and complexity, each with at least 4 turns. These routes are marked in Figure 1 as RX-RY (the red route), BX-BY (the blue route) and GX-GY (the green route). The study area was selected following a series of pilot studies (see Jackson 1994) and was filmed using a Sony Super 8 camera fitted with a wide angle lens.

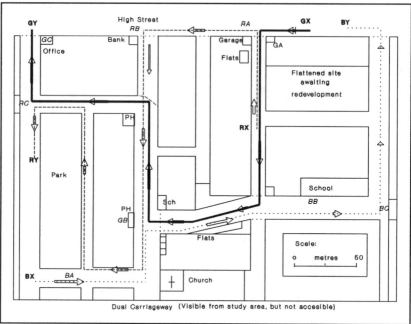

Figure 1 Study area showing routes and landmarks

Procedure

All participants see the same films, but are assigned to one of four experimental conditions which differ in the information heard:

Group 1   see the videos with no additional information (Video condition);
Group 2   hear spoken route guidance instructions (Route guidance condition);
Group 3   hear a news radio broadcast (Radio condition);
Group 4   hear radio, with route guidance information overriding this as required
           (Radio plus route guidance condition).

Subjects are shown the film of the first route. They see two versions of the route: the first with the camera fixed, giving a forward view of the route (the 'route' film); the second with the camera moving left and right to show various landmarks along the route (the 'landmark' film). Having seen both versions participants carry out a series of tasks. These are:

- Sequencing task - subjects arrange randomly shuffled photographs of scenes along the route into their correct sequence (i.e. as they would be seen along the route);
- Distancing task - subjects arrange the sequenced photos along a strip of paper such that their distance apart represents their actual distance apart on the route;
- Single route pointing task - subjects are given photographs of three landmarks on this route. They are asked to imagine themselves standing at the first landmark and must point in the direction of each of the other landmarks, indicating each of these directions in a booklet provided for them. They repeat this for each of the landmarks acting as the 'origin';
- Distance and time estimation - subjects estimate the distance travelled and time taken to travel each route.

This procedure is repeated for the two remaining routes. Having seen the three routes and carried out all the tasks, subjects are tested for their ability to integrate the information acquired from these separate experiences into a configurational whole. This is done by having them carry out the pointing task again, but this time estimating directions from landmarks on one route to landmarks on the other two routes. This is the cross route pointing task. Finally, subjects are asked to draw a map of the area.

Results

The results reported below are based on a sample of 116 subjects (58 males, 58 females) The age range of subjects was 18 to 86 (mean age = 35.1). A complete analysis of results can be found in Jackson (1995), but for the purposes of this summary paper only the cross route pointing task is described as this provides us with a measure of overall performance for each subject. There was also a strong positive correlation between subjects' performance on this task and performance on the other tasks.

Performance of demographic groups
For each direction estimate made by a subject we calculate the deviation from true direction. This enables us to generate a mean error score for each subject for each route. Similarly, the cross route pointing task enables us to produce an overall mean error score for each subject (measured in degrees).

Analysis of variance showed that, of the 4 independent variables - age, gender, driving ability and information group - all but information group were found to be highly significant ($p < 0.0005$). To strengthen the model and to give it more cases to work with the four information groups are collapsed into two by combining subjects who saw the videos in the no information condition with those who saw them under the radio condition (groups 1 and 3). We do the same with the two salient information conditions so that those in the route guidance condition are grouped with those in the radio plus route guidance condition. Thus we create a new variable, group_x, which distinguishes between subjects who are exposed to salient information and those who receive non-salient or no information. With this model the main effects of all four independent variables are significant:

Age	$F(2,113)=15.66, p<0.0001;$
Driving ability	$F(1,114)=25.18, p<0.0005;$
Gender	$F(1,114)= 4.33, p<0.05;$
Information Group	$F(1,114)= 5.62, p<0.05.$

Effects of route guidance information upon spatial knowledge acquisition.
The following is a summary of the main results looking at the effects of route guidance information upon ability to acquire configurational knowledge.

• Subjects who viewed the routes whilst hearing salient (route guidance) information performed significantly worse than those who received no guidance;
• The presence of a non salient source of information (radio), however, did not impair performance. Indeed, the performance of subjects who saw the videos whilst hearing radio was comparable with that of subjects who saw the videos in silence;
• The research design enables a comparison of the performance of similar groups under different information conditions. For example one group of male drivers aged 22-54 saw the videos in the video only condition (mean direction estimate error = 43.28), while a second group saw them in the route guidance condition (mean direction estimate error = 62.98). Similarly, one group of female drivers saw the videos accompanied by radio (mean direction estimate error = 58.08), while a second similar group saw them in the radio plus route guidance condition (mean direction estimate error = 66.92)

The performance of those in the video only and radio conditions is comparable, suggesting that it is not simply the presence of a secondary source of information which is serving to distract subjects. Although subjects in the radio group were being exposed to non-salient information they were able to focus attention upon the primary task of building up a mental representation of the area. Subjects in the route guidance condition, however, were exposed to salient information to which they were required to attend. This sharing of attention between two salient sources appeared to have a detrimental effect upon ability to acquire a mental representation of the area.

Discussion

The four information groups received varying levels of additional information as described above. The performance of the four groups as a whole, judged by their performance of the cross route pointing task, showed that those in the video only (no information) and radio (non-salient information) conditions performed better than those in either of the route guidance conditions. A comparison of like subjects showed that the presence of route guidance information had a detrimental effect upon spatial ability as measured by performance of the various experimental tasks. These are important results because they compare the performances of similar subjects under comparable conditions, the only difference being what was heard.

Contrary to previous research which found no significant differences between age, gender or driving ability groups (eg. van Winsum *et al.*, 1990), the present study suggests that there *are* differences between these groups in their ability to cope with and/or use the additional information provided to them by route guidance and navigation systems. In the under 55's there were no significant differences between male and female drivers. However, both male and female drivers performed significantly better than their non driving counterparts, suggesting that driving ability has more of an effect upon spatial ability than previously supposed. Age had a significant effect upon ability, older subjects performing poorly at most of the tasks. However, this was again mediated by driving experience, most noticeably in the performance of females over 55. The mean error score of this group was 23 degrees lower than that of female non drivers aged 55 and over.

The superior performance of the under 21's as compared with that of subjects aged 22-54 was surprising. We had expected the performance of the youngest subjects to be inferior to that of slightly older subjects, especially if driving experience was found to have a positive effect upon spatial ability. The two groups were closely matched in educational ability and, despite the spread of ages represented by the 22-54 age group, the mean ages of the 8 groups in this age bracket ranged from 26.4 years to 32.5 years. Despite the similarities between the two age groups, male and female drivers in the 21 and under group consistently outperformed all other subjects under all experimental conditions.

Conclusions

It is possible that, for some individuals, the extra cognitive demand associated with attending to an in-car navigation system might detract from the primary task of controlling the vehicle. This issue has received extensive attention over the last decade, with the conclusion that more use should be made of the auditory sense. The results of the present study suggest that whilst this is a successful option for many subjects, certain groups might find spoken instructions as difficult to cope with (and use) as other forms of additional information. Careful consideration needs to be given to the amount of information which can be processed by drivers both when travelling unfamiliar routes on the first few occasions, and also at different stages in the lifecycle. Our information handling and processing capabilities are not a constant across the lifecycle. Future research aims to explore further the interaction between age, gender and driving ability and additional information. In particular, research is now in progress which seeks to identify better ways of expressing route guidance information so as to maximise its utility for a wider range of users.

Acknowledgements

This work is supported by the Engineering and Physical Sciences Research Council. Thanks are due to Professor Kay Axhausen of the University of Innsbruck, Peter Bonsall of the University of Leeds, Dr Mark Blades of the University of Sheffield, Peter Burns and Gary Burnett of HUSAT, Loughborough University of

Technology and my supervisor John Polak for their comments and advice regarding this research.

References

Evans, G.W. (1980) Environmental cognition, *Psychological Bulletin*, **88**, 259-287.

Golledge, R.G. (1987) Environmental Cognition, in D. Stokols and I. Altman (Eds.), *Handbook of Environmental Psychology*, **1** 131-174, Wiley, New York.

Jackson, P.G. (1994) Experimental procedures for studying the updating of spatial knowledge, in J.F. Mortelmanns (Ed.) *Proceedings of the First Erasmus-Network Conference on Transportation and Traffic Engineering*, 4-6 September, Kerkrade, Netherlands, Acco, Leuven.

Jackson, P.G. (1995) Using video to simulate wayfinding and the effects of route guidance information, *Paper presented to the 25th Annual Conference of the Universities Transport Studies Group*, 4-6 January 1995, Cranfield University.

Kitchin, R.M. (1994) Cognitive maps: what are they and why study them? *Journal of Environmental Psychology* **14** (1) 1-19.

Van Winsum, W., Alm, H., Schraagen, J.M., and Rothengatter, T. (1990) *Laboratory and field studies on route representation and drivers' cognitive models of routes*, Drive project V1041 Generic Intelligent Driver Support Systems, Traffic Research Centre VSC, University of Groningen, Netherlands.

STINGRAY: THE DEVELOPMENT OF AN ERGONOMIC KEYPAD FOR USE WITH COMPUTERISED APTITUDE TESTS

Jayne White[1] and Andy Hutchinson

Human Engineering Limited
Shore House
68 Westbury Hill
Westbury-On-Trym
BRISTOL
BS9 3AA

National Air Traffic Services
CAA House
45-59 Kingsway
LONDON
WC2B 6TE

The aim of this programme of work was to review the current man-machine interface (MMI) to the PC based aptitude tests used in the selection of air traffic controller students. The objectives were as follows:

- to gather evidence that problems are experienced with the interface
- to specify new interface designs to overcome the problems
- to evaluate the new designs

This paper reports on the design of new interface options and the experimental study carried out to assess these designs.

Introduction

The National Air Traffic Services use a PC based aptitude test as part of the process for selection of air traffic controller students. The test is in three parts and the candidate makes multiple choice responses using the keys shown in white in Figure 1 on a standard PC keyboard.

Figure 1. Original keyboard layout

The test is necessarily intensive and fast paced, and requires the candidate to concentrate intently throughout the one hour session. There was a concern that use of the current interface was causing discomfort to candidates and was possibly unfair to some candidates. It was feared that candidates were having to pay more attention to how to respond than was desirable. A study was therefore carried out to establish whether the current keyboard was indeed causing these problems.

---

[1] The initial phase of this work was performed by the author at EPP Human Factors.

Evaluation of current interface

It was established that there were a number of problems with the original keyboa
A number of new design options were considered and two designs were proposed. The
first was to modify the original keyboard and secondly to design a custom keypad. Th
aims of the new designs were:

- to increase the accessibility and ease of use of the keys
- to eliminate discomfort during use of the keys
- to increase fairness to left and right handed users and to infrequent users of
  keyboards

Design Rationale

Modified keyboard

The aim of the modified keyboard (Figure 2) was to provide a simple, cost-
effective solution to some of the problems experienced with the original keyboard. Th
provision of a cover protected the redundant QWERTY keys from inadvertent operati
and served as a support for the hands and wrists. The cover also reduced the number of
keys visible thus lessening potential confusion caused by irrelevant keys. In order to
further aid the candidates in locating the appropriate keys, the keys required for the tes
were coloured black and clearly labelled in white text.

Cover protecting
irrelevant keys
and allowing
hands to rest
while using
function keys

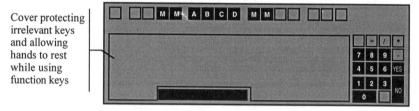

Figure 2. Modified keyboard

Custom keypad

A custom keypad (Figure 3) was designed which took full account of all the usabil
problems identified with the original keyboard.

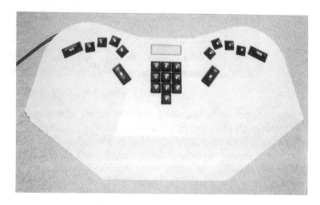

Figure 3. Custom keypad

The problem of inaccessibility of the function keys was resolved by positioning th
keys such that the fingertips "fell naturally" to them. The Match (M) key was positione

so as to be operated by the thumb, and the A, B, C and D keys were positioned so as to be operated by the index, middle, ring and little fingers respectively. This key layout was repeated for both left and right handed operation. The left hand version is a mirror image of the right hand. The labelling of the left hand keys goes against the conventional stereotype of reading from left to right; but is used because left handers when surveyed reported the "A" finger to be the index finger.

The investigation into the problems associated with the current keyboard revealed that the candidates did not like the proximity and orientation of the "Yes" and "No" keys. Therefore the "Yes" and "No" keys were separated so as to be operated by different hands.

The computer keypad/calculator layout was chosen for the number keys as this was considered to be the layout most familiar to potential candidates.

The surface of the keypad was angled such that when using the A, B, C, D, M, Yes and No keys the hands and wrists maintain a neutral posture. This was achieved by referring to the body of competent research reported in the literature (Pheasant, 1988; Sanders & McCormick, 1993).

Experimental Design

Eighteen subjects aged 16-36 (average age 23) took part in the experiment. Each session was videoed, but the experimenter also observed the session real-time via a monitor. Subjects were de-briefed after performing the test and completed a questionnaire designed to gather their opinions about the keyboard design they had used.

Results

The analysis of the data obtained from the experimental study illustrates that the new keyboard designs offer clear advantages over the original keyboard.

Comfort

In terms of comfort, the hypothesis was that discomfort should be reduced for candidates using the modified and custom keyboards compared to the need to support the hands and wrists when using the original keyboard. The experimental results support this hypothesis as there are no cases of reported wrist discomfort for the custom and modified keyboards compared to a moderate amount of wrist discomfort reported for the original keyboard (Figure 4).

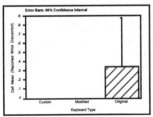

	Mean Diff.	Crit. Diff.	P-Value	
Custom vs. modified	0	.367	•	
Custom vs. original	-.333	.367	.0719	*
Modified vs. original	-.333	.367	.0719	*

(* = significant at 10% level)

Figure 4. Statistical analysis of reported wrist discomfort

The cover on the modified keyboard was regarded by subjects as offering an improvement in comfort compared to the awkward positioning of the hands and wrists demanded by the original keyboard. Subject comments suggest that some modification of the spacing of the A, B, C and D keys on the custom keypad is required, but overall the shape and layout of the custom keypad were considered to be comfortable to use.

Accessibility

It was hypothesised that the accessibility of keys would be considerably improve for the users of the custom keypad and may be slightly improved for the modified keyboard. Analysis of the questionnaire data revealed that the custom keypad was inde perceived as significantly better than the original keyboard for locating the A, B, C, D and M keys (Figure 5) and for reducing the mis-keying of numbers (Figure 6). The modified keyboard was slightly, but not reliably, better than the original keyboard for accessibility of keys and this is in accordance with the hypothesis.

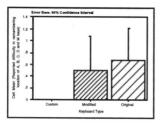

	Mean Diff.	Crit. Diff.	P-Value
Custom vs. modified	-.500	.440	.0648
Custom vs. original	-.667	.440	.0179
Modified vs. original	-.167	.440	.5166

(** = significant at 5% level)

Figure 5. Statistical analysis of reported difficulty in remembering the location of A, C, D and M keys

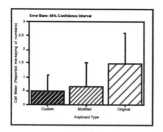

	Mean Diff.	Crit. Diff.	P-Value
Custom vs. modified	-.167	.840	.7328
Custom vs. original	-1.000	.840	.0544
Modified vs. original	-.833	.840	.1025

(* = significant at 10% level)

Figure 6. Statistical analysis of reported mis-keying of numbers

The subjects' comments on accessibility of keys for the original keyboard indicate that the Yes and No keys and number keys had poor accessibility, as did the A, B, C, D and M keys at the top of the keyboard. The accessibility of the keys on the modified keyboard was reported to be improved by their colour coding. The cover on the modifie keyboard was also considered to improve accessibility by reducing the possibility of accidentally hitting irrelevant keys. The advantages reported regarding the custom keypad were the intuitive layout of A, B, C, D and M keys and the provision of only the relevant keys which reduced searching time and eliminated the need to avoid redundant keys.

Subject comments also raised the possible confusion resulting from the translation of ABCD on the screen to DCBA on the left hand custom keys. The accessibility of the Yes and No keys on all three keyboards was considered acceptable although the preference was for the separated custom keypad layout. The numbers on the custom keypad were seen to be more accessible than those on the other two designs due to the central location and hand support provided by the keypad surface.

Fairness

The hypotheses for the effect of the new keyboard designs on fairness were:

- fairness would be improved for inexperienced keyboard users by both the modified and custom keyboards

- the custom keypad would be equally fair to left and right handed users

The custom keypad was perceived by more subjects as fairer to all levels of keyboard experience than the original keyboard. The modified keyboard also improved perceived fairness to infrequent keyboard users compared to the original keyboard.

Statistical analysis of the questionnaire data relating to the perceived fairness of the keyboards is shown in Figure 7.

These results are in accordance with the hypothesis that the custom keypad would be perceived as fairer to infrequent keyboard users than the original design and that the modified keyboard would be seen as slightly fairer than the original keyboard.

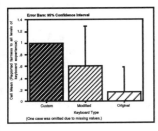

	Mean Diff.	Crit. Diff.	P-Value	
Custom vs. modified	.400	.406	.1050	
Custom vs. original	.833	.388	.0020	**
Modified vs. original	.433	.406	.0814	*

(** = significant at 5% level
* = significant at 10% level)

Figure 7. Statistical analysis of perceived fairness to infrequent keyboard users

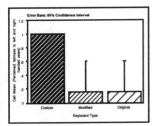

	Mean Diff.	Crit. Diff.	P-Value	
Custom vs. modified	.833	.410	.0006	**
Custom vs. original	.833	.410	.0006	**
Modified vs. original	0	.410	•	

(** = significant at 5% level)

Figure 8. Statistical analysis of perceived fairness to left and right handed users

Fairness to left and right handed people was perceived by subjects as greatest for the custom keypad and indeed significantly more so than the original and modified keyboards.

The modified and original keyboards were predominantly reported as being unfair to left handed users, whereas the custom keypad was considered fair to both right and left handed people (Figure 8).

Summary of evidence

The statistical evidence (as summarised in Table 1) supports the recommendation that the original keyboard should be replaced.

Table 1: Summary of statistically significant evidence

Comfort	Accessibility	Fairness
No reported wrist discomfort for the modified and custom keyboards (p<0.1)	Custom keypad better than original for remembering location of A, B, C, D and M (p<0.05)	Custom keypad fairer to left and right handers than original and modified (p<0.05)
	More mis-keying of numbers for original than for custom keypad (p=.0544)	Custom keypad fairer than original for infrequent keyboard users (p<0.05)
		Modified keyboard fairer than original for infrequent keyboard users (p<0.1)

Modifications

In order to take account of some of the comments received from subjects, a number of modifications were made to improve the custom keypad still further (Figure

The position of the Yes and No keys was modified to avoid any possible confusi between "yes and no" and "left and right" involved in some test items. Yes and No ke are provided on <u>both</u> sides of the keyboard in an orientation such that they are easy to use with either the finger and thumb or index and middle fingers of one hand. Furthermore, colour coding the Yes and No keys green and red respectively provides further information to aid the location of the appropriate key.

A survey of 70 people was conducted to gather more information on Yes and No key position stereotypes and also for left handed stereotypes for the order of A, B, C a D key labelling. The latter resulted in reordering the labelling of the left hand key set.

Some concern was expressed by a few subjects that the spacing of the A, B, C and keys may not be appropriate for users with large hands. Simply increasing the spacing would not necessarily be appropriate as this may create difficulties for users with smalle hands. Therefore a layout was devised in which the key sizes were increased and the orientation of the keys was altered. The keys were angled so that smaller handed users could use the inner arc and those with larger hands use the outer arc formed by the keys

Figure 9: Final custom keypad design

Conclusion

This paper illustrates an ergonomic design project in which a product has been designed using information from existing literature rather than an experimental methodology. Perhaps this is indicative of the progression towards ergonomics becomin a respected engineering discipline with reliable sources of good design guidelines. Furthermore, it is clear that we do not have to rely on lengthy experimentation to obtai the data necessary to produce an ergonomic design. This is good news for ergonomics consultants and in particular for their clients who frequently require reliable solutions to real-world problems without embarking on a major research project.

References

Pheasant, S. 1988, *Bodyspace: Anthropometry, Ergonomics and Design* (Taylor & Francis, London)

Sanders, M. S. and McCormick, E. J. 1993, *Human Factors in Engineering and Design,* 7th Edition (McGraw-Hill International)

White, J. 1995, EPP Report No.: EPP/HF/NATS/9453a/2 *Investigation to Recommend Suitable Man-Machine Interface for Aptitude Tests - Final Report.*

The Effects of Workload on Speaking: implications for the design of speech recognition systems

C. Baber & B. Mellor[†]

Industrial Ergonomics Group,
School of Man. & Mech. Eng.,
The University of Birmingham,
Birmingham.

† Speech Research Unit,
Defence Research Agency,
St. Andrews Rd.,
Malvern.

Speech recognition is often defended in terms of its potential to reduce workload, e.g., by 'relieving' an over-loaded visual channel. However, there is little in the literature on the effects of workload on the production of speech. This is the topic which is addressed in this paper.

INTRODUCTION

One of the prime reasons for considering the use of speech recognition is to reduce the levels of workload experienced by operators. Speech can be said to offer an additional channel of communication which can reduce the attentional demands of manual and visual activity. However, speech recognition technology can place constraints on users, particularly in terms of the permitted vocabulary and speech style. This means that speech recognition need not have a positive effect of reducing workload, nor will it be a neutral component in terms of workload, i.e., the introduction of speech recognition will alter not only the nature of the tasks being performed but also introduce additional attentional demands on the user. Successful applications of speech recognition tend to be in stable environments with little time-pressure or other stress on the speaker. As the range of applications of speech technology continues to move beyond telecommunications to military applications, the question of how speech is effected by workload and stress becomes increasingly important. Ideally, the effect of stress and workload on speech should be of sufficient consistency to allow design modification to accommodate such effects, e.g., if stress led to an increase in speech level, then one could propose a means of tracking speech level in order to modify the speech processing when the level reached a defined limit.

The manner in which workload will lead to changes in speech can be considered from two perspectives. A mechanical perspective would assume that workload would induce measurable physical changes to the speech production apparatus, e.g., muscle tension could increase (resulting in changes to speech level and fundamental frequency), breathing rate could quicken (leading to increased speech rate, reduced inter-vowel spacing, increased glottal pulse rate). Conversely, a psychological perspective would assume that the effects of workload would vary, according to the possible competition between task and speech, and that the competition could be mediated by strategies, e.g., as a result of training or experience. It is assumed that the perspectives are not mutually exclusive. For instance, there might be some mechanical changes which are not subject to mediation, or mediation might remove some of the possible causes of mechanical change.

TIME-STRESS AND THE PRODUCTION OF SPEECH

A characteristic of many high workload situations is an increase in time-stress, i.e., the time-window in which activities may be performed is reduced. Thus, in this section we consider workload simply in terms of the amount of activity a person is required to perform per unit time. A previous study indicated that time-stress impaired performance with a speech recognition system (Graham and Baber, 1993). The performance decrement was related to both reduction in recognition accuracy and to an increase in the number of errors in the vocabulary used to perform the task, e.g., speakers were supposed to use the ICAO alphabet, but under time-stress reverted to other words and expressions. More detailed analysis of the data from this study is currently being conducted. The task involved reading car number plates, using the ICAO alphabet, under slow (19 plates per 90 seconds = 12.7 plates per minute) and fast (33 plates per 90 seconds = 22 plates per minute) conditions. Thus, in the fast condition speaking rate should increase by 69%. The change in speaking rate indicates the increase in proportion of plates attempted from slow to fast condition, and the recognition accuracy is reported at the word level.

It is interesting to note that not all speakers responded to the time-stress in a similar fashion. While the time-stress condition was supposed to lead to an increase in speaking rate, speaker SD showed no change in speaking rate, i.e., he completed 12.7 plates per minute in each condition, and had a slight increase in recognition accuracy, i.e., +2.65%. Thus, speaker SD appeared to deal with the effect of time-stress by ignoring it. This involved a strategy in which plates were sampled from the incoming stream. On the other hand, speaker DS increase speaking rate as expected, i.e., by 51%, and recognition accuracy decreased, i.e., -9.7%.

Previous researchers have concluded that changes in speaking rate result in reduction in pauses between words and argue that this will lead to a reduction recognition accuracy. The implication of this assumption is that the characteristics of the words themselves will remain unaffected; in order for this to be true, one would expect the templates to be static representations (as with template matching devices) rather than dynamic models used in many contemporary speech recognition systems. From this, the problem could be alleviated through the use of connected rather than isolated word recognition. Unfortunately, Graham and Baber (1993) conducted their study on DRA-Malvern's AURIX running as a connected word recogniser.

The recorded speech of two speakers (SD and DS) was analysed using a spectrogram. There was little change between slow and fast conditions for speaker SD, both in terms of speech rate and interword length. Note that, in both conditions speaker SD maintained a speaking rate of 12.7 plates per minute. Comparison of the spectrographs, shown in figure 1, suggest that there is little discernible difference between conditions for this speaker.

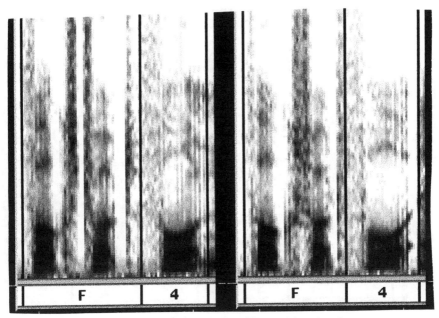

Figure 1: Spectrograph for speaker SD speaking the phrase
"Foxtrot Four", under fast (left hand) and slow (right hand) conditions

Figure 2 shows the spectrograph for speaker DS. There is a discernible increase in speech rate, characterised by a shorter time-frame for the words and a reduction in both vowel space and interword space. This supports the proposal that speaker DS spoke more quickly in the time-stress condition. Furthermore, the high frequency energy in the /F/ of "foxtrot" can be seen to be longer for the fast than the slow condition. This is taken as evidence as increased glottal pulse rate. Thus, one can conclude that the speech of speaker DS was subject to a mechanical change as a result of time-stress; a slight increase in breathing rate led to changes in glottal pulse frequency and speech rate. For the mechanical explanation to be sufficient, one might als anticipate changes in speech level. However, the data from SD shows an increase of <3dB for fast vs. slow conditions, and the data from DS shows a decrease in speech level of around 3 dB from slow to fast conditions. The non-uniformity of changes in speech level suggests that the time-stress was perceived and reacted to differently across individuals. This raises the question of how people deal with stressors, e.g., the fact that some speakers show an increase in speech level and others show a decrease suggests that changes in speech level are not simply physiological reactions to a stressor. In conclusion, the analysis to date suggests that some speakers may exhibit consistent and reliable changes in speech in response to time-stress, while others show no consistent effects, and still others show sporadic effects. Furthermore, there is evidence that speakers can adopt strategies for dealing with the time-stress, other than speaking more quickly, e.g., speaker SD simply maintained a consistent speaking rate in both conditions.

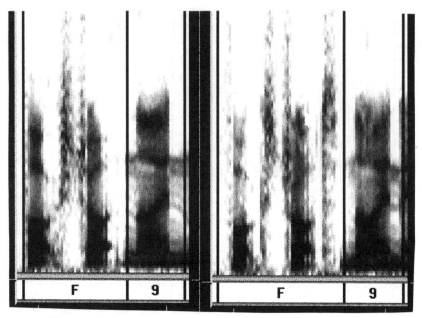

Figure 2: Spectrograph for speaker DS speaking the phrase
"Foxtrot Nine", under fast (left hand) and slow (right hand) conditions

DUAL TASKS AND THE PRODUCTION OF SPEECH

Tunley (1995) found that pairing speech recognition with mental arithmetic led to
worse performance than pairing trackball with mental arithmetic, both in terms of
the mental arithmetic and recognition accuracy. Thus, the additional task led to a
reduction in recognition accuracy. One explanation of this finding is that speech
produced under task loading differs from that under 'normal conditions'. This
issue has been considered above. An alternative explanation is that, as task
demands increase, so the speaker's ability to maintain 'consistent' speech decreases;
in other words, maintaining 'consistent' speech could be proposed to require some
effort, and as effort is shifted to other task demands, so this ability will deteriorate.
For instance, Hapeshi and Jones (1989) have found that recognition performance
decreased when the speaker performs a tracking task, and decreased further when
the cursor moved quickly in the tracking task. Conversely, the tracking task
improved when performed with speech; this could be taken to suggest a strategy by
which effort was directed at the task which participants could explicitly control, and
away from the task requiring more implicit control.

A final explanation appeals to multiple resource theory (Wickens, 1984). In
this explanation, workload would be worse under conditions of competing tasks,
e.g., two verbal tasks, then under conditions of shared tasks. In terms of multiple
resource models, workload leads to competition for processing demands. For
Tunley (1995) the competition was at the processing stage (as the mental arithmetic
task involved the auditory presentation of digits). For Hapeshi and Jones (1989)
both monitoring recognition feedback and a tracking task used a visual display, so
the competition was at the input stage. While this might provide some initial
explanation of deterioration in the performance on non-speech task, it does not
explain why speech recognition performance deteriorates. Interestingly, in his

research Wickens does not tend to report recognition accuracy figures for the speech conditions in his studies (indeed, it is not clear from his papers that the speech of the people is analysed beyond task outcome measures, i.e., whether the speakers were correct or not). The initial assumption in Tunley's (1995) study was that the speech was not necessarily subject to time-stress; consequently, the changes to speech rate ought to be less apparent. However, for several people participating in the study, a limited 'time-window' was perceived between occurrences of the secondary task. This meant that speakers may have been working to self-imposed time-stress, attempting to work through a number of commands prior to the occurrence of the next secondary task stimulus. While the speech data are still being analysed for this study, it is possible to suggest that time-stress is again a variable in the data.

CONCLUSIONS

From these studies, it is proposed that speaking with the intention of clear communication requires some effort on the part of the user. The initial analysis of speech data from these studies reveals two effects of workload on speaking. The first effect can be predicted from considering speech production as a physical and physiological process. If workload, particularly in the form of time-stress, results in a physical response in the speaker, e.g., in terms of increased breathing rate, changes to the operation of the larynx etc., then one would predict specific symptoms such as a reduction in pausing between parts of words and between words, an increased glottal pulse rate. Inspection of figure 2 suggests that these symptoms can be identified in the speaker who responded to time-stress. The lack of consistent changes in speech level suggests, in part, that the time-stress was not of sufficient magnitude to evoke significant physiological changes, but also suggests some possible feedback or control mechanism by which speakers attempt to adapt their speech to the stress demands.

The second effect can be seen as one of strategy, e.g., speaker DS simply ignored the time-stress component and did not modify his style of speaking. At a less extreme level, speech demands can compete with demands on other task components at both input and processing stages of human information processing. Following multiple resource theory, the allocation of resources in such conditions can become a matter of strategy, providing there is sufficient scope to allow either switching between tasks or parallel operation of the tasks. From subjective impressions gathered during the second study, several participants adopted a strategy of attention switching; thus imposing time-stress on themselves, with similar consequences to those observed in the first study.

REFERENCES

Graham, R. and Baber, C. (1993) User stress in automatic speech recognition In Ed. E.J. Lovesey *Contemporary Ergonomics 1993* London: Taylor and Francis 463-468

Hapeshi, K. and Jones, D.M. (1989) Concurrent manual tracking and speaking: implications for automatic speech recognition In Ed. M.J. Smith and G. Salvendy *Work with Computers: Organizational, Management, Stress and Health Aspects* Amsterdam: Elsevier 412-418

Tunley, C. (1995) *A Comparison of Input Devices for the Proposed Future Fighting Soldier System* Birmingham: The University of Birmingham, School of Man & Mech Eng, unpublished MSc thesis

Wickens, C.D. (1984) *Engineering Psychology and Human Performance* Columbus, OH: Charles E. Merrill

COMMUNICATING WITH COMPUTERS: SPEECH AND PEN-BASED INPUT

Jan Noyes

Department of Psychology, University of Bristol
8 Woodland Road, Bristol BS8 1TN

Initial interest in automatic speech recognition and the recognition of handwriting by computers began in the early 1950s when the first serious research attempts to build recognisers to handle speech and handwriting started. However, it is only within the last couple of decades that developments have resulted in speech and character recognition becoming a practical possibility, and several speech and pen-based systems are now on the market. Despite the many commercial-off-the-shelf (COTS) products currently available, there are still a number of issues requiring further human factors engineering research, and these are outlined here. It is concluded that the application of human factors is likely to be a major determinant in deciding the ultimate success and wider use of these technologies in computer communications.

Introduction

Although much research has been conducted into the development of speech recognition and synthesis technologies, it is only within the last couple of decades that the use of speech I/O devices has moved from the experimental laboratories to operational applications. A similar situation exists with pen-based computing and the recognition of handwriting, and it is only within the last decade that developments have resulted in character recognition becoming a practical possibility. In 1993, for example, there were over 30 commercial pen-based systems on the market (Mezick, 1993).

Despite the recent interest in pen-based personal communicators, the concept of pen-based computing is not new. Pen as an input device for computer systems was considered back in the 1950s and 1960s with one of the earliest devices being the Stylator (Dimond, 1957) followed by the RAND tablet in 1963 (Davis and Ellis, 1964). These early digitising tablets employed a variety of electromagnetic techniques to detect the position of a pen point on a writing surface, and although resolution was fairly poor, attempts to program character recognition was carried out with some success. Since then other techniques have been used for detecting pen input including utilisation of electrostatic properties, pen pressure, and ultrasonics (see, Leedham, 1994). Interestingly, the development of speech input devices follows a similar timeline, with work on Automatic Speech Recognition beginning in earnest in the post-war period. Claims to have developed the first speech recogniser, which had the capability of recognising the 10 digits when spoken singly into the machine, were made in 1952 (Lea, 1980). However, with both recognition technologies, it was not until the 1980s that major technological advances were made. More sophisticated algorithms for recognition, plus enhanced computer processing capabilities accompanied by falling costs and greater use of personal computing facilities, all help explain the expanding interest in speech and pen as a means of human-computer communication.

Perceived benefits

One of the most frequently cited benefits of both speech and pen is their perceived naturalness as a mode of human-computer interaction. (In the case of the latter, the 'pen/paper' or 'electronic paper' metaphor is often used.) Learning to speak and to write are skills which the majority of humans master with little effort, and it is with comparative ease that we use these modes of communication throughout adult life. It is therefore not surprising that the development of computer communication using speech and pen is frequently viewed as a natural progression in human-machine interaction. The familiarity, ubiquity and convenience of speech and pen allow communication in the user's language not the machine's, which may be particularly relevant in the case of advanced technologies where control has previously been through the medium of a programming language. The use of a 'natural language' dialogue will reduce the training time needed to interact fully with computer systems, as well as allowing faster response times, which may be particularly relevant in safety-critical systems. However, it could be argued that the naturalness of using speech and pen refers primarily to human to human communications, and may not necessarily be true in the context of human-computer interactions.

When considering the use of speech and pen in the context of system control, other advantages are frequently cited. For example, both recognition modes allow data entry to be direct to a computer system with minimal interruption to the user's primary task (see, Noyes, Baber and Frankish, 1992). Subsequently, data entry can be carried out with greater accuracy, because a step in the data processing cycle has been removed, i.e. collection and input of data occur simultaneously. Speech input is particularly useful in the 'hands/eyes busy' and information overload situations, where other input channels are stretched to capacity or exhausted. In addition, speech recognition devices allow mobility on the part of the user, as it is possible with the use of a radio link to input information to a remote computer system some distance away. Both types of recogniser lend themselves to operation in 'hostile' environments, as they can be portable and engineered for use in the wet or dust.

Current technology

For both speech and pen, there are essentially three types of recogniser: (i) user-independent or user generic, which cannot be trained by the individual so they must adapt to the recogniser; (ii) user-dependent or user specific, which cannot be used until trained by the user, i.e. involving both user adaptation to the recogniser and recogniser adaptation to the user; (iii) hybrid recognisers, which may be trained to a particular user, but may be used by other individuals who have not trained the system.

Speech recognition

Current commercial speech recognisers are predominantly speaker-dependent, i.e. the user must initially train ('enrol') the recogniser with the set of words/utterances they plan to use. This is in contrast to speaker-independent systems, where enrolment is not necessary. The key to successful recognition is 'consistency', since incoming speech (or vocal utterances) is pattern-matched with existing stored templates in order to select the best fit. Problems of detecting the end-points of words and utterances result in 'isolated word' recognition currently being more successful than the recognition of continuous speech, which has been achieved with little success. A final point concerns the vocabulary set being used with the speech recogniser. It is generally accepted that current recogniser technology functions well with vocabulary sets up to 100 words, and some systems will achieve acceptable recognition performance with up to 1000 items. However, this would not be the case, for example, in some word processing applications which would require the use of a natural language vocabulary with the availability of up to 20,000 words.

Pen-based recognition

Pen-based recognisers can generally be divided into two distinct types: those providing on-line recognition and those that operate off-line. On-line recognition results in the handwriting characters being sampled, digitised and processed at the same time as they are being written, i.e. in real-time. Here, an electronic pen or stylus is needed in order to communicate directly with the surface of the display. With on-line recognition, the result of the input will usually appear immediately after completion of a hand-written entry, and consequently, recognition has to match more or less the speed of a person's handwriting. An advantage of this type of recognition stems from receiving writing on-line, as this allows immediate feedback and the acquisition of dynamic information such as pen direction and speed, and the number and sequence of pen-strokes.

In contrast, off-line recognition deals with processing the handwriting after it has been written, and usually involves converting the original source material, e.g. conventionally hand-written documents, to digitised images by means of a scanner. The major difference in terms of processing capabilities is the fact that on-line systems need to operate at the speed of the user, whereas off-line systems are generally intended to operate at speeds determined by the systems in which they are used. In terms of human-computer interaction, on-line recognition is likely to be more appropriate.

Like speech recognition, accuracy will in principle be enhanced by limiting the size of the vocabulary set.

Human factors issues

Recognition performance

The main focus of the human factors considerations of both speech and pen-based computing is concerned with the attainment of acceptable recognition performance. However, it is not realistic to expect recognition systems to achieve perfect recognition, and systems design will always have to take this into consideration. Current recognition technology is not sufficiently advanced to recognise continuous speech or handwriting with any great degree of reliability. This is primarily due to the enormity of the task in the case of speech recognition, and the problems which arise from having to cope with handwriting variability, both within and across the user population, character segmentation, and the general inability of recognition systems to understand context (Martin, Pittman, Wittenburg, Cohen and Parish, 1990). For example, in some speech and automatic handwriting recognition tasks, it is only the context which will allow some spoken items, e.g. 'I scream' and 'ice-cream' and written characters, e.g. 1, I, and l, or o and 0, to be distinguished from others.

Error correction

Given the inevitability of errors, recognition systems need to be designed to be error tolerant with elegant and appropriate error correction procedures (see, Frankish and Noyes, 1990). This includes both errors made by the user for whatever reason, and those created by the system, e.g. time out errors, where the user pauses too long between words or the strokes of a single character leading to misrecognition. In terms of error correction techniques, it is here that the similarity between speech and pen-based systems ends. Due to the ephemeral nature of speech, error correction techniques will be intrinsically linked to the type of feedback provided, whereas in the case of pen-based computing, feedback and error correction will involve identification of incorrectly recognised characters from the feedback display, specification of the corrective action and re-entry of the data (Rhyne and Wolf, 1992).

Feedback

Recognition feedback is necessary in order to help the user identify and understand the cause of errors from their input. In both speech and pen, the timing of the feedback can be either concurrent, i.e. after every item, or delayed, i.e. after a number of items has been entered. The unit of feedback will be dependent upon the task, the unit of recognition, and the probability and criticality of an uncertain result. Concurrent feedback tends to be disrupt carrying out the task, although it does provide immediate information about recognition performance. Terminal feedback, on the other hand, can result in a perceivable delay for the user, whose ability to correlate recognition feedback with input is dependent on either memory, or having the opportunity to view the original input. In the case of terminal feedback, it may also be problematic to make error corrections if the item to be corrected is in the middle of a string, and thus difficult to locate.

Training

Given the need for consistency and the inherent variability in speech and handwriting, this begs the question about whether users can be trained to talk and write in a constrained manner, and in a style that enhances the probability of correct recognition. In the context of character recognition, Ward and Blesser (1985) hypothesised that the requirement for constrained handwriting would result in individuals resorting to old habits sooner than anticipated. Anecdotal evidence also suggests that it would be quite difficult to speak in a constrained manner for any length of time. Looking to the future, the solution may lie in the development of 'intelligent' software, where the system modifies its recognition prototypes over time, as a profile of the individual's speech or handwriting characteristics is collated.

Task design

The nature and characteristics of the task will determine the desired recognition performance. For example, the higher the accuracy requirement the greater the requirement for input feedback, recognition feedback, error correction methods and training. In addition, the individual characteristics of the user, e.g. their experience of using the technology, alternative ways of completing the task, and frequency of use, will all have implications for task design. For example, in a checking/stocktaking task the detection and correction of errors will be essential, since the records need to be accurate and are being used by many individuals. However, personal notes may be more efficiently dealt with using flexible criteria for error recovery as the information is only being used by a single individual, who also generated the items.

Customisation of system interfaces

We know that various types of users require different interfaces, thus the designer is left with a number of decisions, e.g. tailor the interface to the user group, design the interface around all groups, or design customisable interfaces (Potosnak, Hayes, Rosson, Schneider and Whiteside, 1986). Given that no one user group shares the same speech and handwriting patterns, customisable interfaces tailored to meet the needs of individual users would seem to provide part of the solution for coping with the inherent variability exhibited by users of recognition systems.

Conclusions

With judicious and careful task selection, it is possible for both types of recognition systems to be used successfully, and indeed, there are speech recognisers which have been operational in industrial applications since the 1970s (Martin, 1976). However, it is evident that before speech and pen recognition technology will become more widely used in human-computer interaction, further developments are needed. Two major concerns emerge from the preceding discussion: one is the extent to which speech and pen-based recognition map onto user expectations emanating from the perceived 'naturalness' of these devices, and secondly, the extent to which inadequate levels of recognition performance will undermine user interactions with these recognition systems, especially if they have unrealistic expectations and are untrained. Finally, it is evident that speech and pen-based input will not provide the full solution for human-computer interaction, but that there are activities and applications for which they would be suitable and effective. Careful systems' design and human factors engineering would allow some of these to be implemented immediately, e.g. those applications which can utilise the benefits of speech or pen, but are not dependent on the attainment of a very high level of recognition performance.

References
Davis, M.R. and Ellis, T.O. 1964, The RAND tablet: a Man-machine graphical communication device. In *Proceedings of Fall Joint Computing Conference*, 325-331.
Dimond, T.L. 1957, Devices for reading handwritten characters. In *Proceedings of Eastern Joint Computing Conference* , 232-237.
Frankish, C.R and Noyes, J.M. 1990, Sources of human error in data-entry tasks using speech input, Human Factors, **32**, 697-716.

Lea, W.A. 1980, Speech recognition: past, present and future. In W.A. Lea (ed.), *Trends in speech recognition*, (Prentice-Hall, Englewood Cliffs, NJ) 39-98.

Leedham, C.G. 1994, Historical perspectives of handwriting recognition systems. IEE Digest No. 1994/065: Handwriting and Pen-based Input. (Institution of Electrical Engineers, London).

Martin, G., Pittman, J., Wittenburg, K., Cohen, R. and Parish, T. 1990, Sign here, please. Byte, July Issue, 243-251.

Martin, T.B. 1976, Practical application of voice input to machines, In *Proceedings of the IEEE*, **64**(4), 487-501.

Mezick, D. 1993, Pen computing catches on. Byte, October Issue, 105-112.

Noyes, J.M., Baber, C. and Frankish, C.R. 1992, Using automatic speech recognition in industrial applications, Journal of the American Voice Input/Output Society, **12**, 51-68.

Potosnak, K.M., Hayes, P.J., Rosson, M.B., Schneider, M.L. and Whiteside, J.A. 1986, Classifying users: a hard look at some controversial issues, In *Proceedings of CHI '86*, (Human Factors Society, CA) 84-88.

Rhyne, J.R. and Wolf, C.G. 1992, Recognition based user interfaces. Tech. IBM Research Report RC 17637.

Ward, J.R. and Blesser, B. 1985, Interactive recognition of handprinted characters for computer input, Journal of IEEE CG & A, September Issue, 24-37.

General Ergonomics

MEMORY PATCHING AFTER TECHNICAL INSTALLATION MISHAP: AN OPPORTUNITY FOR ERGONOMISTS?

Tay Wilson* and Jennifer MacFarlane**

*Psychology Department
Laurentian University
Ramsey Lake Road
Sudbury, Ontario, Canada
P3E 2C6
fax (705) 675-4823
**Psychology Department
Lakehead University
Thunder Bay, Ontario, Canada

When accident or breakdown afflicts large technical installations, it is often important, for both legal and ameliorative design reasons, to extract the best possible memory record of salient events from diverse groups of involved principals and patch it together in the most accurate and useful manner. Moreover, there is now a considerable body of empirical knowledge concerning memory completeness and accuracy under different evoking conditions in the applied and cognitive psychology literature. It is argued here that a good opportunity exists for some ergonomists to specialize in the provision of memory patching services, in such circumstances, first, by developing recall and patching protocols based on the current state of knowledge in the field; and second, by being ready and willing to go to accident/breakdown sites to carry out the necessary work. An introductory outline of such a protocol based upon some findings in the literature is presented here.

When accident or breakdown afflicts large technical installations, it is often important, for both legal and ameliorative design reasons, to extract the best possible memory record of salient events from diverse groups of involved principals and patch it together in the most accurate and useful manner. Because there exists a considerable body of empirical knowledge concerning memory completeness and accuracy under different evoking conditions, a good opportunity exists for some ergonomists to specialize in the provision of memory patching services, based upon the development of useful recall and memory patching protocols from this knowledge. The problem in developing such protocols is to avoid being drowned in the detail of thousands of empirical findings. To avoid drowning, it is probably better first, that those who engage in such protocol development be arm's length from those engaged in actual experiments in the field and second, that relatively few key memory results be incorporated in early protocols. In order to initiate this development process, there is here presented ten candidate findings about which to develop such a protocol.

Ten Candidate Memory Patching Protocol Building Blocks

There follows ten candidate memory patching building blocks, each accompanied by key reference(s) and/or finding(s) (see Eysenck and Keane, 1993 or Baddeley, 1990 for general discussion). First, consider the so-called amnesic syndrome characterized by normal level intelligence, relatively intact immediate and short term memory, a capability for normal conversation, anterograde amnesia - difficulty in learning new information after onset of amnesia- and, frequently, retrograde amnesia - difficulty in remembering events occurring prior to amnesia (Parkin, 1990). Second, consider chronic alcohol abuse amnesiacs (e.g., Korsakoff's syndrome), who, in addition, may have a poor recognition memory for words and perhaps pictures presented a day before (Crovitz, Harvey, & McClanahan, 1981; Cermak, Talbot, Chandler, & Wolbarst, 1985). Third, consider that, even though amnesiacs can have poor recognition memory, they may have reasonable perceptual memory - ability to detect as a word, a briefly presented display - and, even more interestingly, repetition priming effects - increased speed in finding target figures in complex cartoons in the absence of recognition of having seen such cartoons within the last seven weeks whereas the same finding for "normals" requires an interval of 17 months (Meudell & Mayes, 1981). Fourth, consider episodic and semantic memory differences (Tulving, 1972, 1983). Episodic memory has a autobiographical flavour involving storage of particular events occurring at specific times and places while semantic memory contains organized knowledge about words, rules, their meanings, referents, and relations. Amnesiacs frequently have only fair episodic memory for events occurring before amnesia onset and poor episodic memory for events occurring after onset combined with good pre-onset semantic memory and poor post-onset semantic memory (Gabrielli et al., 1983; Zola-Morgan et al., 1983). Some Korsakoff patients have been shown to have little memory for temporal context of pictorially presented material - they are unable to distinguish pictures were shown to them on day 1 from those shown to them on day 2 - (Huppert & Piercy, 1976). Fifth, consider differences in declarative memory - *knowing that* and procedural memory - *knowing how*. Amnesic patients may have severe impairment of declarative learning/memory combined with intact procedural learning/memory (Squire & Cohen, 1984). Sixth, consider differences in explicit memory - where successful task performance requires conscious recollection of previous experiences - and implicit memory - where successful task performance does not require such conscious recollection

(Schachter, 1987). Interestingly, on word lists, amnesiacs have been found to be much worse that controls on free recall, cued recall, recognition memory but not on word completion tests of memory (Graf, Squire, & Mandler, 1984). Word completion involves presenting the first three letters of a word, asking subjects to say what ever words come into their heads and recording how many emitted words are from the originally presented list.

Seventh consider the memory recency effect - that items occurring at the end of a presented list are remembered better, soon after presentation, than those in the middle of lists. Recency can be truncated in certain amnesic individuals (Shallice & Warrington, 1970); moreover, short term memory deficits in amnesic individuals have often been shown to be modality (auditory or visual) specific (Shallice & Warrington, 1974). Eighth, consider the Baddeley and Hitch (1974) three component model of working memory (modality free limited capacity central executive resembling attention, articulatory loop of about two seconds holding information in a speech based form and a visuo-spatial scratch or sketch pad specialized for spatial and/or visual coding). In particular, articulatory suppression (repeatedly uttering a phrase such as "hi ya" while carrying out a task has little effect on processing time or errors, in deciding truth of simple sentences or in deciding upon essential identity of meaning of two different paraphrases of reading material (Baddeley, 1979; Levy, 1978) but it does diminish performance on tasks involving assessment of correct word order (Baddeley & Lewis, 1981). The important notion that the visual sketch pad is spatial - giving relationships between entities - rather than pictorial is supported by the finding that a specifically visual task (judging light brightness) does not interfere with remembering exactly an easily visualizable auditorily presented matrix of digits, while a spatial-visual task (visually tracking a light moving on a circular track does so interfere (Baddeley, Grant, Wight, & Thomson, 1977).

Eighth, consider effects on memory of type and depth of processing of the material to be recalled during presentation, even when no intention to learn material is evinced (Hyde & Jenkins, 1973; Craik & Lockhart, 1972). In particular, incidental learners (not instructed to memorize material) who examine the meaning (semantics) of material perform as well as intentional learners (instructed to memorize material) who are given no instructions as to how to process material or who are instructed to process material for non semantic aspects (e.g., part of speech or presence of vowels). Memory may also be greater for material presented in complex as opposed to simple sentences. Ninth consider effects of type and context of memory retrieval task. The retrieval of memory from an individual depends upon the nature of the memory task given and in particular how similar the memory task is to learning conditions, i.e., context dependent memory and encoding specificity (Morris, Bransford, & Franks, 1977). Intrinsic context - having direct effect on meaning of to be learned material, e.g., "jam" for strawberry - has different effects on recall and recognition than extrinsic context. In particular, recognition memory may not be affected by similarity of extrinsic context in learning and memory testing whereas, recall is noticeably better when extrinsic conditions are the same in both cases (Godden & Baddeley, 1980). Generation recall techniques - in which individuals are asked to generate as many words as possible to a presented category and then to recognize which words were on an earlier presented list - produce additional correct items compared with mere recall of to be learned material (Rabinowitz, Mandler, & Patterson, 1977). Tenth, consider eye- and ear-witness memory. Turtle and Yuille (1994) report that hypermnesia (recall of more information over successive attempts) is dependent upon recall procedure. In particular, in multiple recall attempts, total number of details recalled at least once, independent of whether subjects are allowed to review their earlier statements increase, total errors increase, but less quickly than total details. However, net number of details correctly recalled decrease. Overall, hypermnesia over trials is seen when subjects are given limited time (seven minutes) to recall on each trial. MacFarlane and Wilson (1995) found, when misleading statements are introduced into the recall process, that confabulation - making up details - increases over later repeated trials. Yarmey (1992), in studying earwitness recall, found that longer description of events when individuals wrote alone after prior two person discussion of events to be recalled; moreover, event duration were relatively over-estimated by individuals whose recall is delayed. In eyewitness research Yarmey (1993) found low correlations between confidence in and accuracy of recall detail. One might conclude by noting that the frequently accepted doctrine of concordance - that there is close agreement between what people know, how they behave and what they experience - is wrong in many instances (Tulving, 1989).

A First Generation Memory Patching Protocol

Here is a generation one memory patching protocol based upon the above material. Let us consider that, for example, a major natural gas plant has exploded and the ergonomists job is, by trying to patch together a collective memory of what happened, to recommend ameliorative practices. Assume, with Meudell and Mayes (1981) that under certain circumstances, normals exhibit many memory problems common to amnesiacs. Get memory statements from as many relevant people as possible, as soon as possible. If practicable, use intuition or other means to sort out key contributors to the story who might exhibit alcohol, stress, or otherwise induced amnesic memory - pay relatively little attention to ability to carry on bantering conversation in this assessment. Provide this group with various pictorial scenarios filled with detail, relevant to the explosion which might be identified more quickly by that group than other neutral details, even though they say they have no memory of it. Give relatively little weight to statements about key episodic details for this group and relatively more weight to their pre-explosion semantic information. If a potential amnesic is judged to be a important contributor to reconstructing memory, try some version of word completion, with a variety of suitably chosen beginning letters. Present material to the "amnesic" group in both auditory and visual form since the deficit could be modality specific. Separately try to obtain contents of spatial and visual memory; note particularly circumstances which might be assumed to interfere with the former. Try to tap procedural memory in a separate manner by getting individuals to carry out key actions without "mouthing" them (without tapping semantic memory). Try the generation technique in which individuals are asked to simply say as many words of a particular category as possible and then later to try to recognize those which are salient to the explosion. Use multiple recall attempts, on a short time line, if possible, to add to the stock of items, while being aware that net recall might decrease, confabulation increase and while discounting individuals confidence that what is reported really happened.

References

Baddeley, A. 1979, Working memory and reading. In P.A. Kolers, M.E. Wrolstad and H Bouma (Eds) *Processing of visible language.* (Plenum, New York).

Baddeley, A. 1990, *Human Memory: Theory and Practice.* (Lawrence and Erlbaum, Hove).

Baddeley A. and Hitch, G. 1974, Working memory. In G. H. Bower (Ed.) *The Psychology of Learning and Motivation, Vol 8.* (Academic Press, London).

Baddeley, A. Grant, S., Wight, E., and Thomson, N. 1975, Imagery and visual working memory. In P.M.A. Rabbitt and S. Dornic (Eds.) *Attention and Performance, Vol V.* (Academic Press, London).

Baddeley A., and Lewis, V.J. 1981, Inner active processes in reading: The inner voice, the inner ear and the inner eye. In A. M. Lesgold and C. A. Perfetti (Eds.) *Interactive Processes in Reading.* (Lawrence Erlbaum, Hillsdale, NJ).

Cermak, L.S., Talbot, N., Chandler, K., and Wolbarst, L.R. 1985, The perceptual priming phenomenon in amnesia. *Neuropsychologia, 23,* 615-622.

Craik, F.I.M. and Lockhart, R.S. 1972, Levels of processing: A framework for memory research. *Journal of Experimental Psychology: General, 104,* 268-294.

Crovitz, H.F., Harvey, M. T., & McClanahan, S. 1981, Hidden memory: A rapid method for the study of amnesia using perceptual learning. *Cortex, 17,* 273-278.

Eysenck, M. W. and Keane, M.T. 1993, *Cognitive Psychology: A Students Handbook.* (Lawrence and Erlbaum, Hove).

Hyde, T.S. and Jenkins, J.J. 1973, Recall for words as a function of semantic, graphic and syntactic orienting tasks. *Journal of Verbal Learning and Verbal Behaviour, 12,* 471-480.

Gabrielli, J.D.E., Cohen, N.J., and Corkin, S. 1983, The acquisition of lexical and semantic knowledge in amnesia. *Society for Neuroscience Abstracts, 9,* 238.

Godden, D. and Baddeley, A. 1980, When does context influence recognition memory? *British Journal of Psychology, 71,* 99-104.

Graf, P., Squire, L.R., and Mandler, G. 1984, The information that amnesic patients do not forget. *Journal of Experimental Psychology: Learning, Memory, and Cognition, 10,* 164-178.

Huppert, F.A. and Piercy, M. 1976, Recognition memory in amnesic patients: Effect of temporal context and familiarity of material. *Cortex, 4,* 3-20.

Levy, B.A. 1978, Speech analysis during sentence processing: Reading and listening. *Visible Language*, 12, 81-101.

Meudell, P.R. and Mayes, A.R. 1981, The Claparede phenomenon: A further example in amnesiacs, a demonstration of a similar effect in normal people with attenuated memory, and a reinterpretation. *Current Psychological Research*, 1, 75-88.

MacFarlane, J. and Wilson, T. 1995, Misleading evidence and confabulation in eyewitness reports. In preparation.

Morris, C.D., Bransford, J.D., and Franks, J.J. 1977, Levels of processing versus transfer appropriate processing. *Journal of Verbal Learning and Verbal Behaviour*, 16, 519-533.

Parkin, A.J. Recent advances in neuropsychology of memory. In J. Hunter and J. Weinman (Eds.) *Mechanisms of Memory: Clinical and Neurochemical contributions.* (Harwood, London).

Rabinowitz, J.C., Mandler, G., and Patterson, K.E. 1977, Determinants of recognition and recall: Accessibility and generation. *Journal of Experimental Psychology: General*, 106, 302-329.

Shallice, T. and Warrington, E.K. 1970, Independent functioning of verbal memory stores: A neuropsychological study. *Quarterly Journal of Experimental Psychology*, 22, 261-273.

Shallice, T. and Warrington, E.K. 1974, The dissociation between long-term retention of meaningful sounds and verbal material. *Neuropsychologia*, 12, 553-555.

Tulving, E. 1972, Episodic and semantic memory. In E. Tulving and W. Donaldson (Eds.) *Organization of Memory.* (Academic Press, London).

Tulving, E. 1983, *Elements of Episodic Memory.* (Oxford U. Press, Oxford).

Turtle, J. W. and Yuille, J. C. 1994, Lost but not forgotten details: Repeated eyewitness recall leads to reminiscence but not hypermnesia. *Journal of Applied Psychology*, 79(2), 260-271.

Yarmey, A. D. 1992, The effects of dyadic discussion on earwitness recall. *Basic and Applied Psychology*, 13(2), 251-263.

Yarmey, A. D. 1993, Adult age and gender differences in eyewitness recall in field settings. *Journal of Applied Psychology*, 23(23), 1921-1932.

Zola-Morgan, S., Cohen, N.J., and Squire, L.R. 1983, Recall of remote episodic memory in amnesia *Neuropsychologia*, 21, 487-500.

ERGONOMI**X** FILES - THE TRUTH IS IN THERE?

Mic L. Porter

Department of Design,
University of Northumbria at Newcastle,
Ellison Place, Newcastle-upon-Tyne,
NE1 8ST

What is captured within a photograph? How, and by whom, should such information be extracted and interpreted? Conference attendees will be invited to consider and discuss a selection of images. They may wish to consider the photographs in terms of the ergonomic questions illustrated, as celebrations of people undertaking useful work, or just simply respond to them as historical records of a moment in the shift of a person at work.

Introduction and context

One hundred and seventy years ago in 1826 Nicéphore Niepce, made the first recorded photograph. The exposure time of 8 hours would, however, limit such images to landscapes and architectural subjects. In 1851 the "wet plate" process made it possible to make negatives and photographs in daylight with exposures of less than a minute. The technology demanded prompt attention to the image before the plate dried and thus mobile coating/developing wagons were developed. Roger Fenton took his horse drawn "Photographic Carriage" to photograph the Crimean War (1854-6) and Mathew Brady recorded the American Civil War (1861-5) with similar equipment. The exposure time, now down to 10/15 seconds, virtually restricted these pioneers to static subjects. However, their staged, behind the lines, images of troops and gun crews or their recording of the aftermath of battle, the dead and the destruction, can be as disturbing as those taken on roll film 80-100 years later.

Photographs - The X factors

The photographers of our twentieth century wars, conflicts and police actions, Robert Capa, Larry Burrow, Lee Miller, Don McCullin, George Roger, Weegee and so many others, had the advantage to stop action as film speed increased and instantaneous lighting became portable. These attributes were then packaged and made available to all. Organizations could record their successes, for example the building of HMS Sans Pareil

(1857) or hand assembling "Standard" motor cars early this century (Royal Commission 1985). The endings of eras were also captured as people observed, and walked by, soon to be forgotten, activities. Macdonald & Tabner (1986) recorded the last days of Smiths Dock and Forsyth (1986) and Konttinen (1983) the last working Tyneside communities. Were these images made and reproduced as a matter of record only or to support debate and argument? This presenter was, by the 1980s, using fast film to support his activities.

"Photographs furnish evidence. Something we hear about, but doubt seems proven when we're shown a photograph of it. In one version of its utility the camera record incriminates... In another version of its utility, the camera record justifies. A photograph passes for incontrovertible proof that a given thing happened. The picture may distort; but there is always a presumption that something exists, or did exist, which is like what's in the picture." (Sontag 1978)

If the photograph captures all that is apparent in an instant in time then the true interpretation of that photograph, to reconstruct the original reality, requires time and experience. Ergonomists adopt a dynamic viewpoint, they need to view the movement frozen in the image. Captions and titles can provide orientation and context but so can the experience that the photographer incorporated into the image. Later the viewer adds their experience and understanding. Ergonomists viewing images made by a Consultant Ergonomist in pursuit of his work can be expected to focus on different aspects than, for example, fellow workers of the subject or consumers. What is the correct interpretation, what is the truth of the situation? These questions are asked when viewing a dramatic newspaper photograph or by managers and judges looking at the work of an ergonomist.

When a continuous cycle of events associated with, for example, the word processing of a document by a keyboard operator, is stopped, split into a series of discrete images and attention focused on one. Is this truth or is the image, shown without context, the sub-truth that we all associate with phenomenon we do not understand. Do we ergonomists have the confidence and ability to interpret images of work or are we just able to understand some of the artefacts of real life? What are the "X" factors? What is truth?

"All photographs are ambiguous. All photographs have been taken out of a continuity.... Yet often this ambiguity is not obvious, for as soon as photographs are used with words, they produce together an effect of certainty, even of dogmatic assertion.... The photograph, irrefutable as evidence but weak in meaning, is given a meaning by the words. And the words, which by themselves remain at the level of generalisation are given specific authenticity by the irrefutability of the photograph. Together the two then become very powerful; an open question appears to have been fully answered." Berger and Mohr (1982).

References

Berger, John and Mohr, Jean, 1982, *Another Way of Telling,* (Writers and Readers Publishing Co-operative Society, London).
Forsyth, Jimmy, 1986, *Scotswood Road,* (Bloodaxe Books, Newcastle-upon-Tyne).
Konttinen, Sirkka-Liisa, 1983, *Byker,* (Jonathan Cape, London).
Macdonald, Ian and Tabner, Len, 1986, *Smith's Dock,* (Seaworks, Newcastle-upon-Tyne).
Royal Commission on the Historical Monuments of England, 1985, Industry and the Camera, (HMSO, London).
Sontag, Susan, 1978, *On Photography*, (Allen Lane, London).

Managing Human Factors

David Collier

GreenStreet Consultancy Ltd.
Bramley Cottages, Claverton Down Rd.
Bath   BA2 7AP
Email 100043.3131@compuserve.com

The UK nuclear industry, in partnership with the HSE Nuclear Installations Inspectorate, funds a joint human factors research programme. This paper briefly summarises some of the results from a study into the management of specialist human factors resources in high hazard industries carried out as part of the 1994/95 programme. It included a literature survey, interviews and an organisational analysis.

Introduction

Human factors professionals have to spend their time on the right tasks and be involved at the most appropriate stage in projects, if they are to be used effectively within the safety management system. They must have the right balance of skills and be adequately resourced. The safety management system must recognise the fundamental importance of human factors issues at all levels, and provide a formal framework for raising awareness, monitoring performance, addressing shortcomings and involving specialist expertise.

This is self-evidently true in principle, but it is not easy to achieve in reality. The UK nuclear industry has a good record in the application of human factors in safety management, but improvement is always possible and the current work was initiated to see whether lessons could be learned from other industries. Over 90 'good practices' were identified under the following headings: human factors input to safety management; developing human factors awareness; research and current awareness programmes; satisfying the regulator; internal consultancy services; professional development; use of external resources and organisational structure. This paper briefly summarises some of the results.

Developing HF Awareness

The development of human factors awareness within the organisation is essential if safety management is to function effectively, and it was seen as a key task by most human factors group heads. A reasonable level of general awareness is necessary to ensure that the need for more detailed human factors work is recognised and human factors professionals involved at the right stage in projects. It is also essential to the survival of human factors as a separate discipline. Some recommendations are given below.

- Prerequisites for a successful human factors awareness-raising programme include a champion at senior management level - perhaps just below Board level; an absence of strong opposition and continuity in case the champion moves on.

- Middle management always has most to lose from change, and must be won over; end users are often more aware of human factors issues than their line managers.

- The difficulties of raising awareness across a broad front need to be balanced against the need to attain a 'critical mass', particularly at management level.

- Ensure that initiatives contribute to the empowerment of the workforce.

- The need for greater awareness must be made obvious to the target audience, in terms they can understand and relate to.

- Without good corporate communications, any awareness campaign will fail.

- Things must be kept simple. Anything that cannot be sustained in the face of competing demands should not be started.

- The infrastructure must be in place to support those putting awareness into practice.

- Crude measures of human factors awareness need to be developed to monitor progress.

It may be better to build human factors awareness into existing culture change initiatives and then explicitly maintain it. Many of the problems tackled by local quality teams will have human factors origins and quality management programmes provide a ready-selected 'sharp end' group of people running problem-solving groups who might benefit from additional knowledge, and often a well organised training framework.

People must be encouraged and empowered to reject work places or equipment with poor human factors. This is difficult to achieve in practice, not least because end users seem to become tolerant of inadequate interfaces (for instance) and do not always appreciate the scope for improvement. They are particularly resistant to improvements that originate within HQ functions. Users can be provided with guidelines or advice to help identify shortcomings and case histories used to illustrate good practice.

Management awareness and support is the key to any successful awareness programme and to the implementation of human factors within the safety management structure. They control the resources and are the dominant influence on organisational culture. The benefits of human factors need selling in terms which the organisation understands and values, so indicators must have face validity and be seen to be relevant to commercial performance.

Internal Consultancy

Whether or not the organisational structure gives them a formal internal consultancy function, human factors groups often have operate in that mode and need the appropriate consultancy and networking skills; just being technically proficient will not be sufficient. Good practices have been identified in: consultancy disciplines; developing human factors as a discipline; internal marketing of human factors services; networking with client groups; quality and customer focus; project and business management.

Products

Many people have difficulty understanding exactly what it is that a human factors professional does, that other engineers cannot. They have difficulty appreciating how the human factors group would fit into the project framework and therefore see the human factors input as being difficult for the project to manage. One way to combat this may be to carefully define a small range of standard 'products' or services whose importance to the organisation is clear. Some organisations have developed checklists which allow people to pick out the human factors issues to be considered and help them decide where they need specialist input. Formal procedures which fulfil the same function exist eg. Def Stan.

Networking

Long term relationships with client groups and related disciplines have to be developed if a human factors group is to make a consistent input to safety management. Successful groups seem to need someone with credibility within the organisation who can build up a reasonably high personal profile and win new internal clients in areas where there ought to be a professional human factors contribution. Charisma is not essential, but vision and enthusiasm probably are. The ability to understand senior management concerns and demonstrate the relevance of human factors is another essential attribute.

Professional Development

Human factors groups need to keep their expertise at a high level, monitor developments in the field and maintain an up to date knowledge base. Survey results reinforce the importance of continuous personal and professional development , but they also reveal that there is scope for improvement. Conferences provide some opportunities, but they are expensive and infrequent, and the topics cannot be chosen to fit in with project or personal needs. A structured professional development programme is essential.

Some human factors professionals are happy to remain within the discipline if the grading structure recognises technical expertise, but others do worry about a lack of wider career opportunities. Valuable human factors skills and experience need to be retained in the company, even if they are being applied in a different context. Reasonable aspirations need to be met if the discipline is to remain healthy. To ensure that professionals are developed, they might be given the option of secondments outside the human factors group, so that in time they can move on within the organisation if they wish.

The insight into human behaviour gained from experience in a broad-scope human factors group should stand the individual in good stead further up the ladder. This suggests giving priority to raising the human factors awareness of managers who act as mentors within the organisation and taking promising operations or safety management staff on secondment.

Organisational Structure

Centralisation

Most human factors groups seem to change models from time to time, but the central group model seems to be the usual starting point, perhaps building from a core area such as MMI, reliability engineering or workplace ergonomics. The centralised model offers benefits in some contexts:

• A multi-disciplinary team can be supported and can address a wide range of tasks;
• Professionals can more easily be recruited and developed;
• Company wide problems can be detected and longer term priorities addressed;
• Solutions can be transferred more easily from one application to another;
• Human factors can be co-ordinated and prioritised more easily;
• Flexibility is improved, and short tasks can easily be accommodated;
• The group is large enough to support a human factors awareness programme;
• Human factors resources is better protected from diversion.

A larger group is likely to have more 'clout' and have higher level representation. It will also be better able to justify the expense of databases and specialist facilities. There seems to be a consensus that the smallest size for a viable central group is three professionals. Below three, the benefits will not be realised; above ten, the group may be at risk of fragmentation or losing its flexibility.

Advantages claimed for a more decentralised model include:

• The work more obviously supports operational activities and ownership is improved;
• Resources and skills can be placed where they are most needed;
• The isolation from the sharp end inevitable in centralised groups is reduced;
• There is more awareness of application-specific issues;
• Tensions within the human factors group are reduced;
• Budgets are small and more easily defended.

Greater productivity is claimed for decentralised arrangements and it is sometimes easier to keep the operational functions and the human factors resources aligned on the same business objectives. Small groups can work together to get most of the benefits claimed for the centralised group, providing that they are willing to co-operate and compromise for the common good. However human factors may be weakened as a discipline under these circumstances. The most stable arrangement in many contexts may turn out to be a strongly networked community with independent embedded groups of 3-4 human factors professionals working in different application areas.

Matrix management attempts to retain the benefits of size and professional interaction, whilst giving functional control to line management to maintain business focus. From a management point of view, there are considerable attractions to wider application of the matrix model, but these advantages again come with matching drawbacks. Some degree of implicit or explicit matrix management will be inevitable, but it requires excellent communications, as well as trust and a culture of cross-boundary working. On large projects, most sources recommend integrating the human factors effort with the project team, supported by central or external resources as needed.

Peer Group Networks

No human factors group can undertake all the human factors work required in a large organisation. A human factors community networking the human factors professionals with groups and individuals doing related work needs to be built up. Building management consensus on the need for action on human factors issues can be difficult, and networking between human factors professionals and related disciplines plays an essential part in the process. Individuals with large personal networks are said to live longer than those with small networks. The same is probably true for human factors groups.

The following principles suggested directly and indirectly by the survey should be borne in mind when trying to establish co-operation between groups which may previously have been competing with each other:

- There must be mutual benefits;
- Repeated interaction breeds co-operation;
- There must be commitment to a long term relationship;
- There must be an expectation of progressively stronger ties;
- 'Business' relationships need to be blended with social and personal ties;
- Relationships must be flexible.

Networks are organic and do not exist to impose co-ordination. A central group may provide a useful focus and help drive a peer group network, but experience from survey companies suggests that dedicated co-ordination groups are a bad idea.

Human Factors 'Champions'

Most human factors groups want to look beyond the peer group networking discussed above, and the concept of volunteer *'human factors champions'* within operations and maintenance functions is attractive. Where a team's role includes human factors related work, at least one team member is trained in the human factors basics relevant to that team. These individuals are then encouraged to promote human factors awareness within their team and provide some routine human factors input.

Human factors professionals act as an involved consultant where specialist input is really required, rather than contributing directly. Many engineers already have an intuitive grasp and enthusiasm for human factors that can be built on, but they have to be identified, trained and supported. There is clearly scope for building on existing quality improvement structures and processes.

Location within the Organisation

All the respondents said that they were happy where they were, whether it was with operations, engineering or safety. Perhaps there really is no one right place to put a central human factors group, or perhaps discretion prevailed. Each location brings advantages and problems, but the organisation has to be designed to reflect customer/client relationships and to optimise communications.

Conclusions

As well as commending the examples of good practice to human factors groups, the study suggested that high hazard industries might consider a number of specific initiatives:

- A systematic review of potential human factors application areas, to determine whether additional human factors input would improve safety management;

- A review of human factors awareness programmes to determine, in particular, whether quality improvement groups should be given additional training;

- Targeted measures intended to raise management awareness of human factors issues;

- A review of the human factors research programme to date and the extent to which results have been implemented - and to what effect;

- A review of the potential for including the need for human factors input in control procedures, and the development of checklists to flag the need for specialist input;

- The development of a defined set of that human factors group products and services;

- An assessment of the extent to which human factors groups should develop an internal consultancy role and whether they should more overtly adopt consultancy disciplines;

- The development of structured human factors professional training programmes;

- A review of options for strengthening peer and extended networking within the organisation, and externally between human factors groups.

The consultants and sponsors would like to take this opportunity to express their gratitude to the survey participants for their support and advice.

NEURAL NETWORK INVESTIGATION OF POSTURE AND MOTION

Samuel Y.E. Lim, S.C. Fok & Irving T.Y. Tan*

School of Mechanical & Production Engineering
Nanyang Technological University
Singapore 639798

* Mobil Oil Singapore Pte Ltd
18 Pioneer Road
Singapore 628498

Neural network has gained much attention in the last decade as an effective artificial intelligent technique. This paper examines the potential of neural network analysis to predict the range of anatomical joint motions for the design/layout of workstation and tasks. Simulated assembling tasks were carried out on a custom-built multi-adjustable workstation, and the posture and motion data were recorded with a flexible electrogoniometric system. Prediction of joint motions was analyzed using a feedforward back-propagation neural network. The trained neural network was capable of memorising and predicting the maximum and minimum angles of joint motions associated with a range of workstation configurations. The average prediction accuracy was found to be around 10°.

Introduction

The conventional approach for improving productivity is through the use of motion time studies. These techniques attempt to maximise the output by using standard time established for each type of work element to determine the layout of the workstation. Although improvements have been made to these assessment methods, such as with anthropometric considerations (Ho and Lim, 1990), the comfort, health and safety of the users are still not adequately addressed.

Motion time studies may allow the workforce to achieve short term productivity goals, however, in the long term it would be detrimental for the operators concerned. They may be forced to adopt undesirable postures, since motion time study is primarily output oriented, with no emphasis on the postural conditions of the users.

User Centred Approach

This paper presents a user-centred approach for the design of workstation. Its aim is to develop a method whereby the design and/or specification of a workstation/task will be based on the comfort, health and safety of the user, rather than on how fast the operator can complete a certain task. This approach is potentially beneficial for operations that require repetitive motions, such as in manipulative and assembling tasks. Undesirable workstation/task configurations can induce awkward postures, which may result in work-related disorders. Jung, Kee and Chung (1995) reported an inverse kinematic method for predicting the reach envelop for the upper limbs. However, the limitation of this approach is that its motions are based on a robotic type of linkage system.

In this approach, layout of workstations and/or configuration of tasks will be based on the criteria of comfortable reach, optimum range of motion, and a balanced posture for the operator. This project explores the ability of neural networks in predicting optimum layout for designated tasks and user profiles.

Neural Network

Artificial Intelligence

Neural network has gained much attention in the last decade as an effective artificial intelligent technique (Simpson, 1990). It uses a system of processing elements (nodes), similar to neurons in the central nervous system, to emulate the biological organization and thinking ability of the brain. These nodes are interconnected to form a network which is capable of identifying patterns of the data by learning from the "experience" that it had been subjected to. A typical neural network is shown in Fig. 1, which consists of a number of input and output channels. In between the inputs and the outputs are the hidden layers, where prediction (output) based on the past experience (input) is processed.

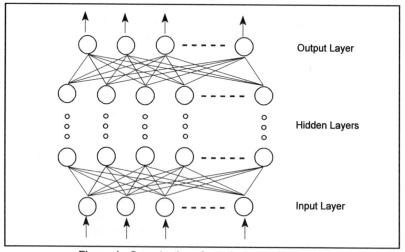

Figure 1. Organisation of a typical neural network.

Each node is fully connected to the nodes in the layers immediately preceding and following it. Thus the output from each node is influenced by the different degree of interactions (weighed sum) among the processing nodes. With this the network learns to reduce its prediction errors based on past events. It can generalize from past experience, recall knowledge from memory, and abstract essential information from noisy inputs.

Neural Network Learning Process

Unlike conventional design methods, neural network does not require a specific "solution" to be composed for each aspect of the problem. The designer only needs to specify a few fundamental variables, and provides it with sufficient raw data associated with the problem.

The behaviour of a neural network depends on its overall architecture and the arrangement of its processing units. "Learning rules" are specified to enable the system to learn and gain knowledge from the characteristics of the input data. In a supervised learning process, the network is trained with both input and output values. Whereas in unsupervised learning, the network receives only the inputs. Back-propagation of error paradigm is the most widely used learning method for the multi-layered feedforward network. A feedforward network consists typically of at least one hidden layer as illustrated in Figure 1.

Application in User-Centred Design

The artificial intelligence of neural networks has wide potential applications in the design of products, workstations, and tasks. The network can be trained to recognise and generalise data which constitute user-centred design characteristics. The training inputs may consists of anthropometric, posture and motion information (Table 1). A trained neural network could assist designers, engineers, and planners to achieve a more operator-oriented working environment, which will lead to higher productivity.

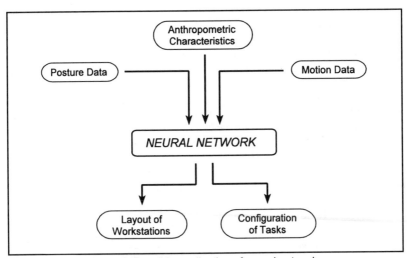

Figure 2. Possible application of neural networks.

Two aspects of user-centred design may be considered for neural network training and applications. Firstly, it may be used in the specification of workstation layout and dimensions. It may also be trained to define configuration of tasks, such as component bin locations, or the sequence of picking and placing operations (Fig 2). For example, if the extreme joint angles associated with a particular configuration can be accurately predicted during the planning stage, the undesirable layout/setup which causes the operators to adopt awkward postures can be eliminated right from the early stages of the design process.

Table 1. Criteria for user-centred design.

Criteria	Application
Reach	Positioning of objects/workpiece from operator.
Range of Motion	Spatial relationship between objects and workpiece.
Posture	Overall layout of task and workstation.

Laboratory Investigation

Experimental Procedure

The potential of neural network in the analysis of posture and motion was investigated with a custom designed height adjustable workstation. A sliding back panel allows the depth of the work surface to be varied from 280 to 750 mm. Horizontal supports on the back panel enable component bins to be positioned within the working envelop. The sitting device provides a vertical adjustment from 435 to 910 mm.

A plain workpiece, which slides along guide rails on the work surface, was used to simulate an assembling task involving simple pick and place operations. Eight component bins were laid out according to 18 data acquisition plans to cover the entire volume of the work space under investigation. Their locations (task layout) are identified with respect to the distance from the origin, which is defined as a point on the near edge of the work surface, facing the mid-sagittal plane.

As the simulated operation may be considered as a somewhat precision task, with interspersed manipulation action, the optimum working height was taken to be 25 mm above the sitting elbow height. For each cycle of the tests, table heights corresponding to high, optimum, and low setup were specified.

Three male subjects (Table 2) between the ages of 23 and 26, were selected from the 5th, 50th and 95th percentile of the Singapore male adult population (Lim, Ngan and Lam, 1989). The subjects are right handed and have no apparent physical impairments. The posture and motion data were collected through two twin-axis flexible electrogoniometers (Penny & Giles 'M' series), linked to a data logger (Penny & Giles DL1001) sampling at 50 Hz, and interfaced with a personal computer. The required training data (Table 3) were subsequently extracted and normalised.

Table 2. Anthropometric data and work surface settings.

	Stature	Percentile	Height of Work Surface		
			High	Optimum	Low
Subject 1	1630 mm	5th %le (1590 mm)	845 mm	735 mm	625 mm
Subject 2	1700 mm	50th %le (1690 mm)	845 mm	755 mm	625 mm
Subject 3	1760 mm	95th %le (1780 mm)	845 mm	775 mm	625 mm

Training of Neural Networks

Normalized experimental data were used to train a feedforward neural network (based on the back-propagation of errors paradigm) to predict the extreme wrist and elbow motions associated with the given bin locations, subjected to table height and anthropometric characteristics (Fig. 3). The networks (NeuralWorks Professional II/Plus) consisting of one hidden layer with 20 to 35 processing elements, were trained to satisfy an RMS error convergence criteria of 0.05. A total of four backpropagation paradigms with 8 input parameters and 6 output parameters were constructed (Table 3). Each were built with different parameters in the number of hidden processing elements and learning schedule (Models A to D).

Table 3. Input and output parameters of the neural network.

Input		Output
Anthropometric Characteristics	• Stature. • Sitting elbow height. • Elbow-fingertip length. • Shoulder-fingertip length.	Wrist: • Ulna/radial angle (max). • Ulna/radial angle (min). • Flexion/extension (max). • Flexion/extension (min).
Workstation	• Height of work surface.	Elbow:
Bin Locations	• x location of bin. • y location of bin. • z location of bin.	• Flexion/extension (max). • Flexion/extension (min).

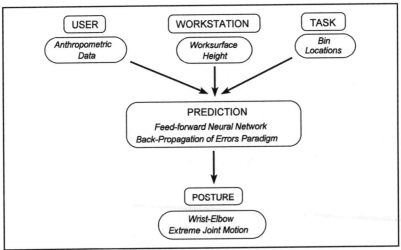

Figure 3. Neural network for prediction of wrist-elbow posture.

Memory of Trained Neural Networks

Memory of the trained networks was assessed by presenting random samples of the training data, and the "predicted" outputs were then verified with the known data. Table 4 shows the results obtained from assessment with 25% of original data. It is evident that the neural networks were quite capable of "predicting" the maximum and minimum joint motions. Model D is the most promising, with an accuracy ranging from 4° to 11° for the wrist, and 11° to 15° for the elbow. The relatively large errors in the elbow angle were due to normalised data at the asymptotic regions of the sigmoid transfer function used in the neural network analysis.

Table 4. Errors of predicted joint angle (in degrees).

Model	A	B	C	D
Wrist: ulna/radial (min)	5.7	5.4	6.4	5.3
ulna/radial (max)	4.2	4.0	5.6	4.0
Wrist: flexion/extension (min)	11.5	11.0	12.7	11.1
flexion/extension (max)	11.9	11.0	13.8	11.0
Elbow: flexion/extension (min)	15.3	15.5	15.7	15.0
flexion/extension (max)	14.2	13.2	14.2	12.6
Average error	10.5	10.0	11.4	9.8

Generalization of Trained Neural Networks

Competency of the trained networks in generalization was tested with data outside the domain of the training set. However, this requires extrapolations in the network analysis, and resulted in larger errors. Errors in the predicted ulna/radial angles were around 25°. Flexion/extension of the wrist had an error of up to 40°, while that for the elbow was less than 20°. The errors are mainly due to the limited training data.

Conclusion

Results of this investigation indicated that neural networks are quite capable of memorizing and predicting the elbow and wrist joint angles for workstation design. The maximum and minimum ulnar and radial deviations of the wrist had an average prediction error of about 5°, while those of wrist and elbow flexion/extension have errors of less than 15°. The capability of neural network to analyse complex relationship between posture, motion and workstation configuration is potentially useful for user-centred approach to workstation and/or workplace design. The neural network analysis is being refined to improve the prediction.

References

Ho N.H. and Lim T.E. 1990, Computer-aided workplace layout and line balancing. In W. Karwowski, A.M. Genaidy and S.S. Asfour (ed.) *Computer-aided ergonomics: a researcher's guide*, (Taylor & Francis, London) 181-193.
Jung E.S., Kee D. and Chung M.K. 1995, Upper body reach posture prediction for ergonomic evaluation models, *Int. Journal of Industrial Ergonomics*, **16**, 95-107.
Lim L.E.N., Ngan L.M.T. and Lam C.Y., 1989, Anthropometric data of the Singapore populace, *The Industrial Engineer*, (Institute of Industrial Engineers, Singapore) 67-76.
Simpson P.K. 1990, *Artificial neural systems: foundations, paradigms, applications and implementations*, (Pergamon Press, New York).

Visual Search As A Function Of Field Of View Size

RDF Harrison* & PJ Barber**

*Defence Clothing & Textiles Agency, Science & Technology Division
Flagstaff Road, Colchester, Essex, CO2 7SS

**Department of Psychology, Birkbeck College
Mallet Street, London, WC1E 7HX

This investigation contributes new data about the consequences for work performance of a restricted field of view (FOV). A visual search task was used, with *targets* designated by a stationary or slowly moving location *signal*. Potential '*targets*' were arrows pointing vertically or horizontally. The task was to locate and report the direction of a designated *target* using a joystick. Subjects performed under four viewing conditions (individually calibrated): an unrestricted FOV, and FOVs limited to 45°, 90° and 135°. Reaction times and subjects' head movements were recorded. The restrictions were imposed in the vertical-lateral plane only. Search was affected by FOV size, *signal* type (stationary vs. moving), and *target* location. It was slowest for the two heaviest FOV restrictions, slower for the static *signal*, and it decreased with increasing *target* displacement from the centre of the field of view. Each pair of the three independent variables interacted significantly. Subjects made more and faster head movements under greater restrictions. A simple additive model of the performance costs of these variables would not be appropriate and design guidelines need to allow for the implications of compounding performance limiting factors.

Introduction

Certain types of protective equipment (e.g. helmets, goggles and respirators) may have visual effects, such as restricting the use of peripheral vision. This has particular significance in military contexts, where a restriction of this sort on the operator's field of view (FOV) may result in increased risk and impaired performance both on the battlefield and in training. Little work appears to have been undertaken by military agencies or others to quantify performance reductions. Many current items of equipment have largely been designed as stand-alone items with cursory acknowledgement of other items with which they might have to interface. Nevertheless there is an increased interest in the systems aspects of protective equipment and how this will affect the design of future solutions. Elements of protection technologies may have profound or unforeseen affects on the wearer and their performance. The principal motivation for the present research was to establish the possible effect on a vision-based task of a range of FOV restrictions.

This study examines the possible effect of restricted FOV on performance on a visual search and detection task. The eye has developed in such a way that different regions are relatively specialized for dealing with different types of stimuli. For instance, the fovea supports the processing of fine detail, while the extrafoveal part of the retina signals which areas of the visual field may contain useful information and require foveal processing, achieving this by flagging properties such as size and brightness directly. Moreover different parts of the retina are better at perceiving moving and non-moving stimuli

(Gregory, 1990). For instance, the edge of the retina is more sensitive to large fast moving stimuli, while the fovea can detect slowly moving targets more easily than static targets. Hence restricting an operator's FOV is likely to affect performance by impeding access to the information received peripherally that may be used to guide visual search. While the full range of relevant stimulus properties needs to be fully investigated, the present study makes a start in contrasting moving and stationary targets.

Research on restrictions of an observer's FOV has mainly concentrated on head up displays (HUDs) for pilots, although there is also some work on motorcycle safety helmets. Venturino and Wells (1990) investigated the performance and head movements of subjects wearing helmet mounted displays and found that subjects hit fewer targets and were threatened by the presence of potential enemies for longer under greater FOV restrictions. Wells, Venturino and Osgood (1989) investigating the effect of FOV size on performance relation to simple air to air missions, reported that the FOV required for acceptable performance to be maintained increased as the number of targets presented increased. Szoboszlay, Haworth and Reynolds (1995) applied restrictions identical to those imposed by night vision HUDs on pilots flying rotor wing aircraft. Preliminary results indicate that reduced FOVs increased the difficulty of controlling the aircraft and altered the pattern of the pilots' head movements. Reduced situational awareness was reported, which led to pilots being unable to provide accurate reports of their flying performance and unable to monitor warning indicators in the cockpit effectively. Gordon and Prince (1975) also found significant performance decrements as FOVs were reduced by increased motorcycle safety helmet coverage. Reduced performance effects were illustrated by longer reaction times to target detection, increased head movement, increased speed of head movement and reduced eye movement. The performance effects are generally very similar across different investigations.

A simple framework for construing the possible effects of restricted FOV was described by Sanders (1970) in his distinction between an observer's stationary, eye and head fields. This is a functional classification, and was based on studies using a variety of visual tasks in which performance decreased as the visual angle from fixation increased, with marked drops at about 20-35° (the "boundary" between the stationary and the eye fields), and at about 75-90° (the "boundary" between the eye and the head fields). In the stationary field performance nothing more than peripheral vision is needed, while in the eye field, eye movements are required for the task to be carried out, and in the head field, head movements too are needed. While this may be a simple and obvious distinction, an important qualification is that the boundaries were affected by task difficulty, the transition between fields being displaced further into the periphery for simpler tasks. An extensive study would require performance to be assessed comprehensively, including eye and head movements. In the present case, it was possible only to record visual search times and the incidence of head movements. The latter are of course also relevant to the issue of physical fatigue, especially given the load imposed by wearing protective headgear.

Method

A repeated measures design was used with three within subjects factors: FOV size (unrestricted, 135°, 90° and 45°), *signal* type (stationary vs. moving), and *target* location. There were 24 separate locations, bracketed into eight screen segments for the purposes of analysis and presentation (centred at 12.5°, 37.5°, 62.5° and 87.5° to the left and right of screen centre). Twelve volunteer subjects (including one female) were recruited from military units in Colchester garrison. There were eleven men and one woman, and their average age was 23 (range 20-27); none wore glasses or reported being aware of any visual defects. A Goldman Perimeter on loan from the Defence Engineering and Research Agency, Centre for Human Sciences, was used for measuring angles of peripheral vision.

Subjects wore MK 6 General Service combat helmets with full face visors. Subjects were strapped into the perimeter device and a 5 mm diameter light dot was displayed at 300 mm (subtending a visual angle of 2.5°). The light was presented at eye level, at 67.5°, 45° and 22.5° relative to each eye. The visors were marked with vertical graduations at 5 mm intervals and parallel lines scored into the surface at 50 mm intervals. The visors were only masked in the lateral plane.

Special purpose display equipment was constructed at the Defence Clothing and Textiles Agency, Science and Technology Division, Colchester (DCTA, S&TD). This equipment consisted of a screen of boards, 5 m in diameter. A horizon was marked at 1.35 m above the floor. A central axis was marked which equally divided the screen into two sections, 125° in arc. Subjects sat in the centre of the screen area, on a height-adjustable swivel seat. This was to allow vertical adjustment of the subject's eye height so that it matched the level of the horizon. A joystick was mounted vertically from the chair. Above and below the horizon was marked a band which correspond to 10° above and below the horizon from the central eye height point.

Targets were presented as arrows which pointed up, down, to the left and to the right. The *targets* were indicated by a 1 mW, 670 nm, continuous wave, laser diode (Class II), which produced a spot 8 mm in diameter at the screen. The *signal* dot was either still or oscillating (1 Hz) with an amplitude of 0.25 m. The *signal* dot fell within the bordered area which contained the *target* arrow. The laser was directed by a small robot arm. Software was generated at the DCTA, S&TD to control the robot arm and the sequence of *target* presentations. The laser only illuminated when it reached its designated *target* and was cancelled when the subject moved the joystick. Subjects were briefed to sit facing the central axis, and to monitor the green LED set into the joystick. Once the LED illuminated the subjects were to search for the *signal* which was randomly presented. Once the subject detected the *signal* and identified the *target*, they had to move the joystick in the direction that the arrow was pointing. This action cancelled the laser, stopped the timer, recorded the subject's search time and the direction that the joystick was moved. Following each presentation of the laser dot, the subjects were instructed to return to face the centre of the screen and monitor the LED for the next exposure. For each of the eight FOV x *signal* type conditions, there were 48 individual presentations of the laser dot, corresponding to two visits to each of the *targets* on the screen. The order in which the subjects served in these eight conditions was chosen at random. Each condition took about 6-8 minutes to complete, with breaks between conditions.

A video-recorder was used to monitor head movements, using a camera positioned behind the subject, aimed at a mark of the rear of the helmet. This was done for sixteen of the 96 sessions (12 subjects x 8 conditions). Gross eye movements were similarly video-recorded for eight sessions, using a camera positioned directly in front of the subject.

Results

There were 85 incorrect responses made by the 12 subjects in the course of the 384 each completed. The error rate of 1.84% is too low for meaningful analysis. Only search times for correct responses were entered into the following analyses. A three factor repeated measures analysis of variance (ANOVA) was used to analyse search times, with a pooled error variance. The main effect of FOV was significant, $F (3,4459) = 47.61$; $p < 0.001$, as was that of *signal* Type, $F (1,4459) = 25.80$; $p < 0.001$, and *target* location, $F (7,4459) = 116.39$; $p < 0.001$. All of the two way interactions were also significant: that of FOV x *signal* type was highly significant, $F (3,4459) = 13.83$; $p < 0.001$, as was that of *target* location x *signal* type ($F (7,4459) = 4.84$; $p < 0.001$), while that of FOV x *target* location was less so, $F (21,4459) = 2.048$; $p = 0.003$. All but the last of these effects are

significant even when conservative degrees of freedom are applied. The three way
interaction between FOV x *signal* type x *target* location was not significant (F < 1).

Overall search times for the four FOV conditions indicated that performance in the
unrestricted and 135° fields was the same (1.98 s), but performance with greater restriction
was impaired, 2.27 s for the 90° field, and 2.44 s for the 45° field. The dynamic *signal*
produced faster performance overall than the stationary *signal* (2.08 s vs. 2.25 s). The
interaction between these two factors suggested that performance was faster when the *sign*
designating the *target* was moving, except when the FOV was most severely limited,
although it should be noted that the performance advantage for the 135° restriction was ve
small.

Search times increased as the *target* was progressively displaced from screen cent
with some sign of a field asymmetry; the biggest decrements in performance accompanied
the shifts in from screen segments 1 to 2 in the left field (from 12.5° to 37.5°) and from
segment 2 to 3 in the right field (from 37.5° to 62.5°). The interaction between *signal* type
and location appears to arise from the advantage to the dynamic signal being limited to the
most central *target* locations. The remaining significant interaction, between FOV and
location (see Figure 1), suggests that the unrestricted and 135° field were essentially
equivalent, and that performance was not substantially affected by FOV in the most centra
locations (segments 1 and 2). However, there is a clear differentiation for outer locations,
with performance suffering when the two most severe FOV restrictions were imposed.
There is some evidence too that performance for the 45° restriction was worse than that wi
the less severe 90° restriction (with the exception of segment 4 in the right field). The
search time profile again suggests a minor asymmetry between left and right fields.

Figure 1 Mean RTs for Screen Segments and
Different FOVs - Static and Dynamic Conditions Combined

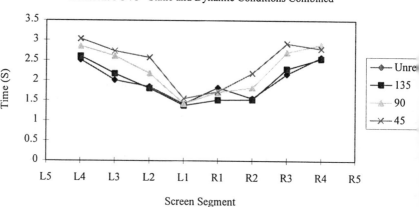

Video recordings taken of the subjects' head and eye movement suggested marked
changes in search strategies, and some common measures. All subjects undertook one of
three strategies involving head movement and/or torso movement with the chair. Four
subjects used their head only when scanning with their body remaining facing forward
during this, two moved only their bodies and not their head, and six used a combination of
head and body movement. The ratio of head and body movement was generally equal, with
the technique being a half move to the left or right with the body, with head movement to
compensate for the rest of the distance in the case of extreme *targets*.

Subjects who used head movement only generally made what seemed to be a
ballistic movement in either direction. Subjects using axial rotation about the chair

generally had a slower search strategy, while subjects who combined methods made sweeping torso movements and sweeping head movements rather than the ballistic head movements made by the subjects moving the head only. Eye movement by the subjects followed a similar pattern, and depended on the level of restriction. All subjects undertook a phase of eye movement which involved scanning the visible area to their front, prior to any other physical movement. The greater the restriction, the shorter the period of eye movement.

Discussion

The present experiment produced results which support other findings of a similar nature where reduced FOVs lead to performance decrements. By using a range of FOV restrictions it seems that appreciable effects on a search task such as that used in this study would be liable to occur when the FOV is reduced to below $135°$, and that any impairment is likely to increase as the restriction is intensified. A task in which the target location was specified intrinsically rather than by a separate signal would serve to test the generalizabilty of the findings. It seems unlikely, however, that the use of a peripherally salient cue as in the present study would exacerbate the deleterious effect of FOV restriction on search time. It seems reasonable to conclude that a task in which the target location was less clearly designated would be no less prone to the effect of FOV restriction. This would follow from Sanders's (1970) findings on the effect of task complexity on the transition between functional visual fields but it would need to be confirmed experimentally.

A moving *signal* used to identify the position of the *target* was generally somewhat more effective than a stationary *signal*. Technical difficulties limited the speed of movement of the cue to 1 Hz, but even at this slow rate performance benefited relative to the static signal. Faster rates of oscillation might well be more effective further into the periphery, as would a flashing signal. Further research would be needed to establish the extent to which FOV restriction effects on search performance may be offset by properties of the *target* or any *signal* used to specify it.

As might be expected search time declined as *target* location became more displaced from the central starting point. There were no obvious signs of stepwise increases in the search times that would be suggestive of a relatively abrupt shift between regions of the functional visual field, although close analysis of individual data may reveal this kind of evidence, particularly when the search times are linked to the eye and head movement data. It is of interest, however, that these data suggest that subjects did attempt to compensate physically for restrictions in their peripheral vision. The data shows that as restrictions increased, head movement increased and eye movement decreased. It seems that the $45°$ and $90°$ restrictions impaired the subjects' use of their eye field, and that they compensated with increased head and body movement. The heaviest restriction ($45°$ FOV) would also have affected their use of their normal head field and also required compensation by head and body movement. Some subjects spontaneously resorted to search mediated by body movement, using axial rotation about the chair; they generally had a slower search strategy than those who used head movement only.

It has been reported that non-hearing individuals rely on their peripheral vision in using sign language (Loke & Song, 1991). It has been suggested that their peripheral vision does not posses greater acuity than in others, however they are better able to attend to it through practice. Although it is debatable whether the motivation of soldiers to develop peripheral viewing skills could ever be comparable with that of people with impaired hearing, experience is evidently a factor that needs to be investigated further in connection with FOV research.

The specific working context that is of concern here is a military one but the findings may have more general relevance, given the wide range of everyday circumstances in which an individual's visual field may be limited. Restrictions of varying severity are routinely placed on one's FOV by wearing goggles, spectacles, rainhoods, motor-cycle helmets, and so forth, and the effects on hazard detectability and visual search generally need to be understood and their consequences minimized by sensitive design. In those case where a range of protection measures have to be adopted, interactions between such measures must be considered before the final design of a protection system is decided. Knowledge of the effect of restrictions, including those imposed on FOVs, and their interactions offers the ability to predict performance during different tasks. Further work i needed to investigate a range of protection measures and their interactions, to add to the da base of information available to design engineers. The present findings show that the FOV restriction effect is not additive with other relevant target properties including target location, and this needs to be allowed for in design guidelines.

References

Barbur, J.L. 1979. Visual periphery. In J.N. Clare and M.A. Sinclair (Eds.), *Search and the human observer*. London: Taylor and Francis.

Gordon, M., & Prince, J. 1975. Field of view with and without motorcycle helmets. *DOT - NHTSA - Safety Research Laboratory, USA* - Technical Report.

Gregory, R. L. , 1990 , *Eye and Brain: The Psychology of Seeing*. (London: George Weidenfeld & Nicholson Ltd)

Loke, W.H., & Song, S. 1991. Central and peripheral visual processing in hearing and non-hearing individuals. *Bulletin of the Psychonomic Society*, 29, 437-440.

Sanders, A.F. 1970. Some aspects of the selective process in the functional visual field. *Ergonomics*, 13, 101-117.

Sivak, B., & MacKenzie, C.L. 1990. Integration of visual information and motor output in reaching and grasping: The contributions of peripheral and central vision. *Neuropsychologia*, 28, 1095-1116.

Szoboszlay Z., Haworth, L., & Reynolds, T. 1995. Effect of field of view restriction on rotorcraft pilot workload and performance - preliminary results. *Society of Photo-Optical Instrumentation Engineers*, 2465, 142-153.

Venturino, M. & Wells, M.J. 1990. Head movements as a function of field of view size on a helmet mounted display. *Proceedings of the Human Factors Society 34th Annual Meeting*, 1572-1576.

Wells, M.J., Venturino, M., & Osgood, R.K. 1989. The effect of field of view size on performance at a simple simulated air-to-air mission. *Society of Photo-Optical Instrumentation Engineers*, 1116, 126-137.

NON-PRESCRIPTION DRUGS IN THE WORKPLACE: ANTIHISTAMINES

Sona J Toulmin, Susan Teasdale and Anthony Wetherell

Chemical and Biological Defence Establishment
Porton Down
Salisbury
Wiltshire SP4 0JQ

People rarely perform at peak capacity all the time, and one of the reasons is that they take drugs. Most people will take drugs at some time, and some people might have to take them for long periods of time. All drugs have side effects which may impair performance directly by effects on the central nervous system, or indirectly through peripheral effects. A vast number of drugs is available over the counter without prescription, and this paper considers antihistamines, which are widely taken, mainly for allergic reactions, and upper respiratory diseases such as coughs, colds and influenza. Antihistamines have a variety of side effects, which vary considerably between individuals, but most people will experience sedation, and impairment of cognitive and psychomotor performance and driving behaviour. Even though "daytime" antihistamine preparations, which purport to be less sedative, are now available, many people still use traditional antihistamines, and their use should be taken into account by ergonomists and designers.

Introduction

Ergonomists generally assume that the people for whom they design are fully functional and operating at peak capacity. However, it is common experience that people often operate at much less than peak capacity for a variety of reasons, eg physical disability, fatigue, and illness. However, one reason that ergonomics has so far tended to ignore is drugs. Almost everyone will take drugs at some time, and some people may have to take them for long periods of time. All drugs have side effects which can impair performance directly by actions on the central nervous system, or indirectly by affecting mood or other subjective state, or through peripheral actions which may worry or distract the patient. Thus, people who are taking drugs may not be able to cope with equipment or procedures that have been designed with healthy people in mind. There are far too many drugs to cover in one paper, and we have concentrated here on antihistamines that are widely available from pharmacies, without a prescription, and which are widely known to cause sedation and impair performance.

Pharmacology

Antihistamines antagonise, or block, the action of histamine, a chemical which is widely synthesised and stored in mast cells throughout the body, and in circulating blood basophils. Mast cell histamine is involved in immediate hypersensitivity reactions, whereas basophil histamine is thought to contribute more to delayed hypersensitivity reactions. Histamine can be released by several agents, including antigens and simpler substances that can combine with proteins to form antigens, mechanical trauma, stings and venoms, surface active agents such as detergents, high molecular weight compounds such as dextrans and polyvinylpyrrolidine used as plasma substitutes, and many drugs.

The actions of histamine depend on the type of receptor involved - H_1: reddening and itching of the skin, bronchial constriction and mucus secretion, oedema owing to increased capillary permeability, hypotension owing to dilation of arterioles, cardiac arrhythmia, and contraction of intestinal smooth muscle; H_2: bronchodilation and mucus secretion, cardiac arrhythmia and increased gastric acid secretion; H_3: inhibition of histamine and neurotransmitter release. Thus, although histamine is presumably involved in the healing process, its actions can cause more problems than the original injury.

Histamine blocking activity was first discovered in 1937, and the first clinical use of an antihistamine, pyrilamine maleate, was in 1944, but it was not until the 1950s that a selection of antihistamines became available for clinical use [Bovet 1950]. Now, a very large number of antihistamines is available. Most of the older drugs are non-specific in that they act on more than one type of histamine receptor, but recently, specific antihistamines have been developed, such as the H_2 blockers cimetidine and ranitidine, used to treat gastric ulcers. These drugs were originally available only on prescription, but can now be bought over the counter.

Even more recently, a new subclass of H_1 antihistamines has been developed, which do not cross the blood-brain barrier so easily as do the traditional H_1 drugs, and so are reputed to be less sedative (see below). Examples include terfenadine, astemizole, loratadine and cetirizine (a metabolite of hydroxyzine). These drugs are being used increasingly in "daytime" preparations for upper respiratory tract infections, but traditional H_1 antihistamines are still very widely used, and even the newer H_1 drugs may not be totally non-sedative.

H_1 antihistamines are classified into six groups on the basis of their chemical structure: a) amino alkyl ethers, eg diphenhydramine and carbinoxamine, b) ethylenediamines, eg antazoline and pyrilamine, c) alkylamines, eg chlorpheniramine, d) piperazines, eg cyclizine and chlorcyclizine, e) piperidines, eg cyproheptadine, and f) phenothiazines, eg promethazine. All of the drugs are well absorbed in the gastrointestinal tract, with effects appearing within 15 to 30 minutes, reaching full activity within one hour, lasting about three to six hours (although some last much longer), and disappearing after about 24 hours. H_1 antihistamines interact with alcohol and other central nervous system depressants.

Use

H_1 antihistamines are used widely for a variety of conditions involving histamine release, but their main use is in the symptomatic treatment of various allergic diseases such as seasonal hay fever, and in the symptomatic treatment of upper respiratory tract infections such as coughs, colds and influenza, but they are very widely used in proprietary preparations for these purposes. A recent development has been to capitalise on the side effect of sedation; antihistamines are now available over the counter to treat insomnia. Antihistamines are also sometimes used to treat pruritis, motion sickness, and nausea during pregnancy, and, less commonly, in Meniere's and Parkinson's diseases and petit mal epilepsy.

The exact scale of use in the United Kingdom is not known, but in the United States, about 30 million people took single H_1 antihistamines in 1988, and total sales exceeded \$500 million. If antihistamine-decongestant combinations are included, total sales almost reached \$2 billion [Meltzer 1991]. Also in the United States, non-prescription drugs appear to be used more by women than by men, but there seems to be no difference in use with age [Gerda et al 1993].

Preparations containing antihistamines normally come with instructions on how to take them, to avoid alcohol, and to avoid driving or operating machinery. However, many people do not comply, either deliberately, or unwittingly. They might deliberately change the dosing regimen according to their perceived symptoms, or in anticipation of exposure, eg to pollen, or they might misunderstand the dosing instructions. Bradley et al [1994] found that the reading age values in a sample of 50 drug information leaflets were well above the average reading age of the general adult population. People might drink alcohol because they simply forget, or because they underestimate the effects: a common failing.

People might drive or operate machinery unwittingly because they are unaware that they are affected, because they have misjudged the onset, duration, type, or severity of effects, or because they have misunderstood the instructions. People might not comply deliberately because they want to make a journey, or do not want to lose money or time off work. Here, they might consciously decide to take a risk in full knowledge of the effects and warnings, or they might rationalise that their trip is not really "driving", their vehicle is exempt, or their equipment is not really "machinery".

Side Effects

The side effects of traditional H_1 antihistamines vary considerably between individuals. Central nervous system effects include sedation, dizziness, tinnitus, lassitude, incoordination, fatigue, blurred vision, diplopia, euphoria, nervousness, insomnia and tremors. Gastrointestinal system effects include loss of appetite, nausea, vomiting, diarrhoea, dry mouth. Genitourinary system effects include urinary retention and frequency, dysuria, impotence. Other effects that have been reported include palpitations, hypotension, headache, tightness of the chest, and tingling, weakness and heaviness of the hands.

H_1 antihistamines also have some anticholinergic actions which can cause other problems [Toulmin & Wetherell 1995], and some H_1 antihistamines such as promethazine and pyrilamine, can have local anaesthetic activity in high doses. It is even possible for antihistamines taken for allergic reactions to cause allergic reactions of their own.

Performance Effects

Several studies have shown that H_1 antihistamines impair cognitive and psychomotor performance. Hindmarch [1976; 1980] reported that chlorpheniramine impaired critical flicker fusion, and that ketotifen, mebhydrolin and promethazine impair both critical flicker fusion and choice reaction time. Nicholson et al [1982] reported that triprolidine impaired dynamic visual acuity and critical flicker fusion, but terfenadine and astemizole did not. Nicholson [1985] reported that promethazine impaired visual-motor coordination, but terfenadine did not. Seidel et al [1987] reported that hydroxyzine impaired reaction time, and the subjects were unaware that they had been affected. Gengo et al [1990] reported that diphenhydramine impaired symbol substitution performance, and reaction time in a simulated driving test, but cetirizine did not. Gengo & Gabos [1987] reported that traditional H_1 antihistamines such as hydroxyzine impaired performance on the Stroop colour-word naming test, and that the effects were related to the drugs' blood levels and their ability to cross the blood-brain barrier.

Several studies have been reported concerning the effects of antihistamines on driving. Warren et al [1981] reported that Canadian drivers who had been killed in accidents attributable to their own error were 1.5 times more likely to have been using traditional H_1 antihistamines than were drivers who were not responsible for the accident in which they had been killed. Aso and Sakai [1988] studied vehicle steering, and found that, of ten subjects taking chlorpheniramine, only four were able to complete the test, and the other six were judged to be incapable. Subjects taking placebo or terfenadine were not affected. Volkerts et al [1991] found that triprolidine impaired drivers' ability to maintain speed and lateral position, compared with terfenadine and cetirizine.

O'Hanlon et al [1988] studied drivers' weaving behaviour, and found that triprolidine caused impairments, similar to those caused by a blood alcohol concentration of 0.05%, compared with terfenadine and loratadine. This effect of triprolidine persisted for up to four hours, but drivers only reported subjective sedative effects for one to two hours. Thus, the drivers thought they were not impaired when, in fact, they were.

Betts et al [1986] studied the effects of triprolidine, a traditional H_1 antihistamine, with those of terfenadine, a non-centrally acting antihistamine, on women drivers performing slalom-type and gap acceptance tests. On the slalom test, drivers were significantly slower, struck more bollards, and made more mistakes while taking triprolidine than when taking either terfenadine or a placebo. On the gap-acceptance test, there was no difference between the drugs in terms of the risk that drivers were prepared to take, but drivers taking triprolidine struck the sides of the gap significantly more than did drivers taking either terfenadine or placebo.

There seems to be little relationship between the objective and subjective effects on performance. Seidel et al [1987] could not correlate their objective and subjective effects. Nicholson [1985] found that his subjects reported little subjective sense of impairment when, in fact, the antihistamines were exerting their maximum objective effect. O'Hanlon et al [1988] found that the effects of triprolidine lasted for up to four hours, but his drivers only reported subjective effects for one to two hours.

The performance impairment caused by H_1 antihistamines can be exacerbated by alcohol [Moser et al 1978; O'Hanlon et al 1988; Hindmarch & Bhatti 1987] and diazepam [Moser et al 1978].

Conclusions

People who take antihistamines will experience several side effects that could impair performance directly or indirectly. Most people will experience sedation which will impair cognitive and psychomotor performance, including driving. The impairment may not always be what is expected since subjective sensations of sedation do not always match objective performance impairment, and people may not understand or comply with instructions on dosing regimens or advice not to drive or operate machinery. Even "non-sedating" antihistamines may cause some impairment, particularly if dosing regimens are not followed, which could be more dangerous because it is not expected.

Whatever the reason, people taking antihistamines will experience performance impairment for some or all of the time, causing at best loss of production, and at worst, injury or death to themselves or others. Almost everyone will take antihistamines at some time, and some people may have to take them for long periods of time. Thus, the problem is considerable, and deserves some attention. As ergonomists, we are taught not to regard ourselves as representative of the target population, so we should not think that our target population will be healthy just because we are. In the case of non-prescription drugs, however, we probably are representative of the target population; we should remember that even ergonomists catch colds and suffer from allergic reactions.

References

Aso, T. Sakai, Y. 1988, Effects of terfenadine on actual driving performance, *Jpn J Clin Pharmacol Ther*, 19:681-688.

Betts, T. Mortiboy, D. Nimmo, J. Knight P. 1976, A review of research: the effects of psychotropic drugs on actual driving performance. In J.F. O'Hanlon and J.J. de Gier (eds), *Drugs and Driving*, (Taylor and Francis, London) 83-100.

Bovet, D. 1950, Introduction to antihistamines and antergan derivatives, *Ann NY Acad Sci*, 50:1089-1126.

Bradley, B. Singleton, M. Po, A.L.W. 1994, Readability of patient information leaflets on over-the counter (OTC) medicines, *J Clin Pharm Ther*, 19:7-15

Gengo, F.M. Gabos, C. 1987, Antihistamines, drowsiness and psychomotor impairment: central nervous system effects of cetirizine, *Ann Allergy*, 59:53-57.

Gengo, F.M. Gabos, C. Mechtler, L. 1990, Quantitative effects of cetirizine and diphenhydramine on mental performance measured using an automobile simulator, *Ann Allergy*, 64:520-526.

Gerda, G. Fillenbaum, G.G. Hanlon, J.T. 1993, Prescription and nonprescription drug use among black and white community-residing elderly, *Am J Public Health*, 83:1577-1582.

Hindmarch, I. 1976, The effects of the sub-chronic administration of an antihistamine, clemastine, on tests of car driving ability and psychomotor performance, *Curr Med Res Opin*, 4: 197.

Hindmarch, I. 1980, Psychomotor function and psychoactive drugs, *Br J Clin Pharmacol*, 10: 189.

Hindmarch, I. Bhatti, J.Z. 1987, Psychomotor effects of astemizole and chlorpheniramine, alone and in combination with alcohol, *Int Clin Psychopharmacol*, 2:117-119.

Meltzer, E.O. 1991, Comparative safety of H_1 antihistamines, *Ann Allergy*, 67:625-633.

Moser, L. Huther, K.J. Koch-Weser, J. 1978, Effects of terfenadine and diphenhydramine alone or in combination with diazepam and alcohol and psychomotor performance and subjective feelings, *Eur J Clin Pharmacol*, 14:417-423.

Nicholson, A.N. 1985, Central effects of H_1 and H_2 antihistamines, *Aviat Space Environ Med*, 56:293-298.

Nicholson, A.N. Smith, P.A. Spencer, M.B. 1982, Antihistamines and visual function: studies on dynamic visual acuity and the pupillary response to light, *Br J Clin Pharmacol*, 14:683-690.

O'Hanlon, J.F. 1988, Antihistamines and driving performance: the Netherlands, *J Respir Dis*, 9:S12-S17.

Seidel, W.F. Cohen, S. Bliwise, N.G. 1987, Cetirizine effects on objective measures of daytime sleepiness and performance, *Ann Allergy*, 59:58-62.

Toulmin, S.J. Wetherell A. 1995, Some effects of anticholinergic drugs on performance. In S.A. Robertson (ed), *Contemporary Ergonomics 1995*, (Taylor & Francis, London) 505-510.

Volkerts, E.R. van Willigenburg A.P.P. van Laar, M.W. 1991, A comparison of the acute and subchronic effects of two "non-sedating" antihistamines (cetirizine and terfenadine) versus those of an older antihistamine (triprolidine) upon on-the-road driving performance, *Schweiz Med Wochenschr*, 121: 34.

Warren, R. Simpson, H. Hilchie, J. 1981, Drugs detected in fatally injured drivers in the province of Ontario. In L. Goldberg (ed), *Alcohol, Drugs and Traffic Safety, vol 1*, (Almquist and Wiksell, Stockholm) 203-217.

VARIABILITY IN FORCE AT A HAND-HELD JOYSTICK DURING EXPOSURE TO FORE-AND-AFT SEAT VIBRATION

G.S. Paddan and M.J. Griffin

Human Factors Research Unit
Institute of Sound and Vibration Research
University of Southampton
Southampton
SO17 1BJ England

Force in three orthogonal directions was measured at the right hands of seated subjects holding a joystick during exposure to fore-and-aft whole-body vibration. The aim of the experiment was to determine the variability in the force during repeated measures for a single subject and the variability in force between individuals. One male subject took part in the intra-subject variability study; twelve male subjects took part in the inter-subject variability study. The subjects sat on a rigid flat seat with their backs in contact with the seat backrest. The subjects held their arms such that the forearm was horizontal and in front, with the upper arm vertical (i.e. an elbow angle of 90°). Subjects were exposed to random vibration in the fore-and-aft axis at frequencies below 10 Hz with a vibration magnitude of 0.5 ms^{-2} r.m.s. Transmissibilities were calculated between acceleration at the seat and force at the hand in each of the three translational axes. Higher transmissibilities occurred for force in the fore-and-aft axis of the hand compared with the forces in the other two axes, with a peak fore-and-aft response occurring at about 5 Hz. Methods of comparing subject variabilities are defined and show that the variation in transmissibility between subjects was significantly greater than the variation within the single subject.

Introduction

A knowledge of the variability in response within and between individuals is essential when predicting human responses. In the context of human response to vibration, the variability in the transmission of vibration to the head has been demonstrated by Paddan and Griffin (1994), the variability in apparent mass has been reported by Fairley and Griffin (1989) and the variability in subjective responses by Griffin and Whitham (1978).

In this paper the forces at the hand when holding on to an isometric joystick during exposure to whole-body vibration in the fore-and-aft axis are presented. Variabilities within an individual and between individuals (i.e. intra- and inter-subject variability) have been investigated.

Equipment and Procedure

Forces were measured using a joystick mounted on a Kistler 3-component quartz force transducer (type 9251A). The joystick consisted of a wooden dowel 28 mm in diameter and 130 mm long. The force signals were amplified using a Kistler charge amplifier type 5041C. Seat acceleration was measured using an Entran type EGCS-240-B-10D accelerometer.

One male subject took part in the intra-subject variability experiment. He was 27 years old, weighed 76 kg and had a stature of 1.78 m. A group of twelve male subjects took part in the inter-subject variability study; their physical characteristics are summarised in Table 1. All subjects were fit and healthy, and were required to comply with the medical contraindications specified in British Standard BS 7085 (1989). The subjects sat on a rigid seat (seat surface 480 mm above the moving footrest) in an upright comfortable posture with their backs in contact with the seat backrest (i.e. a 'back-on' posture). The seat backrest extended up to the subject's shoulders. The subjects kept their upper arms vertical and their forearms horizontal (i.e. the included elbow angle was 90°). Forces at the right hands of the subjects were measured using the joystick. Figure 1 shows a subject sitting in a 'back-on' posture and holding on to a force joystick. The subject's arms were not in contact with the seat or the body.

Table 1. Physical characteristics of subjects taking part in the inter-subject variability experiment.

	Age (yr)	Weight (kg)	Stature (m)
Minimum	24	52	1.65
Maximum	39	76	1.88
Mean	30.0	67.0	1.76
Standard Deviation	5.30	6.69	0.07

Vibration was generated using an electro-hydraulic vibrator capable of producing horizontal displacements of 1 metre. The vibration at the seat covered a frequency range up to 10 Hz and the vibration magnitude was 0.5 ms^{-2} r.m.s. The subject who participated in the intra-subject variability study had twelve repeat exposures to fore-and-aft seat vibration with no support for the arm. The other subjects were exposed to the same vibration once in the same posture. Each vibration exposure lasted 120 seconds; all signals were acquired into a data acquisition and analysis system, *HVLab*, at a sample rate of 65 samples per second. The four signals (one of acceleration and three of force) were low-pass filtered at 15 Hz.

Analysis

Transfer functions between seat acceleration and force at the joystick have been calculated using the 'cross-spectral density function method'. The transfer function, $H(f)$, was determined as the ratio of the cross-spectral density of acceleration at the seat and force at the hand, $G_{af}(f)$, to the power spectral density of acceleration at the seat, $G_{aa}(f)$:

$$H(f) = G_{af}(f)/G_{aa}(f).$$

Figure 1. Subject sitting in a 'back-on' posture and holding on to a joystick.

Transfer functions were calculated between fore-and-aft seat acceleration and each of the three axes of force at the hand. Frequency analysis was carried out with a resolution of 0.127 Hz, giving 64 degrees of freedom.

Results and Discussion

Figure 2 shows transmissibilities between fore-and-aft seat vibration and triaxial force at the joystick with the subjects sitting in a 'back-on' posture. The top row in Figure 2 shows the twelve transmissibilities for the single subject exposed to vibration twelve times (i.e. intra-subject variability); the bottom row shows the transmissibilities for the 12 subjects (i.e. inter-subject variability).

Transmissibilities for force in the fore-and-aft axis show higher values than those in the lateral and the vertical axes, and exhibit one main peak at about 5 Hz. Some large differences between subjects are seen in transmissibilities in Figure 2. For example, although the variability at 5 Hz for the single subject fell in the range 7.2 N/ms^{-2} to 9.1 N/ms^{-2}, the variability between subjects fell in the range 4.2 N/ms^{-2} and 8.5 N/ms^{-2}.

The phase between fore-and-aft seat vibration and fore-and-aft force at the hand is shown in Figure 3 for both intra-subject variability and inter-subject variability.

Median and interquartile ranges of the transmissibilities in Figure 2 are shown in Figure 4. The median transmissibility for the twelve subjects shows a value of about 3.5 N/ms^{-2} at frequencies below about 2 Hz for the force in the fore-and-aft axis. It is seen that the variability in transmissibility between the different vibration exposures for the intra-subject study is small; the interquartile range curves are almost coincident at some frequencies.

Two different measures of variation that have been calculated for the transmissibility data are normalised variability and relative variability (see Paddan and Griffin, 1994). These are defined as: normalised variability = (interquartile range)/median, and relative variability = (inter-subject interquartile range)/(intra-subject interquartile range).

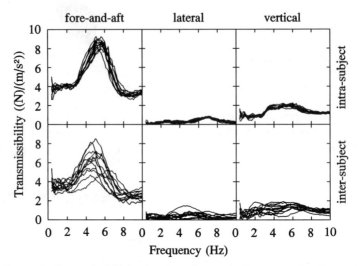

Figure 2. Transmissibilities between fore-and-aft seat acceleration and triaxial force at the joystick for subjects sitting in a 'back-on' posture.

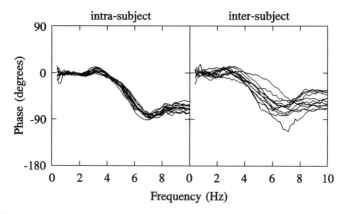

Figure 3. Phase between fore-and-aft seat acceleration and fore-and-aft force at the joystick for subjects sitting in a 'back-on' posture.

'Normalised variability' can be used to compare intra-subject variability and inter-subject variability irrespective of the magnitude of the transmissibility. The variability may visually appear greater for higher median transmissibilities compared to smaller transmissibility values. Normalised variability for the force transmissibilities presented in Figure 1 is shown in Figure 5. The normalised variability is least for the data obtained from the single subject compared to the twelve subjects, and less for the fore-and-aft axis than the lateral or vertical axes.

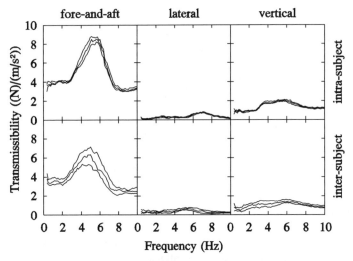

Figure 4. Median and interquartile range of transmissibilities between fore-and-aft seat acceleration and triaxial force at the joystick.

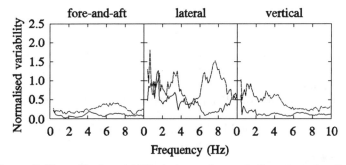

Figure 5. Normalised variability between fore-and-aft seat acceleration and force at the joystick (——— intra-subject; — — inter-subject).

Relative variability shows the variability in transmissibility between subjects *relative* to the variability within an individual. The relative variabilities in Figure 6 show that over most of the frequency range, inter-subject variability is two or more times greater than the variability of the single subject. The relative variability is as high as 7 for some frequencies (see data for lateral and vertical axes).

It is seen from Figures 2 and 4 that even though there is large variability in transmissibility between subjects for force in the fore-and-aft direction, all subjects show a transmissibility of approximately 4 N/ms^{-2} for frequencies below about 2 Hz. This is thought to be related to the mass of the hand and the arm. The peak in the fore-and-aft transmissibility at about 4.8 Hz corresponds to a resonance of the arm in this posture.

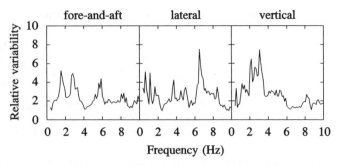

Figure 6. Relative variability in the transmission of fore-and-aft seat acceleration to force at the hand-held joystick.

Conclusions

During exposure to fore-and-aft seat vibration, force at the joystick occurred mainly in the fore-and-aft direction. Variations in the force measured at the joystick were significantly greater between subjects (i.e. inter-subject) compared with the variation measured for an individual during repeat measures (i.e. intra-subject). A peak in transmissibility for fore-and-aft force at the hand occurred at about 5 Hz.

References

British Standards Institution 1989, Guide to safety aspects of experiments in which people are exposed to mechanical vibration and shock. BS 7085. London: *British Standards Institution.*
Fairley, T.E. and Griffin, M.J. 1989, The apparent mass of the seated human body: vertical vibration. *Journal of Biomechanics,* **22**(2), 81-94.
Griffin, M.J. and Whitham, E.M. 1978, Individual variability and its effect on subjective and biodynamic response to whole-body vibration. *Journal of Sound and Vibration,* **58**(2), 239-250.
Paddan, G.S. and Griffin, M.J. 1994, Individual variability in the transmission of vertical vibration from seat to head. ISVR Technical Report No. 236, Institute of Sound and Vibration Research, University of Southampton, Southampton, Hampshire.

Acknowledgements

This work has been carried out with the support of the UK Ministry of Defence under an extramural research agreement with the Centre for Human Sciences, DRA Farnborough.

ERGONOMIC RE-DESIGN OF AIR TRAFFIC CONTROL FOR INCREASED CAPACITY AND REDUCED STRESS

T. B. Dee

7 High Wickham, Hastings
Sussex, TN35 5PB, England
Tel/Fax +44.1424.430919
tbdee@ibmpcug.co.uk

The contemporary Air Traffic Control (ATC) system evolved under severe practical constraints, which place heavy mental loads on the controllers. Recent technical developments provide far more precise data on the situation in the air, and on its expected near-term developments. Existing systems do not make effective use of this information. Although very advanced (and very expensive) display systems are now being installed in ATC centres, they are being used to emulate obsolescent manual systems, using 'windowing' systems that are ill-adapted to continuous dynamic control processes. Examination of ATC Tasks suggests that much routine workload can be eliminated, and a survey of remaining tasks shows that a re-designed display system and control interface can relieve controllers of much of their existing mental strain, improve safety, and increase sector capacity, while enhancing the more satisfying aspects of Air Traffic Control.

Introduction

Air Traffic Control (ATC) is interesting to the ergonomist as a domain in which human judgement is continuously intimately engaged with a complex dynamic system. High levels of safety and reliability are traditional in this field, and have always been considered as the primary task of the system. Air traffic has grown steadily since the Second World War and shows no sign of stabilising. The traditional response to increased traffic is to reduce the 'sectors' (volumes of airspace) for which controllers are responsible, but increasing the number of sectors increases the amount of coordination necessary, and reduces the room for manoeuvre within the sector. A point of diminishing returns is approaching, and qualitative rather than quantitative changes are necessary.

For readers unfamiliar with ATC a brief, necessarily simplified, description may help in understanding the problem and the proposed solution. In most contemporary systems, a Planner (PC) compares data abstracted from flight plans and presented on printed strips, primarily as times and heights over beacons. The PC decides on height changes for aircraft to avoid dangerously close approaches between aircraft (conflicts), measured as time and height differences over beacons. The Executive (EC) watches the radar (or SDD - System Data Display), identifies aircraft, refers to the strips for details, and for the PC's changes, and communicates with the aircraft. The EC also watches for unexpected behaviour by aircraft, for unplanned aircraft, or failures to follow planned trajectories. Because the SDD shows the position, direction and speed of aircraft directly, with height shown only as a three digit code, the EC tends to change aircraft headings to maintain separation, measured as a minimum direct distance between aircraft. After the EC has changed an aircraft's direction, (s)he must remember to return it to track after the problem has passed. Depending on the complexity of the air-routes in the sector, the EC may have to cope with from eight to twenty aircraft at the same time. Many of these aircraft may require no intervention, but the EC must be aware of exactly where they are and what they are going to do, all the time. This is a stimulating but stressful task. Controllers in busy sectors work for about two hours, including half an hour of handing over to the next team.

Current Developments

The growth in the number of strips required is approaching the physical limits of capacity. Handling, sorting, comparing and marking all these strips simply takes too much time. The immediate response is to provide electronic displays representing the strips, and give the controllers facilities to sort or mark the electronic strips. This solution requires large, very high quality, screens to present sufficient data in characters of readable quality. Colour codings of various sorts can be automatically applied, and, in principle, the electronic system can monitor the controller's actions and call attention to unresolved conflicts. However, the limits of technology and human vision combine to require physically large, heavy and clumsy screens. (The current favourite displays a 50 cm square image, with 40 pixels to the cm, and weighs about 100 kilos.) Even so, they require significant reductions in the data directly visible to the controller.

At the same time, since the information is stored electronically, it is technically feasible to provide much more information directly to the EC via the SDD. The label attached to each plot can be enlarged to show what is virtually a full 'strip', in place of the height-and-identity data currently provided. Providing all this data for all aircraft would cover the screen with labels. A basic label is normally displayed, which can be expanded by 'clicking' using a mouse and pointer system, to open a 'window' with a full flight plan. Orders to the aircraft can be passed to the system by clicking on appropriate items and selecting appropriate orders. (However, these are then conveyed to the aircraft by the traditional 1930's style R/T.)

Other windows have been suggested. One particularly interesting one is the Conflict (And) Risk Display (C(A)RD) (Graham, Young, Pichancourt, Marsden and Ikiz,1994), modified from a suggestion by Falzon (1982) which presents a two dimensional picture, where the vertical represents minimum separation and the horizontal time from the present. Points on the display represent potential conflicts, at their minimum distance. Sometimes a line indicates the duration of the infringement of separation standards. Each point is numbered, and a separate table lists the callsigns of the aircraft involved. The EC can then locate the aircraft on his SDD and estimate what their relative positions will be when the conflict occurs, and decide what action to take. (The system has, in fact, calculated these future positions, but this information is not made available to the EC.) Because conflicts are relatively rare at current traffic levels, the C(A)RD is usually blank. This is probably its most useful state, since it reassures the controller that (s)he has not forgotten or failed to see a conflict - a constant source of stress in current systems. This display is greatly appreciated in study simulations.

Other windows have been proposed to show the traffic over a specific entry or exit point; the positions of aircraft at a future time; or enlarged sections of the current picture. More radical innovations include proposals to show, for a particular aircraft, the 'forbidden zones' where manoeuvres at a particular point would lead into potential conflicts.

All of these innovations, although promising, welcome and possibly helpful to hard-pressed controllers are, to garble a phrase of Brian Shackel, 're-designing the steam engine'. They are attempting to use the power and flexibility of the computer to emulate a system developed to surmount physical constraints that no longer apply. Even the ODID study simulations and the currently planned PHARE demonstrators are essentially 'horseless carriages' rather than the fast and safe airliners we need for our own and our childrens' futures.

Analysis

Several extremely detailed and systematic descriptions of the full task of the controller are available, but for this demonstration a fairly general description is sufficient.

Controllers (in general, we abandon the PC/EC distinction) : -
1) plan the future 'streams' of aircraft in their sectors,
2) check that they will not come into conflict within the sector.
3) determine how to resolve conflicts.
4) acknowledge aircraft coming on to their frequency as they enter the sector.
5) instruct aircraft to manoeuvre to avoid conflicts.
6) assure the arrival of aircraft at the planned sector exit point,level and time
7) coordinate their actions with adjacent sectors
8) monitor aircraft behaviour for deviations from plan
9) hand aircraft to the next sector
10) comply with special requests and handle emergencies.

Ergonomists will not need to be reminded that the strengths of human beings as operators of complex systems are their pattern-recognition and problem-solving ability, flexibility and initiative, nor that their weaknesses are poor recall, low reliability, tendency to boredom and inability to judge complex probabilities. Some of these tasks (4,7 and 9) are essentially routine, and although at present they are justified as aids to controller memory, could easily be automated. Tasks 2 and 8 are ill suited to humans, and can be relatively easily automated. Tasks 3,5 and 10 are most interesting and satisfying to controllers, and those at which they are remarkably efficient. Task 1 is inherent in the manner in which the task is currently performed, and Task 6 is the natural and irksome consequence of task 5.

The major cause of strain on controllers (other than their awareness of the sheer importance of the task they perform), is the amount of information they must retain in their memories at all times (Stein and Bailey, 1994). This information, known colloquially as "the picture", is necessary if they are to carry out their scheduling and separation tasks as they do at the moment. Exactly how this information is structured mentally is not well understood, although there are indications (Lafon-Millet 1978) that it changes in nature with the controllers experience. The considerable time taken to learn any sector, and the lack of transfer from sector to sector, suggest that controllers tend to rely on recognition rather than recall or conscious logic. The use of pre-defined route structures and height conventions appears to assist in the compilation of a mental data-bank of situations, by restricting the possible cases to remember.

A radical proposal, originally due to Maignan is that those aircraft which will pass through the controller's airspace without any intervention should not be shown to the controller. The difficulty arises that aircraft not shown to the controller may come into conflict as a consequence of his intervention to resolve conflicts he can see.

Another source of disquiet is the channel of communication between controller and pilot. At present this is a classical VHF radio link. Identification, communication and acknowledgement are purely manual operations. Even with the most disciplined and consciencious controllers and pilots, mistakes are not unknown (Mell,1992, Cardosi 1994). Most of these have no practical consequences, and result only in a few more nightmares for the controller. Some have tragic consequences (Cushing,1994). This problem will be compounded in some future systems, where the controller must also tell the computer what he is asking the aircraft to do. In this demonstration system, it is supposed that the controller will instruct the system what manoeuvres he wishes the aircraft to undertake, and that the system will send this information to the aircraft by data-link, using an automatic read-back protocol to verify the transmission. A voice synthesiser could be used to convey the instructions on the traditional R/T to maintain the 'party-line' effect for other pilots, and an on-board synthesiser could inform linguistically-challenged pilots of the manoeuvres required in their own language. The R/T channel can never be totally suppressed, if only for Task 10 (although the appearance of a human voice may become a general alerting signal)

Re-design

We will now re-design the ATC display to suit a system where full contemporary automation is available, adopting the ergonomic principles that : -
- displays should not be loaded with unneeded detail,
- display features should be intuitive,
- and that information should be easily accessible.

We have available one VGA display, capable of showing 640 x 350 pixels in 8 colours, and a keyboard for input, and an 286 CPU to run the demonstration engine.

Our main image is a plan display, on which we show individual aircraft as below:-

Where is it?	Position on display
What is it?	Ikon Size
How fast is it?	Sweep back
Where is it going?	Direction ikon points
How high is it?	Colour Code
Is it climbing or descending?	Colour Combination

Height has always been a problem on ATC displays. The current solution (three digits representing hundreds of feet) is a relic of early SSR systems, and is, ergonomically, a sin. Most aircraft cruise at pre-determined flight levels, of which there are only a few in regular use in a given sector. We mostly compare aircraft at the same level. We follow the current European strip convention, coding :

Westbound	31,000 ft (FL310) =	Blue
Eastbound	33,000 ft (FL330) =	Yellow
Westbound	35,000 ft (FL350) =	Cyan
Eastbound	37,000 ft (FL370) =	Brown (dark yellow)
Westbound	39,000 ft (FL390) =	Magenta.

We identify aircraft changing levels by showing their front in the colour they are going to, and their rear in the colour they are coming from.

The most important point about an aircraft is whether it will be in conflict. We indicate aircraft that will not be in conflict by not showing them. We show which aircraft will be in conflict by linking them with lines that become more intense as the conflict is appproached. This draws the eye to aircraft that are involved in multiple conflicts, so that several problems may be resolved by one manoeuvre.

To solve problems ,we choose an aircraft, or allow the system to select an aircraft from the most urgent conflict, and examine its future trajectory and actual or potential conflicting traffic. The system presents a secondary height by time/distance image on the same scale as the main display. Blocks of solid colour show the times and heights for which other aircraft will be in conflict with this one. Hollow blocks show aircraft separated by height only. (This display alone is sufficient to resolve most problems by height changes). The future trajectory of the aircraft is superimposed on the main display in green, with dashed green lines for the trajectories of height-separated aircraft, and solid green lines for potentially conflicting aircraft. These tracks end at the 'point of closest approach', which is linked by a solid line (white for conflicting, green for non-conflicting aircraft) to the corresponding point on the track of the original aircraft.

Control orders can be constructed (because our system is limited to keyboard input) as a series of keystrokes. For example "S+H+++T+++F--T+++++R?" would mean "Increase speed ten knots,turn right thirty degrees. In three minutes time,descend four thousand feet. Maintain level, speed and heading for five minutes. Then proceed to exit flight level, adjusting speed and heading to arrive at original exit point exactly at the planned time." The final ? instructs the system to try out the sequence, then present the image of the result. If it is satisfactory it can be accepted, if not another solution can be tried. The essence of this system is that a controller can designate a complex system of manoeuvres, to take place at a future time, use the aircraft systems to return the aircraft to its required course, check his proposed solution, and, having solved the problem, forget it.

Although formal evaluations have not been made, unskilled trainees can handle at least 45 aircraft simultaneously - 180 per hour, for indefinite periods.

Discussion

This proposal rejects the 'evolutionary' approach to ATC. This traditional approach minimises short-term risks, but can no longer accomodate the speed and size of modern technical change. Airlines are subject to fierce evolutionary pressure in the market place. ATC services have been essentially civil service monopolies, usually associated with safety and regulatory organisations, and concerned more with minimising costs than with maximising 'profit'. This is now changing, for better or worse.

The type of system foreshadowed here requires changes, not merely to displays, but to ways of thought. Ironically many of the physical elements are already in place, and considerable efforts are under way to integrate them into the traditional system. This system requires a far greater trust by controllers and pilots in the automatic systems, similar to the 'glass cockpit' revolution introduced by the Airbus 320 and its successors. The introduction of the 'glass cockpit' was the outward sign of a radical change in the way civil airliners are flown, and although some unexpected snags are still showing up, the change is now irreversible.

This short demonstration does not solve all the problems of ATC. Monitoring for deviations may require a separate display, or a separate controller. Approach Control,

Flow Control, Strategic control of traffic flows and control manning, require entirely different displays.

Although most commercial aircraft are already equipped with much of the necessary equipment, some would require extensive reequipment, and all would need some modifications. Light aviation would certainly require relatively expensive additional equipment, although this would have considerable concomitant advantages.

The transition to such a system would require careful planning, considerable delicacy, and possibly some brutality. It involves alterations in the controllers' way of thinking, and could be regarded as 'de-skilling' rather than 'skill-enhancing'.

Conclusions

An in-depth review of Air Traffic Control suggests that a re-designed system can provide considerable additional capacity using relatively inexpensive displays while reducing the strain on the controllers.

More generally, it has been demonstrated that substantial contributions can be made to innovative display design using inexpensive personal computing equipment.

The general principles remain :

- Humans are responsible, therefore must be in control
- They should do the tasks they do well, and enjoy doing them
- They must have equipment suited to their mental (and physical) abilities
- Computer based systems should be designed around the human operators
- Emulation of existing systems is a blind alley, because it fails to make use of the potential of modern computer systems.

References

Cardosi, K. M. 1994, *An Analysis of Tower (Local) Controller-Pilot Voice Communication* FAA, Washington DC, USA (DOT-VNTSC-FAA-94-117)

Cushing, S. 1994, *Fatal Words: Communication Clashes and Aircraft Crashes*, Univ.of Chicago Press, Chicago, USA.

Falzon, P. 1982, Display structures: Compatibility with the Operator's Mental Representation and Reasoning processes, in Johannsen G and Boller H E, *Proceedings of the Second European Annual Conference on Human Decision Making and Manual control*, Bonn, Germany.

Graham, R. V., Young, D., Pichancourt, I., Marsden, A. and Ikiz, A., 1994, *ODID IV Simulation Report*, EUROCONTROL Experimental Centre, Report No. 269, Bretigny, France.

Lafon-Millet, M-T, 1978, *Observations en trafic reel de la resolution des conflits entre avions evolutives*, Raport INRIA N0 CO/R/55, INRIA, France.

Mell, J. 1992, *Etude des Communications Verbales entre Pilote et controllleur en Situation Standard et Non-Standard*, Ecole Nationale de l'Aviation Civile,Toulouse,France

Stein E S and Bailey J, 1994,*The Controller Memory Guide: Concepts from the Field*. DOT/FAA Technical Centre, Atlantic City, New York,USA.

(The demonstrator interface is freely available in QuickBASIC 4.5 source code and compiled form on a 3.5 inch disc for PC compatibles with EGA or better colour interfaces. The disc contains full documentation in WordPerfect 6.0 and ASCII code.Colour images and a fuller description are also available from the author.)

Late Submissions

THE CRISIS IN EUROPEAN AIR TRAFFIC CONTROL: AN ASSESSMENT OF CONTROLLER WORKLOAD MODELLING TECHNIQUES.

Arnab Majumdar

PhD. Student,
Centre for Transport Studies,
Imperial College, University of London,
London SW7 2BU

The design, planning and management of European airspace is a highly complex task, involving numerous issues. After briefly noting the current problems of air traffic control (ATC) in Europe, the central role of controller workload and the difficulties in its understanding the paper then examines the salient features of several controller workload models. Featured are those used by ATC authorities in the UK i.e. the DORATASK and PUMA; by Eurocontrol,i.e. EAM and RAMS; and also the SDAT model of the USA.

Introduction: Europe and Workload

Air traffic in Europe doubled over the last decade, much in excess of predictions upon which developments in the national air traffic control (ATC) systems were based. It is forecast (ATAG, 1992) that the total number of flights in Western Europe will increase by 54% in the period 1990-2000, and by 110% in the period 1990-2010 leading to more than 11 million flights/year. Furthermore, these flights are unevenly spread throughout the region, with a core area where the density of traffic is highest, leading to a situation of an ever-increasing workload being placed on the ATC network in an area where current traffic levels often exceed the capacity of system. The cost of the inefficiencies related to the ATC system has been estimated to be as high as US$ 5 billion in 1987 (Lange 1989).

In 1990, the Transport Ministers of the European Civil Aviation Conference (ECAC) States, covering then Western Europe but now also Central and Eastern Europe, responded by adopting an ATC strategy for the 1990s which would harmonise and progressively integrate the patchwork of different ATC systems in Europe at present. The overall objective of the strategy is to improve airspace and control capacity whilst maintaining safety (Marten 1993). Eurocontrol - the European Organisation for the safety of air navigation - was chosen to manage the implementation of the strategy. It established the European Air Traffic Control Harmonisation and Integration Programme (EATCHIP) to implement the strategy for the en-route environment, comprising of four overlaping phases, the last of which relies upon major technical innovations to provide an ATC system for Europe for the future (Majumdar 1994). In most Western European en-route airspace sectors currently, the limiting factor on capacity is the workload of air traffic controllers, i.e. the nature, complexity and duration of their tasks. This likely to remain so in the short to medium term (the next 25 years). Therefore controller workload will remain the dominant factor in determining sector and system capacity, and any modification of the airspace structure (air routes re-organisation, re-sectorisation, etc.) which reduces controller workload should increase airspace capacity.

Workload is a construct, i.e. processes or experiences that cannot be seen directly, but must be inferred from what can be seen or measured, and despite numerous theories about what generates human workload, by 1980 there was no generally accepted definition of the construct. Definitions and models of workload in the literature are numerous and varied in form and workload is used to describe various matters. The most common conception of workload as used by system designers is akin to "task difficulty" or "complexity" which simply replaces one ill defined concept for another. Stein and Rosenberg (1983) at the FAA Technical Center proposed the following definition of workload, conceived of as a phenomenological or subjective experience, which is more than the sum of all the demands placed upon the operator from the environment and from within. In this holistic, unified model, the formal definition is as follows:

"the experience of workload is based on the amount of effort, both physical and psychological, expended in response to system demands (taskload) and also in accordance with the operator's internal standards of performance."

Workload definitions and measurement will only have credibility if they fit within how well the system performs towards its objectives. Stein (1990) notes the common misconception in that taskload in the form of system demands and workload are synonymous. While workload, in general, may be predictably based on a knowledge of what is going on in the system, it is important to realise that individual differences do exist and that there is no one idealized model of a controller.

It is generally held that workload within the ATC system has dramatically risen in the past few years. With further growth predicted in the near future, there has been an increase in complexity of traffic. Various strategies needed to overcome these problems that have been proposed, e.g. increasing the number of controllers or increased training to enable are either impractical or infeasible. Given the difficulties in understanding workload, the following sections outline various models of controller workload.

The UK Models - DORATASK and PUMA.

DORATASK developed by the Civil Aviation Authority (CAA) in the UK, is a fast-time simulation for evaluating the capacity of a sector (as limited by controller workload) by sytematically summing up the the time spent by the controller on observable and non-observable tasks for each category of traffic traversing the sector.The objectives of the model are to predict the capacity changes which result from: changed manning levels; changed route structure or relative traffic loadings; changed ATC procedures or equipment and airspace re-sectorisation. The methodology defines the capacity of the sector as that traffic flow rate which creates a level of workload equal to a level that has been specified to be acceptable by ATC managers. For the en-route sectors, the components of controller workload modelled by DORATASK are:

- Routine actions: these are actions taken by the controller as a matter of routine to control each aircraft as it proceeds through the sector, e.g. heading and altitude instructions
- planning actions: this models the actions of the controller upon receiving a new flight strip. It involves checking the strip afgainst every strip on the relevant boards;
- conflict resolution actions: this component of the workload represents the actions taken by controllers to resolve potential conflicts between pairs of aircraft. It is defined in terms of the number of heading and altitude instructions which need to be given to each of the aircraft involved in the potential conflict, and the number and type of co-ordinations with other controllers. Conflicts are categorised under a number of headings: conflict geometry, combination of controllers, climb/level/descend status of each aircraft, flight category of each aircraft, the current (waypoint) destination of each aircraft, stage of resolution (avoidance/completion).
- recovery time: a proportion of the controller's time is allowed to represent the time needed to recover from active work. This represents a safety factor allowing the controller to work at a higher rate for a short period.

For terminal manoeuvring area (TMA) sectors, the workload is modified. In particular, monitoring workload is represented by two elements consisting of: making checks of recently arrived aircraft against all others on the radar screen and regular monitoring of the radar screen to check each aircraft against all others for potential conflicts. The model has been calibrated against known capacities of many sectors in the UK, known either from the use of other capacity assessment methods, or from empirical evidence.

The PUMA model developed by Roke Manor in the UK and also used CAA is different from the other models outlined in the paper in that it attempts to model the human information processing of controllers. The method in PUMA involves (Day and Hook 1995):

- establishment of a base-line of controller activities by analysis (or reference to pre-existing analysis of) ATC activities as they are currently performed;
- fragmentation of those ATC activities into those fundamental components which impose a predictable loading on the controller;
- establishment of what new circumstances or procedures are to be examined using the toolset, which might for instance involve introduction of changes to the fine task structure (perhaps associated with the use of new computerised support tools), and then setting that in the context of a scenario of aircraft movements within a sector;
- calculation of workload on the basis of the Wickens' "multiple resources theory" (Wickens 1992). This involves the concept of multiple channels within the user, upon which demands are made when tasks are undertaken, and which may cause conflict.

PUMA consists of a family of tools which can exchange data between them readily. Its starting point is the Definition of the Operational Concept- a process defining and linking together the roles, tasks, actions and events involved in the area of ATC under study. These are defined carefully. A role is seen as associated with the performance of particular duties. Different roles exist and it is necessary to be able to associate a person with a particular task which consists of a number of actions. An action places an unvarying demand on the user's cognitive processing channels, whilst a task- probably consisting of a number of actions, some overlapping- is a recognisable ATC duty, e.g. giving an aircraft a clearance. An event is an externally generated phenomenon that causes a controller to take some action, either overt, observable or covert, cognitive. Many sources define an operational concept, e.g. the published literature and the descriptions of their given duties and tasks by the controllers. In the use of PUMA to examine the workload implications new work practices, the operational concept differs in some way from the baseline that has been established by these means. The Membership Editor (ME) allows the user to represent in the form of lists the roles, events, tasks and actions, and define and display the links between them. The Operational Concept (OC) File stores all the information concerning events, tasks, actions and associated information on channel loadings and conflict matrix.

An initial Observational Task Analysis (OTA) builds up the definition of the operational concept and involves observing the controllers performing their duties, and then relating observed actions to tasks. The controller then talks through a videotape of his actions immediately after his spell on duty has ended, allowing a good insight into what the controller did/ did not do, and also why. This "cognitive walkthrough" interview is recorded on video. Subsequent video analysis results in the expression of the operator's activities in terms of actions, and their time and duration. This is a time-consuming activity and is conducted to obtain a baseline of task detail within the definition of the OC. Given that operational concepts explored by PUMA typically involve some departure from current practice, this will involve modifications to tasks, etc. Where such departures are small, confidence in the predictive capacity of PUMA is high. A further OTA on a high fidelity man-in-the loop simulation of some future system concept must be conducted to provide an update to the baseline and thereby reduce the uncertainty in predictions. As no known technique can access those cognitive functions that the controller is unaware of, even though these may be of importance, and possibly prone to disruption by certain automation approaches, is one reason why PUMA should serve as only a "coarse filter" in evaluating future scenarios.

From the ME, the user can call two editors to allow a complex operational concept to be explored from different perspectives, in terms of the workload involved in the tasks and the workload associated with individual roles. The Scenario Builder/Editor (SBE) supports the process of creating an ATC scenario which typically involves defining a sector of particular dimensions, with reporting points and standard routes and a number of aircraft types with realistic flight plans. The Workload

Assessment Tool (WAT) plays the scenario through observing the curve of workload against time and also allows the user to see the workload data expressed in a histogram form, with the amounts of time spent at each workload level graphically displayed. **Conflict Matrix Editor (CME)** allows the conflict matrix to be defined, in terms of the channels involved in the actions that make up the tasks, and the conflicts between each channel and every other.

The EUROCONTROL Models - EAM and RAMS.

The method used at Eurocontrol to determine the capacity of air traffic control sectors is the ATC Capacity Analyser. This uses the EUROCONTROL Airspace Model as a simulation tool to generate the workloads on the simulated controller positions for a given traffic sample. On completion of a run, the Capacity Analyser analyses the loading recorded on the controller positions of the sector whose capacity is being determined. Comparison of these results with pre-defined thresholds for workload then leads to a re-run of the simulation with either an increase or decrease in the traffic sample of the sector, a process which continues until the loading on the simulated working positions has reached the heavy load threshold.

The Eurocontrol Airspace Model (EAM) is a critical event model which during the simulation treats a number of defined events in the life cycle of a simulated flight, e.g. entry into the first simulated sector, exit from a sector, conflict search and resolution etc. It simulates the effect of the controller's actions on the air traffic. On completion of the simulation, an analysis package examines the resulting profiles of each aircraft and determines a defined number of tasks required of the controllers to process the flight. As each task has a defined execution time and working position(s), the amount of work required to handle a given traffic sample can be determined. Due to the level of the detail in the simulation, significantly greater than in DORATASK, a typical simulation run consists of assessing about four airspace organisations, under the conditions of about six different traffic scenarios.

There are four main types of control and input data into the model:

i) airspace structure and route network; simulates en-route, TMA and approach sectors, as well as superimposed and overlapping sectors. Parallel, unidirectional, specialised or non-ATC routes can be simulated.

ii) traffic samples; Traffic data is simulated for a three hour period and can be broken down to the level of individual aircraft trajectories. Projected increases in future traffic, which can occur uniformly or non-uniformly throughout the airspace can be obtained. The simulation contains a detailed aircraft performance model, recognising more than 200 different aircraft types which are grouped into 35 aircraft performance categories. Information from aircraft performance tables determines cruise speed, rates of climb and descent, etc. Three types of flight are distinguished; long haul, short haul and local.

iii) ATC logic and procedures. The model can simulate procedural, radar or mixed control with varying separation minima to take into account the procedures in operation for co-ordinations, data transfer and the functions of each controller.

iv) controller task definitions. Up to 110 different ATC tasks are recored and identified, divided into various categories, e.g. flight data management tasks, external and internal coordination tasks. For each task, information is input into the model on task duration, the number of times it is performed and under what conditions, the controller position performing the task and any exceptions to this.

Iterative applications of EAM are performed with different traffic levels until sector capacity is reached. An one hour rolling average of the EAM workload is calculated and capacity defined as that level of traffic flow producing a peak in the rolling average of 70% - based on experimentation and corresponds to 42 minutes measured working time in one hour, with 18 minutes left for undefined tasks in the model and recuperation.

The EAM is over twenty years old, and is now being updated to be replaced by the Reorganized ATC Mathematical Simulator (RAMS), a fully integrated simulation tool to which most features of the EAM have been transferred. Its sophisticated conflict detection/resolution mechanisms and highly flexible user interfaces and data preparation environment, allows users the possibility of carrying out planning, organisational, high-level or in-depth studies of all of the wide range of ATC concepts (Eurocontrol Experimental Centre 1995).

Each air sector has two control elements associated with it: RAMS Planning Control, and RAMS Tactical Control, which maintain information regarding the flights which wish to penetrate them, and have associated separation minima and conflict resolution rules. As aircraft fly, they are scheduled to enter a sector at the physical sector pierce point. On the basis of this calculation, they are added to the planning information at a user specified time offset, prior to entering the sector. As the flight progresses, it enters the RAMS tactical information area- an user defined extended boundary surrounding the sector. On leaving the sector, the flight information is removed.

Conflict detection in RAMS are based upon existence of the two control elements, which have their own user defined separation standards. Conflict search is performed on the occurrence of a simulation event that causes a change in state for the a particular control area. Having identified potential conflict situations, RAMS applies state of the art rulebased conflict resolution algorithms, to decide on the most appropriate action based on the control area. The recommendation is based on a set of ATC rules and heuristics developed in conjunction with Eurocontrol's own operational ATC experts. At present, the conflict resolution rulebase comprises a set of over 100 rules which are grouped into one or more of 19 rulegroups. This is the major difference between RAMS and the EAM. In the EAM this is a cumbersome process and the model does not attach priorities as is the case in reality. The EAM examines the traffic sample and there is a timetable of entry into the model which examines one, and only, aircraft at a time. The flight profile of this aircraft is examined and then compared with those of other aircraft which have previously flown through the sector. If EAM recognizes a conflict with other aircraft, it always applies a level change to aircraft concerned- not necessarily the appropriate solution as other aircraft may change their profile i.e. a penalty on these other aircraft may be the realistic strategy. In RAMS, each aircraft is surrounded by a box, which if infringed leads to a conflict which is solved by a rules databas obtained from the experience of controllers. This rules-database is not yet extensive, solving upto 70% of conflicts. The hope is that eventually upto 90% of conflicts.

The Sector Design Analysis Tool - SDAT

The Sector Design Analysis Tool (SDAT) is a workload model developed by the Federal Aviation Agency (FAA) in the USA, to estimate the controller workload generated by air traffic at any specified ATC sector (Geisinger and MacLennan 1994). The workload here is a surrogate for true controller workload, useful for estimating the differences in workload between sectors or such differences from changes in airspace design or traffic loading but is not meant to measure individual controller performance or establish staffing standards. The data sources are workload actions noted in the HOST computer and input/output (I/O) messages, together with the actions specified by the user. For SDAT purposes, sector workload is characterised by the data source about an action, whether the action is required or optional and the control position in the sector. All workload is related to specific actions, each with a time requirement and is measured by the total time required to perform the actions.

The actions performed by controllers - physical, communications and mental - are estimated from the available data, easier for the observable tasks of the controller, e.g. computer entry and response. The total time for each type of workload action is then the sum of the time for processing an I/O message and the time required to perform the reported work action. For the unobservable mental tasks, SDAT calculates the strategic planning time by estimating the number of flight strips active in the sector and the tracks being monitored on the radar screen, and allocating a time to these activities. The critical separation assurance component of the workload, is also not identified from the HOST data and the number of controller actions has to be statistically estimated from by determining the potential conflicts (at present, crossing and merging) and the time required to resolve them.

In SDAT, controller workload refers to the workload at a particular position to which the model allocates the actions, e.g. radar or interphone controller. The more useful measure is sector workload since current ATC practices make workload interchangeable between positions to a large extent. The HOST computer data indicates when I/O measures occur, and is used by the model to identify when a workload action occurred. This information allows for the calculation of sector workload and display of workload for a sector on a time basis.

For each aircraft in the traffic sample, SDAT develops a flight history. This timeline reflects if and when the flight progress strip was generated for the sector and any subsequent modifications, and when the sector observed the flight track as a result of a handoff, request data block or point out action. This information is then used to calculate workload. An unique feature of SDAT is its attempt to determine sector complexity measures to try and explain the vagaries between sectors and controllers. Given the category of sector, e.g. en-route, departure, the model measures the mixture of three types of data for a sector to determine complexity:

- types of flight operation - departure, arrival or en-route
- types of users operating, e.g. air carrier, air taxi, military or GA
- aircraft speed mix - fast, slow moderate.

To assess sector complexity, each data type provides information to be considered in assessment. At present, SDAT does not provide any composite analysis of this information, though with experience of use this should be available. Instead an empirical measure of complexity based on numerical values and intuition is used: complexity = workload/traffic. However, this value can change for different levels of traffic loads and mixes within the same sector, and offers no explanation for the value of complexity but will help in developing a theoretical measure. The model allows the user to interactively modify both airspace features and traffic loading.

Conclusions

The rapid growth in air traffic in Europe since the 1960s has placed ever-increasing demands on the ATC system as designed. Despite various technology programmes to increase airspace capacity, controllers remain vital to the system. Their existence leads to variability in workload and variability in performance. Given this the elusive construct of human workload in complex person-machine systems must be understood.

Most of the models outlined are based upon experience of observing controllers. This has helped define the reference scenarios and feedback has enabled the models to reflect reality as far as possible. Both DORATASK and SDAT attempt to model the tasks of the controller which cannot be observed, on the basis of various assumptions, whilst the EAM and RAMS covers around 70% of the controllers total work, i.e. the observed tasks in much greater detail, but not the cognitive element. However, all models tend to overestimate the conflict resolution workload, e.g. busy periods, these models do not take account of the rationalisation in the controller's workload,e.g. he does not give weather forecasts to the pilots, features are taken into account in the models, thereby deviating from. In slacker periods, the models do not estimate appropriately the cognitive workload involved in monitoring. PUMA differs in trying to model the difficult cognitive processes. Note that all these models were designed with specific needs of their client authorities in mind and apart from exceptions, notably RAMS, are based on historical criteria and are of limited use in future ATC planning.

References
ATAG (1992) European Traffic Forecasts, ATAG, Geneva.
Day, P.O. and M.K. Hook (1995) PUMA, a description of the PUMA method and toolset for modelling air traffic controller workload, Roke Manor Research Limited, Romsey, UK.
Eurocontrol Experimental Centre (1995) RAMS User Manual Version 2.0, EEC/RAMS/UM/OO13, Eurocontrol, Breitgny-sur-Orge, France.
Geisinger, K. and B. MacLennan (1994) Sector Design Analysis Tool (SDAT) Workload Model Design Document, Operations Research Service, AOR-200, Federal Aviation Authority, USA.
Lange, D.G.F. (1989) The crisis of European air traffic control: costs and solutions, Wilmer, Cutler & Pickering, London/ Brussels/ Washington DC.
Majumdar, A. (1994) Air traffic control problems in Europe - their consequences and proposed solutions, Journal of Air Transport Management 1(3) 165-177
Marten,D. (1993) European ATC harmonization and integration programme, Journal of Navigation 46(3) 326-335.
Stein, E.S. and B. Rosenburg (1983) The measurement of pilot workload, DOT/FAA/CT-82-23, NTIS No.AD A124 582)
Stamp, R.G. (1992) The DORATASK method of assessing ATC sector capacity - an overview, DORA Communication 8934, Issue 2, Civil Aviation Authority, London.
Wickens, C.D. (1992) Engineering psychology and human performance, Harper Collins, USA.

Author Index

Subject Index